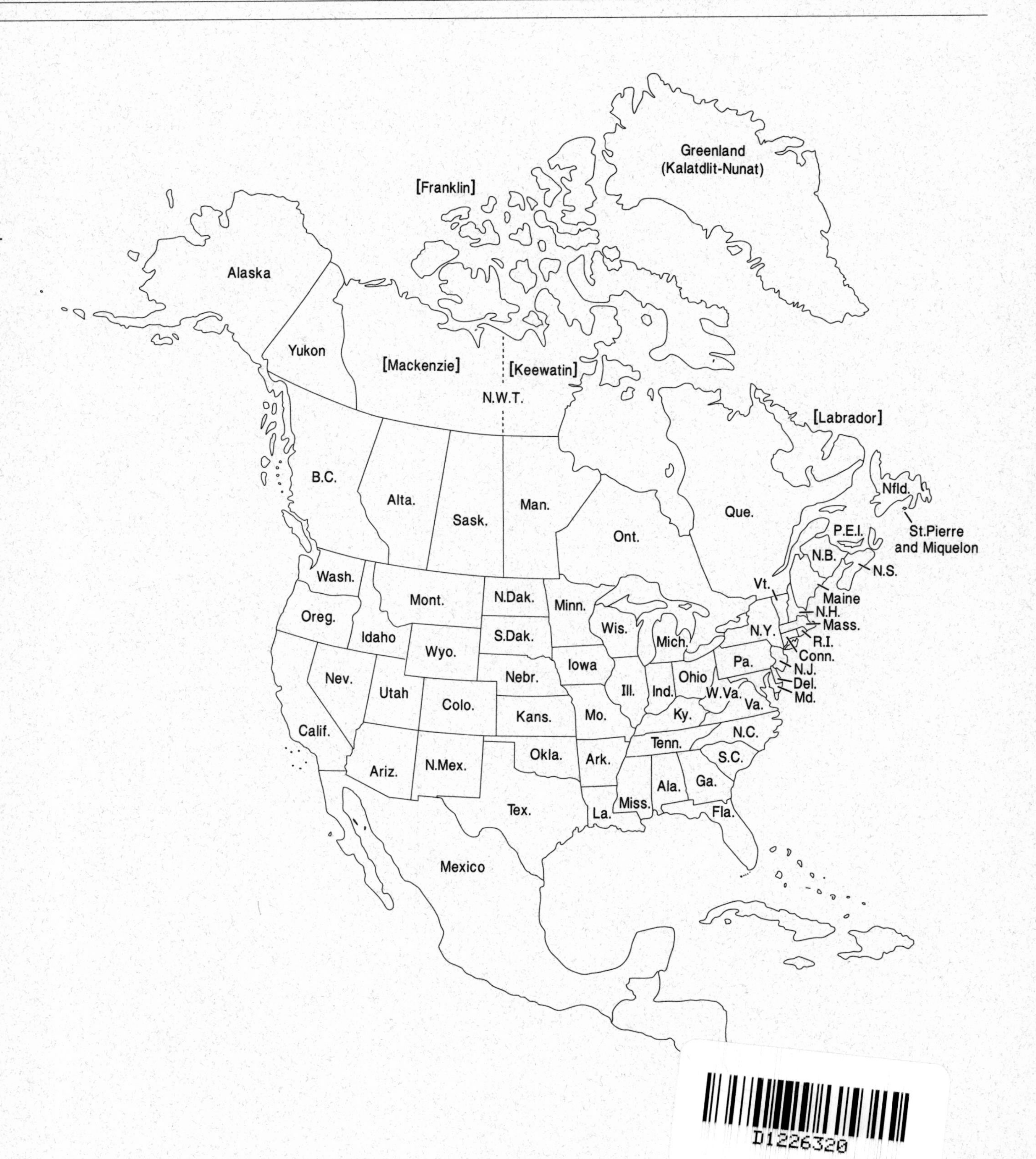
Greenland
(Kalatdlit-Nunat)
[Franklin]
Alaska
Yukon
[Mackenzie]
[Keewatin]
N.W.T.
[Labrador]
B.C.
Alta.
Sask.
Man.
Ont.
Que.
Nfld.
St.Pierre
and Miquelon
P.E.I.
N.B.
N.S.
Wash.
Oreg.
Mont.
Idaho
N.Dak.
Minn.
S.Dak.
Wyo.
Wis.
Mich.
N.Y.
Vt.
Maine
N.H.
Mass.
R.I.
Conn.
Pa.
N.J.
Del.
Md.
Iowa
Nev.
Utah
Nebr.
Ill.
Ind.
Ohio
W.Va.
Va.
Colo.
Kans.
Mo.
Ky.
Calif.
N.C.
Tenn.
Okla.
Ark.
S.C.
Ariz.
N.Mex.
Ga.
Ala.
Miss.
Tex.
La.
Fla.
Mexico

Flora of North America

FLORA OF NORTH AMERICA EDITORIAL COMMITTEE

Flora of North America

North of Mexico

Edited by FLORA OF NORTH AMERICA EDITORIAL COMMITTEE

VOLUME 1

Introduction

NEW YORK OXFORD · OXFORD UNIVERSITY PRESS · 1993

Oxford University Press

Oxford New York Toronto
Delhi Bombay Calcutta Madras Karachi
Kuala Lumpur Singapore Hong Kong Tokyo
Nairobi Dar es Salaam Cape Town
Melbourne Auckland Madrid

and associated companies in
Berlin Ibadan

Published by Oxford University Press, Inc.,
198 Madison Avenue, New York, New York 10016-4314

Oxford is a registered trademark of Oxford University Press

Library of Congress Cataloging-in-Publication Data
Flora of North America north of Mexico
edited by Flora of North America Editorial Committee.
p. cm.
Includes bibliographical references and indexes.
Contents: v. 1. Introduction.
ISBN 0-19-505713-9 (v. 1)
1. Botany—North America.
2. Botany—United States.
3. Botany—Canada.
I. Flora of North America Editorial Committee.
QK110.F55 1993 581.97–dc20
92-30459

9 8 7 6 5 4

Printed in the United States of America
on acid-free paper

The Flora of North America Editorial Committee
takes pleasure in dedicating this volume
to Dr. Peter H. Raven,
Director of the Missouri Botanical Garden
and ardent supporter of this project.

IN MEMORIAM

Arthur Cronquist
1919–1992

The Editorial Committee notes with deep regret the passing of Dr. Arthur Cronquist on 22 March 1992. Dr. Cronquist was without doubt our generation's foremost student of the temperate North American flora, as was Asa Gray in the previous century. For more than four decades Dr. Cronquist was associated with the New York Botanical Garden, where he wrote or variously participated in most of the major floristic works of the time. He was a prominent student of the family Asteraceae, and the author of numerous influential texts and reference books. His monumental system of classification of the flowering plants is widely respected, and it is followed here in the *Flora of North America North of Mexico*. Dr. Cronquist was a giant among botanists, and he will cast a long shadow into the future.

Contents

PART IV HUMANKIND AND THE FLORA

PART V CLASSIFICATIONS AND CLASSIFICATION SYSTEMS

For their partial funding
of the Flora of North America Project,
we gratefully acknowledge and thank:

National Science Foundation
The Pew Charitable Trusts
The Caleb C. and Julia W. Dula Foundation
The Surdna Foundation
The Robert and Lucile Packard Foundation
National Fish and Wildlife Foundation
ARCO Foundation
The William and Flora Hewlett Foundation

MEMBER INSTITUTIONS

Flora of North America Association

Arnold Arboretum
Jamaica Plain, Massachusetts

Biosystematics Research Institute
Canada Agriculture
Ottawa, Ontario

Canadian Museum of Nature
Ottawa, Ontario

Carnegie Museum of Natural History
Pittsburgh, Pennsylvania

East Central University
Ada, Oklahoma

Field Museum of Natural History
Chicago, Illinois

Fish and Wildlife Service
United States Dept. of the Interior
Washington, D.C.

Harvard University Herbaria
Cambridge, Massachusetts

Hunt Institute for Botanical Documentation
Carnegie Mellon University
Pittsburgh, Pennsylvania

Jacksonville State University
Jacksonville, Alabama

Jardin Botanique de Montréal
Montréal, Quebec

Kansas State University
Manhattan, Kansas

Missouri Botanical Garden
St. Louis, Missouri

New Mexico State University
Las Cruces, New Mexico

New York State Museum
Albany, New York

Northern Kentucky University
Highland Heights, Kentucky

Royal Ontario Museum
Toronto, Ontario

Southern Illinois University
Carbondale, Illinois

The New York Botanical Garden
Bronx, New York

The University of British Columbia
Vancouver, British Columbia

The University of Texas at Austin
Austin, Texas

Université de Montréal
Montreal, Quebec

University of Alaska
Fairbanks, Alaska

University of Alberta
Edmonton, Alberta

University of California
Berkeley, California

University of California
Davis, California

University of Idaho
Moscow, Idaho

University of Illinois
Urbana-Champaign, Illinois

University of Iowa
Iowa City, Iowa

University of Kansas
Lawrence, Kansas

University of Michigan
Ann Arbor, Michigan

University of Oklahoma
Norman, Oklahoma

University of Ottawa
Ottawa, Ontario

University of Southwestern Louisiana
Lafayette, Louisiana

University of Tennessee
Knoxville, Tennessee

University of Western Ontario
London, Ontario

University of Wyoming
Laramie, Wyoming

Utah State University
Logan, Utah

Foreword

These essays provide information on the physical and biological factors that have affected the evolution and distribution of the plant life of North America. They also reflect how we humans have developed our understanding of the flora. The essays are written to summarize current knowledge in each of the topics, and to provide access to the relevant literature.

The reader will note that the several continent-wide detail maps share more in common than mere geographic outlines. The major boundaries within the maps for climate, vegetation, soils, and phytogeography were derived from different kinds of information, yet they are strikingly similar. The clear conclusion is that these several topics are intimately related, and that the nature of the vegetation is a product of complex, long-term interactions among these factors.

The drawings of plants in the essays were taken from various nineteenth- and twentieth-century works. They were chosen in part to illustrate some of the plants mentioned in the text, and in part as examples of the style and qualities of botanical illustration from an earlier era to the present.

Theodore M. Barkley

Special Thanks

Many people have worked countless hours to bring volumes 1 and 2 of *Flora of North America North of Mexico* to publication simultaneously. I particularly thank some of the key people involved in this gargantuan undertaking.

Editors for volume 1 were Theodore M. Barkley, Luc Brouillet, and Richard W. Spellenberg. Work on these introductory essays was coordinated by Luc Brouillet.

Preparation of manuscripts for publication has been an immense task. All technical editing and most of the word processing has been done by Helen Jeude, the project's technical editor. Bruce Ford, Denis Kearns, Bruce Parfitt, and Alan Whittemore helped check many aspects of the manuscripts. Sara Jenkins assembled the illustrations and maps and helped with other details of the project.

Editorial Committee members who voluntarily reviewed every manuscript in its final stage of preparation for publication—and who became known as the "final five"—were Theodore M. Barkley, John G. Packer, Richard W. Spellenberg, John L. Strother, and R. David Whetstone.

I am personally grateful to project staff, editorial committee members, authors, and reviewers for the tremendous amount of work they have done, and for their enthusiasm for and support of Flora of North America.

Nancy R. Morin
Convening Editor

Acknowledgments

Flora of North America North of Mexico is the product of creative thinking and hard work on the part of many people. This project owes its start to those who attended an organizing meeting on 30 April and 1 May 1982 at the Missouri Botanical Garden: G. Argus, D. Bates, W. Burger, S. Hatch, N. Holmgren, T. Jacobsen, M. Johnston, R. Kiger, J. Massey, J. McNeill, R. Mecklenburg, N. Morin, J. Phipps, D. Porter, J. Reveal, S. Shetler, F. Utech, and G. Webster.

The project has been designed, in large part, by the Editorial Committee. This committee has changed its makeup slightly over the years, but overall it has been a remarkably stable group. By 1986 the Editorial Committee consisted of G. Argus, T. Barkley, D. Boufford, L. Brouillet, D. Henderson, R. Kiger, B. MacBryde, J. McNeill, N. Morin, J. Packer, J. Phipps, L. Shultz, R. Spellenberg, J. Strother, J. Thieret, G. Webster, and D. Whetstone. Since then, Henderson and MacBryde have resigned from the committee and have been replaced by R. Hartman and J. Fay. J. Estes, M. Johnston, D. Murray, G. Straley, and R. Thompson were added to the committee as permanent members, and J. Eckenwalder, A. Smith, and W. H. Wagner Jr. joined as special editors for volumes 1 and 2. In 1991, bryophytes were added to the project, and the following bryologists were added to the Editorial Committee: W. Buck, M. Crosby, J. Engel, M. Hicks, D. Horton, N. Miller, B. Murray, W. Reese, R. Stotler, B. Thiers, and D. Vitt.

The Flora of North America project is coordinated from the Organizational Center, located at the Missouri Botanical Garden. Support by the Missouri Botanical Garden for this center has been instrumental in the success of the project, and we are grateful to Peter H. Raven, Marshall R. Crosby, Enrique Forero, and W. D. Stevens for committing institutional resources and for contributing their own time and energy to Flora of North America. The institutions at which editors work provide more than half of the overall cost of the project through in-kind support. We are grateful to all of the foundations and individuals who have financially supported Flora of North America. In particular, the Pew Charitable Trusts and the National Science Foundation have supported the project through its very early, developmental phases. We are particularly grateful to program directors at the National Science Foundation for their help and ideas.

Coordination of the activities of the project's many editors and hundreds of authors and reviewers has, in itself, been an enormous task. Project coordinators, who have also been co-editors of the Flora of North America Newsletter, have been, in succession, Carol L. Blaney, Kay Tomlinson, and Judith Unger. Lois Ganss, Eloise Halliday, Mary Lawrence, Nan Winkelmeyer, and Eleanor Zeller have provided secretarial support. Editors at other institutions have been assisted by Gavin D. R. Bridson, Donna M. Connelly, Darlene O'Neill, Carolyn Parker, Kimberley Perez, Robert Rhode, and Charlotte A. Tansin. Undergraduate interns, some of whom were funded by the Research Experiences for Undergraduates program of the National Science Foundation, have been a tremendous help in all aspects of the project. They have been: Carol Davit, Sandra S. Howell, Jennifer R. Milburn, Lisa Nodolf, Stacy A. Oglesbee, Susan Reiter, Viviana M. Tharp, and Larry A. Turner.

For volume 1, Carol Baskin, Joseph Ewan, Priscilla Fawcett, Gail Fritz, Dale Johnson, John Montre, Robbin Moran, Roger Sanders, Mary Smith, and George Yatskievych helped with the illustrations. Ann Dillon, at the Washington University Medical Center Computer Graphics unit, oversaw preparation of the maps. The editors thank Gregg Bogosian, Jerry Brown, Larry Davenport, Roger Etcheberry, Joseph Ewan, Stuart Hay, Harold Hinds, David Hopkins, and Pierre J. H. Richard for reviewing parts of the Introductory Essays, and Linda Oestry for providing assistance.

The Flora of North America taxonomic database is an integral part of Missouri Botanical Garden's database system. This system was designed by Marshall R. Crosby and Robert E. Magill, and the programs were written by Magill. The computer staff at the Garden, M. Christine McMahon, Deborah Kama, and John Satterfield, have been helpful with many aspects of Flora of North America. The bibliographic database, held at the Hunt Institute for Botanical Documentation, was designed by Robert W. Kiger. Development of the Flora of North America database was initially overseen by G. E. Gibbs Russell and is now being designed and managed by Deborah Kama. Character lists have been compiled by A. Whittemore and J. Bruhl. Data entry has been done by Kimberly Lindsey and more recently by Kathleen Janus and Judith McMurtry. Gibbs Russell, with Flora of North America undergraduate intern Larry A. Turner, designed and analyzed a questionnaire on floristic databases, the report of which has been an important reference for determining the status and potential compatibility of floristic databases.

The following institutions signed a Memorandum of Cooperation as members of the Flora of North America Association: Arnold Arboretum, Jamaica Plain; Biosystematics Research Institute, Canada Agriculture; Carnegie Museum of Natural History; Field Museum of Natural History; Harvard University Herbaria, Cambridge; Hunt Institute for Botanical Documentation; Jacksonville State University, Alabama; Kansas State University, Manhattan; Missouri Botanical Garden, St. Louis; Jardin Botanique de Montréal; National Museum of Natural Sciences, Ottawa; New Mexico State University, Las Cruces; Northern Kentucky University; Office of Scientific Authority, U.S. Fish and Wildlife Service; The New York Botanical Garden, Bronx; Université de Montréal; University of Alaska, Fairbanks; University of Alberta, Edmonton; University of California, Berkeley; University of California, Davis; University of Idaho, Moscow; University of Illinois at Urbana-Champaign; University of Kansas, Lawrence; University of Ottawa; University of Western Ontario; and University of Wyoming, Laramie.

Members of the Flora of North America Project Advisory Panel were: F. A. Almeda Jr., California Academy of Sciences; D. M. Bates, Liberty Hyde Bailey Hortorium; T. P. Bennett, Academy of Natural Sciences; A. Bouchard, Jardin Botanique de Montréal; W. C. Burger, Field Museum of Natural History; T. Duncan, University of California, Berkeley; C. G. Gruchy, National Museum of Natural Sciences, Canada; V. L. Harms, University of Saskatchewan; J. Hickman, University of California, Berkeley; A. G. Jones, University of Illinois; R. W. Kiger, Hunt Institute for Botanical Documentation; M. M. Littler, National Museum of Natural History; J. Massey, University of North Carolina, Chapel Hill; G. A. Mulligan, Vascular Plant Herbarium, Ottawa; G. B. Ownbey, University of Minnesota; J. G. Packer, University of Alberta; D. H. Pfister, Harvard University Herbaria; G. T. Prance, New York Botanical Garden; R. F. Scagel, University of British Columbia; R. L. Shaffer, University of Michigan; R. F. Thorne, Rancho Santa Ana Botanica Garden; B. L. Turner, University of Texas.

Members of the Flora of North America Project Database Consulting Group were: Guy Baillargeon, Biosystematics Research Centre, Ottawa; Chris Beecher, NAPRALERT, Chicago; Warren Brigham, Illinois Natural History Survey; Christian Burks, Genbank, Los Alamos; Theodore J. Crovello, California State University, Los Angeles; Thomas Duncan, University of California, Berkeley; Janet Gomon, National Museum of Natural History, Washington D.C.; Ronald L. Hartman, Rocky Mountain Herbarium; Maureen Kelley, BIOSIS, Philadelphia; Robert W. Kiger, Hunt Institute for Botanical Documentation; Kenneth M. King, EDUCOM, Washington, D.C.; Robert Magill, Missouri Botanical Garden; Jim Ostell, National Library of Medicine, Bethesda; David Raber, Case Ware Inc., Costa Mesa; Beryl Simpson, University of Texas; Frederick Springsteel, University of Missouri, Columbia; Kerry S. Walter, Center for Plant Conservation, Jamaica Plain.

We are deeply grateful to these people and institutions for their hard work and continuing support on behalf of the Flora of North America project.

N. R. M.

Contributors

Michael G. Barbour
Botany Department
University of California
Davis, California

Theodore M. Barkley
Division of Biology
Kansas State University
Manhattan, Kansas

Luc Brouillet
Université de Montréal
Montreal, Quebec

Norman L. Christensen
Botany Department
Duke University
Durham, North Carolina

Arthur Cronquist (deceased)
The New York Botanical Garden
Bronx, New York

Paul A. Delcourt
Center for Quaternary Studies of the Southeastern United States
Department of Geological Sciences
University of Tennessee
Knoxville, Tennessee

Hazel R. Delcourt
Center for Quaternary Studies of the Southeastern United States
Department of Geological Sciences
University of Tennessee
Knoxville, Tennessee

James E. Eckenwalder
Department of Botany
University of Toronto
Toronto, Ontario

Alan Graham
Department of Biological Sciences
Kent State University
Kent, Ohio

Charles B. Heiser
Department of Biology
Indiana University
Bloomington, Indiana

Nancy R. Morin
Missouri Botanical Garden
St. Louis, Missouri

James S. Pringle
Royal Botanical Gardens
Hamilton, Ontario

James L. Reveal
Department of Botany
University of Maryland
College Park, Maryland

Alan R. Smith
University Herbarium
University of California
Berkeley, California

Richard W. Spellenberg
Department of Biology
New Mexico State University
Las Cruces, New Mexico

G. Ledyard Stebbins
Department of Genetics
University of California
Davis, California

Donald Steila
Department of Geography and Earth Sciences
University of North Carolina
Charlotte, North Carolina

Ronald L. Stuckey
Herbarium
Ohio State University
Columbus, Ohio

Robert F. Thorne
Rancho Santa Ana Botanic Garden
Claremont, California

Warren H. Wagner Jr.
Department of Biology
University of Michigan
Ann Arbor, Michigan

R. David Whetstone
Department of Biology
Jacksonville State University
Jacksonville, Alabama

George Yatskievych
Natural History Division
Missouri Department of Conservation
St. Louis, Missouri

Flora of North America

Introduction: History of the Flora of North America Project

Nancy R. Morin
Richard W. Spellenberg

The Flora of North America project has grown out of a long history of interest in the flora area and has built on the results of previous efforts to produce a comprehensive flora. It has been designed to draw on the expertise of the entire systematic botany community and to make the best possible use of the literature and herbarium specimens that are essential resources for the project. Despite current interest in and urgency of botanical studies of many other kinds, there continues to be a need for information about the plants of North America on a continental scale.

Historical Background

The Flora of North America project derives from the deep interest that botanists have in studying and characterizing the plants of the region and the need for authoritative information for basic and applied research, conservation, and resource management. Ever since explorers first sent North American plants to Europe, the world has known that the flora is rich and interesting. The first accounts of plants from the area were published in Europe by European botanists (e.g., in Robert Morison's *Historiae* [1680–1699], Leonard Plukenet's *Phytographia* [1691–1705], and John Ray's *Historia Plantarum* [1686–1704]). Fredrick Pursh's *Flora Americae Septentrionalis* (1814) was the first flora of continental scope. Treatises on North American plants were written by resident botanists in the early 1800s, and the first attempt at production of a comprehensive flora of the continent was undertaken by John Torrey and Asa Gray in the 1830s (see J. L. Reveal and J. S. Pringle, chap. 7). Subsequently, botanists focused on producing regional floras, until 1905 when N. L. Britton began publishing the monographic series *North American Flora* at the New York Botanical Garden.

Flora North America, 1965–1973

Because regional floras do not provide fully comparative information on plants throughout their ranges, the potential utility of a continental flora is self-evident. Recognizing the need and spurred, at least in part, by the announcement at the Tenth International Botanical Congress in Edinburgh in 1964 that volume 1 of *Flora Europaea* (T. G. Tutin et al. 1964–1980) was soon to be published, North American botanists considered launching a comprehensive flora project at home. At the annual meeting of the American Institute of Biological Sciences (AIBS) in August 1964, the Council of the American Society of Plant Taxonomists (ASPT) appointed a committee to study the feasibility of undertaking the production of a flora of North America.

This committee of about 25 interested botanists was chaired by Robert F. Thorne, and it met in Washington, D.C., in May 1966. The committee concluded that production of a flora of North America was both desirable and feasible. Thorne, on instructions from the committee, corresponded with many botanists about their willingness to serve in some capacity on the flora project. The response was overwhelmingly positive and left no doubts as to the advisability of proceeding.

In August 1966, during the meeting of AIBS at the University of Maryland, the Council of ASPT voted to sponsor the flora project, to be called "Flora North America," and designated an organizational committee under the chairmanship of Mildred E. Mathias. Mathias drew up an organizational structure in which the Smithsonian Institution served as the project headquarters. Steering and editorial committees were appointed, and the organizational committee was disbanded.

In 1967 sponsorship of the project was transferred to AIBS. The National Science Foundation awarded two grants—one from the Systematic Biology Program and one from the Information Systems Program, Office of Science Information Service—to AIBS in July 1969 to support planning for the project. The grants provided for a small central staff at the Smithsonian Institution and for travel by members of the planning committees. By 1969 automated data processing had assumed major importance, and part of the planning grants funded a systems development manager. Peter H. Raven, chairman of the editorial committee, was program director. Stanwyn B. Shetler was program secretary and devoted nearly full time to the daily activities of the program.

Pilot projects were developed to define the Flora North America databasing approach. The biological community was canvassed by use of several questionnaires designed to identify potential users of the database, to inventory plant systematists, and to evaluate systematists' interests and potential as contributors. Pilot databases included the Type Specimen Register, which comprised 7000 records at the end of 1971, and "authority files," such as an author file, a higher taxon list, a species name list, a contributor file, and a morphological/ecological vocabulary. B. R. Rohr et al. (1977) provided a comprehensive account of publications resulting from the Flora North America program.

In 1971 a program council was formed to act as the single body charged with advising the administering institution (then AIBS) regarding policy. The program council recommended that the Smithsonian Institution become the administering institution, taking that responsibility from AIBS with the latter's full approval. This transfer took place in July 1971. The Secretary of the Smithsonian Institution made a commitment to the National Science Foundation that the Smithsonian Institution would share in the costs for the project and would maintain the database indefinitely.

A proposal was submitted to the National Science Foundation in 1971 requesting funds for the implementation of the project. Similar proposals were submitted to the Canada Department of Agriculture and the National Research Council of Canada. The funding, which was approved, was to support seven regional editorial centers in Canada and the United States, and the headquarters and systems development center at the Smithsonian Institution.

After six years of planning, the project became operational in the fall of 1972. National Science Foundation support was made contingent on the willingness of the Smithsonian Institution to increase its funding year by year, and their commitment to do so. Examining its overall priorities, the Smithsonian Institution decided that it was unwilling to make such a guarantee, unless Congress would in turn guarantee that the institution's total request would be funded each year. Consequently, the National Science Foundation felt obliged to withdraw its support, and the project was suspended in the summer of 1973. Attempts to revive the project through collaboration with the Man in Biosphere project and the National Park Service were not successful.

The Flora of North America Project

Beginnings

In spring 1982, 22 botanists from throughout the United States and Canada met at the Missouri Botanical Garden to consider a plan proposed by the Hunt Institute for Botanical Documentation and the Carnegie Museum to resume collaborative preparation of a flora of North America north of Mexico. In the new plan, a significant portion of the project costs, particularly staff time and use of facilities, would be contributed by participating institutions, and the flora would consist of traditional treatments prepared by members of the systematic botanical community. Information from the treatments would be used to develop a database. That meeting was the basis for the formation of the current Flora of North America project.

A steering committee was appointed by those attending the 1982 meeting in St. Louis. The steering committee met at the Hunt Institute in January 1983 and determined that a proposal detailing the project should be written and submitted to the American Society of Plant Taxonomists and the Canadian Botanical Association/L'Association Botanique du Canada for endorsement.

At the annual ASPT Council meeting in August 1983, the Missouri Botanical Garden offered to serve as the organizational center for the multi-institutional, collaborative project and offered staff assistance for preparation of the proposal and for continued organization of the project. Administrators from several of the larger North American botanical institutions were subsequently invited to serve on an advisory panel, and an editorial committee was established. Representatives of the home institutions of the editors, and of other institutions supporting the project, each signed a memorandum of cooperation and together they established the Flora of North America Association.

To help establish the Flora of North America project, the advisory panel counseled project members about organization and implementation from the standpoints of both the systematic botanical community at large and their own institutions. The original advisory panel fulfilled its purpose and was disbanded in November 1989.

Community support for the project was voiced in 1982, when the Canadian Botanical Association/L'Association Botanique du Canada passed a resolution reaffirming its commitment and approving the steps being taken to produce a flora of North America. Further, at its annual meeting in August 1983, the American Society of Plant Taxonomists passed a resolution affirming its commitment for production of such a flora.

The Organizational Years

From 1983 to 1987, when outside funding was first acquired, the project was supported entirely by participating institutions. Missouri Botanical Garden contributed staff time and facilities for the organizational center and sought funding on behalf of the project. Proposals for funding were submitted to the National Science Foundation and to private foundations. The editorial committee used this time to identify and contact potential contributors and reviewers for taxa in the first volumes, to determine what should be included in the treatments and database, to write sample treatments, and to develop a "Guide for Contributors." In February 1988 the Pew Charitable Trusts awarded a matching grant to the project; that funding made it possible for the editorial committee to meet and for the organizational center to hire support staff for the project. A year later, matching funds were received from the National Science Foundation.

Filling the needs of users who are not botanists or systematists for authoritative information on plants has been a major goal of the project. A workshop, "Floristics for the 21st Century," was organized to define these needs and to discuss how these needs could be met. The workshop, supported by the National Science Foundation, met in Alexandria, Virginia, in May 1988, with some 70 biologists from around the world and from a variety of disciplines in attendance. The proceedings were published the following year (N. R. Morin et al. 1989).

Until 1990 Flora of North America was scheduled to include only vascular plants. At the annual meeting of the American Society of Bryologists in December 1990, members agreed to prepare treatments for a bryophyte volume of the flora. Bryologists were added to the editorial committee for this purpose and started work in 1991. *Flora of North America,* therefore, will treat all embryophytes growing without cultivation in North America north of the United States–Mexico boundary.

Project Organization

In order to synthesize existing knowledge concerning the vascular plants of North America, the project has attempted to identify and draw on the expertise of the entire systematic botanical community. In addition to plant taxonomists in North America and other parts of the world, biologists in the various governmental agencies help the project directly by reviewing and sometimes contributing treatments and indirectly by providing their expert knowledge to those who are preparing treatments.

The editorial committee is the governing body of the Flora of North America project. It consists of (1) taxon editors, each of whom is responsible for soliciting and editing treatments of blocks of families; (2) regional coordinators, each of whom acts as liaison between the project and the botanical community in his or her region; (3) a bibliographic editor; (4) a nomenclatural advisor; and (5) a convening editor, who has ultimate responsibility for the coordination, quality, and completion of the project. Special advisors with expertise in specific areas, such as economic uses or conservation, have been asked to join or assist the editorial committee, and others may be added before the project is completed.

Almost all of the treatments are written by specialists in the groups. Treatments for which no specialists are available are written by project staff or other botanists. Reviewers who are particularly knowledgeable about given geographic regions or taxa critique treatments of species that occur in their areas or that fall within their taxonomic expertise.

Flora of North America Databases

The original Flora North America project had the clear vision that computerization could make floristic work easier and more valuable. In addition, participants at the "Floristics for the 21st Century" workshop, held in 1988, identified many potential users of Flora of North America information. In order for floristic information to be useful to this large audience, however, it must be easily accessible in a variety of ways. Computerized databases allow access to and analysis of the many kinds of information contained in floristic treatments. A botanical database for the Flora of North America project

Members of editorial committee, Missouri Botanical Garden, October 1990. Front row: Rahmona Thompson, George Argus, Luc Brouillet, Leila Shultz, R. David Whetstone, Nancy Morin, Robert Kiger. Middle: Marshall Johnston, John Packer, Alan Smith, John Thieret, Ronald Hartman, Theodore Barkley, David Boufford, Grady Webster. Back row: Richard Spellenberg, John Strother, Gerald Straley. Not pictured: James Estes, John Fay, John McNeill, David Murray, James Phipps, W. H. Wagner Jr.

was established at the Missouri Botanical Garden using the TROPICOS system developed at the Garden. In 1989 a computer databases consultant group was instituted to discuss the long-term computerization needs of the project, and its recommendations helped to improve the system. The TROPICOS database uses plant names to link information on features such as relationships, distribution, characteristics, and specimens. Data from other wide-area floristic projects, such as Flora of China and Flora Mesoamericana, connect to Flora of North America data, allowing comparisons of distributions and features on a broad geographic scale.

Production of the database proceeds simultaneously with the preparation of the hard-copy flora. The database will eventually contain all the information found in the published volumes as well as additional information, such as more detailed morphological data, information on type specimens, and documentation for chromosome counts. Furthermore, it will also incorporate information on habitats, precise distributions, populations, and plant communities. Morphological descriptions are recorded as characters and character states in a way that allows searches for character states or suites of character states, and permits various other kinds of analyses. The database is online, and additional information about it and access to it are available from the Flora of North America organizational center at Missouri Botanical Garden.

All references used in the published volumes are entered into a computerized database at the Hunt Institute for Botanical Documentation. This database will allow searches of published sources and will be used to

Editors working at Missouri Botanical Garden Library. Theodore Barkley, Nancy Morin, Luc Brouillet, Richard Spellenberg.

Deborah Kama, Programmer/Systems Analyst, demonstrates TROPICOS for Scientific and Managing Editor Bruce Parfitt.

Author George Yatskievych collecting plants. [Photograph by Kay Yatskievych.]

produce bibliographies in the individual volumes and to produce the cumulative bibliography.

In order to provide comprehensive information to Flora of North America contributors, reviewers, and editors, the project maintains files on plant systematists, floristic and taxonomic databases, relevant references, and unpublished studies. These resources, coupled with the ever-building database on North American plants, give the project the ability to act as a referral service providing information on North American floristic studies, references, and specialists.

Resources for Flora of North America

Flora of North America treatments are based on critical evaluation of the literature, examination of specimens, and field and laboratory experience. Most contributors and reviewers have studied the plants in the field, and such firsthand experience adds to the quality of the treatments. Information derived from recent experimental work and numerical taxonomy is also incorporated into the treatments.

In the preparation of manuscripts, authors have either examined specimens from herbaria to determine morphological variation and geographic distribution, or they have used recent monographs and revisions, which in themselves are the results of critical examination of specimens. Whether a manuscript has been generated from a fresh look at specimens or primarily from review of critical revisions, specimens have been used by authors to test the work, and they have been examined by reviewers to determine accuracy of descriptions, usability of keys, and correctness of stated geographic ranges.

Literature

Taxonomic work depends to a great extent on information recorded in the literature. Original descriptions of taxa, many of them dating from the early 1700s, must be consulted in order to determine the correct application of scientific names. Monographs and revisions contain information about the taxa themselves and about the taxonomic opinions of those who studied them. Shorter reports contain additional information about the biology and distribution of the taxa. Local and regional floras may contain detailed information about the morphology and distribution of taxa within the specific regions that they cover.

Many major regional floristic works have been published in the past three decades. For Canada, examples include: H. J. Scoggan, *The Flora of Canada* (1978–1979); A. E. Porsild and W. J. Cody, *Vascular Plants of Continental Northwest Territories, Canada* (1980); and E. H. Moss and J. G. Packer, *Flora of Alberta* (1983).

Similar projects in the United States include: *Vascular Flora of the Southeastern United States* (A. E. Radford et al. 1980+); *Flora of the Great Plains* (Great Plains Flora Association 1986); *Intermountain Flora: Vascular Plants of the Intermountain West, U.S.A.* (A. Cronquist et al. 1972+); *Vascular Plants of the Pacific Northwest* (C. L. Hitchcock et al. 1955–1969); *Manual of the Vascular Plants of Texas* (D. S. Correll and M. C. Johnston 1970); and *The Jepson Manual: Higher Plants of California* (J. C. Hickman, ed., 1993).

This literature has been made available to Flora of North America participants by university, museum, and botanical garden libraries either directly or through interlibrary loans. Without this considerable assistance from libraries and librarians, the project would be virtually impossible to complete.

Herbaria

Flora of North America is primarily specimen-based. Plant specimens are generally pressed and dried after they have been collected. Information is recorded about where the plant was collected, and additional information about the habitat or species associated with the plant may be included. The pressed specimens are mounted on paper with a label containing the recorded information. They are then stored in herbaria, where they are given protection from pollutants, moisture, destructive insects and rodents, and careless handling. Given proper protection, such specimens last indefinitely and provide a wealth of historical information. Specimens have been made available to Flora of North America participants directly, or as loans. Thus it has been possible for an author to study together all collections of a particular taxon made through time and throughout its range. In some cases curators have provided information recorded on labels to confirm reported distributions, morphological features, and dates of flowering or fruiting. Much of what is known about the historical floristic diversity of North America is based on information held in herbarium specimens. Specimens have provided a picture through time of the changes in distribution of a taxon, sometimes of the rate and pattern of spread of a noxious plant, but more often of the collapse of the geographic range of a taxon through the influence of human activity. In a few instances, specimens are the only indication that an extinct taxon once existed.

TABLE 1.1. Large Herbaria in the United States and Canada.

	Date Established	*Number of Specimens*
New York Botanical Garden	1891	5,300,000
Harvard University Herbaria	1864	4,858,000
Smithsonian Institution	1848	4,368,000
Missouri Botanical Garden	1859	3,700,000
Field Museum of Natural History	1893	2,415,000
University of California, Berkeley	1872	1,785,000
University of Michigan	1837	1,613,500
California Academy of Sciences	1853	1,600,000
Academy of Natural Sciences, Philadelphia	1812	1,590,000
Bailey Hortorium, Cornell University	1871	1,092,000
Biosystematics Research Center, Agriculture Canada	1886	1,050,000
U.S. Department of Agriculture	1869	1,010,000
University of Texas	1900	1,006,000
University of Montreal	1920	730,000
Canadian Museum of Nature	1882	540,000

[1]Compiled from data in P. K. Holmgren et al. 1990
U.S. herbaria with more than 1,000,000 specimens and Candian herbaria with more than 500,000 specimens are listed.

Herbarium specimens document much of the information reported in *Flora of North America North of Mexico* and the database. Authors have annotated specimens they examined in preparation for writing the treatments. Documentation is maintained on specimens from which illustrations were drawn. Herbarium specimens serve as vouchers for chromosome reports. Much of the experimental work mentioned in the discussion sections is documented by herbarium specimens or has used herbarium specimens as a source of material.

There are more than 60 million specimens in 628 herbaria in the United States and nearly 7 million specimens in 110 Canadian herbaria; a small local collection of specimens exists in the herbarium in Greenland.

Author Robbin Moran in the herbarium.

The oldest existing herbarium in the United States is at Salem College, Winston-Salem, North Carolina, established in 1771; the oldest in Canada is the Royal Ontario Museum, Toronto, Ontario, established in 1838 (P. K. Holmgren et al. 1990). Table 1.1 lists the largest herbaria of the United States and Canada, with the date of establishment and number of specimens. Of the nearly 273 million herbarium specimens in the world, 25% are held in institutions in the United States and Canada. Many specimens from North America are located in European herbaria, particularly the Natural History Museum in London; Royal Botanic Gardens, Kew; the Muséum National d'Histoire Naturelle, Paris; Conservatoire et Jardin botaniques, Geneva; Liverpool Museum, Liverpool; and the University of Copenhagen, Copenhagen.

Flora of North America North of Mexico— A Flora for the 21st Century

Flora of North America is an enormous undertaking, but the accumulation of data from monographic studies and regional floras has contributed so much that the time to attempt the kind of overall synthesis envisioned by Torrey and Gray more than 160 years ago is now clearly at hand. Flora of North America distills and synthesizes information found in monographs and revisions, and makes it more easily accessible to ecologists, plant morphologists and physiologists, zoologists, conservationists, and workers in many other scientific disciplines. The flora also brings together information about the characteristics, relationships, distribution, and biology of all of the vascular plants and bryophytes of

Illustrator Laurie Lange drawing ferns from herbarium specimens.

At the Organizational Center: Helen Jeude, Technical Editor; Judith Unger, Project Coordinator; and Eleanor Zeller, Office Clerk.

North America north of Mexico so that it can be used as a basis for further studies of many kinds. Such inmation is indispensable for the proper management, scientific study, and conservation of the plants of North America.

Identification of individual plants is essential to most kinds of botanical research, including agriculture, horticulture, forestry, and plant conservation. Local and regional floras and other specialized guides will continue to be the principal means of identifying plants, but the taxa they treat can be understood in a regional context only by means of a work such as the *Flora of North America North of Mexico*. Such a flora can also serve well for the identification of plants found for the first time beyond their known ranges, and thus not included in regional works. Flora of North America also synthesizes the important regional floristic work and results of exploration of little-known areas currently underway in North America.

A continental flora provides a catalog of the plant resources available within the area and their status—it can be used as a gauge to assess environmental change and as a tool to help manage resources in the future. Flora of North America, with its printed volumes and online database, will provide information for many essential purposes.

PART I PHYSICAL SETTING

1. Climate and Physiography

Luc Brouillet
R. David Whetstone

The area covered in this book includes North America north of Mexico plus Greenland (Kalâtdlit-Nunât). It encompasses about 21.5 million km^2, between latitudes 26° and 85° N, and longitudes 15° W and 173° E, and it stretches from the Florida Keys northward to Ellesmere Island, and from Greenland westward to Attu Island in the Aleutian Archipelago. Widest in the north, the continent narrows sharply at the Gulf of Mexico. South of the United States border with Mexico, it tapers gradually to the Isthmus of Panama. It is surrounded by three oceans—the Arctic, Pacific, and Atlantic, respectively to the north, west, and east—and by the Gulf of Mexico to the south. It is separated from northeast Asia by the Pacific Ocean, and by the epicontinental Bering Sea, the Chukchi Sea, and the connecting Bering Strait. The Greenland and Norwegian seas, as well as the North Atlantic Ocean, separate North America from Europe and link the Atlantic to the Arctic Ocean; the Denmark Strait divides Greenland from Iceland. The Strait of Florida divides North America from the West Indies (Cuba).

Climate, physiography, and geology play major roles in determining the distributions of present-day soil classes, vegetation types, floras, and faunas. Biogeographers agree that climate is the primary factor in the control of these distributions (e.g., R. Good 1974; G. T. Trewartha and L. H. Horn 1980; F. I. Woodward 1987; E. O. Box 1981; M. J. Müller 1982; A. L. Takhtajan 1986; R. Hengeveld 1990; see also D. Steila, chap. 2; A. Graham, chap. 3; P. A. Delcourt and H. R. Delcourt, chap. 4; R. F. Thorne, chap. 5; M. G. Barbour and N. L. Christensen, chap. 6). Climate determines the erosional and soil-forming processes that occur, and the life forms that are able to survive at a given locale, all of which may be affected secondarily by the types of bedrock and surficial deposits encountered in the area. In turn, relief influences climatic patterns through elevation above sea level and its effects on wind patterns and rainfall (R. A. Bryson and F. K. Hare 1974b).

Geoclimatic changes that occurred throughout Earth history have affected the distribution of biotas through time. Climate has changed under cosmic influences, such as the Milankovitch cycles (discussed by P. A. Delcourt and H. R. Delcourt, chap. 4; see also F. I. Woodward 1987; H. E. Wright Jr. 1989). The climate has also been affected by the relative position of the drifting continents, because drift implies latitudinal shifts, changes in the distribution of landmasses relative to oceans and oceanic currents, and modifications in the position of mountain ranges relative to airflow patterns. For instance, the Tertiary opening of the Atlantic onto the Arctic Ocean, and the establishment of the circumantarctic current with the opening of the Drake Passage between South America and Antarctica, played a significant role in subsequent climatic cooling (e.g., M. E. Collinson 1990).

The deep oceanic conveyor belt (a bottom sea current that links all the oceans) was presumably modified by changes in continental distribution and may have

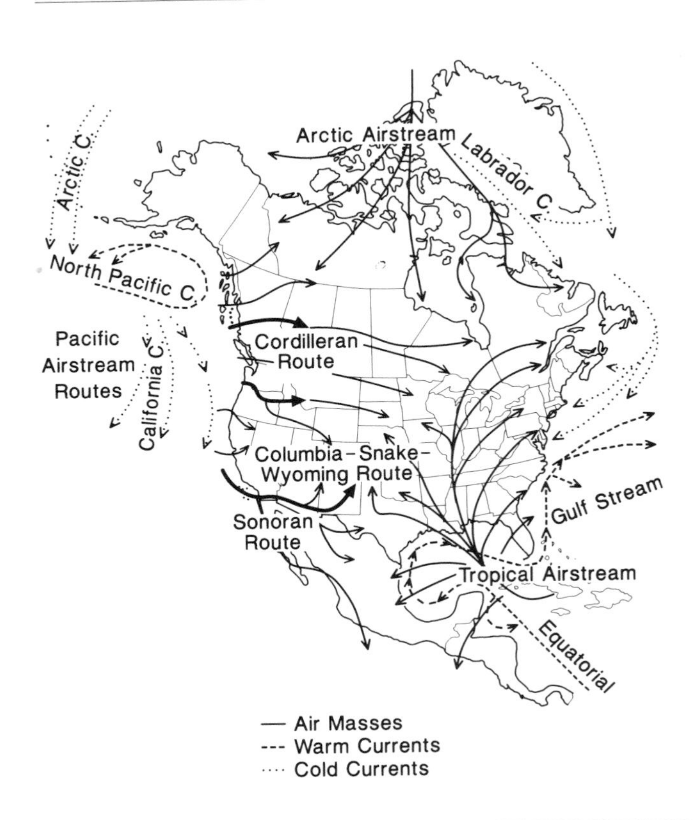

FIGURE 1.1. Primary airstream routes over North America and sea currents in adjacent oceans. [Modified from R. A. Bryson and F. K. Hare 1974b, with permission from Elsevier Science Publishers.]

affected climate (H. E. Wright Jr. 1989; J. Gribbin 1991). W. F. Ruddiman and J. E. Kutzbach (1991) proposed that the 3-km uplift of the high plateaus in Tibet and in western North America in the Pliocene-Pleistocene were instrumental in provoking the late Tertiary trend of climatic cooling. Finally, the pathways by which biotas have been able to spread between continents were also affected by the existence of bridges. Such dynamic factors influenced the evolution of life on the North American continent.

Climate

Climate may be defined in a simplified manner as the composite weather of a region (G. T. Trewartha and L. H. Horn 1980; J. Gribbin 1991). Factors that affect climate are discussed in climatology texts (e.g., G. T. Trewartha and L. H. Horn 1980; P. Pagney 1973; F. W. Cole 1980; J. R. Eagleman 1985; F. K. Lutkens and E. J. Tarbuck 1989; J. Gribbin 1991), and the effects of these factors on plant distribution are described in ecological and biogeographical treatises (e.g., S. Tuhkanen 1980; E. O. Box 1981; R. Hengeveld 1990; F. I. Woodward 1987). Changes in climate and vegetation through time are addressed by A. Graham (chap. 3) and by P. A. Delcourt and H. R. Delcourt (chap. 4) and are not discussed further here. R. A. Bryson and F. K. Hare (1974b) provided a thorough discussion of North American climates. T. W. Böcher et al. (1968) and G. Halliday (1989) described the climate of Greenland, although in very general terms.

Three major air masses affect the climate in North America (fig. 1.1). The source of the cold, continental polar air mass is the frozen Arctic and, in winter, snow-covered northwestern Canada and Alaska. The moist, mild, maritime Pacific air mass originates in the North Pacific and is ushered eastward by the midlatitude westerlies. The Gulf of Mexico, Caribbean Sea, and subtropical western Atlantic are the source areas for the moist, warm, tropical air mass. Climate in northeastern North America is also affected, especially in winter, by a fourth air mass: the cold, moist, oceanic North Atlantic air mass that finds its source over the cold waters off Newfoundland, Labrador, and Greenland. The influence is sporadic along the coast, not penetrating beyond the Appalachians or Cape Hatteras.

Air masses converge along fronts. The Polar Front is the fundamental climatic phenomenon at midlatitudes, where temperature contrasts are sharpest. It consists of the zone of convergence between the polar and tropical air masses, and coincides with a pressure trough with abundant cyclonic storms. The Polar Front intersects the earth slightly south of the high-altitude, high-velocity jet stream, and it migrates seasonally in relation with the latter. The jet stream thus represents an important element in the control of weather. This wind is circumpolar and has a meandering course around 30°–35° N, farther north in winter (35°–45° N), and south in summer (20°–25° N). Poleward of the jet stream, air is colder, and equatorward, warmer.

Precipitations are concentrated in areas south of the stream, along the zones of convergence. In the subtropical zone (20°–30° N), an area dominated by anticyclones, a minimum of precipitation is encountered; the great deserts are located in this zone. Between 45° and 55° N, a maximum of precipitation occurs in the region of convergence associated with the Polar Front and the belt of maximal cyclonic activity. From 50° to 55° N poleward, precipitation declines sharply to an absolute minimum in the cold regions at very high latitudes (greater than 75° N). Seasonal variations occur in this pattern: the subtropics are mostly dry all year, and the Mediterranean climate at the western margin of the continent has rainy winters and dry summers. In temperate latitudes, precipitation is present in all seasons, but it is most abundant in summer over the continents. Polar regions have slight precipitations in all seasons.

Annual cycling in the position of the meteorological equator (the latitude at which the noon sun is vertical to the earth surface at a given time of the year) causes pressure and wind belts to migrate latitudinally. This shifting of wind belts is significant in regions intermediate between different air mass systems; these transition zones experience contrasting weather conditions at opposite seasons. In North America, transition zones exist: (1) in the area between latitudes 30° and 40° N, (a) over the eastern parts of the Pacific and adjacent coastal California (part of the western facade of the continent), a region located between the subtropical anticyclone and the westerlies that experiences summer drought and winter rain, and (b) on the subtropical east side of the continent, along the Gulf of Mexico, where onshore monsoon winds favor maximum summer rainfall; and (2) in the area of subpolar lows between 60° and 70° N, intermediate between stormy westerlies and polar easterlies, a region of alternating colder and warmer air, as is encountered in Hudson Bay, northwestern Canada, and Alaska.

Oceanic currents affect the heat balance of the continental margins along which they flow (fig. 1.1); winds must be onshore, however, for currents to affect land climates. The major currents flow clockwise both in the Atlantic and in the Pacific. Because of the direction of the prevailing westerlies, the east coast of North America is less affected by the temperature of the sea currents than the west coast. The currents flowing along the southeastern sides of North America are warm. Thus the warm Equatorial Current enters the Caribbean Sea and the Gulf of Mexico from the tropical Atlantic, before returning to the North Atlantic through the Florida Strait as the Gulf Stream (fig. 1.1). The Gulf Stream then passes up along the east side of North America, transferring heat from subtropical to cold latitudes, but it leaves the shore at Cape Hatteras. Cold currents either rise from oceanic depths at subtropical latitudes (e.g., California Current) or are polar in origin (e.g., Labrador Current from the Arctic Sea on both sides of Greenland). On eastern seaboards, warm and cold currents converge at about 40° latitude. Thus the Labrador Current and Gulf Stream meet off Cape Hatteras (fig. 1.1). At such sites, temperature gradients are sharp, and fog is frequent. Therefore at subtropical latitudes the east coast has a warm and rainy climate; at lower middle latitudes it has a modified continental climate with relatively cold winters and warm to hot summers; and at higher midlatitudes, where cold currents flow offshore, cool summers are characteristic.

On the west coast of North America, cool currents flow equatorward from about 40° latitude, warm ones poleward, and they diverge from each other. When the prevailing westerlies blow over cool offshore water onto land that is warmed in summer, as in coastal California, the summers are dry; the moisture-holding capacity of the air increases when it passes onto the land. Such climates, known as Mediterranean climates, are found at similar latitudes along the western side of the other continents as well. Warm currents at midlatitudes result in maritime climates with cool summers and mild winters, characterized by small annual ranges in temperature and adequate precipitation. Such climates, which are highly equable, are encountered along the temperate Pacific Coast of North America.

R. A. Bryson and F. K. Hare (1974b) described the climate of North America north of Mexico through a topographic model that was correlated to the global atmospheric circulation between subtropical latitudes at 26° N, which experience an excess of radiation, and the polar latitudes at 85° N, where net radiation is in deficit. They stated that two geographical features have major effects on North American climates: the western Cordillera, trending north-south, and the Interior Plains to the east. The former constitutes a major obstacle to westerlies and trade winds, while the latter provides an uninterrupted path for the flow of the Arctic and tropical air masses (fig. 1.1).

The major effect of the high Cordillera on climate is in intercepting large-scale patterns of air circulation. The prevailing, low-altitude winds at midlatitudes (30°–65° N) are from the west (westerlies, fig. 1.1), due to the

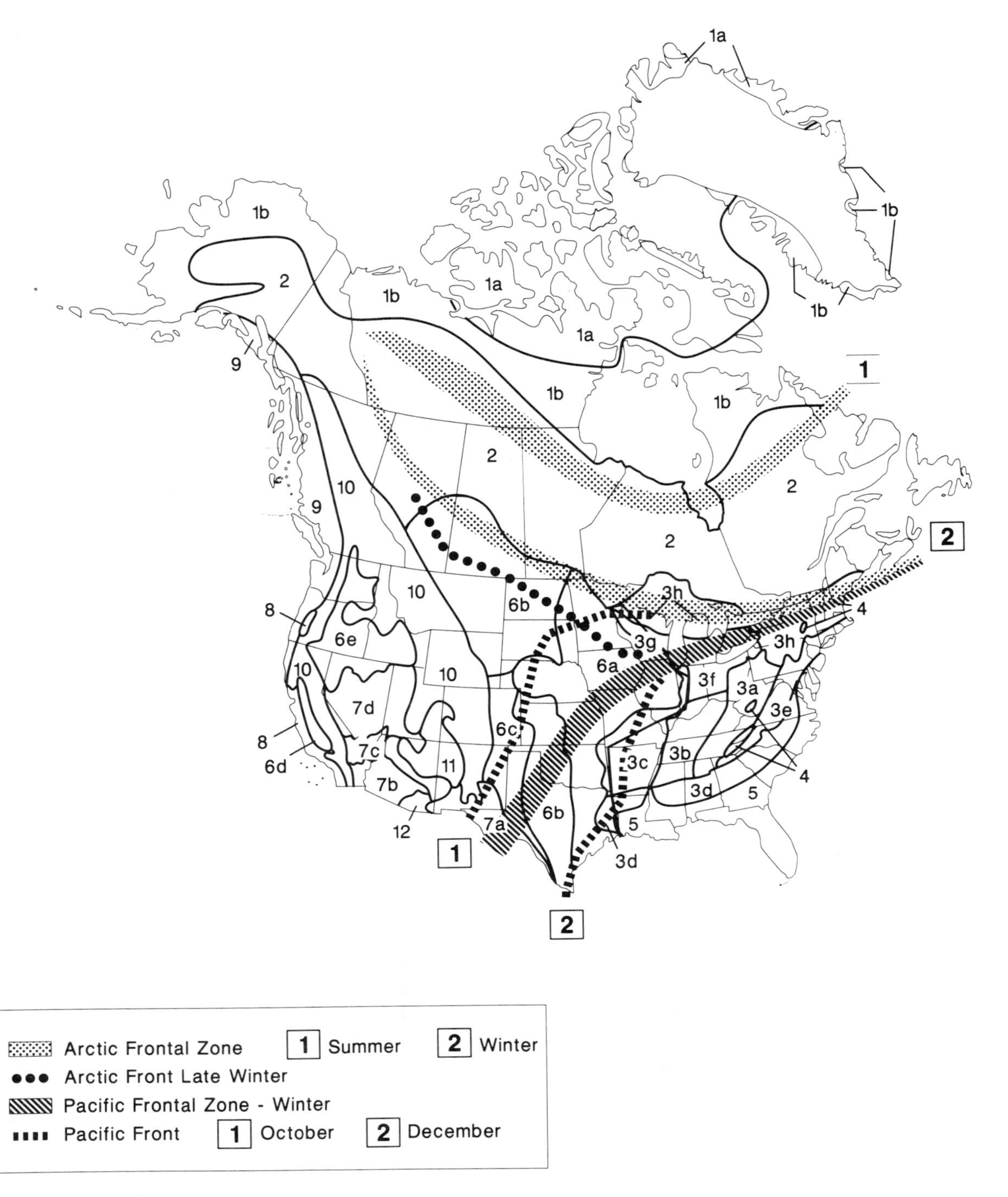

FIGURE 1.2. Relation between vegetation boundaries and average limits of major air masses in North America. [Modified from R.A. Bryson et al. 1970, with permission from the University Press of Kansas, using the vegetation map of M. G. Barbour and N. L. Christensen, chap. 5, fig. 5.1.]

rotation of the earth, i.e., Coriolis force. Equatorward (0°–30° N), easterlies or trade winds predominate; poleward, winds are also easterlies. Where air masses are forced above extensive mountain ranges, they discharge much of their water vapor windward as orographic precipitations, while leeward, a rain shadow is formed. Consequently westerlies that blow onshore from the Pacific must pass over the Cordilleran high plateaus (1500–2000 m) and mountains (to more than 4000 m) before reaching the interior. Near the region where westerlies are strongest (45°–50° N), surface winds are deflected both north and south, while north of the main flow, they are deflected northward.

Few low passes in the Cordillera provide direct routes for surface winds through the Cordillera (fig. 1.1). The major passes are: (1) near 45°–50° N (maximum westerlies); (2) the Columbia River–Snake River–Wyoming Basin gap; and (3) the region of lower relief near the Mexican border. Most of the Pacific air mass that penetrates to the interior is from the upper part of the air mass that enters north of the Cordillera. Along the eastern slopes, the air descends and is warmed in the process, i.e., adiabatically, with a concomitant decrease in relative humidity. As a result the Pacific air mass has a mild, dry character east of the Cordillera. By this process, the moist Pacific air mass provides precipitations on the windward side of the ranges, most notably the Coastal Ranges, and leaves a rain shadow on the leeward side. Over the plains, the Pacific air forms a wedge that invades eastern North America at midlatitudes, dragging along air from both the Arctic and tropical masses (fig. 1.1).

East of the Cordillera, no significant obstacle impedes the southward flow of the Arctic air mass or the northward flow of the tropical air mass. Arctic air may rapidly cross the Interior Plains and reach Mexico, with little temperature change. Tropical air rarely extends beyond the Canadian border on the continent, although it may reach farther north over the Atlantic or the Pacific. On average, Arctic air reaches south into mid-Canada in summer, and into the northern United States in winter. Tropical air stretches to the northern United States in summer, but in winter it is restricted to the extreme southern United States. In winter, an anticyclonic eddy often develops in the southeastern United States as a relict of Arctic or Pacific air masses, at a latitude where subtropical anticyclones normally develop.

The behavior of air masses produces discrete patterns recognized as climatic regions. The position of fronts at various seasons coincides with the limits of major biomes, as was shown for eastern North America by R. A. Bryson et al. (1970; fig. 1.2). Historical shifts in the boundaries between air masses help explain the biogeographic history of species and biomes (P. A. Delcourt and H. R. Delcourt, chap. 4). R. Hengeveld (1990, p. 213) stated this concept:

> Global climate can thus be considered as a process of regional atmospheric warming and cooling and, related to this, of regional humidity variation.... The resulting, highly dynamic spatial patterning of climatic conditions accounts for relatively sharply, though temporarily defined regions. Biogeographical processes, both present and past, should be viewed against this background of regional and global climatic patterns, shifting on various time scales.

In the following paragraphs, the climatic classification of North America developed by C. Troll and K. H. Paffen (M. J. Müller 1982; fig. 1.3) is presented. The descriptions are from M. J. Müller (1982) unless otherwise indicated. The system by Troll and Paffen is used here because, among other reasons, the level of resolution is comparable to that of the soil map by D. Steila (chap. 2) and of the vegetation map by M. G. Barbour and N. L. Christensen (chap. 5). Other systems of classification and a discussion of their relative value may be found in climatology or biogeography textbooks (e.g., F. W. Cole 1980; J. R. Eagleman 1985; F. K. Lutkens and E. J. Tarbuck 1989; P. Pagney 1973; G. T. Trewartha and L. H. Horn 1980).

Much of the northern and eastern North American climatic pattern is latitudinal; it integrates both the average seasonal positions of air masses and the temperature and moisture gradients. In the boreal region, air masses and gradients tend to slope eastward to lower latitudes, so that a particular climatic regime is found farther south in the east than in the west. Patterns in the west are more complex due to the impact of the Cordillera on air movement and on temperature.

Polar and subpolar climates occur north of a line running from northern Alaska to northern Quebec-Labrador. They are dominated by the Arctic air mass. The areas of icecaps and glaciers of Greenland and the Arctic Archipelago have a high polar icecap climate. Adjacent areas of the Arctic islands have a polar climate, with the warmest month averaging below 6° C; this is the High Arctic climate or belt (G. Halliday 1989; T. W. Böcher et al. 1968). Along the coasts of the Arctic mainland and of Hudson Bay, in the southern Arctic Archipelago and in southern Greenland, a subarctic tundra climate occurs, characterized by cool summers (warmest month 6°–10° C) and intensely cold winters (coldest month below −8° C); this is the Low Arctic climate or belt (G. Halliday 1989; T. W. Böcher et al. 1968).

The Aleutians and the Alaska Peninsula have a highly oceanic subpolar climate with cool summers (warmest month 5°–12° C), moderately cold winters (coldest

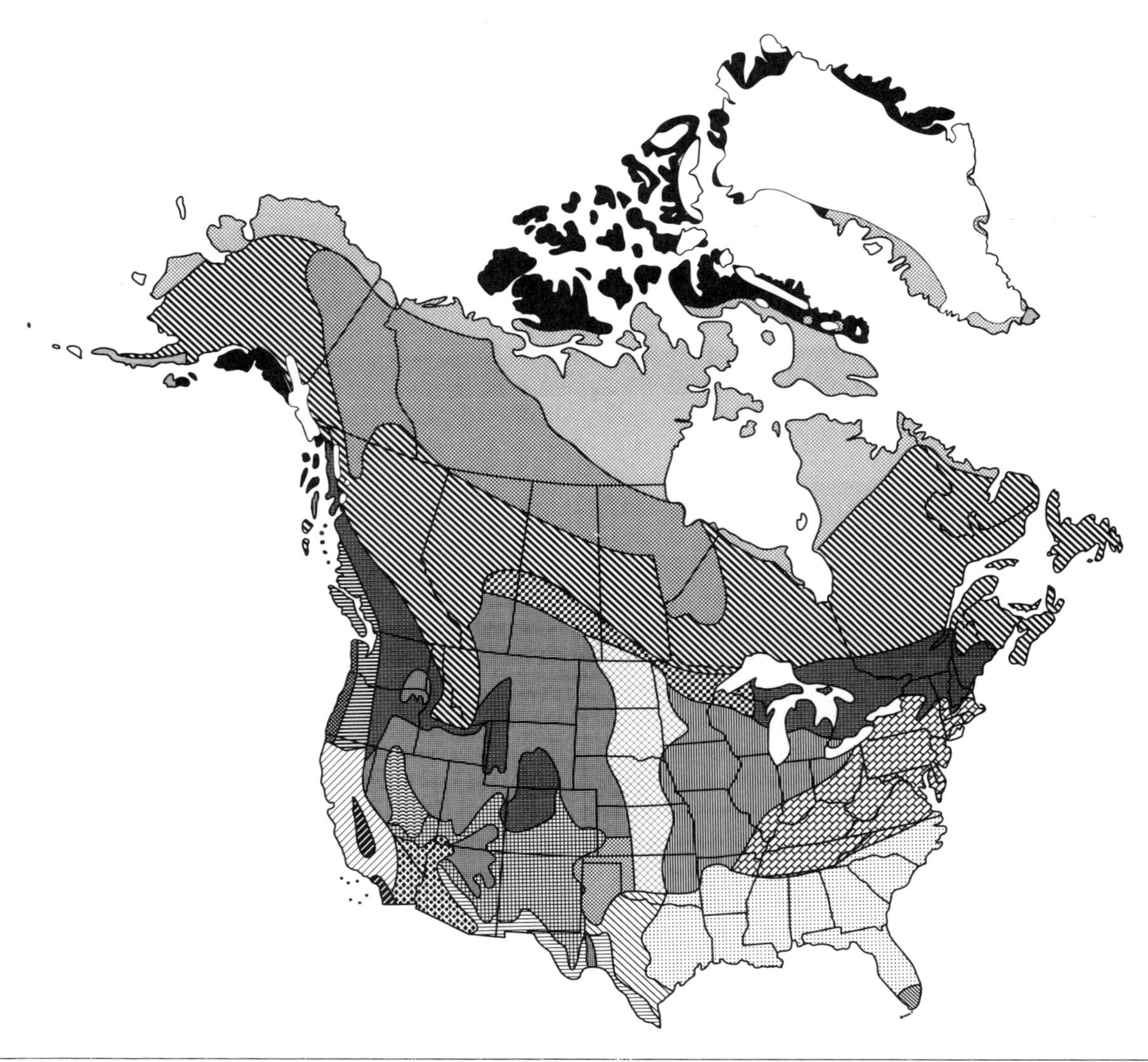

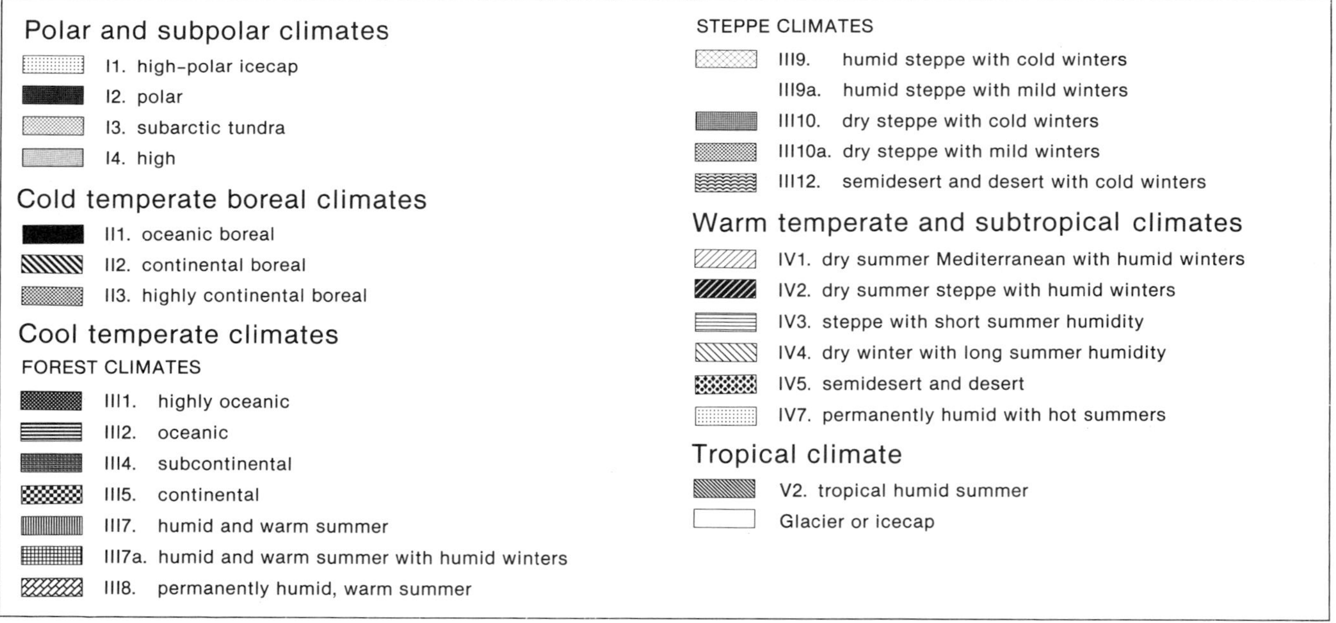

FIGURE 1.3. Climatic regions of North America, according to Troll/Paffen. [Based on data from H. Walter and H. Lieth 1967, and Climate Zones after C. Troll/K. H. Paffen in M. J. Müller 1982, with permission from Kluwer Academic Publishers.]

month −8° to 2° C) with little snow cover, and annual fluctuations in temperature of less than 13° C. Such a climate type may also occur at Cape Farewell, Greenland, an area of high precipitation (H. Walter and H. Lieth 1967; the maps published by M. J. Müller [1982] do not cover Greenland).

The cold-temperate boreal climates stretch across the continent south of the Polar Zone. This belt barely reaches south of the Canadian border; its southern limit corresponds to that of the average winter position of the Arctic air mass. The west coast from Kodiak Island (Alaska) to Prince Rupert (British Columbia) has an oceanic boreal climate with moderately warm summers (warmest month 10°–15° C; growing season of 120–180 days), moderately cold winters (coldest month −3° to 2° C) with abundant snow (winter maximum in precipitation), and annual fluctuations in temperatures between 13° and 19° C. A broad, extensive band running from the Bering Sea to the North Atlantic is subject to a continental boreal climate with short, warm summers (warmest month 10°–20° C; growing season of 100–150 days) and long, cold winters with abundant snow, and annual fluctuations in temperature of 20°–40° C.

A zone of highly continental boreal climate occupies interior Alaska and northwestern Canada eastward to Hudson Bay in northern Manitoba. This region is part of the source area for the Arctic air mass in winter. The summers are short but relatively warm (warmest month 10°–20° C), permafrost is prevalent, the winters are long, extremely cold (coldest month below −25° C) and dry, and annual temperature fluctuations exceed 40° C.

Cool temperate climates span about one-third of the continent. Two groups of climates are recognized: woodland (forest) and steppe.

Woodland climates are classified as highly oceanic, oceanic, subcontinental, continental, humid and warm summer, humid and warm summer with humid winters, and permanently humid with warm summer. Climates of the west coast benefit from the abundant precipitation brought by the warm, moist Pacific air mass. A narrow band along the west coast from Washington to northern California has a highly oceanic climate characterized by cool to moderately warm summers (warmest month below 15° C), very mild winters (coldest month 2°–10° C) with high precipitation, and annual temperature fluctuation of less than 10° C. From the Queen Charlotte Islands to Vancouver Island, and a short distance inland from Washington to northern California, an oceanic climate dominates. Summers are moderately warm (warmest month below 20° C), winters are mild (coldest month 2°–10° C), maximum precipitation occurs in fall and winter, and annual fluctuation in temperature is less than 16° C.

Subcontinental climates are found on both sides of the continent. A sizable area extending from the Great Lakes–St. Lawrence region to the St. Lawrence estuary and the coast of Maine has this climate. Water vapor from the Great Lakes humidifies the Pacific air mass traversing the region (fig. 1). This vapor and the cyclonic storms of the Polar Front contribute to the precipitation regime of this zone. A second area of subcontinental climate occurs near the west coast, inland from the oceanic climate zone, in the form of a narrow north-south band with an eastward extension to the mountains of Montana and Wyoming and a small outlier in the middle Rocky Mountains. This extension reflects a greater penetration of moist westerlies at that latitude through the Columbia River–Snake River–Wyoming Basin gap. Summers are moderately warm and receive the maximum precipitation (growing season 160–210 days), winters are cold (coldest month −3° to −13° C), and annual fluctuation in temperature is 20°–30° C. A narrow band north of the prairies, from Minnesota northwestward toward Edmonton (Alberta), has a continental climate characterized by moderately warm and humid summers (warmest month 15°–20° C; growing season 150–180 days), cold, slightly dry winters (coldest month −10° to −20° C), and annual temperature fluctuations of 30°–40° C.

A broad triangle, with its base between Minneapolis (Minnesota) and Oklahoma City (Oklahoma) and its apex at the west end of Lake Erie (Michigan and Ohio), corresponds with the eastern limit of penetration of the dry Pacific air mass, and it is strongly influenced by the tropical air mass in summer and the Arctic air mass in winter. Summers are warm and humid (warmest month 20°–26° C), winters are moderately cold and dry (coldest month −8° to 0° C), and annual temperatures fluctuate by 25°–35° C. In the southern Rocky Mountains and at the margins of the Colorado Plateau, a similar climate prevails, but with mild to moderately cold, slightly humid winters (coldest month −6° to 2° C). A band stretching from southern New England to northeastern Arkansas has a permanently humid, warm summer climate. It straddles the southern part of the Polar Frontal Zone, although the influence of the tropical air mass is more prevalent than in the subcontinental climate to the north. The climate exhibits warm and humid summers (warmest month 20°–26° C), mild to moderately cold winters (coldest month −6° to 2° C), and annual temperature fluctuations of 20°–30° C.

Steppe climates occupy the western interior half of the continent at cool temperate latitudes. The prairies in a narrow north-south band along the eastern edge of the region are designated humid steppe with cold winters. Temperature averages below 0° C in the coldest month. More than 6 months are humid (months without water deficit), and the period of vegetative

FIGURE 1.4. Physiographic map of North America by Erwin Raisz. [Reprinted with permission from Raisz Landform Maps.]

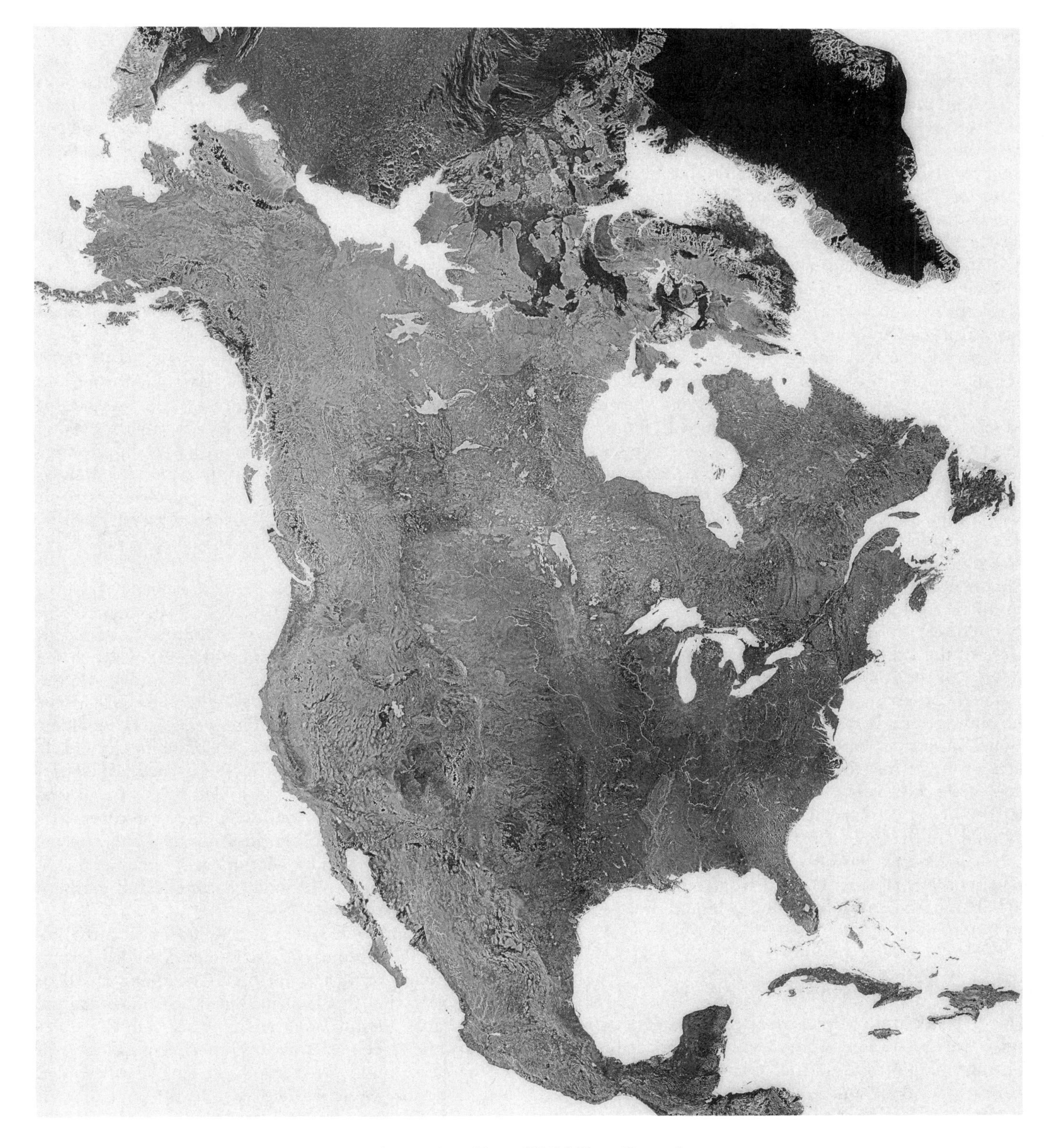

FIGURE 1.5. Satellite photograph of North America. [From EROS Data Center.]

growth lasts from spring to early summer. At the southern end of this zone, winters are milder (coldest month above 0° C). West of this band is a broad area extending westward to the Sierra Nevada, dominated by a dry steppe climate. Summers are arid, and winters are cold (coldest month less than 0° C). Less than 6 months of humidity occur. A small area to the southeast, in Colorado–New Mexico, has milder winters (coldest month 0°–6° C). Some basins of the Basin and Range physiographic province in Utah, Nevada, and Arizona experience a semidesert and desert climate with cold winters (coldest month less than 0° C).

Except for south Florida, the southern United States has warm temperate and subtropical climates. Much of California has a dry summer Mediterranean climate with humid winters (more than 5 humid months). The Central Valley and the southwestern coast of California experience a dry summer steppe climate with humid winters (fewer than 5 humid months). An irregular band along the Rio Grande in southern Texas, extending into New Mexico and parts of mountainous Arizona, has a steppe climate with summer humidity of short duration (fewer than 5 humid months) and dry winters. East of this zone in Texas, a narrow region has a dry winter climate with long-lasting summer humidity (8–9 humid months).

Semidesert and desert climates are encountered in two areas of the United States: in the Mojave and Sonoran deserts of Arizona and California, and in the Chihuahuan Desert reaching extreme southern New Mexico and western Texas. They are characterized by fewer than 2 humid months; winters are not hard, but transient or night frosts occur. Southeastern North America east of the Edwards Plateau, Texas (fig. 1.4), is characterized by a permanently humid climate. Summers are hot and are the time of maximum precipitation.

North America has only a small tropical zone, which is in southern Florida. It has a tropical humid summer climate, characterized by 7–9 1/2 humid and 3–5 1/2 arid months.

Physiography and Geology

The diversity and complexity of North American geology is well illustrated by the volumes produced for the Geology of North America Décade program (see the list in A. W. Bally and R. A. Palmer 1989). The classical map of landforms by Erwin Raisz (fig. 1.4) expresses this diversity. The map was an attempt to reproduce realistically the continental landscape from information gathered on the ground. Today, in contrast, satellites with remote-sensing devices help us to explore the globe, and from the vast amounts of data, computers produce stunning images and maps (fig. 1.5). In both Raisz's map and the mosaic image, diversity and complexity are revealed. Before reviewing the physiography and geology of North America, we shall examine the conditions that prevailed in the past and the processes responsible for the patterns observed today.

Tectonic Geology of North America

The movement of continents as a result of plate tectonics has led to their present configurations and has resulted in most mountain-building (orogeny) around the world. A chronology of the geologic periods mentioned below is provided in figure 1.6.

Most of North America is on a single major plate, the North American Plate, while Baja California and California west of the San Andreas Fault are part of the Pacific Plate. These two plates are colliding along the deep Aleutian Trench, whereas strike-slip motions (i.e., lateral motions along a fault line) dominate along the west coast of the United States (San Andreas Fault and associated faults, fig. 1.7). North America is drifting away from Europe with the widening of the Atlantic Ocean.

The major tectonic and geomorphic elements of North America are given in figure 1.7 (A. W. Bally et al. 1989). The craton (the stable core of a continent) of North America includes a Precambrian basement, of which the Canadian and Greenland shields are outcrops, and the sedimentary platforms that overlie it. The shields have been coherent for almost 2 billion years, and the oldest part (Greenland) is more than 3.5 billion years old. It comprises some of the oldest rocks on Earth. The basement extends underneath the folded belts of much of North America. In the southwest, the craton was deformed and uplifted during the Paleozoic, either in the formation of the Wichita Mountains and of the ancestral Rocky Mountains or in the Laramide and southern Rocky Mountains orogenies.

Three major folded belts are found around the periphery of the craton: (1) the Innuitian folded belt in the Canadian Arctic and northern Greenland, of Paleozoic age; (2) the Appalachian folded belt (including the Ouachita Mountains) of the east coast and the Caledonian folded belt of northeastern Greenland, also of Paleozoic age; and (3) the younger Cordilleran folded belt along the whole western edge of the continent, of Mesozoic-Cenozoic age. Superimposed on the Cordilleran folded belt are the Tertiary extensional systems of the Basin and Range Province and the widespread Tertiary volcanics encountered from the Basin and Range Province into southern British Columbia. Quaternary volcanoes of the circum-Pacific "ring of fire" are found

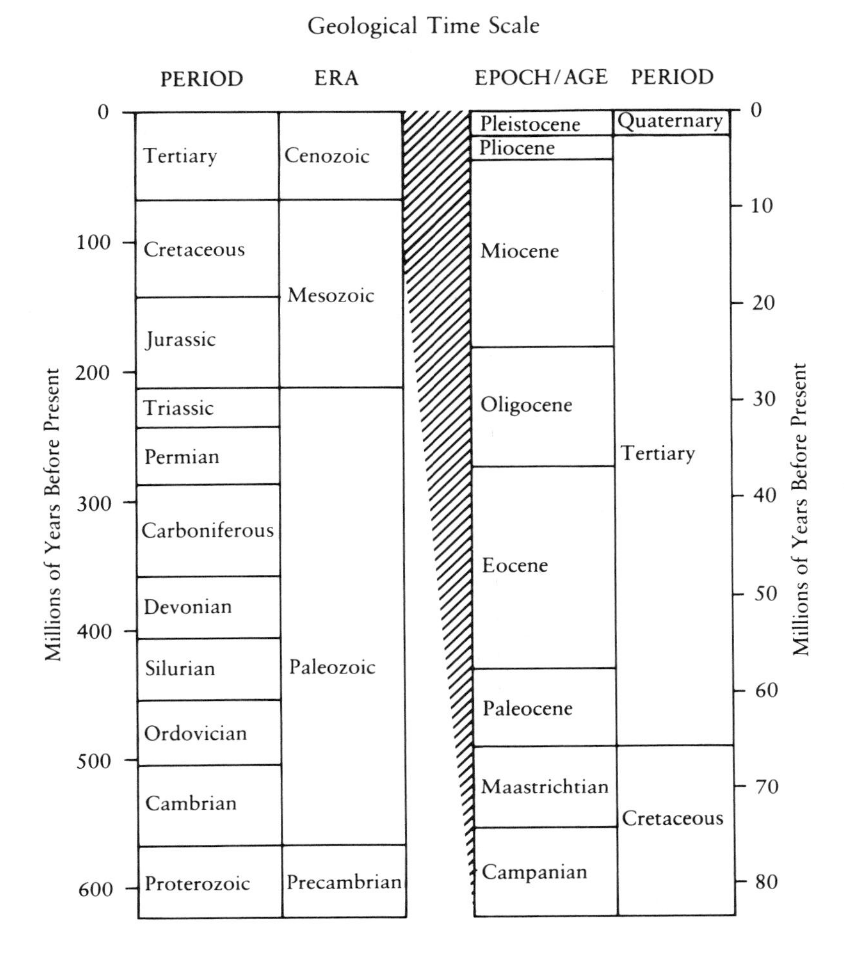

FIGURE 1.6. Geological time scale.

from Central America to the Aleutians. The Atlantic and Gulf coastal plains and the adjacent continental shelf are a passive continental margin. The west coast of North America is an active continental margin that coincides with the western zone of the Cordillera (A. W. Bally et al. 1989).

A. W. Bally et al. (1989) provided a description of events and reconstructions of the relative position of North America in the context of the global development of oceans and continents since the Precambrian. The flowering plants—the plant group dominant today—had arisen near the beginning of the Cretaceous (e.g., E. M. Friis et al. 1987b), but probably not much earlier. Contemporary families of angiosperms mostly originated between the mid-Cretaceous (100 Ma, or million years ago) and the Paleocene (65–58 Ma), so that it is the last 100 million years of Earth history that is of prime importance in explaining their contemporary distributions, and those of their constituent genera and species. The contemporary representatives—families and genera—of modern vascular plants have also developed largely during this period.

To help understand the present-day distribution of vascular plants, therefore, we describe and illustrate the tectonic events that have occurred since the Cretaceous, using maps published by A. G. Smith et al. (1981; fig. 1.8). Epicontinental seas are not represented (see A. Graham, chap. 3, fig. 3.1), although they represented significant barriers to dispersal at various times. The paleogeography and paleoclimates of the late Cretaceous and early Tertiary in relation to plant geography and evolution, and the land bridges that linked North America to Europe and Asia, have been thoroughly discussed and summarized (P. H. Raven and D. I. Axelrod 1974; D. I. Axelrod and P. H. Raven 1985; J. T. Parrish 1987; M. E. Collinson 1990; B. H. Tiffney 1985b; D. W. Taylor 1990; D. I. Axelrod et al. 1991; A. Graham, chap. 3).

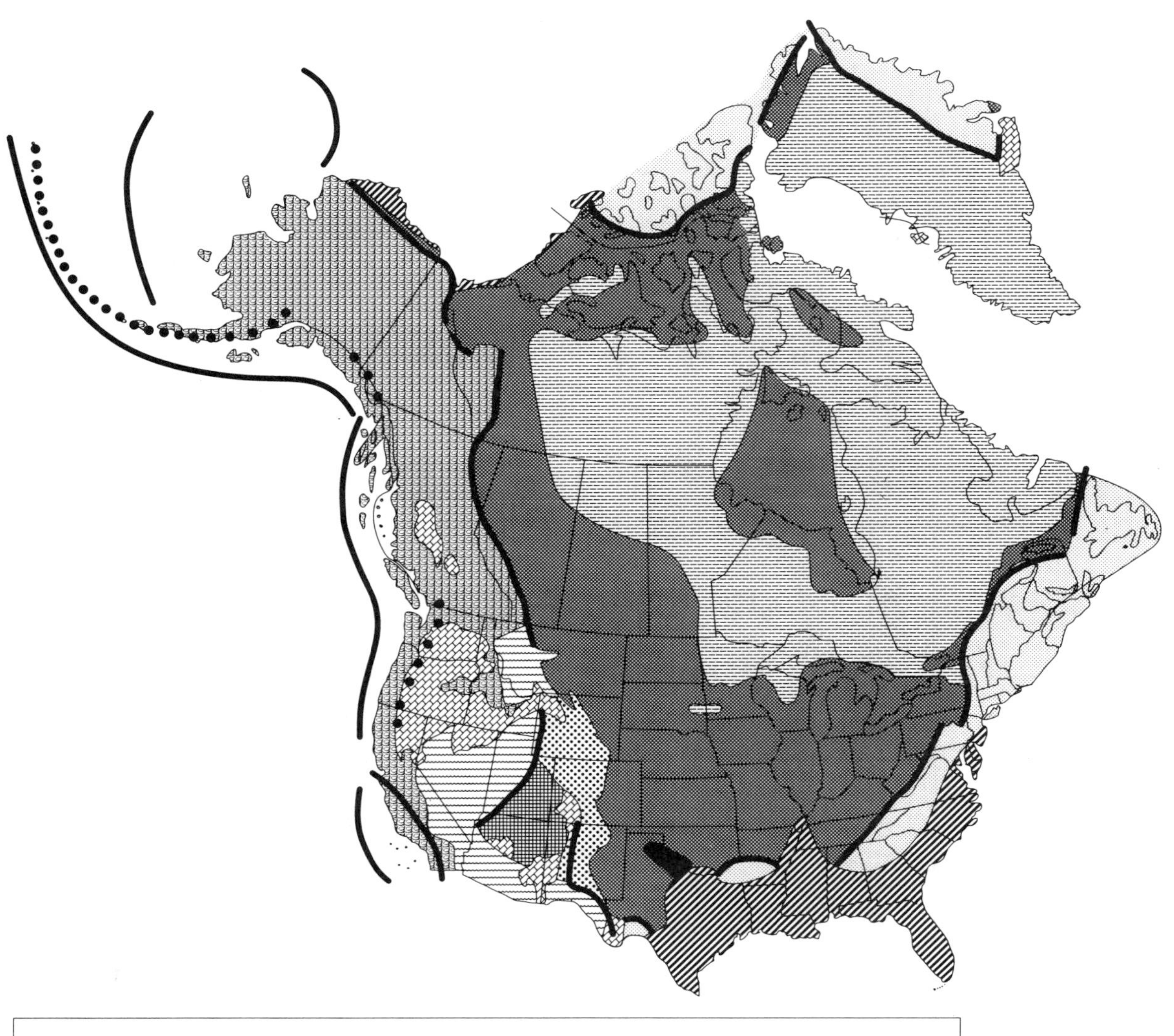

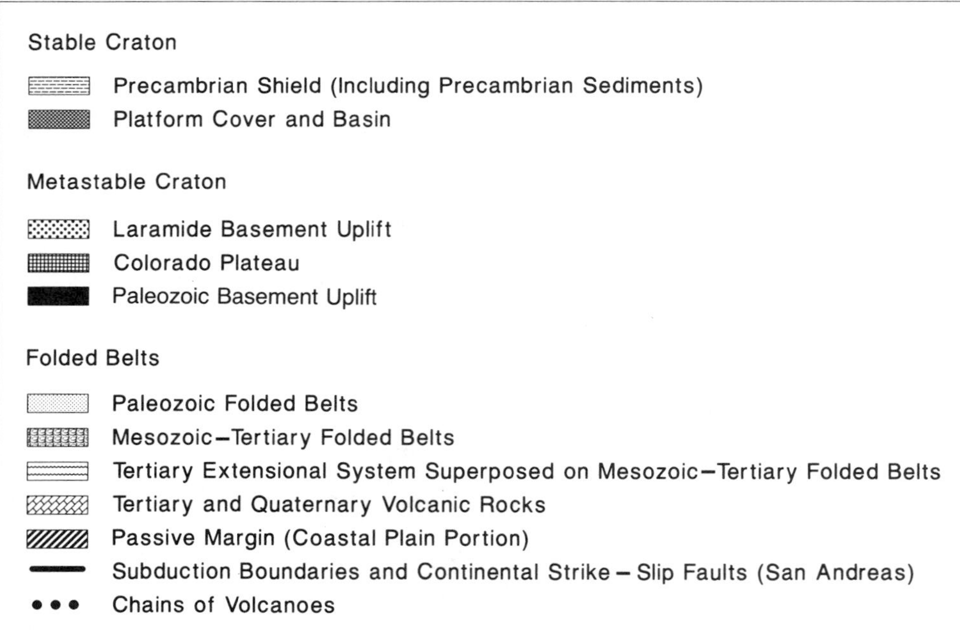

FIGURE 1.7. Simplified tectonic map of North America. [Modified from A. W. Bally et al. 1989, with permission from the author.]

In the Jurassic, the continents were close together and formed the super-continent Pangaea. In the Cretaceous, they became increasingly separate due to the expansion of new oceanic basins. In the early Cretaceous, a narrow oceanic basin formed in the position of the central Atlantic Ocean between South America and Africa, and possibly also in the Gulf of Mexico. South of the Gulf, the Proto-Caribbean Sea formed in the widening gap between North and South America (fig. 1.8a). Expansion of the central Atlantic and Pacific oceans occurred. In the late Cretaceous, the spreading of the seafloor continued between North America and Africa, and a new rift system began between North America and Europe (fig. 1.8b). The Proto-Caribbean stopped widening, and the Greater Antilles arc (including Cuba) began its collision with the Bahama Platform, causing the initiation of the Panamanian arc system.

The south Atlantic and the west Pacific oceans expanded during that time. Intensive volcanism occurred, associated with the subduction (phenomenon whereby the lighter oceanic plate is driven underneath the denser continental plate along their collision front) of the Pacific Plate. A large meteoric impact may have occurred at the end of the Cretaceous. This cataclysm may have been responsible for mass extinction of living groups (volcanism is given as the alternative explanation of this phenomenon; M. E. Collinson 1990). Although Europe and North America formed a single geological unit, sea level was high during the Cretaceous, and shallow continental seas between Europe and eastern North America and between eastern and western North America probably prevented direct overland migration until the beginning of the Tertiary (A. Graham, chap. 3, fig. 3.1). The relative position of North America and Siberia had been stable since the end of the Mesozoic (fig. 1.8c–f).

In the early Oligocene (the period immediately following that illustrated in fig. 1.8d), Baffin Bay and the Labrador Sea opened, and Greenland separated from Norway. Strike-slip motions along the Nares Strait occurred between Greenland and Ellesmere Island. Formation of the Eurekan folded belt in eastern Ellesmere Island resulted in the opening of northern seaways. Seafloor spreading continued in the central and southern Atlantic. In the Caribbean, the eastern island arc system of the Caribbean Plate began to approach its modern location. The leading edge of the Pacific Plate impinged on the margin of North America during the Oligocene. Consequently, the San Andreas strike-slip fault system developed, and the Gulf of California opened as present-day Baja California commenced moving northward from approximately the location of the Mexican state of Jalisco.

The San Andreas Fault runs through the Gulf of California to the vicinity of Cajon Pass north of San Bernardino, California, turns west and runs along the San Bernardino and San Gabriel mountains to the vicinity of Mount Pinos, and then turns again, running north-northwest, leaving the continent at the north end of the Point Reyes peninsula just north of San Francisco. All of the lands west of the fault are on the Pacific Plate and have moved to their present positions during the past 35 million years. The Transverse Ranges of southern California, which have formed along the east-west trending section of the fault, are a zone of violent tectonic activity and uplift, and movements along the fault may well also be related to the development of the Basin and Range physiographic province. Changes in plate motion and stress systems may have been responsible for the development of the Basin and Range physiographic province.

North America and Europe remained in contact via Greenland until the Miocene (fig. 1.8e), when the mid-Atlantic Ridge extended to the Arctic Ocean. Before the Miocene, migration of terrestrial organisms could have occurred intermittently, hindered at times by shallow seaways. A narrow seaway existed between Europe and North America from the Cretaceous rise in sea level onward; during the Eocene, high sea levels contributed to submersion of continental regions. Bridges may have emerged for short times during drops in sea level in early Paleocene and early Eocene. Direct overland migration between North America and Europe could not have occurred through eastern Asia during that period.

In Paleocene and early Eocene, eastern Asia was isolated from Europe by the Turgai Strait, a shallow seaway. This strait was closed by the early Oligocene, but the European landmass was then separated from the Russian platform by the Polish Seaway. The land between Alaska and northeastern Siberia was continuous throughout late Cretaceous and early Tertiary. Submergence of this region occurred in the late Tertiary. The Isthmus of Panama was closed in the late Pliocene, allowing the direct overland interchange of terrestrial biotas between North and South America for the first time during the history of the flowering plants.

Continental Glaciation in North America

The beginning of the Quaternary is set by convention at 2 Ma (H. E. Wright Jr. 1989). During the Quaternary major glaciations that strongly affected North American biotas repeatedly occurred. Glaciations resulted from both climatic and geological phenomena. Glaciers affected regional weather (P. A. Delcourt and H. R. Delcourt, chap. 4), and they modified the landscape, scouring the surface or producing surficial deposits of considerable extent. These deposits can result

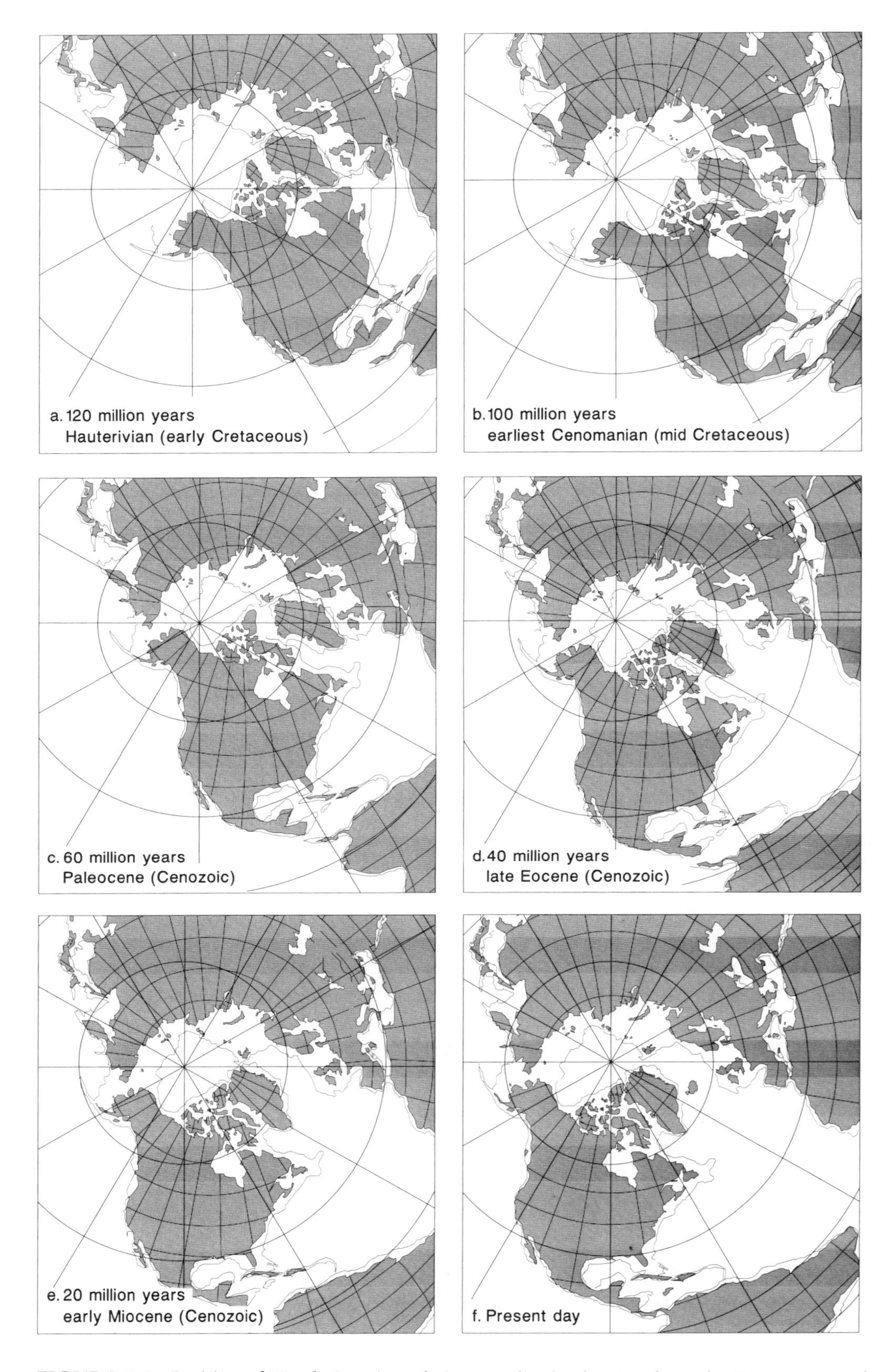

FIGURE 1.8. Position of North America relative to other landmasses from the Cretaceous to the present, as viewed in a polar projection. [Adapted from A. G. Smith et al. 1981, with permission from Cambridge University Press.]

directly from glacial erosion, or less directly from proglacial lacustrine deposition or from transport of silt and sand by wind along modified wind circulation patterns.

Several episodes of glaciation occurred during the Quaternary, and much debate surrounds the interpretation of regional and local events. Geoclimatic events prior to the late Wisconsinan glaciation (also called the Wisconsin glaciation) doubtless affected the distribution of plants and animals, but their reconstruction is uncertain. Also, the last glacial maximum, the Wisconsinan, is the episode that contributed most to the definition of the distribution of modern North American biotas, and to the distributions of the taxa treated in this flora. Therefore, we have limited our summary of physical glacial events (i.e., position and contour of glaciers, proglacial lakes, and marine invasions) to the late Wisconsinan, from maximum glaciation to total deglaciation, i.e., 18,000–ca. 5000 yr B.P. (fig. 1.9). These physical factors directly determined the areas where living organisms could survive (refugia; see R. W. Mathewes 1989) or migrate (e.g., C. E. Schweger 1989) during the Pleistocene. (For discussions of various reconstructions and models proposed by Quaternary geologists, see R. F. Flint 1971; A. Weidick 1976; H. E. Wright Jr. 1989; and authors in W. F. Ruddiman and H. E. Wright Jr. 1987 and in R. J. Fulton 1989. Vegetation and climatic changes during the late Wisconsinan are discussed in P. A. Delcourt and H. R. Delcourt, chap. 4.)

Glacial maximum occurred at about 18,000 yr B.P. (fig. 1.9a), but it was not reached simultaneously all along the ice limits. At that time, ice covered nearly the whole of Canada (except for much of Yukon) and lowland areas in the Arctic. The extent to which the Gulf of St. Lawrence was ice-free is controversial. Glaciers in Alaska were restricted to the Alaska Range and adjacent areas, and to parts of the Brooks Range. The Bering Strait area was dry land, allowing migration between eastern Siberia and Alaska. Greenland was completely ice-covered, and pack ice reached to the southern end of the island. Two major ice masses, confluent at that time, constituted the ice cover: the Cordilleran Ice Sheet in the Canadian Cordillera, and the Laurentide Ice Sheet that covered half of the continent east of the Rocky Mountains. Separate, smaller ice sheets, also confluent, occupied the High Arctic and Newfoundland. Discontinuous mountain glaciers were present in several high ranges of the western United States, and pluvial lakes (P. A. Delcourt and H. R. Delcourt, chap. 4, fig. 4.9) developed at various times during glacial retreat. Sea levels lowered as much as 200 m, and broad expanses of the continental shelf were exposed and available for plant colonization and migration. Meltwater discharge was mainly to the south of the ice, notably via the Mississippi River system.

At 14,000 yr B.P. (fig. 1.9b), the ice sheets had receded perceptibly, particularly in the east. An ice-free corridor was gradually developing between the Cordilleran and Laurentide ice masses. The Greenland sheet had receded, but the island was still fully covered. Mountain glaciers were disappearing. Proglacial lakes were forming at the margins of the retreating ice sheets between morainic deposits and the ice sheet; some of them were forerunners of the Great Lakes. The Des Moines lobe reached its maximum in Iowa, contributing to the delimitation of the Driftless Area of Wisconsin. Between 18,000 and 12,000 yr B.P., strong winds from the northwest deposited sand and loess over the Great Plains. Drainage to the ocean was largely to the south.

At 12,000 yr B.P. (fig. 1.9c), ice was globally receding. The Cordilleran and Laurentide ice sheets were separated in Alberta by a corridor through which migration could occur between Beringia and the southern, unglaciated part of the continent. Mountain glaciers in the United States became more restricted. Major proglacial lakes were found all along the southern and western margins of the Laurentide Ice Sheet, notably Lake Agassiz in the northeastern prairies and the precursors of the Great Lakes. In the east, the ice was retreating from the St. Lawrence Valley. A local dome existed over Newfoundland. Sea levels were slowly rising, inundating the continental shelf and reducing the area of the coastal plain.

Around 10,000 yr B.P. (fig. 1.9d), in response to the warming of the climate, recession of the ice accelerated. The Cordilleran Ice Sheet was broken into several domes; mountain summits became ice-free before lowlands. Ice had disappeared from most mountain areas of Alaska and the western United States, except in the St. Elias and Alaska ranges; isolated glaciers still persisted locally. Much of the Arctic was also ice-free, and the coastal areas of Greenland were uncovered. Retreat of the Laurentide ice mass north of the St. Lawrence Valley allowed marine invasion by the Champlain Sea to occur. Large proglacial lakes existed along the western margin, notably Lake Agassiz. The Great Lakes Basin was occupied by smaller water bodies. Drainage could now occur to the east, via the St. Lawrence estuary, as well as to the south. The Laurentide ice mass started receding over the Torngat Mountains of Labrador, and little ice was left on Newfoundland.

By 8400 yr B.P. (fig. 1.9e), ice remained in the Canadian Cordillera and Arctic only in areas where glaciers persist today. The Laurentide ice mass was centered on Hudson Bay, having receded from west of the

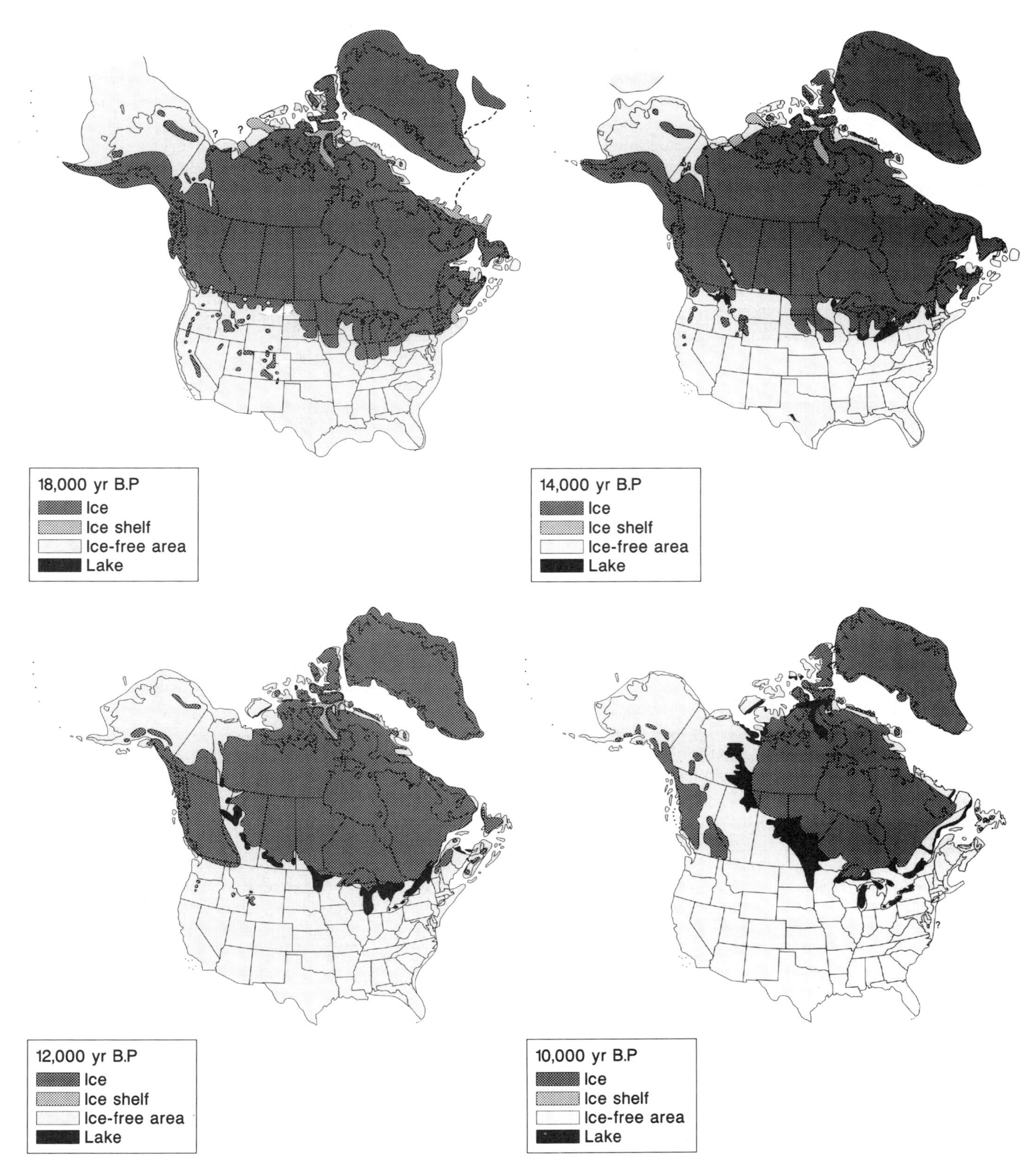

FIGURE 1.9. Deglaciation of North America from the maximum of the Wisconsinan glaciations (18,000 to 5000 yr B.P.). [Based on A. S. Dyke and V. K. Prest 1987; modified with data from D. M. Hopkins 1982; T. D. Hamilton et al. 1986b; T. Hughes 1987; S. Funder 1989.]

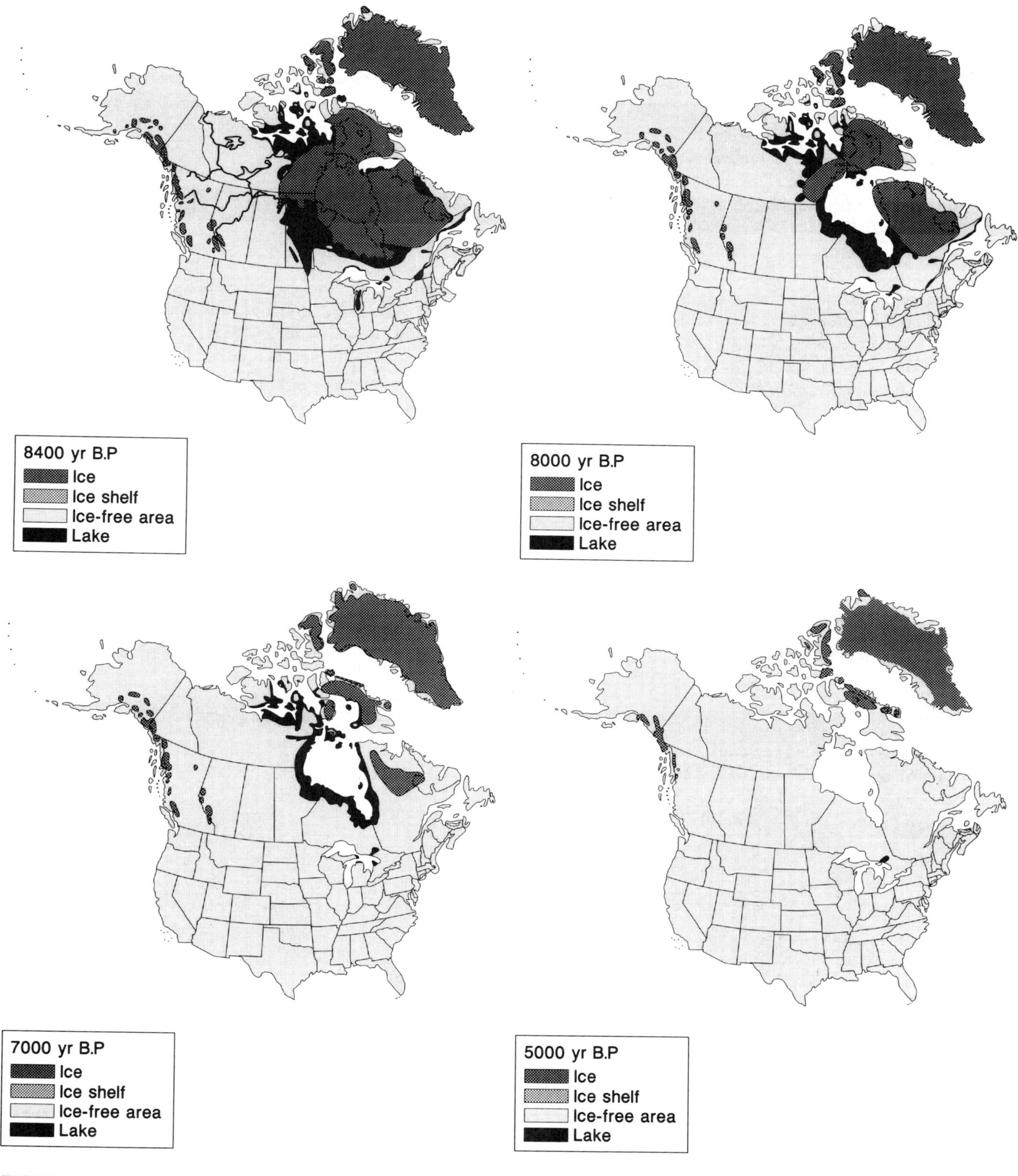

FIGURE 1.9 (Continued).

Torngat Mountains in the east, the Keewatin District in the west, and the south of James Bay. A separation was forming between the Keewatin and Laurentide domes. Lake Agassiz occupied nearly all of Manitoba and extended in the east to join Lake Ojibway–Barlow, responsible for the deposition of the Clay Belt of northern Ontario and western Quebec. The Champlain Sea was receding from the St. Lawrence Valley and the Laflamme Gulf occupied the Saguenay–Lake St. John area. The Atlantic Coastal Plain was nearly as it is now.

Soon after this time, a sudden marine invasion occurred around the periphery of Hudson Bay, provoking a fast recession of the ice and its separation into the Keewatin and Labrador domes by 8000 yr B.P. (fig. 1.9f). By then, the Tyrrel Sea was transgressing over much of the Hudson Bay Lowlands, Hudson Strait was open, and the Champlain Sea was nearly gone. The Great Lakes slowly achieved their present extent. By 7000 yr B.P. (fig. 1.9g), the ice had disappeared from Keewatin, but a dome was left on Baffin Island. The Labrador dome slowly melted in central New Quebec and adjacent Labrador. Through that whole period, ice receded from the periphery of Greenland, and the Greenland Ice Sheet reached a minimum at about 5000 yr B.P. (fig. 1.9h), followed by a readvancement to the position it occupies today. No ice was left on the continent at that time, except in the glacier areas of the Cordillera. The domes on Baffin Island receded to the areas where glaciers are found at present.

Geographical Features

River systems, surficial geology, bedrock types, and permafrost are features of North American geography that transgress the boundaries of the physiographic units described below. Thus a general description of these features is provided here.

River Systems

North America has several major river basins (fig. 1.4). The St. Lawrence River drains the Great Lakes Basin and opens into the Gulf of St. Lawrence. The Mississippi River system drains the center of the continent between the Appalachian and Rocky mountains from near the Canadian border southward, and terminates at the Gulf of Mexico. Flowing at the border of Texas and Mexico, the Rio Grande system drains the southern Rocky Mountains and reaches the Gulf of Mexico. The Colorado River Basin drains a part of the Basins and Ranges and the Colorado Plateau, and finishes at the Gulf of California. The Columbia River and its tributaries, which drain the Columbia-Snake plateaus, and the Fraser River, which drains the interior of British Columbia, both empty into the Pacific at midlatitudes.

The mighty Yukon River drainage system covers the interior of Yukon and Alaska, and terminates at the Bering Sea. The Mackenzie River Basin takes its source in the Canadian Rocky Mountains and in the great northern lakes (Great Bear, Great Slave, Athabaska), and empties into the Beaufort Sea. The Canadian Prairies and adjacent areas of Minnesota and North Dakota are drained by several rivers (Saskatchewan North, Saskatchewan South, Souris, Red) that flow from the Rocky Mountains or the border area, near or through Lake Winnipeg, north of which they merge into the Nelson River, which reaches Hudson Bay. Other rivers on the continent are smaller, drain relatively small areas, and empty directly into the adjacent oceans (fig. 1.4).

Surficial Geology

The following description is based on the map by R. C. Heath (1989b). The diversity of surface deposits encountered on the continent reflects the erosional and depositional processes that have predominated in various parts of North America at various times (fig. 1.10). The northern half of the continent is covered with deposits (e.g., tills, moraines) that reflect the action of the ice sheets, especially during the Wisconsinan glaciation (fig. 1.9). Some areas were laid bare by glacial action, notably mountainous British Columbia, ice-free Greenland, parts of the Arctic Archipelago and of the Ungava Peninsula, the north shore of the Gulf of St. Lawrence, and much of Newfoundland (fig. 1.10). Where proglacial lakes or marine inundations occurred (fig. 1.9), extensive silts and clays were left; these are prominent in central Canada and adjacent United States, south and west of Hudson Bay, in the Arctic Coastal Plain, in the Arctic Lowlands, south of the Bay of Ungava, and in the St. Lawrence Valley. Lake deposits are also encountered in areas where pluvial lakes (P. A. Delcourt and H. R. Delcourt, chap. 4) existed within the Great Basin physiographic province during the time of glaciation (fig. 1.10).

The Central Valley of California was also inundated by sea at that time. Alluvial plain and loess deposits in the Great Plains and in the Central Lowlands are associated with fluvial and aeolian action during the Wisconsinan glaciation, particularly in the central United States, east of the Rocky Mountains (fig. 1.10). Surficial material in much of the area outside of the limits of the Laurentide and Cordilleran ice sheets, in Alaska-Yukon and in the conterminous United States, results mostly from in situ weathering. In the mountainous, arid western United States, water and wind erosion played a major role, leaving bare rock and alluvial fans

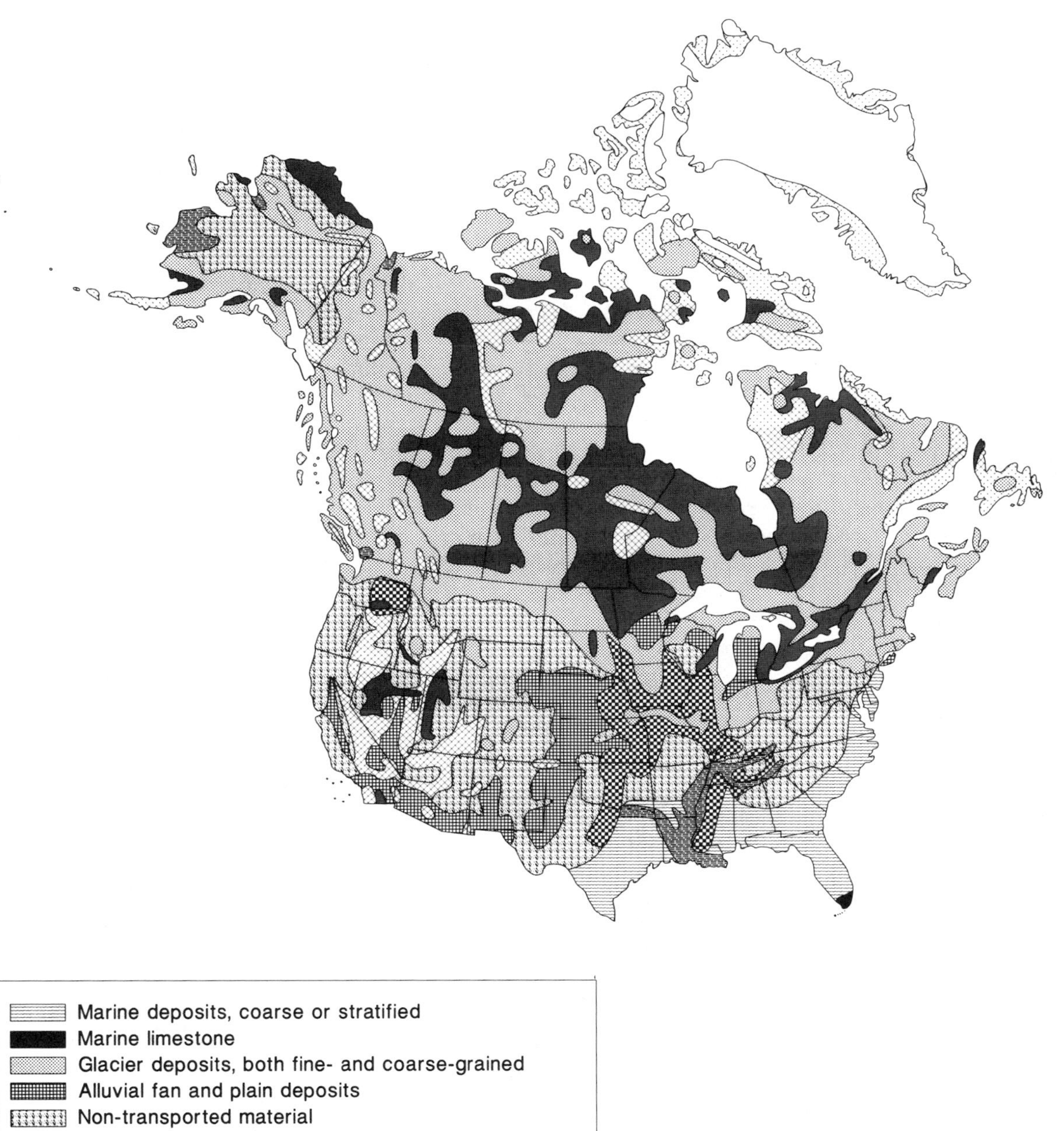

FIGURE 1.10. Surficial geology of North America. [Simplified from R. C. Heath 1989b, with permission of the author.]

and plains (fig. 1.10). The Atlantic and Gulf coastal plains are covered by coarse, stratified marine deposits of Cretaceous to Quaternary age, with marine limestone prominent in southern Florida. Finally, the alluvial plains and deltas of the Yukon, Mackenzie, and Mississippi rivers have fine-grained deposits (fig. 1.10).

Bedrock Geology

The following description is based on the map by R. C. Heath (1989). The rock formations that underlie surficial deposits and that are also sometimes exposed (fig. 1.11) are diverse and reflect tectonic processes. Unconsolidated deposits (see also fig. 1.10) predominate in the Atlantic and Arctic coastal plains, in the western Great Plains, in intermontane basins and valleys, in the Central Valley of California, and in parts of Alaska (fig. 1.11). Intrusive igneous and metamorphic rocks (e.g., granites, gneiss) dominate the Canadian and Greenland shields, Newfoundland, and parts of the Appalachian Mountains and adjacent Piedmont; and several batholiths are found in the western Cordillera, including a large part of British Columbia and Yukon. Sedimentary rocks predominate on the stable platform surrounding the craton (fig. 1.7) and cover the Great Plains and Central Lowlands, the St. Lawrence Lowlands, the Hudson Bay Basin, the Colorado Plateau, and the Wyoming Basin, and parts of Alaska, the Arctic Archipelago, and northern Greenland. Metamorphic formations have been folded notably in the Queen Elizabeth Archipelago, along the Appalachian Mountains and in the Ouachita Mountains, in the Rocky Mountains and Brooks Range, and in the Pacific Coast Ranges. Volcanic rocks are encountered mainly in the Aleutian Archipelago, in the interior of British Columbia, in the Columbia-Snake plateaus, and in various areas of the Basins and Ranges physiographic province (fig. 1.11).

Glaciers and Permafrost

Because of its high latitude, the Arctic Ocean is almost completely covered by a permanent sea pack of ice. In winter, the continuous sea pack extends southward to cover the Bering Sea as far south as the Alaska Peninsula in the west, and reaches south of Greenland and along the Labrador Coast in the east. Looser pack ice also covers the northeastern coast of Newfoundland and the Gulf of St. Lawrence. The Great Lakes, the large northern lakes (fig. 1.4), and the St. Lawrence River also freeze during the winter months. Ellesmere Island and northern Greenland are the landmasses closest to the North Pole (figs. 1.4, 1.12), and the whole Arctic Archipelago and parts of the continental Northwest Territories and of Alaska are located north of the Arctic Circle (fig. 1.4). An icecap covers about 90% of Greenland (figs. 1.4, 1.10). The highlands of the Queen Elizabeth Islands and of Baffin Island (Arctic Archipelago), and the St. Elias Range (Alaska-Yukon) have glaciers (fig. 1.10). In the western Cordillera, smaller glaciers are found farther south along the Coast Ranges and in the Rocky Mountains.

Permafrost is ground that remains permanently frozen to a certain depth; in summer only the surface thaws. Figure 1.12 illustrates the distribution of permafrost in North America (T. L. Péwé 1983b). Continuous permafrost is found in the northern two-thirds of Greenland, the Canadian Arctic Archipelago, the Ungava Peninsula and the Torngat Mountains, and the south coast of Hudson Bay, northwestwardly to Great Bear Lake, northern Yukon, and the Brooks Range area. Discontinuous permafrost occurs in the southern third of Greenland, Alaska, and the Northwest Territories, and the northern third of British Columbia, then southeastward across the northern Prairie Provinces to James Bay and southern Labrador. South of this area, only alpine permafrost is encountered. In the west, it is found in the Coast Ranges of British Columbia, the northern Cascade Mountains, the Sierra Nevada, and along the Rocky Mountains (fig. 1.12). Small areas of alpine permafrost also occur in the east in the Shickshock Mountains of the Gaspé Peninsula and on the highest mountains of New England.

Physiographic Regions

The division of North America north of Mexico into physiographic units (fig. 1.13) has been addressed in detail by several authors (notably N. M. Fenneman 1931, 1938; C. B. Hunt 1974; J. B. Bird 1980; M. J. Bovis 1987; W. D. Tidwell 1972; R. F. Madole et al. 1987; and E. C. Pirkle and W. H. Yoho 1982). This section is a synthesis of their information. Absolute consistency does not exist concerning the circumscriptions or nomenclature of physiographic units among the sources consulted. The following classification reflects these uncertainties, and it should not be construed as definitive. The formal physiographic units (usually provinces) of the various authors were grouped into informal "systems." These systems are aggregations of physiographic units that have similar geologic histories and maintain geographic continuity and distinctness. The systems are described in a clockwise sequence, from the northeast to the northwest.

Arctic System

The Arctic system (fig. 1.13) comprises the extreme north of the continent and is fronted on the north by the Arc-

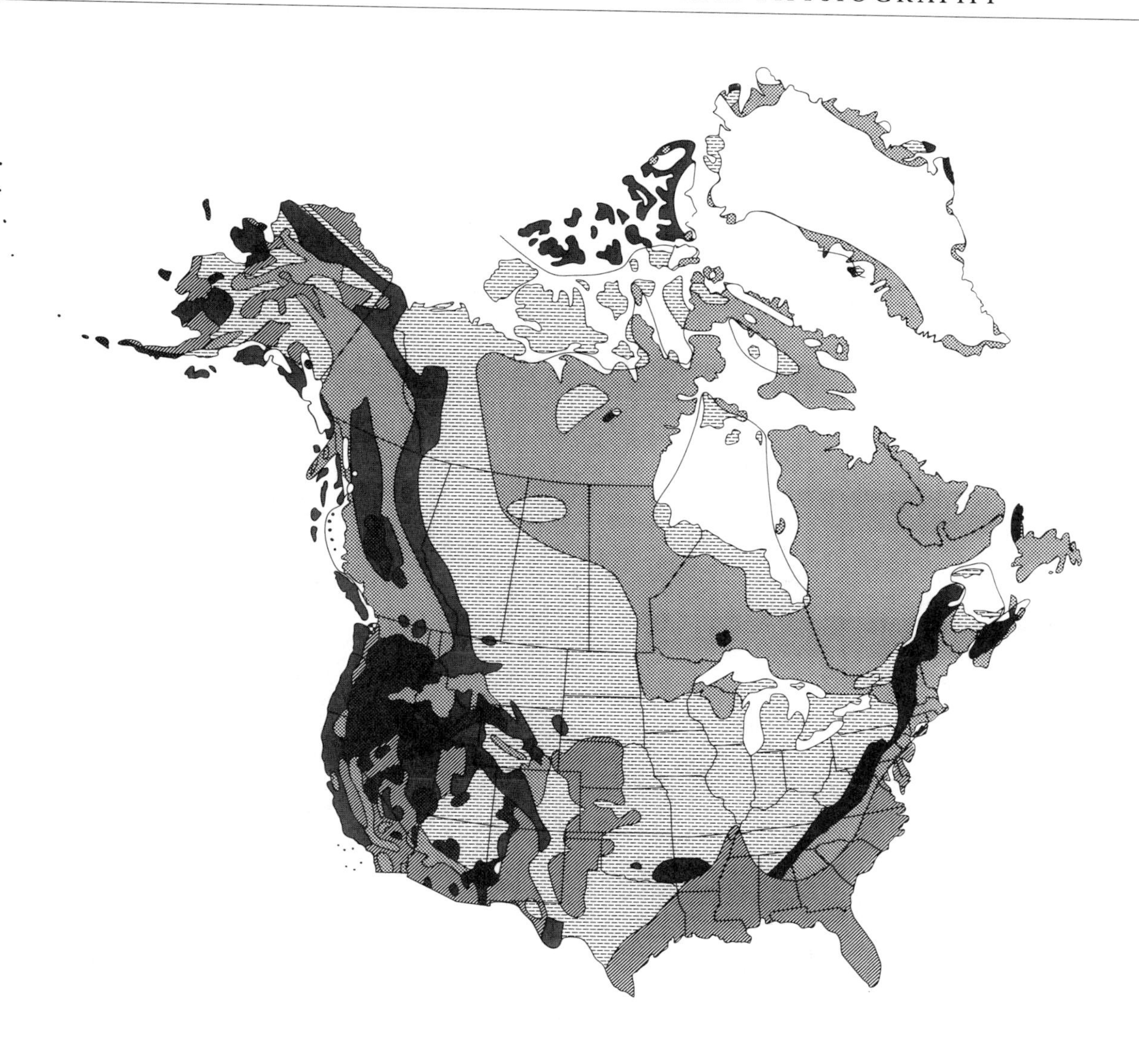

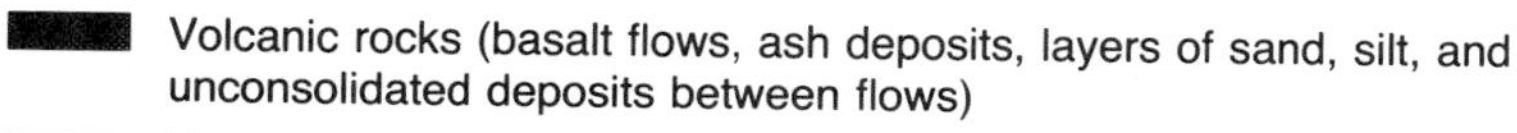

Volcanic rocks (basalt flows, ash deposits, layers of sand, silt, and unconsolidated deposits between flows)

Unconsolidated deposits (gravel, sand, clay, and semi-consolidated limestone of Coastal Plains, Great Plains, and intermontane valleys)

Flat-lying to gentle-dipping, consolidated and semi-consolidated rocks (sandstone, shale, limestone, dolomite)

Intrusive igneous and metamorphic rocks (granite, gneiss, schist, phyllite, etc.)

Folded consolidated rocks (sandstone, shale, limestone, dolomite)

FIGURE 1.11. Bedrock geology underlying surficial deposits in North America. [Simplified from R. C. Heath 1989, with permission of the author.]

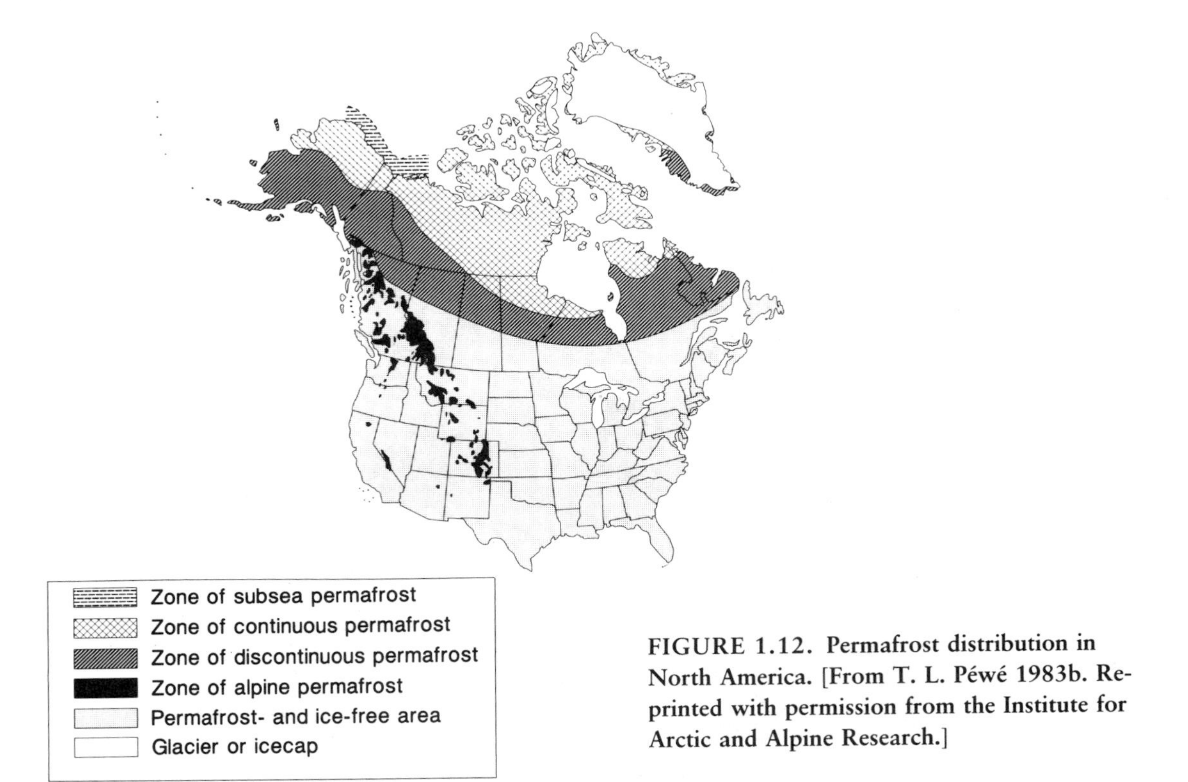

FIGURE 1.12. Permafrost distribution in North America. [From T. L. Péwé 1983b. Reprinted with permission from the Institute for Arctic and Alpine Research.]

tic Ocean. The system encompasses about 1.3 million km^2, including intervening waters. Included are the Caledonian Uplands, High Arctic Uplands, Arctic Lowlands, and Arctic Coastal Plain provinces.

Quaternary glaciers covered most of this system (fig. 1.9); they created straight, deep channels and numerous fjords, and removed most of the weathered mantle from the islands, often leaving bare rock (fig. 1.10). Marine transgressions over the Arctic Coastal Lowlands and Arctic Coastal Plain were extensive after glaciation (fig. 1.9). Glaciers remain on promontories of Axel Heiberg, Ellesmere, and Devon islands, and on Greenland (fig. 1.13).

The rugged North Greenland Mountains and the High Arctic Uplands are geologically related (figs. 1.7, 1.13). Both were uplifted during the Innuitian (Paleozoic) and Eurekan (Cretaceous) orogenies (H. P. Trettin 1989). Subsequently, they were separated by the movements of the Greenland Plate. The rocks are mostly folded sedimentary formations (fig. 1.11). Elevations on Ellesmere Island reach 2604 m (Barbeau Peak). The Paleozoic Caledonian orogeny uplifted the mountain ranges of northeastern Greenland that constitute the Caledonian Province (A. Escher and W. S. Watt 1976). The topography of the western part of the system includes plateaus, lowlands, and coastal plains dissected by channels and fjords. The Mackenzie Delta comprises fine-grained deposits of sand, silt, and clay (fig. 1.10). The Arctic Coastal Plain (figs. 1.4, 1.11), also referred to as the Arctic Slope, is low and dotted with ponds and lakes. It stretches 1200 km from a point 160 km east of Point Hope (Alaska) to the Mackenzie Delta. At its widest, it is about 160 km across.

Canadian Shield System

The Canadian and Greenland shields form the core of the continent (figs. 1.7, 1.13). The Greenland Shield rifted from the Canadian Shield during the Tertiary and now is separated from the mainland by Davis Strait (fig. 1.4); it constitutes most of Greenland. The north-south axis of the Greenland Shield is about 2700 km, and the east-west axis, 1200 km. Most of it lies below sea level because of the weight of the 3-km-thick icecap of the Greenland Icecap Province. This ice occupies almost 90% of the surface of Greenland (G. Halliday 1989). A narrow, discontinuous, coastal band, mostly along the west side, is ice-free. This zone comprises deep fjords and a rocky coast (Western Mountain Fjords Province).

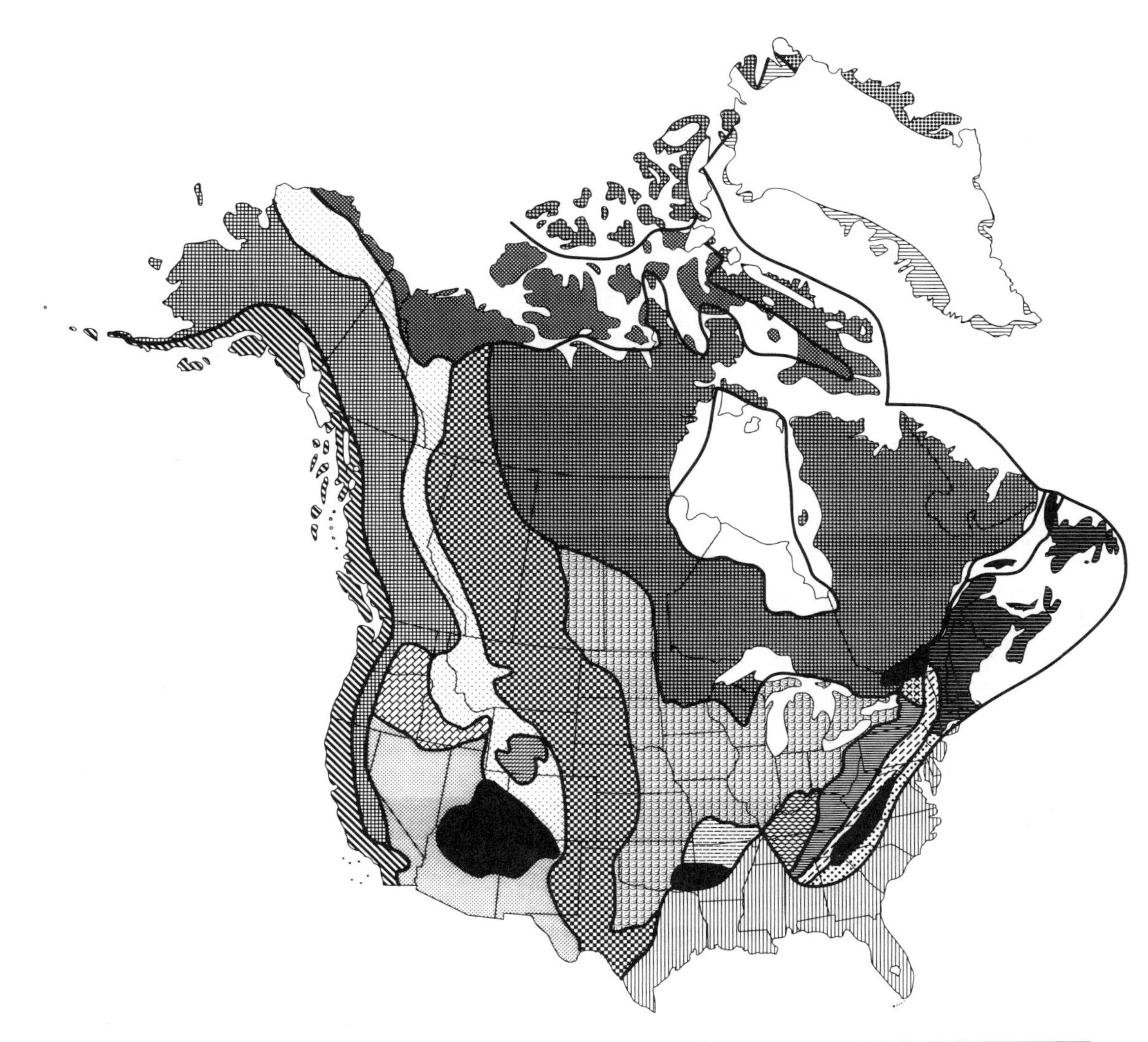

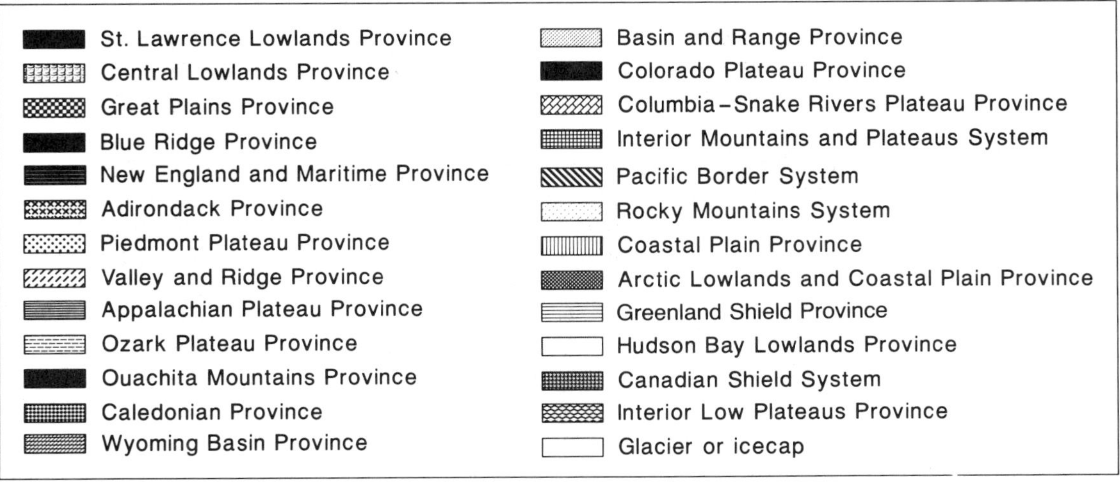

FIGURE 1.13. Physiographic regions of North America. [Modified from C. B. Hunt 1974, with permission from W. H. Freeman and Company, using data from M. J. Bovis 1987 and A. Escher and W. S. Watt 1976b.]

The Canadian Shield is a gently rolling peneplain, about 3400 km long and 3000 km wide, covering over 5 million km^2. The elevation is mostly below 600 m, with half lower than 300 m; some peaks exceed 1500 m. It comprises mostly acidic, granitic uplands separated by igneous and sedimentary metamorphic formations (fig. 1.7). At one time, sedimentary rocks covered it, but subsequent erosion exposed Precambrian formations and only left remnants of the sedimentary formations. A series of major lakes is found along its western margin with the Interior Plains: Winnipeg, Athabasca, Great Slave, and Great Bear lakes. The entire region was glaciated, and the cores of the continental glaciers were found in this system, one in Franklin District and the other on the Quebec-Labrador peninsula. Eastern and western portions of this shield are covered by coarse-grained tills. Areas peripheral to Hudson Bay were inundated during glaciation by extensive proglacial lakes and seas that left behind silts and clays (fig. 1.6).

The Canadian Shield has been divided into a series of provinces based on geological history, but they are not particularly relevant to modern-day plant distributions. Mountainous highlands with eroded summits are found in several belts on the eastern half of the shield. These highlands include the Davis Mountains (1500 m; maximum 2591 m) on Baffin Island, which are partially glaciered, the Torngat Mountains (1200 m; maximum 1595 m; fig. 1.4) in northern Labrador, and the Otish Mountains (1135 m) in the Laurentian Highlands of Quebec. Thelon Plains (Northwest Territories) and Athabasca Plains (Alberta-Saskatchewan), two extensive low-lying plains of sandstone, limestone, and dolomite, are found on the western half of the shield (fig. 1.11). Sand dunes occur on the south shore of Lake Athabasca. In the middle of the shield, Hudson Bay north to Southampton Island, James Bay, and the Hudson Bay Lowlands form a shallow basin that is over 1600 km across from north to south (figs. 1.4, 1.11, 1.13). These lowlands are limestone overlain by marine sediments of Quaternary age (figs. 1.10, 1.11).

Appalachian System

The Appalachian system consists of a series of roughly parallel plateaus, valleys, and ridges that extend for more than 3000 km, from Newfoundland southwestward to central Alabama (fig. 1.13). The system resulted from Paleozoic orogenic events followed by erosion (fig. 1.7). The central ranges are thrust-fault mountains flanked by plateaus. This system includes several physiographic provinces: Maritime and New England, Adirondack Mountains, Piedmont Plateau, Blue Ridge, Valley and Ridge, and Appalachian Plateau.

The Maritime and New England Province is made of hilly uplands ranging from sea level to about 450 m, with a few peaks exceeding 1800 m (Mt. Washington, 1907 m). In New England, rocks are mostly gneiss and granite, whereas in the Gaspé Peninsula and in the Maritimes, rocks are mostly sedimentary (sandstones) and igneous intrusions (fig. 1.11). Within this province, substrates are mainly acidic. Pleistocene glaciers covered the entire area, leaving till as they receded; much of Newfoundland is bare rock (fig. 1.10).

The Adirondack Mountains Province is a nearly circular dome of Precambrian acidic rocks surrounded by Paleozoic sedimentary formations (figs. 1.11, 1.13). The highest peak, Mt. Marcy, is 1629 m. The Adirondacks were completely glaciated and are now covered by coarse till (fig. 1.10).

The Piedmont Plateau Province extends 1600 km from southern New York to central east Alabama (fig. 1.13). It is a rolling plateau tilted eastward from 450 m to about 150 m at the Fall Line, which represents the boundary with the coastal plain. Rocks are metamorphic and diverse, but most are acidic gneiss and schist (fig. 1.11). Along the eastern edges of the plateau, expanses of granites are exposed as flatrocks.

The Blue Ridge Province consists of a narrow range extending from southeastern Pennsylvania southward to northern Georgia (fig. 1.13). The ridge system is 900 km long and 8 to 80 km wide. Elevation is from 750 to 1950 m, and Mt. Mitchell (2326 m) is the highest point of eastern North America. Substrates are mostly Precambrian granitic rocks (fig. 1.11).

The Valley and Ridge Province is 2000 km long and 40 to 120 km wide, a complex of ridges and valleys extending from southern Quebec to central Alabama (fig. 1.13). Sandstones, shales, and limestones of varying degrees of metamorphosis constitute major rock formations (fig. 1.11). The area is noted for its fold mountains and broad flat valleys.

The westward sloping Appalachian Plateau Province stretches from the Adirondacks south to central Alabama and covers a distance of 1500 km (fig. 1.13). The plateau surface is mostly at 600 to 900 m. The Catskill Mountains, at the northern end, reach 900 m. The surface comprises mostly resistant, acidic sandstones deeply incised by streams. Underlying the sandstones are shales and limestones, which lie exposed along the edges of the plateau (fig. 1.11). The glaciated, northern portion has lakes and till (fig. 1.10).

Coastal Plains System

The Atlantic and Gulf Coastal Plains Province forms a belt along the Atlantic Ocean and the Gulf of Mexico, extending from Cape Cod in Massachusetts south to the Mexican border (fig. 1.13). The coastal plain is an elevated sea bottom of low relief that rises inland from below sea level to an elevation of about 150 m. The submerged portion forms the Continental Shelf of eastern North America, which slopes gently to a depth of about 200 m. The coastal plain dips to the north and east; this slope results in extensive tidal inlets, such as Chesapeake Bay (fig. 1.4), and submerged banks, such as George Banks and the Grand Banks. The 1800-km shore consists of sandy beaches, notably barrier islands (outer banks, such as those at Cape Hatteras; fig. 1.4). Behind these beaches, low-lying areas are occupied by sounds, estuaries, swamps, or wetlands.

The province comprises mostly unconsolidated marine sediments (figs. 1.10, 1.11), eroded from the Appalachian Mountains, of Cretaceous, Tertiary, and Quaternary ages, and occurring in this sequence from inland to the coast. These formations form belts of low ridges and valleys roughly parallel to the coastline. Peninsular Florida is underlain by limestones, and the area is characterized by sinkholes with numerous wet prairies and lakes. Stratified marine limestones, including coral reefs, are exposed extensively in the Everglades and Florida Keys (fig. 1.10).

The Mississippi alluvial plain (fig. 1.4) divides the province into three areas: eastern (Atlantic and eastern Gulf coasts), central (Mississippi flood plain), and western (coastal band of Texas). This alluvial plain extends from the confluence of the Ohio and Mississippi rivers, where the coastal plain penetrates far inland, to the extensive Mississippi Delta (fig. 1.13). The Mississippi River meanders through the low-lying plain for about 300 km, and drains mostly through its delta. It comprises fine-grained floodplain and delta deposits of silts, sands, and clays (fig. 1.10).

Interior Highlands System

This system comprises two nearly disjunct areas (figs. 1.4, 1.11): the Interior Low Plateau Province east of the Mississippi River, and the Ozark Plateau Province and the Ouachita Mountains Province west of it. The Ozarks and Ouachitas are very old (Paleozoic). The Interior Highlands lie south of the glacial limits.

Elevations of the Interior Low Plateau Province are generally less than 300 m and are characterized by moderate, rolling relief, sloping from the Appalachians toward the Mississippi River. The province is 470 km long and 400 km wide. The Lexington Plain and the Nashville Basin (fig. 1.4) are mostly limestone exposures with sinkholes and underground drainages.

The Ozark Plateau Province (figs. 1.4, 1.13) is primarily a rolling upland of moderate elevation, mostly above 300 m. The province is about 470 km north-south and 270 km east-west. Limestone and dolomite are the most abundant substrates (fig. 1.11). Some of the higher summits have sandstone caps, such as the Boston Mountains in the south that reach more than 600 m (maximum elevation 781 m).

The Ouachita Mountains Province is approximately 360 km long and 100 km wide and is mostly separated from the Ozarks by the Arkansas River Valley. The Ouachita Mountains arose from the same uplift as the Appalachians. They comprise folded linear ridges and valleys ranging east to west, with a maximum elevation of 839 m (Magazine Mountain). Predominant rocks are sandstone and shale (fig. 1.11).

Interior Plains System

The Interior Plains system stretches from the Appalachian Mountains, Interior Highlands, and Canadian Shield on the east to the Rocky Mountains on the west, and from southern Texas northward to the Arctic Lowlands. It comprises three provinces: St. Lawrence Lowlands, Central Lowlands, and Great Plains (figs. 1.4, 1.13). The Interior Plains constitute the stable platform bordering the Precambrian Shield (fig. 1.7). They cover one-third of the continent and are the most extensive system. An epicontinental sea (A. Graham, chap. 3, fig. 3.1) linked the Gulf of Mexico and the Beaufort Sea across this area during the Cretaceous and early Tertiary. It deposited extensive sediments on the Great Plains and the western part of the Central Lowlands (fig. 1.11).

The St. Lawrence Lowland Province follows the course of the St. Lawrence River from the Frontenac Axis (a narrow outcrop of Precambrian rock linking the Canadian Shield to the Adirondacks, just east of Lake Ontario) to western Newfoundland on the Gulf of St. Lawrence (fig. 1.13). This narrow lowland is constricted between the Canadian Shield and the Appalachians. It is oriented southwest to northeast and stretches 1500 km. Elevations vary from sea level to 150 m near the Frontenac Axis. The province consists of three disjunct land areas: the St. Lawrence Valley, Anticosti Island and the Mingan Archipelago, and northwestern Newfoundland. The valley is 500 km long, about 150 km at its widest bulge, and it terminates near Quebec City. In the Gulf of St. Lawrence area, the lowlands are submerged except at Anticosti Island and the Mingan Archipelago, and on the coastal plain of northwestern Newfoundland. The entire area was glaciated. During ice retreat, much of the St. Lawrence River Valley was inundated by the Champlain Sea (fig. 1.9), which left behind marine sediments, mainly clay (fig. 1.10). The province is characterized by flat limestone and sandstone outcrops (fig. 1.11). The Monteregian Hills are igneous intrusions and represent a conspicuous feature on the flat valley. In many respects, the St. Lawrence Province can be considered an extension of the Central Lowland Province.

The Central Lowland Province (figs. 1.4, 1.13) is a vast plain oriented north-south just west of the Mississippi River Valley; it projects into the south central Canadian Prairies, and a broad wedge extends eastward to the Great Lakes. It is 2400 km long (north-south) in the west and 2200 km at its greatest width (east-west). Elevations range from 150 m along the Mississippi River to 600 m in the west. The northern part was glaciated, except for the Driftless Area of southwestern Wisconsin and adjacent Minnesota and Iowa. The shallow proglacial Lake Agassiz covered much of the northern extension of the province, mostly in Manitoba, and left a flat plain of clay and silt sediments (fig. 1.10).

Similar deposits in southern Ontario resulted from various developmental phases of the proglacial Great Lakes. Gravel and sand were deposited on glacial outwash plains in the Lower Peninsula of Michigan and southwest of Lake Superior. Other glaciated areas were covered by fine-grained tills (figs. 1.9, 1.10). The southern portion, the Osage Plain, was not glaciated; it is located south of the Missouri River. Loess deposits (fig. 1.10) cover much of the surface, and they are deeply dissected by steep-sided river valleys; there is little relief otherwise. The Red Hills of south Kansas are noteworthy features caused by stream erosion.

The Great Plains Province is a western continuation of the Central Lowlands but is distinguished by having Tertiary deposits washed eastward from the Rocky Mountains. It stretches from southern Texas to the Arctic Lowlands, east of the Rocky Mountains (figs. 1.4, 1.13). Elevations range from 600 m in the east to 1500 m at the foot of the Rocky Mountains. Most rivers run transversely across the province and drain to the Mississippi River in the United States, and to Hudson Bay and the Beaufort Sea in Canada (fig. 1.4). Since aridity increases toward the west, soils are more alkaline than in the Central Lowlands.

The portion located in Canada and the adjacent United States was glaciated and is covered by fine-grained till and proglacial lake deposits (figs. 1.9, 1.10). The area is flat with rolling hills, the most conspicuous being the gravel-topped, rocky Cypress Hills (Alberta-Saskatchewan; fig. 1.4), which reach 300 to 600 m above the surrounding plains (maximum elevation about 1400 m). The unglaciated southern half lies within the United States.

Most of the western side is an elevated plain with varying degrees of dissection due to stream cutting. Extensive erosion sculpted impressive canyons, hills, and bluffs, referred to as the Badlands, that are located mostly in South Dakota on the Missouri Plateau. Several dome mountains rise 500 to 600 m above the plain; the most prominent are the Black Hills of South Dakota (1000 m above the adjacent plain; highest elevation is Harney Peak, 2207 m; figs. 1.4, 1.11). The Black Hills are an uplift of Precambrian rock topped with limestone eroded in the eastern half, exposing granite and schist. Extensive areas of sand sheets formed by wind from alluvial deposits are distributed over large portions of this section (fig. 1.10). Striking results of this process are stabilized dunes such as the Nebraska Sandhills (fig. 1.4), which occupy 62,000 km^2. At the southern end is an extensive limestone tableland, the Edwards Plateau (fig. 1.4), demarcated from the coastal plain by an escarpment.

Intermountain System

The Intermountain system derives its name from being located between the Rocky Mountains and the Sierra Nevada–Cascades axis. It extends south of the Mexican border (fig. 1.13). The system comprises four distinct and geologically diverse provinces: Basin and Range, Colorado Plateau, Wyoming Basin, and Columbia River plateaus. It was uplifted during the Laramide orogeny (Cretaceous and Tertiary) along with the flanking mountain ranges (fig. 1.7). The intermountain region consists mainly of a series of plateaus that are the highest on the continent.

In arid and semiarid areas, winds cause sheet erosion of cliff faces, valley floors, and plateau surfaces, as opposed to gully erosion by water. At the bottoms of basins that lack external drainage are playas, i.e., mud flats that often have accumulation of salts. These playas derive from lakes that evaporate in times of limited precipitation (P. A. Delcourt and H. R. Delcourt, chap. 4, fig. 4.9). Other characteristic features are alluvial fans that are built from sediments eroded from the adjacent mountains, rock pediments that are sloped and eroded by stream floods, desert pavements, and wind-accumulated sand dunes (fig. 1.10).

The Basin and Range Province is bounded to the north by the lava flows of the Columbia Plateau and extends south into Mexico. The province occupies about 780,000 km^2. Elevations vary from below sea level (Death Valley, -86 m; Salton Sea, -71 m) to nearly 4000 m above sea level at the highest peaks; most of the plateau is at about 1500 m. Relief from basin floor to adjacent mountain tops is from 900 to 1500 m. The province is characterized by broad, level desert basins and narrower, elongate, isolated, parallel mountain ranges trending north to south (fig. 1.4). In the south, this pattern becomes irregular. In the north, most basins lack external drainage. This topography originated by block faulting in the Oligocene, accompanied by extensional phenomena (fig. 1.7). Paleozoic formations predominate; they consist of limestone, siltstone, shale, and sandstone (fig. 1.11). The province is geologically very diverse, and several sections may be recognized (described below).

The Sacramento Mountains Section adjoins the Great Plains. It is mostly plateau with diverse topography and geology. The western and eastern rims are montane, with a basin between, which is occupied by salt basins, clay hills, and dunes rising to 30 m. Erosion of gypsum deposits by water and wind led to the formation of alkaline flats and dunes, such as those at White Sands (New Mexico). The Mexican Highland Section, from southern Arizona to trans-Pecos Texas, is topographically varied. Basins decline westwardly from 1500 m to 600 m, and ranges are 900 to 1500 m higher. The New Mexico and Texas portion drains into the Rio Grande and to the Gulf of Mexico, while the Arizona portion drains to the Gila River and to the Gulf of California (fig. 1.4). Lava flows are found in this section (fig. 1.11).

The Sonoran Desert Section includes the Mojave Desert of southeastern California and the Arizona deserts. Elevations there are mostly below 600 m and rarely exceed 900 m. Mountains are mostly granite and volcanics. Basins are drained, except in the Mojave; aridity, however, often prevents drainage to the sea. The Salton Trough Section is a continuation of the Gulf of California Trough. Imperial Valley is south of the Salton Sea, which occupies the lowest part of the trough.

The northern Great Basin Section is north of the Mojave Desert and is delimited by the Garlock Fault. Centered on Nevada, it has topography typical of the Basin and Range Province: isolated mountain chains oriented north to south, with broad, intermontane basins. John C. Frémont gave the area the name "Great Basin." Only small portions have external drainage. The lowest and highest elevations of the province occur in this section.

In the central area of elevated basins and ranges, valley floors are at 1500 m and have no perennial lakes. Pluvial Lake Bonneville formerly occupied the entire Bonneville Basin and left behind alluvial fans, salt playas, and salt lakes such as Great Salt and Sevier lakes in Utah (figs. 1.4, 1.10; P. A. Delcourt and H. R. Delcourt, chap. 4, fig. 4.9). Similar to the Bonneville Basin is the Lahontan Basin to the west, except that volcanism is more prevalent in the latter. The southern area is similar to the central area. A noteworthy feature of the section is Death Valley, which includes a dry, salt-encrusted playa, the lowest point on the continent.

The Colorado Plateau Province is delimited by the southern and middle Rocky Mountains to the north and east, and by the Basin and Range Province to the south and west (figs. 1.7, 1.13). It covers 340,000 km^2 and is roughly circular, with a diameter of about 470 km. Uplifted to its present level during the Pliocene, it is the highest plateau on the continent: most of it is above 1500 m, some tablelands and several peaks reaching to 3300 m. The surface is deeply dissected by hundreds of steep-walled canyons that expose an impressive display of geological history. Among them, the Grand Canyon is the most famous. The province consists of high plateaus of nearly horizontal sedimentary formations of sandstone, shale, and limestone, volcanic mountains, lava plateaus, and intrusive dome mountains, sand deserts, and shale deserts with badlands (figs. 1.10, 1.11). Rainfall charges streams that drain externally, such as the Colorado River. Glaciation was slight in the area and occurred mainly in south central Utah. Several sections are recognized.

The Grand Canyon Section constitutes the high southwestern part of the province. Rock formations there are Carboniferous, and the northern third of the section is covered by volcanics. The Grand Canyon (fig. 1.4) exposes these formations. Several discrete plateaus are found within this section, such as the Kaibab Plateau, the highest of them (Pt. Sublime, 2273 m). The Datil Section, at the southern end of the Province, is covered by thick lavas (fig. 1.11). North of these two sections is the Navajo Section, a depression with broad flats on shaley formations separated by sandstone cuestas. The Painted Desert is located in this region. Further northward, the Canyon Lands Section is characterized by gorges carved in sandstone and by badlands in areas of thick shale capped by sandstone. The Uinta Basin Section is a deep bowl in Tertiary formations that rises steeply northward onto the flank of the Uinta Mountains (fig. 1.4). Along the western side, the High Plateau Section comprises tablelands higher than 2700 m, with some exceeding 3300 m. Mesozoic and Tertiary formations are often capped by lava flows. Where exposed, the Tertiary formations are severely eroded into badlands, such as Bryce Canyon.

Disjunct from the bulk of the Intermountain system, the Wyoming Basin Province (fig. 1.13) is surrounded by the three Rocky Mountain ranges and opens to the Great Plains along a northeast corridor. Elevations are between 1500 and 2400 m. The province comprises semiarid basins and isolated low mountains.

The Columbia–Snake River Plateau Province occupies the northern end of the Intermountain system (figs. 1.4, 1.13). It is a series of semiarid plateaus of rolling, mostly laminar, basaltic lava flows (figs. 1.7, 1.11). Much of the surface is covered by loess (fig. 1.10). Along with the extensive basalts, huge amounts of sand, gravel, and clay occur in alluvial fans and washes. Average elevation is about 900 m. The plateau is deeply dissected by the Columbia and Snake rivers. Substrates exposed in gorges are mostly igneous with lowest walls of granite and schist. Several sections are recognized.

The Snake River Plain Section (fig. 1.4) of southern Idaho is covered mostly with Quaternary basalts (figs. 1.7, 1.11). Along the northern border is a series of volcanic craters, such as Craters of the Moon, and numerous sinks and ponds. The Snake River flows along the southern side of the section. The Payette Section to the west is generally lower in elevation. Former lake beds of basalt overlain by lake sediments constitute much of its surface, now dissected by stream valleys. Two batholiths frame this basin traversed by the Snake River, which, after flowing through a gorge, meanders through extensive lava plains. The Owyhee Mountains rise sharply above the lake beds to 2400 m.

The Harney Lake Section is a volcanic plain at the southwestern corner of the province with little local relief except at centers of volcanism, where cones rarely exceed 60 m above adjacent floors. Much moisture percolates through the porous surface: Harney (alkaline) and Malheur (fresh) lakes, included within a marshy tract, fluctuate greatly throughout the year. The Blue Mountain Section (fig. 1.4) is an uplift of Paleozoic and Mesozoic sediments surrounded by elevated plateaus of basalt. Here the Snake River carves Hell's Canyon, a gorge deeper than Grand Canyon. The Blue Mountains rise 900 m above the dissected surface to an elevation of 2700 m.

The Walla Walla Section is underlain by basalts and is covered by lake sediments and loess (fig. 1.10). The eastern side has gently rolling relief and canyons to 600 m deep. The Columbia and Yakima rivers cut gorges in the east, which become shallower to the west. The Spokane River flows at the northern edge of the basalt. After confluence with the Spokane River, the Columbia River is joined by the Snake River in the Pasco Basin. The northern part of the plateau is the Scabland, an area stripped of loess by a catastrophic flood as the ice dam holding a proglacial lake broke during the late Wisconsinan. This flood also cut the Grand Coulee and other large valleys.

Rocky Mountains System

The Rocky Mountains system occurs along the western edge of the stable craton of the continent and is part of the folded belt uplifted during the Laramide orogeny of the early Tertiary (fig. 1.7). The system runs northward some 5000 km from the vicinity of Santa Fe (New Mexico) to Cape Hope (Alaska), with a bend westward near the Beaufort Sea (figs. 1.4, 1.13).

The Southern Rocky Mountains Province extends from central Wyoming south to north central New Mexico. Mountain ranges in this province are mostly on a north to south axis, sometimes exceeding elevations of 4200 m, such as Pike's Peak (4301 m) in Colorado. The main ranges are arrayed in eastern and western parallel series, separated in the north by the North Platte River Basin and the Arkansas River Valley, and in the south by the San Luis Valley and the Rio Grande (fig. 1.4). Two-thirds of the drainage is eastward of the Continental Divide. Underlying rocks are granitic formations flanked by sedimentary rocks (fig. 1.11). The San Juan Mountains of Colorado are mainly formed by volcanics.

The Middle Rocky Mountains Province extends nearly 600 km from near Payson (Utah) almost to Livingston (Montana). The ranges have diverse orientations, and they are separated by semiarid, intermontane basins (fig. 1.4). Elevations vary from 1200 to 3000 m. All the major ranges were glaciated during the Wisconsinan (figs. 1.9, 1.10), and snow and ice still cover most of the mountains above 2400 m, particularly in the northern part. The Yellowstone Plateau is a high (2300–2600 m) volcanic tableland contiguous with the Intermountain system, demarcated by the Snowy and Absaroka ranges. The plateau was also glaciated. Numerous alkaline and calcareous hot springs and geysers occur in the area, among which is Old Faithful. Another noteworthy feature is Bighorn Basin, nearly surrounded by the Bighorn Mountains and connecting with the Great Plains (fig. 1.4). This whole area was extensively glaciated, and the glaciers left impressive cirques and U-shaped valleys. In western Wyoming, the Teton Range rises majestically to 4197 m (Grand Teton). In the south, the Uinta Mountains, the largest east-west trending chain in North America (fig. 1.4), form the southern border where the province abuts the Colorado Plateau.

The Northern Rocky Mountains Province is a narrow, north-trending chain of mountains running 1700 km from Boise (Idaho) to the Liard River in British Columbia (figs. 1.4, 1.13). The area is characterized by linear block-faulted mountains of mostly igneous and metamorphic rocks separated by long straight valleys of Tertiary sediments (fig. 1.11). This province includes the Rocky Mountain Trench, the longest terrestrial trench in the world, which begins near Flathead Lake (Montana) and terminates 1200 km to the north at the Liard River. The range is 300 km wide at its southern end, narrows to less than 120 km near the Canadian border, and flares north of the Liard River into the Liard River Plateau and Plain. Elevations range from 1200 m to 3000 m.

In Montana, most peaks above 2400 m were centers of Pleistocene glaciers (figs. 1.9, 1.10). The Canadian Rockies were covered by the Cordilleran Ice Sheet that extended eastward on the Great Plains to meet the Laurentide Ice Sheet. In Montana, numerous Tertiary lake basins and mountains occur. Much of central Idaho comprises separate mountain groups that were formed by dissection of the Idaho batholith during the Cretaceous (fig. 1.4). This area is characterized by block faulting. Most valleys are deep gorges; few are broad basins. Some peaks harbor mountain glaciers, such as in the Lewis Ranges (Glacier National Park). The Canadian Rocky Mountains are a series of long, sharp-crested ranges. They consist mostly of Paleozoic limestones, with some older rocks thrust into them. To the east, the foothills are underlain by Cretaceous rocks. The highest peak of the Canadian Rockies is Mt. Robson (3954 m).

The Brooks Range Province is delimited from the northern Rocky Mountains by the Liard Plateau and Plain. It is bordered to the north by the Arctic Coastal Plain. The province forms an arch 2200 km long from the northern extremity of the Northern Rocky Mountain Province to Cape Hope. The area comprises a series of ranges and plateaus along the northern side of the Yukon Basin. This province was strongly folded, thrust-faulted, intruded by molten masses, and metamorphosed. It was last uplifted during the Tertiary and Quaternary. Two major ranges compose the Brooks Range Province: the Mackenzie Mountains and adjacent mountain chains in Northwest Territory and Yukon, and the Brooks Range of Yukon and Alaska (fig. 1.4). The Brooks Range is rugged with large areas above 1500 m and some reaching above 2400 m near the Mackenzie Delta (Mt. Chamberlin, 2749 m; Mt. Isto, 2761 m). The Mackenzie Mountains and associated ranges rarely exceed 2100 m (although Keele Peak is 2972 m). The province was glaciated during the Pleistocene, and some glaciers persist today. The ranges are interrupted by lowland plateaus and plains.

Pacific Interior System

The Pacific Interior system extends 4900 km from the southernmost Sierra Nevada in California to the mouth of the Yukon River in Alaska. Narrow in the south, it broadens near the Canadian border, and it arches and further widens in Yukon. It is characterized by belts of rugged mountains and dissected uplands. Some geographers subdivide this complex system into numerous regions, corresponding to terranes, i.e., distinctive geological formations, associated with the Cordilleran orogenies. Several episodes of uplift occurred in late Cretaceous and early Tertiary, with renewed uplift and erosion during the Pliocene-Pleistocene (fig. 1.11).

The Sierra Nevada–Cascades Province is located between the Intermountain system and the Pacific Border system (fig. 1.4). It consists of narrow (80–100 km wide) mountains trending north-south for a distance of more than 1600 km. The province is underlain by a great granitic batholith (fig. 1.11). Elevations are generally highest along the east side of the Sierra Nevada, where many peaks rise to more than 3800 m (Mt. Whitney, 4418 m, highest peak in the conterminous United States). These summits are flanked to the west by a large upland above 3000 m that slopes gently westward to the Central Valley of California. Mountains of the Sierra Nevada are blocks of granite that range for more than 600 km north and south. Rivers cut deep valleys, some to 1500 m, that drain mostly to the west. Most peaks and higher valleys were glaciated, but glaciers affected little terrain south of Owens Lake (California).

To the north, the Cascade Mountains extend over a distance of 1100 km from Lassen Peak (northern California) to Meager Mountain (southwestern British Columbia). The Cascade Mountains include 12 major stratovolcanoes, some of which approach the elevation of peaks in the Sierra Nevada, such as Mount Shasta (4317 m), Mount Adams (3751 m), and Mount Rainier (4392 m). The only rivers to breach the Cascades are the Columbia, Klamath, and Pit, and all flow westward.

Three sections are recognized within the Cascade Mountains. The Southern Section is a lava-covered sag in the granite, mostly less than 1500 m high. Numerous volcanic cones are encountered, including Mount Shasta and Mount Lassen (3187 m; an eruption in 1914–1915, and some activity in the 1970s). The Middle Section is an uplift of middle Tertiary lavas dominated by Quaternary volcanic cones. The east side is higher, overlooking the Columbia Plateau; the crest is marked by the High Cascades. The Western Cascades are so severely eroded that no trace of the original landscape persists. Crater Lake (2486 m) was formed after the eruption of the high volcanic cone of Mount Mazama, which occurred during the late Pleistocene. The cone subsequently collapsed and filled with water. Mount St. Helens (now 2550 m) violently erupted in 1980, losing its top 400 m. Glaciers formed on peaks above 2700 m during the Pleistocene, and some still persist.

The Northern Section of the Cascades is composed of two kinds of mountains, volcanic and granitic (e.g., Mount Baker, volcanic, and Glacier Peak, granitic). These mountains comprise mostly folded, metamorphosed sediments and intruded granites. The peaks and ridges are approximately uniform in elevation, although they are rugged and steep-sloped. None of these mountains rises above 2400 m. In British Columbia, the range reaches its northern limit south of the lower Fraser Valley (fig. 1.4). The northern Cascades were glaciated extensively, and the effect of the ice stops near their southern end, except for higher elevations. Several small glaciers persist today.

The northern part of the system is wider than the Sierra Nevada–Cascades Province (fig. 1.13) and consists of a series of plateaus bordered to the east and west by the northern Rocky Mountains and the Coastal Mountains, respectively. In British Columbia two mountain chains invade the plateaus. Interior British Columbia was fully glaciated during the Pleistocene by the Cordilleran Ice Sheet, but the Yukon-Alaska segment was not.

The Columbia Mountains Province is comparable in elevation to the Sierra-Cascades. Two peaks exceed 3500 m elevation. The province consists of four parallel ranges, the Purcell, Selkirk, Monashee, and Cariboo mountains (fig. 1.4). It forms a triangle with the narrow side abutting the Columbia Plateaus on the south, the east side delimited by the Rocky Mountain Trench, and the western edge delimited by the northern Cascades and the Fraser Plateau. Geologically, the granitic Nelson batholith is combined with metamorphic and sedimentary formations of older ages (fig. 1.11). Orogenic activity ended here long before it did in the Rocky Mountains.

The Fraser Plateau Province is delimited to the east by the Columbia and northern Rocky mountains, to the west by the northern Cascades and Coastal Mountains, and to the north by the Cassiar–Skeena Mountains Province. The Fraser Plateau is a rolling upland of 1000 m elevation tilted toward the north, mostly higher than the Columbia Plateaus to the south. Isolated ranges reaching to 800 m cross the semiarid plateau, which comprises Tertiary lava flows and intrusions (fig. 1.11). The Fraser River and tributaries traverse the plateau (fig. 1.4); in the south, they have cut gorges 500 to 1000 m deep; in the north the plateau is less deeply incised.

The Cassiar–Skeena Mountains Province lies north of the Fraser Plateau. It is bordered on the west by the Coastal Mountains, on the north by the Stikine Plateau, and on the east by the Liard Plateau and the northern Rocky Mountains. Two rugged mountain belts comprise the province: the Cassiar-Omineca mountains in the east and the Skeena-Hazelton mountains in the west. The eastward-flowing Peace and Liard rivers cut the Cassiar-Omineca mountains, while the westward-flowing Skeena (fig. 1.4), Nass, and Iskut rivers dissect the Skeena-Hazelton mountains. Most of the area is below 2000 m, but some peaks reach 2400 m. The province developed on a granitic batholith (fig. 1.11).

The Stikine Plateau Province (fig. 1.4) is bordered by the Cassiar Mountains to the east, by the Skeena Mountains to the south, by the Coastal Mountains to the west, and it extends to the Yukon Plateau on the north. Elevations are mostly from 600 to 1200 m. On the western edge, volcanic Mount Edzina towers over the plateau, reaching a height of 2700 m. The Stikine River drains the plateau to the west.

The Yukon Plateau Province is delimited on the east by the Mackenzie Mountains (fig. 1.4), on the south by the Cassiar Mountains and the Stikine Plateau, on the west by the Shakwak Trench that borders the St. Elias Mountains, and on the north by the Yukon-Tanana Upland. The Yukon Plateau averages 1000 m, but it has high points at almost 2000 m on the plateau proper and in the Ogilvie Mountains to the northeast. Part of the eastern side is occupied by the Selwyn Mountains. The plateau is deeply dissected into hills and lowlands by the Yukon River and other rivers that have cut the surface by 600 m. Rock types are diverse, including granitic intrusions and metamorphic formations (fig. 1.11). The plateau is bisected east of the Pelly Mountains by the Tintina Trench, which runs northwestward parallel with the Shakwak Trench; Kluane Lake lies in the latter.

The Porcupine Plateau Province is a lowland wedged between the Brooks Range (fig. 1.4) and Richardson Mountains on the north and east. It is bounded on the south by the Yukon Plateau, and by the Yukon-Tanana Upland on the west.

The Alaskan portion of the Pacific Interior system consists of a series of uplands separated by large basins trending east to west. The area lies between the Brooks Range to the north and the Coastal Mountains to the south (figs. 1.4, 1.13); it is bounded to the east by the Porcupine and Yukon plateaus. It includes 11 physiographic provinces (M. J. Bovis 1987), which are, from east to west: the Yukon-Tanana Upland, the Yukon Flats, the Tanana-Kuskokwim Lowland, the Ray Mountains (fig. 1.4), the Kuskokwim Mountains, the Nushagak Lowland, the Yukon-Kuskokwim Lowland, the Kobuk-Selawik Lowland, the Nulato Hills, the Ahklun Mountains, and the Seward Peninsula. Most of the area was unglaciated during the Pleistocene.

Most of the uplands of central Alaska are drained by the Yukon River and its tributaries, and generally they are lower than 1500 m. Highlands include the hills of the Seward Peninsula and the Nulato Hills (450–600 m) between the Yukon River and Norton Sound. The uplands fronting the Bering Sea consist of a rugged, dissected plateau of 300 to 750 m with some low mountains. The Kuskokwim Mountains (Kiokluk Mountain, 1248 m) east of the Yukon delta extend into central Alaska; they are part of a Quaternary uplift that extends into Siberia. The vast Yukon-Tanana Upland is located in eastern Alaska. The arched Tanana-Kuskokwim Lowland lies just north of the Alaska Range, and runs from Canada to the Kuskokwim Mountains (fig. 1.4). The Yukon River flows through a complex of lowlands, the most extensive being the Yukon Flats with an area of 23,300 km^2, located near Porcupine Plateau where the Yukon River changes its course to the west. Flowing 3200 km, the Yukon River is one of the longest in the world. It has several major tributaries: the Koyukuk, Tanana, Porcupine, and Klondike rivers. The Kuskokwim River, which originates within the Alaska Range, converges with the Yukon River to form a low deltaic flat more than twice the size of the Mississippi Delta.

Pacific Border System

The Pacific Border system fronts the Pacific Ocean. It is the longest North American system, extending for more than 7000 km from western Alaska to Baja California (fig. 1.13). In the conterminous United States, a series of low ranges (mostly less than 600 m) with an inland trough constitutes the Pacific Border Province. The system continues in British Columbia and Alaska as the Coastal Mountains Province, a series of high, rugged, coastal and insular mountains with glaciers and fjords. Where the coast curves to the west, the mountains are heavily glaciated and form the Glaciered Coast Province. Farther west, an inland arcuate belt of mountains constitutes the South Central Alaska Province, which is succeeded westward by the Alaska Peninsula and Aleutian Islands Province.

The Lower California, or Peninsular Range, Province represents the north end of the Baja California peninsula. It is a plateau tilted westward, divided into ranges oriented northwest by faults, and with an east-facing escarpment. San Jacinto Peak (3293 m) at the north end towers above the Salton Trough. The province comprises a granitic batholith of Cretaceous age that is intrusive into Cretaceous sedimentary formations.

The Pacific Border Province comprises a series of coastal mountains separated from the Sierra-Cascades mountain chains (Pacific Interior system) by a trough that is partly submerged in the north (Puget Sound) and elsewhere is less than 150 m in elevation. At the southern end of the province, the Transverse Ranges Section consists of ranges and basins trending east to west, perpendicular to the orientation of adjacent mountain ranges (Sierra Nevada, Coastal Ranges) (fig. 1.4). The northern Channel Islands (Santa Barbara, Santa Cruz, and Santa Rosa) represent peaks of the submerged part of the Santa Monica Mountains (maximum elevation 861 m); the southern islands are not part of this chain, however. The basins have thick Tertiary deposits overlain by mostly marine sediments; the Ventura Basin is submerged in the Santa Barbara Channel; and the Los Angeles Basin is the only coastal plain of the province.

The western mountains of the Transverse Ranges are mostly marine, Tertiary formations; the higher eastern ranges comprise older rocks, including much granite (fig. 1.11). The San Bernardino (maximum elevation 3506 m) and the San Gabriel mountains (Mt. San Antonio, 3068 m), the highest ranges, are separated by the San Andreas Fault (fig. 1.7). They consist of granitic rocks and volcanics. The Central Valley of California Section (figs. 1.4, 1.10, 1.11, 1.13) is a trough between the coastal ranges and the Sierra Nevada. Most of the valley is below 150 m, and one-third is lower than 30 m. It is mostly filled with recent sedimentary deposits from the Sierra Nevada that bury the western edge of the underlying granitic batholith (figs. 1.10, 1.11). The Sacramento River drains the northern part, and the San Joaquin River the southern part, both of them reaching the Pacific through San Francisco Bay. The southernmost part of the valley is a closed basin containing two playas, Tulare and Buena Vista lakes. Two intrusive, domed mountains, the Marysville or Sutter Buttes, rise 600 m above the valley floor.

The California Coast Ranges Section (fig. 1.4) consists of a series of four ridges and three valleys that parallel the coast. Ridges rarely exceed 1500 m, with many below 900 m. Rocks are mostly sedimentary. Numerous faults, including the San Andreas Fault (fig. 1.7), among others, are involved in the geological history of the Coast Ranges and continue to affect them. The Pacific Plate to the west of the faults is moving north relative to the American Plate and has provoked the displacement northward of the Coast Ranges area by more than 160 km since the Cretaceous, relative to the area just east of the faults.

Near the Pacific Coast of northern California and southern Oregon, the Klamath Mountains Section is an upland deeply incised (450–750 m) by the Rogue and Klamath rivers. Isolated peaks rise to elevations 1500 to 2000 m above the upland. The rock formations are deformed, metamorphosed, and intruded by granite. They were uplifted during the Cretaceous and again during the past 2 million years. The Oregon Coast Range Section is a gently rolling plateau of Tertiary rocks along the coasts of northern Oregon and southwestern Washington. The highest summits are mostly less than 900 m, the hills are rounded, and the valleys are open. The Willamette Valley forms the eastern side of the section and is the southern end of Puget Trough.

The Puget Trough Section is a lowland partially submerged in the north, and it does not exceed 150 m in elevation at its southern end. This lowland was submerged during the Pleistocene. To the north, the trough extends 2500 km along the coast of British Columbia and Alaska, creating the Inside Passage. The Olympic Mountains Section (fig. 1.4) of northwestern Washington is a domed uplift that reaches an elevation of 2408 m (Mt. Olympus). The core of resistant rocks is surrounded by Tertiary formations. The Olympic Mountains were glaciated during the Pleistocene, and some cirques still contain glaciers.

The Coastal Mountains Province extends from just north of the Fraser River (British Columbia) to the coastal glaciers of the St. Elias Mountains (southeast Alaska). The province is about 1600 km long and sometimes reaches a width of 160 km (figs. 1.4, 1.13). The rugged coastal and insular mountains reach 3900 m. The greatest batholith of the Pacific Coast underlies the province (fig. 1.11). The batholith was uplifted during the Columbian orogeny (Tertiary). The province is divided longitudinally into two ranges separated by the coastal trough and associated lowlands: the Coastal Mountains on the mainland, and the Queen Charlotte Island and Vancouver Islands seaward (fig. 1.4). Portions of the islands escaped glaciation during the Wisconsinan glaciation. The interior ranges are associated with lava plateaus and three large shield volcanoes, including Mount Garibaldi (southeastern British Columbia). The Fraser River (fig. 1.4) cuts through this plateau and reaches the Strait of Georgia in a valley that curves around the south end of the Coastal Mountains. Most peaks in the Coast Range are about 2400 m above sea level, but a few reach 2700 m. Several rivers cross the range, notably the Skeena (fig. 1.4) and the Stikine rivers, and some of the valleys are very deep. Glaciers are present along inlets, where they calve icebergs into the sea.

The Glaciered Coast Province lies between the Coastal Mountains and the South Central Alaska provinces, where the coast bends westward. The St. Elias Mountains reach elevations near 6000 m, including Mount Logan (Yukon, 5951 m), the highest peak in Canada. Glaciers cover about 13,000 km^2 in the province; perpetual snow exists above 750 m. Along the coast, formations are sedimentary.

The South Central Alaska Province includes the Aleutian Range, the inland Alaska Range, and other belts of mountains and lowlands (troughs) that curve around the Gulf of Alaska. The mountains were uplifted during the Tertiary, and in the late Tertiary the Aleutian Trench developed (fig. 1.7). The Chugach and Kenai mountains are composed of metamorphics and volcanics (fig. 1.11), and they are separated from the Alaska and Aleutian ranges by lowlands (fig. 1.4). The Alaska Range is nearly 1000 km long. Elevations reach 6000 m, including Mt. McKinley (Denali) (6194 m), the highest peak in North America. The range is mostly sedimentary with extensive intrusions of granite (fig. 1.7). Pleistocene glaciations were extensive, and many glaciers persist today.

The Aleutian arc represents the northern segment of the "ring of fire" (fig. 1.7), the volcanically active margin of the Pacific Plate. The arc is highest at its eastern end, where it begins at Mt. Spurr (3330 m), and sweeps westward as the Aleutian Range. As the range dips beneath oceanic waters, the tops of the volcanoes (maximum elevation 2857 m, Shishaldin Volcano, Unimak Island) form a 2500-km, narrow, rugged archipelago that ends at Attu Island, the westernmost part of North America (ca. 173° E). Glaciation was heavy in the eastern portion; icecaps and glaciers persist on high peaks. Geological evidence indicates that the Aleutian Islands formed a corridor of land south of the Bering Land Bridge during the Wisconsinan, when sea level was about 50 m lower than today.

2. Soils

Donald Steila

Soil is a medium wherein plants are anchored and from which they draw water and mineral nutrients. Reciprocally, plants protect the soil from erosion, they increase its moisture-holding capacity via the incorporation of organic matter, and they recycle solid elements important to sustain biomass production. Soils are derived from complex interactions of geologic, biotic, and climatic factors, acting over time.

There are many kinds of soil in North America, and soil scientists group them among several categories called "orders." The geographic distribution of these soil orders is correlated to regional vegetation types and to climate, and it has been described by several biogeographers (W. E. Akin 1991; P. W. Birkeland 1984; S. W. Buol et al. 1980; S. R. Eyre 1971; D. Steila and T. E. Pond 1989).

Soil classification provides a framework whereby the relationships among the kinds of soils may be compared and patterns among them studied. As with most fields of science, a specialized nomenclatural scheme has been developed, but the details of its applications are not universally accepted. The soil classification scheme used here follows *Soil Taxonomy*, prepared by the U.S. Soil Survey Staff (1975). This is a hierarchical model that employs six ranks of soil taxa: order, suborder, great group, subgroup, family, and series. Soils are grouped within a given taxon on the basis of sharing certain features. The order represents the greatest degree of generalization, and 10 orders are recognized here, each with an identifying name ending in *sol* (Latin *solum*, soil). These are listed in table 2.1 with their approximate equivalents in the soils classification scheme of Canada (Canada Soil Survey Committee 1978).

At the order level, the position of each of the mineral soils is based largely on the degree of weathering, i.e, the degree of mineral alteration and profile development (fig. 2.1). The continental distribution of the soil orders is illustrated in figure 2.2.

As a result of macroscale soil formation, or pedogenesis, regional homogeneity occurs in diagnostic soil characteristics. A detailed examination of figure 2.2 reveals a broad pattern in the distribution of soil orders that can be readily associated with relatively extensive climatic and vegetation realms (see L. Brouillet and R. D. Whetstone, chap. 1; M. G. Barbour and N. L. Christensen, chap. 5). Yet each order includes hundreds of subtypes, each of which has attributes associated with its unique environment.

Soil orders are differentiated into suborders on the basis of chemical and/or physical properties related to drainage conditions, or major parent material, because they possess genetic differences due to climate and vegetation. Suborders are separated into great groups on the basis of the kinds and array of diagnostic horizons, base status, and regimes of soil temperature and moisture. The great groups are divided into subgroup categories that indicate the extent to which the control concept of the great group is expressed. The criteria for recognition of families stress features of importance to plant growth, such as texture, mineralogy, reaction class,

TABLE 2.1. **Soil Orders Name Derivation and Canadian Taxonomy Equivalents**

U.S. Soil Orders	*Derivation of Root Word*	*Canadian Soil Orders**
Entisols	(Artificial syllable)	Regosolic, some gleysolic
Vertisols	Latin: *verto,* turn upward	Some chernozemic
Inceptisols	Latin: *inceptum,* inceptive	Brunisolic, some gleysolic and podzolic
Aridisols	Latin: *aridus,* dry	Some solonetzic
Mollisols	Latin: *mollis,* soft	Chernozemic, some brunizolic, gleysolic, and solonetzic
Spodosols	Greek: *spodes,* wood ash	Podzolic
Alfisols	(Artificial syllable)	Luvisolic, some gleysolic and solonetzic
Ultisols	Latin: *ultimus,* ultimate	No equivalents
Oxisols	Greek: *oxy,* sharp, acid	No equivalents
Histosols	Greek: *histos,* tissue	Organic soils

* These are general equivalents, as the definitions of horizons and classes differ between the U.S. and Canadian systems.

and soil temperature. The individual soils are called series and are distinguished on the basis of the kinds and textures of horizons and on their chemical and mineralogical properties. They are named after a natural feature or place where the soil was first recognized as distinct.

Clearly, the soils at the series level are myriad and cannot be adequately treated within the scope of this chapter. Therefore, the remainder of this essay is directed toward the soils of North America at the order level.

North American Soils

Entisols

The entisols are true soils and should not be confused with geologic materials incapable of supporting plant life. An unstabilized sand dune, for example, does not constitute soil, but rather a mineral deposit. Once stabilized and supportive of vegetation, however, it represents a recent soil. The key criterion for entisols is that they lack diagnostic horizons or features that would place them in another order. In effect, soil-forming processes either have not been operative on the parent material for a sufficient period of time, or some aspect of the physical environment (e.g., wind erosion) has prevented development of traits characteristic of mature soils within the bioclimatic regime of their location. Typically they lack an E- or B-horizon and have a profile sequence of an A-horizon either overlying a C-horizon or resting on unaltered geologic material (see fig. 2.3 for soil profile and explanation of horizons). They may be found in any moisture or temperature regime, on any type of parent material, and under widely varying vegetative covers. Thus their geographic distribution is exceedingly diverse. The best means to understand the rationale for their location is by referring to figure 2.2, as each of the five suborders is described below.

Aquents (from the Latin *aqua,* water) are the wet entisols. They are commonly found in tidal marshes, in deltas, on margins of lakes where the soil is continuously saturated with water, in flood plains of streams where soil is saturated at some time of the year, or in very wet, sandy deposits. These soils are bluish or gray, and mottled. Temperature does not restrict the occurrence of aquents, except in areas constantly below freezing. The soil moisture regime is primarily a reducing one, virtually free of dissolved oxygen due to lengthy saturation by ground water. Aquents are generally found in recent, usually water-deposited sediments. They may support any form of vegetation that will tolerate prolonged periods of waterlogging.

Arents (from the Latin *arare,* to plow) are entisols that lack horizons, normally because of human interference. They have been deeply mixed by plowing, spading, or moving. Some of these soils are the result of deliberate human attempts either to modify soil or to break up and remove restrictive pans; in other instances, they result from cut and fill operations intended to reshape a surface. An uncommon few are formed from the natural effect of mass movement in

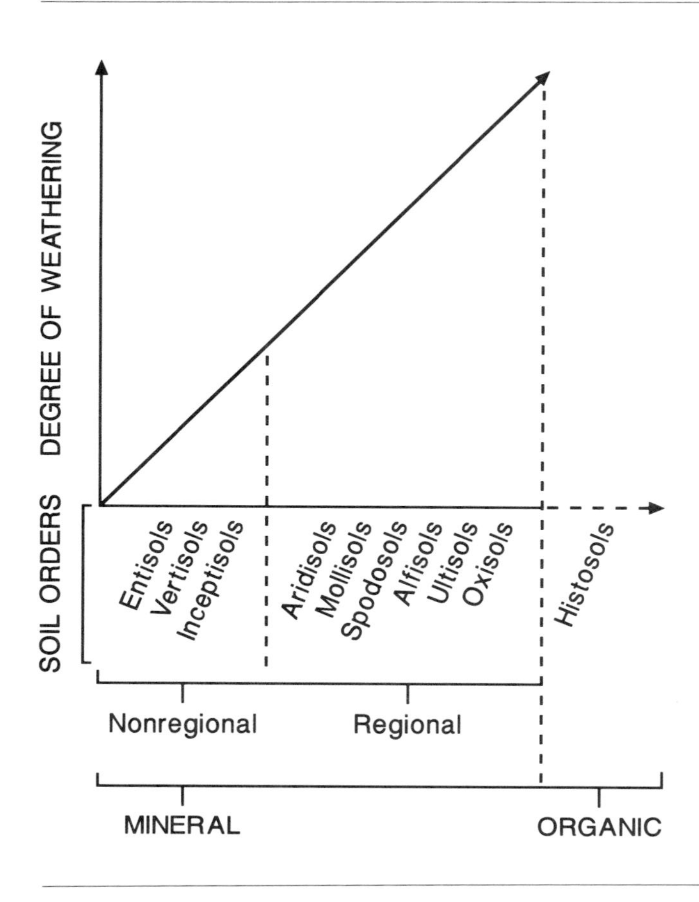

FIGURE 2.1. Soil orders organized by degree of weathering. [From D. Steila and T. E. Pond 1989. Reprinted with permission from Rowman & Littlefield.]

earth slides. Arents are not extensively developed, and their characteristics vary from place to place.

Fluvents (from the Latin *fluvius,* river) are brownish to reddish entisols. They have formed in recent water-deposited sediments—primarily floodplains, fans, and deltas of rivers and small streams, but not in back swamps where drainage is poor. The age of sediments in which these soils form is usually very young, only a few years or decades. Under normal conditions fluvents are flooded frequently, and deposited materials show signs of stratification—layers of a given texture alternated with layers of other textures. Most alluvial sediments, coming from eroding surfaces or stream banks, include appreciable amounts of organic carbon that are dominantly associated with the clay fraction. These soils do not occur under any specific vegetation type and may be found in any moisture or thermal regime, except those that are subject to temperatures constantly below freezing.

Orthents (from the Greek *orthos,* true) are entisols occurring primarily on recently eroded surfaces created by geologic factors or by cultivation. The basic requirement for recognition of an orthent is that any former existing soil has been either completely removed or so truncated that the diagnostic horizons typical of all orders other than entisols are absent. Such formations normally support only scattered plants. If they do not support plants, they are considered to be rock rather than soil. A few orthents occur in recent loamy or fine wind-deposited sediments, in glacial deposits, in debris from recent landslides and mudflows, and in recent sandy alluvium.

Psamments (from the Greek *psammos,* sand) are sandy entisols that lack pedogenic horizons. They include sandy dunes, cover sands, and sandy parent material produced in an earlier geologic cycle. Others are found in water-sorted sands, like those along natural levees or beaches. Occurring under any climate or vegetation, they may be located on surfaces of virtually any age. The older psamments are usually dominated by quartz sand and cannot form subsurface diagnostic horizons.

Vertisols

By definition these soils must be at least 50 cm deep, have 30% or more smectite clay in all horizons, and

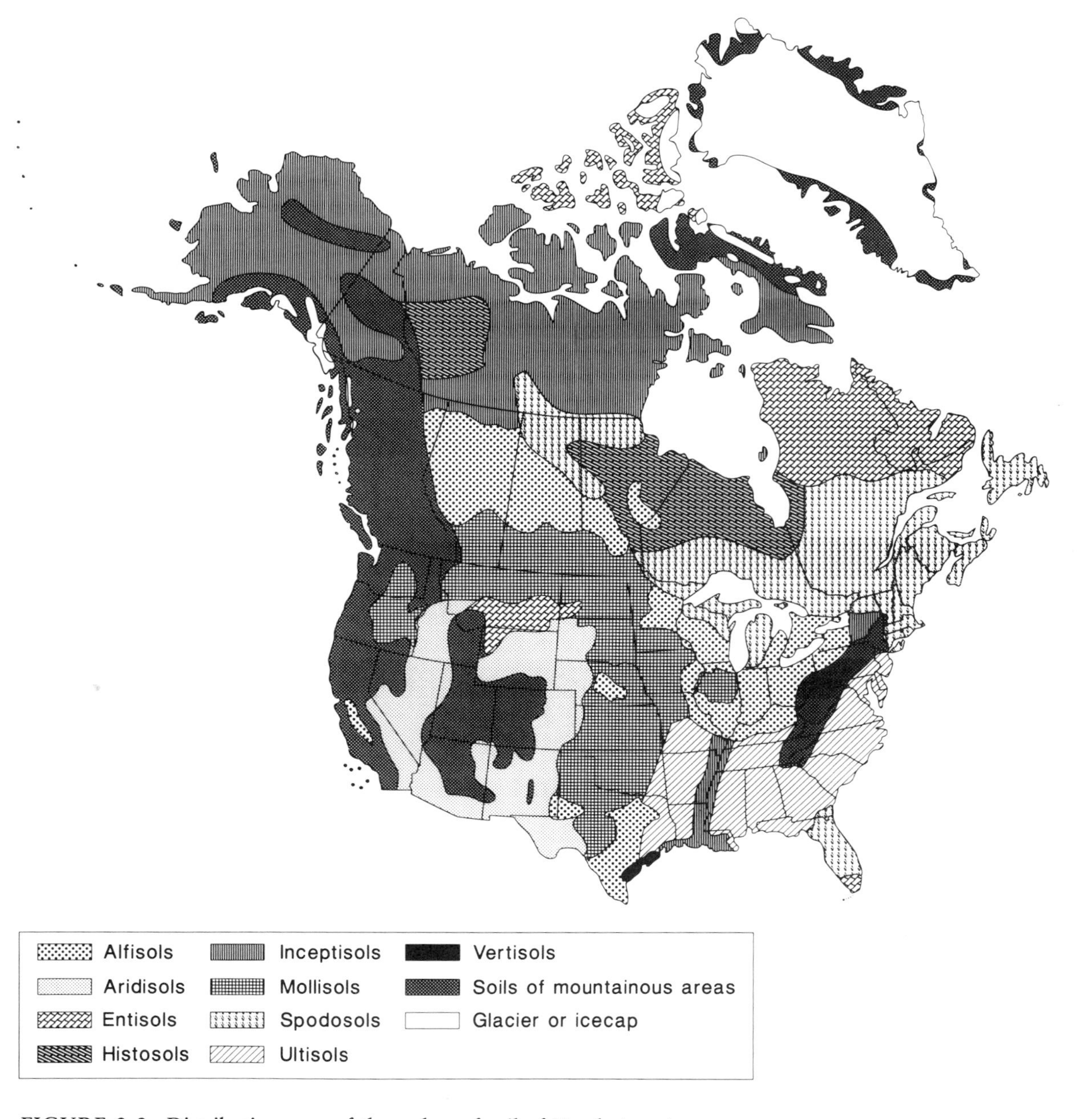

FIGURE 2.2. Distribution map of the orders of soil of North America.

have cracks at least 1 cm wide at a depth of 50 cm during part of the year. Indeed, the cracks that occur during the dry season are their most conspicuous feature. Frequently the cracks extend deeper than 1 m and are very wide.

Vertisols are mostly found in regions with distinct wet-dry seasons. They are of minor extent in North America, being found mainly in east central and southeastern Texas, west central Alabama, and east central Mississippi.

Their development takes place on alkaline parent materials wherein clays form that have high shrink-swell properties. When the soil dries, clay shrinkage produces the characteristic deep cracks, which are frequently arranged in polygonal patterns. During this time, surface soil material collects in the cracks by several processes: deposition of wind-borne materials, surface-washing and dislodgment by rainfall, and grazing of animals. When wetted, the clays expand and the cracks close, trapping the displaced particles below the sur-

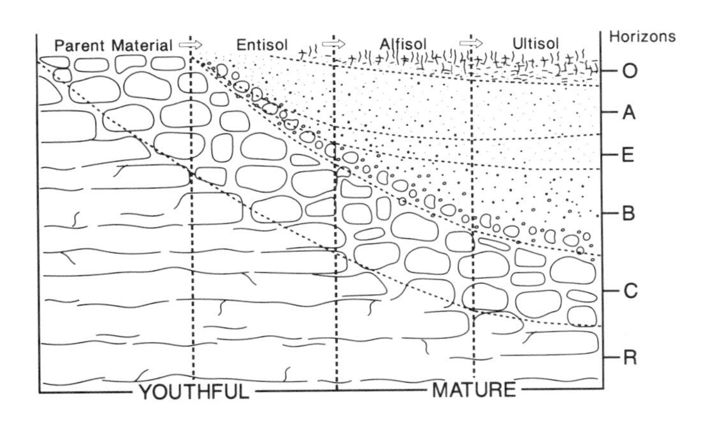

FIGURE 2.3. Soil horizons are layers of soil or soil material lying approximately parallel to the land surface. They differ from adjacent related layers in physical, chemical, and biological properties or characteristics, such as color, structure, texture, consistency, kinds and numbers of organisms, and degree of acidity or alkalinity. The number and variety of horizons indicate stage of soil evolution. Well-developed (mature) and undisturbed soils normally exhibit a sequence of horizons from the surface downward that are classified (in the pedologist's short-hand) as O, A, E, B, C, and R.

face. This extra volume at lower levels prevents the expanding clay from occupying its former space, thus exerting pressure on adjacent soil particles that forces them to move laterally and upward. Consequently, lower soil material moves upward, toward the surface. This continuous cycle of overturning has earned these soils the reputation of being self-plowing or inverted, hence the name vertisols. The entire landscape of a vertisol habitat may become mottled with a mosaic of micromounds and depressions, referred to as "gilgai topography."

Inceptisols

Similar to the entisols, inceptisols may be found in almost any bioclimatic regime. Some are weathered sufficiently to form slightly altered horizons, yet they normally lack a well-developed B- or E-horizon. In brief, inceptisols exhibit traits associated with their soil-forming environment, but the features are too weakly developed for them to be considered mature soils. Generally they bear close relationship to their parent material.

The primary difference between entisols and inceptisols is that the direction of soil development is evident in the latter, suggesting that it is possible to predict their development into a mature form, such as an alfisol or ultisol. Typically they have altered horizons that have been somewhat leached of either bases or iron and aluminum. At the same time, they usually lack subsurface horizons that have been enriched with silicate clays. The diversity of inceptisols creates problems with attempts to generalize their geographic distribution. It is safe to state only that they may be found almost anywhere and simply represent an intergradational state between the immature soil of regionally dominant entisols and the mature soil of the dominant bioclimatic regime.

Aridisols

North America's aridisols are found primarily west of longitude 100° W in areas of desert where water is lacking for long periods. During the time when these soils are warm enough for plant growth, the climate is moisture-deficient. Aridisols have surface horizons that normally are light in color and low in organic matter. They also may contain pedogenic horizons derived from the translocation and accumulation of carbonates, salts, or silicate clays, or that have formed by carbonate or silica cementation.

Aridisols are associated with climatic regimes that have meager precipitation and relatively high potential evapotranspiration, meaning that certain types of chemical weathering are limited. Hence excessive leaching of soluble minerals is prevented. Although many sets of soil-forming processes predominate in arid regimes, two dominate: calcification and salinization. The former is extensive within moisture-deficient climates, and the latter occurs in more localized sites.

Calcification is the process by which calcium carbonate or calcium and magnesium carbonate accumulate in the soil. The primary genetic factor has been associated with limited rainfall (S. W. Buol et al. 1980; D. Steila and T. E. Pond 1989). In some desert areas, these processes are sufficiently active to cement the carbonates into a solid rocklike mass, commonly called "caliche" in the Southwest.

Salinization leads to the accumulation of mineral salts in the soil in sufficient concentration to limit plant production to halophytes. This occurs in localized areas of entrapped surface drainage. Such conditions are common in valley basins with no exterior drainage (bolsons), where rainfall can become imponded. Water descending from surrounding valley slopes assimilates soluble minerals and transports them into the basin, where an intermittent lake may be formed during precipitation events. Between periods of precipitation, the entrapped water evaporates, leaving behind a residue of mineral salts encrusted on the surface and/or precipitated in the soil. The principal salts are cations of calcium, magnesium, and sodium, and the chloride and sulfate anions. Also occurring, in smaller quantities, are potassium, bicarbonates, and nitrate ions, and, even less commonly, borates. Borates have received considerable attention because of their high toxicity to plants.

The aridisols have a sparse growth of plants. The organic matter within the surface soil is low, as are carbon/nitrogen ratios and microbial populations. Barren areas are not uncommon, and plant life is dominated by annuals and by xerophytic perennials.

Mollisols

Mollisols are relatively fertile, dark-colored, humus-rich soils with a soft surface horizon. They are thought to be formed primarily by the underground decomposition of organic matter, related to the rooting and life cycle of grasses. They occur in regions transitional between the arid and humid climates of North America. Natural vegetation ranges from sparse shortgrass prairie along the arid margins to a rich, luxuriant tallgrass prairie along the humid boundaries.

Grasslands differ from forests in both the total amount of organic matter incorporated within the soil and in its distribution throughout the profile. In forested areas, the major source of humus is leaf-fall that accumulates on the surface. Grasses also produce an organic mat as their decaying aerial parts accumulate. In contrast to the roots of trees, however, the dense, fibrous masses of grass roots permeate the soil profile. As the plants die and decay continuously, a large quantity of humus is incorporated within the soil.

In general, grasses require greater quantities of basic mineral nutrients, particularly calcium, than do trees. As a result, their organic remains are richer in base nutrients, which upon decay are returned to the soil. The released bases are available for use by the plants in a continuous cycle that under natural conditions maintains a relatively high degree of soil fertility. The less soluble calcium salts have a tendency to accumulate in lower horizons because normal rainfall, which carries them down in solution, reaches only a few feet into the subsoil; they may later become solidified into hard nodules.

Spodosols

The spodosols of North America are sporadically found in scattered parts of the Atlantic Coastal Plain and in high mountainous areas, both east and west, but they are most extensively developed around the Great Lakes region, eastern Canada, and the northeastern United States. Typically they are associated with coniferous vegetation or heath plants. The dominant climatic regime in which they occur is cool and humid (humid continental and subarctic realms).

The classic spodosol profile is an ashy gray sand overlying a dark sandy loam; an abrupt boundary separates the horizons. The soil reaction is acidic, the parent material is quartzose sand, and a spodic horizon is present. Spodic horizons are illuvial subsurface horizons in which amorphous materials composed of organic matter, aluminum, and/or iron have accumulated.

In regions of spodosol occurrence, surplus precipitation infiltrates organic litter to form organic acids. The acid solution percolates through the soil, leaching lime and/or salts to allow clays to disperse. The dispersion (called peptization) speeds the movement of clays from the surface horizon. As a consequence, in time, the soil develops strata of different texture and degree of mineral alteration. The surface horizon is depleted of fine-sized particles, leaving a coarse texture through which organic colloids may be transported downward. Calcium and magnesium are leached from the topsoil, resulting in a horizon dominated by acidic, silicate minerals. Simultaneously, aluminum and/or iron may be liberated from their mineral bonds and migrate downward from the surface horizon. Subsequently these precipitate in the subsoil to form the spodic horizon.

Alfisols

Alfisols contain significantly more clay in the B-horizon than in the A-horizon and have a base-saturation greater than 35%, indicating that they are relatively fertile. In addition, most have a pale-colored surface horizon. They form in many bioclimatic regimes, but they are most extensive in humid and subhumid temperate climates. They occur widely on lands where deciduous forest is present or where it formerly occurred, on some prairie lands, on calcareous glacial drift, and within loess deposits.

The high clay content of the alfisol's B-horizon is attributed to substantial translocation of clay downward from the soil's upper A- and E-horizons. This soil feature is referred to as an argillic horizon. Clays migrate downward rather than being synthesized in situ.

Ultisols

Ultisols are deeply weathered, acidic soils that are confined largely to the southeast and eastern seaboard of the United States. They are found on old surfaces that lie south of the advance of the last glaciation. The climatic regime is one of both relatively high precipitation and high summertime potential evapotranspiration. In many respects, ultisols are morphologically similar to alfisols. Processes leading to the formation of a clay-enriched subsurface horizon are common to both. And both can be found in adjacent landscapes within a uniform climatic regime, but on different parent materials. The distinct difference between the two is that the alfisols have moderate to high fertility, whereas the ultisols are less fertile and are acidic for a considerable depth, particularly when the parent materials are deficient in bases, as are siliceous crystalline rocks, for example.

Ultisols are considered the most weathered of all midlatitude soils, hence their name. The key to understanding their traits lies in the extremely long period they have been subjected to chemical and physical weathering and their occurrence in a climatic regime propitious to rapid disintegration and decay of parent material. The soils are deep and leached of soluble bases and frequently have colors associated with oxidation of their iron and magnesium compounds. Consequently, reddish, orange, and mahogany colors indicate free drainage, ample oxidation, and the presence of residual iron (red soil) and magnesium (mahogany soil). Other ultisols have yellow to gray colors that reflect poorer drainage conditions. The yellow results from hydration of iron whereas the gray colors occur under reducing conditions in water-saturated soils.

Oxisols

Oxisols are soils found on ancient landscapes within the humid tropics and consequently are not represented within the flora area. These are deep, thoroughly weathered soils, with few remaining, weatherable minerals and containing a horizon of hydrated oxides of iron and/or aluminum along with kaolinite clays. They are mentioned here solely to provide a complete discussion of all soil orders.

Histosols

Unlike the other soil orders, the histosols are not considered primarily mineral but organic. These soils are commonly called bog, moor, peat, or muck. They are rather extensively developed along the southern margins of Hudson Bay, in the Northwest Territory, and occur in smaller areas in Minnesota and within the Gulf and Eastern coastal plains. Isolated pockets can, however, occur within any bioclimatic regime, as long as water is available. They occur in the tundra and on the equator, and their vegetation consists of a wide variety of water-tolerant plants.

The histosols contain organic materials that are either more than 12% to 18% organic carbon by weight or well over half organic matter by volume. The presence of water is the common denominator in all histosols, regardless of location. Unless drained, most histosols are saturated or nearly saturated most of the year.

In addition to water, the controlling factors that regulate the accumulation of organic matter include the temperature regime, the character of the organic debris, the degree of microbial activity, and the length of time in which organic accumulation has taken place. The deposition of organic material within a water-saturated environment rapidly leads to the depletion of oxygen in the wet zone through decomposition by aerobic microorganisms. This eventually results in an anaerobic milieu, in which the rate of organic matter mineralization is considerably reduced. As a consequence, organic materials accumulate. The development of these soils is, therefore, characterized by a deepening of the organic layer.

The major distinction among the suborders of histosols relates to the degree of decomposition of organic remains and the soil's moisture. There are four suborders: fibrists, folists, hemists, and saprists.

Fibrists (from the Latin *fibra,* fiber) are histosols of relatively unaltered plant remains. The reasons for the preservation of plant remains vary, but the absence of oxygen is probably the most important factor. The rate of mineralization varies according to the thermal regime and the nature of the vegetative debris.

Folists (from the Latin *folia,* leaf) are rather freely drained histosols consisting of an organic horizon derived from leaf litter, twigs, and branches (but not sphagnum), resting on rock or on fragmental materials made of gravel, stones, and boulders, with the interstices filled or partially filled with organic materials. There is very little evidence of mineral soil development, and plant roots are restricted to the organic materials.

Hemists (from the Greek *hemi,* half) are histosols in which the organic material is strongly, but not completely, decomposed. Enough, however, has been broken down to the point that its biologic origin cannot be determined. These soils are normally found where the ground water is at or very near the surface most of the year. The groundwater levels can fluctuate, but they seldom drop more than a few centimeters below the surface tier. These soils were once classified as bog. They are usually found in closed depressions and in broad, very poorly drained flat areas such as coastal plains.

Saprists (from the Greek *sapros,* rotten) consist almost entirely of decomposed plant remains. Their color is usually black. The botanic origin of the materials is, for the most part, obscure. They occur in areas where the groundwater level fluctuates within the soil, and they are subject to aerobic decomposition.

Acknowledgments

I am indebted to Drs. Tyrel G. Moore and James F. Matthews of the University of North Carolina at Charlotte for their helpful comments on this manuscript, and to Andrew C. Nunnally for his aid in drafting the Soils Map of North America.

PART II VEGETATION AND CLIMATES OF THE PAST

3. History of the Vegetation: Cretaceous (Maastrichtian)–Tertiary

Alan Graham

A discussion of the history of the vegetation of North America most logically begins with the events of the late Upper Cretaceous epoch, 70–60 Ma (million years ago). By then, the angiosperms and other major present-day groups were clearly established as dominant in the world's terrestrial flora. The continents were closer together than they are at present, and indeed, Eurasia and North America were still conjoined across the northern Atlantic. The plate tectonic forces that have placed the continents in their present configurations, however, were already in motion.

Our knowledge of the botanical events of the past rests on an interpretation of the fossil record, which for vascular plants occurs in two forms. Macrofossils are structures such as leaves, stems, fruits, seeds, wood, and flowers, whereas plant microfossils representing terrestrial or freshwater aquatic macrophytic vegetation include pollen grains, spores, and phytoliths (crystals formed within living plants). Paleobotany (including specialized approaches such as dendrochronology and analysis of pack-rat middens) has come to imply the study of plant macrofossils, and paleopalynology designates studies concerned with plant microfossils.

Experience has shown that most elements comprising a fossil assemblage are broadly consistent in terms of habitat preference, or they can be sorted into subsets reflecting habitat diversity (viz., elevational gradients). This organization gives rise to the concept of paleocommunities from which it is possible to deduce past climates, paleophysiography, and biogeographic patterns. Such reconstructions are based on a direct comparison and presumed general equivalency of most members of a fossil flora with modern analogs (composition of the flora), on the observation that present-day plants with certain morphological attributes (e.g., leaf physiognomy) are found in certain habitats, and on the assumption that most fossil plants with similar morphological attributes occurred in comparable habitats. For example, modern plant assemblages containing many large-leaved, entire-margined species with drip-tips typically occur in humid tropical habitats; therefore, a fossil flora with many similar leaf types is taken to indicate a humid tropical paleoenvironment. The composition of a fossil flora, based on the combined inventories provided by macro- and microfossil remains, leaf physiognomy, and dendrochronology, are all valuable methods for studying vegetational history and reconstructing the environments that influenced the development of North American vegetation through time.

The following discussion begins with the floras of the Maastrichtian stage of the Upper Cretaceous, 70–65 Ma, and progresses forward in time through the Tertiary to 2 Ma, the end of the Pliocene epoch, i.e., to the advent of the Pleistocene, the "Ice Ages." Within each section, the fossil floras are discussed in a sequence that begins with the southeastern corner of the continent and proceeds westward and around the con-

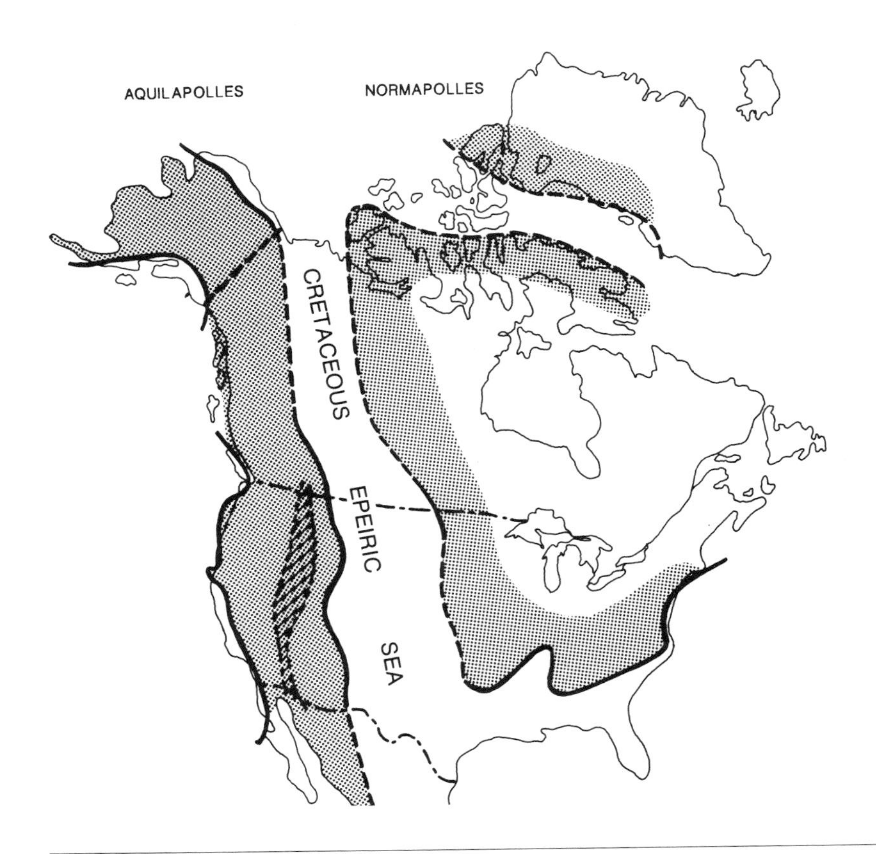

FIGURE 3.1. Map showing extent of Cretaceous epeiric sea in Maastrichtian time and Aquilapolles/Normapolles provinces. Shading shows presumed land areas; hachured area shows highlands during late Cretaceous time. [From E. B. Leopold and H. D. MacGinitie 1972. Reprinted with permission from the authors.]

tinent in a clockwise fashion. Events from the Pleistocene to the present are covered by P. A. Delcourt and H. R. Delcourt (chap. 4).

In describing paleoevents, degrees of latitude and longitude, unless otherwise noted, are given in terms of present-day locations of the poles and continents, even though the North American continent has moved slightly relative to the poles during the Tertiary and to the present.

Late Cretaceous

As the late Cretaceous came to a close (Maastrichtian times, 70–65 Ma), an epeiric (epicontinental) sea extended from the Gulf of Mexico through the western interior lowlands to the Arctic Ocean (fig. 3.1). The Appalachian Mountain system, uplifted at the end of the Paleozoic, had undergone 185 million years of erosion. In the western portion of the continent, the Rocky Mountains were locally of moderate relief, the Sierra Nevada was only a series of low hills until the Mio-Pliocene, and the High Cascades would not appear until the Pliocene. As a consequence of widespread inundation, low physiographic relief, and high temperatures (fig. 3.2), equable subtropical, warm temperate, and temperate climates extended in successive zones along low thermal gradients from the southern present-day United States to the Arctic regions. Across the latitudinal zones of late Cretaceous vegetation in North America, a temperature gradient of only about 0.3° C/1° latitude is inferred from leaf physiognomy of Maastrichtian megafossil floras, compared to just under 1° C/1° latitude at present. As noted in E. M. Friis et al. (1987), the period from the Cretaceous through the middle Eocene (70–35 Ma) included some of the warmest temperatures in all of Phanerozoic time, perhaps 5°–10° C warmer than at present at low midlatitudes (30° N) and 30° C warmer at high latitudes (80° N) (J. T. Parrish 1987).

With the continent physically divided by the epeiric sea into western and eastern regions, two principal floristic provinces characterized the late Cretaceous vegetation of North America (fig. 3.1; D. J. Batten 1984; N. O. Frederiksen 1987). In the west, and extending into Asia, palynofloras are distinguished by the presence of Aquilapolles (*Aquilapollenites* and other *Triprojectacites* types; M. J. Farabee 1990), while in the east, and extending into Europe, the Normapolles group is prominent. Suggested affinities of *Aquilapollenites* include the Loranthaceae and Santalaceae (D. M. Jar-

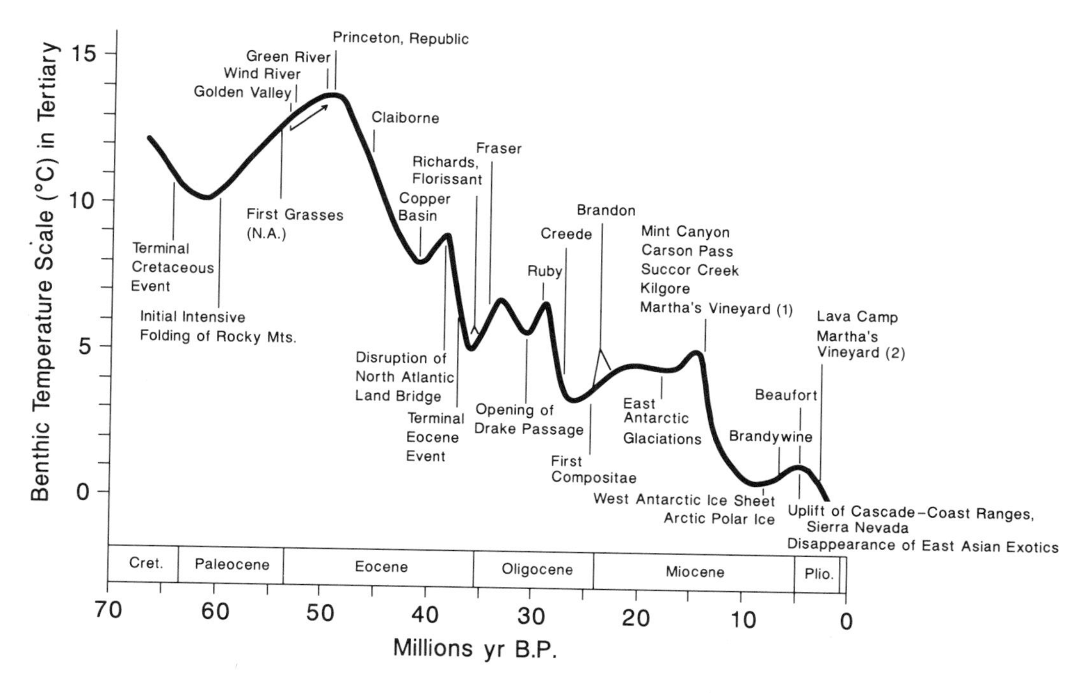

FIGURE 3.2. Tertiary paleotemperature curve of benthic temperatures with principal events and floras cited in text. [Based on fig. 4 in D. I. Axelrod and P. H. Raven 1985, with permission from the Journal of Biogeography.]

zen 1977), but the Aquilapolles in the broad sense are likely polyphyletic (M. J. Farabee and J. J. Skvarla 1988). The Normapolles are a group of Cretaceous and lower Paleogene colpate and porate (usually triporate) pollen with a complex pore structure of thick protruding ektexine (A. Traverse 1988). Suggested affinities of the polyphyletic Normapolles include various Hamamelidae, the Juglandaceae, the related family Rhoipteleaceae, or an extinct family of the Juglandales.

At the end of the Cretaceous, much of southeastern North America (30°–45° N) was covered by megathermal (mean annual temperature above 20° C) evergreen tropical woodland (estimated mean annual temperature above 25° C), and in the west by broad-leaved evergreen forests of dicotyledons and conifers (paratropical vegetation, estimated mean annual temperature 20°–25° C) (fig. 3.1; G. R. Upchurch and J. A. Wolfe 1987; J. A. Wolfe and G. R. Upchurch 1987). Fossil woods show little evidence of seasonality.

In the west, between latitudes 45° N and 65° N, mesothermal vegetation (estimated mean annual temperature 13°–20° C) was prominent (C. N. Miller Jr. et al. 1987; J. A. Wolfe 1987b). Leaf size was commonly of the notophyll class (maximum size 45 cm^2), intermediate between the microphylls (20 cm^2) indicative of drier and/or colder climates and the macrophylls (1640 cm^2) and megaphylls (no maximum size) indicative of more tropical climates. The canopy was less open in the notophyll forest than in the paratropical woodland. Middle latitude forests contained araucarian, rosid, platanoid, hamamelid (D. L. Dilcher et al. 1986; P. R. Crane and S. Blackmore 1989), and trochodendroid elements, as well as Betulaceae, Ulmoideae, Tiliaceae, Flacourtiaceae, Juglandales (possibly represented by some Normapolles), and Loranthaceae-Santalaceae (some *Aquilapollenites-Triprojectacites*). At the southern limits of this western zone (ca. 45° N), growth rings in fossil woods are absent or poorly developed, whereas at the northern limits (56°–60° N) they are clearly defined. Evergreen conifers were common, and woody dicots were represented mostly by small trees and shrubs.

Deciduous angiosperm vegetation is first recorded in high northern latitudes in the Cenomanian. Above 55° N latitude, late Cretaceous macrofossil floras occur in central Alberta (W. A. Bell 1949; C. G. K. Ramanujam and W. N. Stewart 1969; J. A. Wolfe and G. R. Upchurch 1987b), and slightly older ones (Campanian to Maastrichtian, 82–65 Ma) are known from Alaska (R. A. Spicer et al. 1987). The vegetation was a mosaic of deciduous angiosperm and gymnosperm forests in a climate where temperatures ranged from mesothermal to microthermal (estimated mean annual temperature at or below 13° C). Macrofossil remains include platanoids, trochodendroids, hamamelids, and *Viburnum*-like leaves representing vegetation from middle latitudes and extending into these northern regions primarily along streams. Away from the streams the vegetation included *Fagopsis,* a genus of the Fagaceae that became extinct at the end of the Eocene (S. R. Manchester and P. R. Crane 1983; W. L. Stern et al. 1973), Flacourtiaceae, Aceraceae, *Alnus,* and possibly *Betula.* Fossil woods show well-defined growth rings, an indication of contrasting seasons.

The estimated 1° C mean temperature isotherm for the coldest month separated the zone of predominantly broad-leaved, evergreen, subtropical vegetation from the more northern broad-leaved deciduous vegetation. Regions with coldest month means between −2° C and 1° C had notophyll broad-leaved evergreens as understory vegetation. In modern climates with coldest month means lower than −2° C, notophyll broad-leaved evergreens are lacking (J. A. Wolfe 1978).

At the close of the Cretaceous two events occurred that influenced subsequent development of the North American flora. The first, regression of the western interior sea, reduced the principal physical barrier to dispersal between the eastern and western portions of the continent. In geological strata of this age, diversity in Normapolles increased in California and the Canadian Arctic; *Aquilapollenites* has been reported from the Atlantic Coastal Plain.

The second event marks the end of the Cretaceous. Major extinctions are noted for marine plankton and certain large terrestrial animals and lesser ones for terrestrial vegetation. These extinctions may have resulted from climatic changes immediately following the impact of a large asteroid. A relatively brief lowering of temperatures (winter effect), followed by a rise in temperature (greenhouse effect), is plausible.

Such an event also may have set into motion, or intensified, selection favoring the deciduous habit at the expense of evergreens. J. A. Wolfe (1987) described a change from dominantly broad-leaved evergreen megathermal forest to dominantly broad-leaved deciduous mesothermal forest in the high middle latitudes of North America. Soon afterward temperatures and vegetation resumed the trends of pre-impact times under primarily climatic and physiographic influences.

Tertiary: Paleocene through Middle Eocene (65–35 Ma)

Generally warm temperatures, and possibly increased but seasonal (winter-dry) rainfall regimes, characterized the Paleocene and early and middle Eocene (J. A. Wolfe and G. R. Upchurch 1987b).

Several events important to the history of North American vegetation occurred during the period from late Cretaceous through middle Eocene times. Tropical dry-season vegetation first appeared; the Betulaceae, Fabaceae, Fagaceae, Juglandaceae, and Ulmoideae diversified in the middle and higher latitudes. Araucarian conifers became extinct in the Northern Hemisphere at the end of the Cretaceous. *Glyptostrobus* and *Metasequoia* became abundant, the latter the most frequent and widespread member of the Taxodiaceae (Cupressaceae in the flora) in North America from Cretaceous to Miocene times. *Momipites* (*Alfaroa-Engelhardia-Oreomunnea* type), *Carya, Platycarya* (first appearing in the early Eocene), *Cyclocarya-Pterocarya* of the Junglandaceae, lianas (Vitaceae), Anacardiaceae, and Rutaceae became abundant. *Acer, Ailanthus, Celtis, Chaetoptelea, Comptonia, Eucommia, Hydrangea, Liquidambar, Populus,* and *Rhus* made their first appearance. The family Poaceae was present in North America by the Paleocene/Eocene (W. L. Crepet and G. D. Feldman 1991). The midlatitude mesothermal, broad-leaved deciduous vegetation also extended across the midcontinent region that now supports prairie vegetation. At high elevations in the higher latitudes in western North America, coniferous forests of Pinaceae and Cupressaceae first appeared. In the middle and late Eocene, the Rosaceae, Juglandaceae, and Aceraceae underwent rapid adaptive radiation. By the end of the Paleocene, the epeiric seaway had retreated to a position some 300–500 km inland from the present Gulf and Atlantic coasts.

Interpretations based on fossil pollen (N. O. Frederiksen 1980) and macrofossils (D. L. Dilcher 1973; J. A. Wolfe 1985) suggest that the climates of southeastern North America during the lower and middle Eocene varied between seasonally dry tropical and humid subtropical. In all probability, the flood plain at the northern end of the Mississippi Embayment in Kentucky and Tennessee experienced occasional frosts, but the climate near the sea was frost-free, winter-dry tropical (N. O. Frederiksen 1980).

Tropical forests (Annonaceae, Lauraceae, Menispermaceae) in the Southeast reached their maximum northern expansion (to ca. 50°–60° N). The middle Eocene Claiborne flora of western Kentucky and Tennessee (fig. 3.3) is regarded as preserved in oxbow lake sediments several miles inland from the embayment.

These paleocommunities were a complex assemblage that have no exact modern analog. In addition to neotropical elements, they had at least several paleosubtropical taxa introduced into North America from the Old World tropics over the North Atlantic land

FIGURE 3.3. Location of Tertiary plant fossil formations in North America. 1 = Copper Basin, 2 = Ruby, 3 = Wind River, 4 = Florissant, 5 = Creede, 6 = Kilgore, 7 = Golden Valley, 8 = Princeton, 9 = Republic, 10 = Succor Creek, 11 = Green River, 12 = Mint Canyon, 13 = Lava Camp, 14 = Beaufort, 15 = Richards, 16 = Carson Pass, 17 = Claiborne, 18 = Brandon, 19 = Martha's Vineyard, 20 = Brandywine. ◇ = several sites in close proximity.

bridge during the late Paleocene and early Eocene (N. O. Frederiksen 1988; M. C. McKenna l983; D. W. Taylor 1990; B. H. Tiffney 1985b). The macrofossil components of the lowland flora have primarily tropical affinities. The nearby Appalachian and Ozark highlands, however, added a substantial temperate pollen component to this lowland paleoflora. Pollen types include *Pinus, Alnus, Betula, Castanea, Corylus, Nyssa* (also present in Eocene megathermal assemblages), and *Tilia*, as well as the tropical Annonaceae (*Annona*, possibly *Cymbopetalum*), Bombacaceae (oldest records are in the eastern United States [J. A. Wolfe 1975]), *Engelhardia* (also winged fruits and catkins of *Engelhardia* and *Oreomunnea-Alfaroa* [W. L. Crepet et al. 1980]), Euphorbiaceae (*Retitricolporites* = ? *Amanoa* group), and Sapindaceae *(Cupanieidites, Duplopollis)*. Macrofossils include *Nypa* (from the Laredo formation of the Claiborne group in South Texas [J. W. Westgate and C. T. Gee 1990]), *Podocarpus, Ceratophyllum* (P. S. Herendeen et al. 1990), *Dendropanax*, Fagaceae (Castaneoideae), *Ficus*, Euphorbiaceae *(Hippomaneoidea)*, Gentianaceae (flowers with *Pistillipollenites* pollen), *Hura*, Lauraceae (*Androglandula* flowers with *Ocotea-Cinnamomum* affinities), Fabaceae (flowers *Eomimosoidea, Protomimosoidea*; *Crudia* [P. S. Herendeen and D. L. Dilcher 1990]), Malpighiaceae (pollen, flowers—*Eoglandulosa*, affinities possibly with subfamily Byrsonimoideae [D. W. Taylor and W. L. Crepet 1987]), *Philodendron* and *Acorites* of the Araceae, *Sabal*, Ulmaceae *(Eoceltis)*, and Zingiberaceae (D. W. Taylor 1988).

The Golden Valley formation of North Dakota is of late Paleocene and early Eocene age (L. J. Hickey 1977). The macrofossils in it include *Lygodium, Hemitelia, Onoclea, Salvinia, Glyptostrobus, Metasequoia, Pinus, Acer, Betula, Carya, Cercidiphyllum, Cornus, Corylus, Dombeya, Meliosma, Palmae, Platanus, Platycarya, Pterocarya, Salix, Stillingia*, and Zingiberaceae. The flora and associated fauna are interpreted to indicate a warm temperate climate (mean annual temperature near 15° C) in the late Paleocene, and warmer subtropical conditions in the early Eocene (ca. 18.5° C). The different composition of the sequential florulas suggests the region may have been within a broad ecotonal belt between a predominantly deciduous forest to the north and a broad-leaved evergreen forest to the south (L. J. Hickey 1977, p. 72).

Another Paleocene flora from the Sentinel Butte formation near Almont, North Dakota, is under study (P. R. Crane et al. 1990), and when completed, the results will provide additional information on the Paleocene vegetation and environments of the region. In general, the tropical to warm temperate component of Paleocene and early Eocene floras from the eastern Rocky Mountain region shows affinity with eastern Asia, while by the middle Eocene the affinity is with seasonally dry vegetation of northern Latin America (E. B. Leopold and H. D. MacGinitie 1972; C. N. Miller Jr. et al. 1987; S. L. Wing l987; J. A. Wolfe 1987b).

In western North America, lower to middle Eocene floras include the Republic flora of eastern Washington (J. A. Wolfe and W. Wehr 1987), the Princeton flora of British Columbia (S. R. S. Cevallos-Ferriz and R. A. Stockey 1989, 1990, 1990b; R. A. Stockey 1984, 1987), the Wind River flora of northwestern Wyoming (H. D. MacGinitie 1974), and the Green River flora of Colorado and Wyoming (H. D. MacGinitie 1969). The paleolatitudes of the locations and the proposed stratigraphic relationships of these and other Eocene floras in the region are summarized by D. I. Axelrod and P. H. Raven (1985, p. 23), S. L. Wing (1987, pp. 752–753), and J. A. Wolfe and W. Wehr (1987, pp. 5, 7). Collectively they reflect a seasonally dry, subtropical to warm temperate megathermal vegetation at the lower elevations, a deciduous broad-leaved mesothermal vegetation in the uplands, and locally at higher elevations a dominant coniferous element (D. I. Axelrod 1990), suggesting microthermal climates.

The Wind River flora includes *Acrostichum, Lygodium, Acalypha, Aleurites, Apeiba, Canavalia, Cedrela, Dendropanax, Eugenia, Ilex, Luehea, Machilus-Persea, Platanus, Populus, Sabalites, Saurauia, Schefflera, Sterculia,* and *Symplocos*. The Green River flora includes a more prominent coniferous element with *Picea, Pinus,* and *Tsuga*. The Princeton flora includes *Equisetum, Azolla, Glyptostrobus, Metasequoia, Pinus, Alnus,* Araceae (tribe Lasioideae), *Betula* (also from the nearby middle Eocene Allenby formation [P. R. Crane and R. A. Stockey 1987]), *Cercidiphyllum, Comptonia, Platanus, Sassafras,* and Sapindaceae (Allenby formation [D. M. Erwin and R. A. Stockey 1990]).

The Republic flora was a low montane mixed coniferous forest with the coniferous element dominant, with some lakeside or streamside broad-leaved deciduous and deciduous gymnospermous trees, plus a minor forest element of broad-leaved evergreen and broad-leaved deciduous small trees and shrubs (J. A. Wolfe and W. Wehr 1987). Among the plants were *Abies, Chamaecyparis, Ginkgo, Metasequoia, Picea, Pinus, Pseudolarix, Thuja, Acer, Cornus, Crataegus, Itea, Phoebe, Photinia,* aff. *Potentilla, Prunus, Pterocarya, Rhus, Tilia,* and *Ulmus*. The estimated elevation of the Republic flora is 700–900 m, and at that level the mean annual temperature was below 13° C, the mean of the warmest month was below 20° C, the coldest month mean was between 1° C and 2° C, and the warmest month mean was more than 15° C.

Late Paleocene macrofossil floras from Alaska are stratigraphically complex (J. A. Wolfe 1972, 1977) and include leaves similar to *Anemia, Dioön, Zamia, Glyptostrobus, Metasequoia, Pseudolarix, Acer, Carya, Corylus, Cercidiphyllum, Cocculus, Comptonia, Hypserpa, Macaranga, Palmacites, Planera, Pterocarya,* and *Sabalites*. The varied biogeographic affinities suggested by these assemblages include the Sino-Japanese region, Atlantic North America, to a lesser degree the Neotropics, and possibly Malaysia and continental Southeast Asia. The leaves are mostly evergreen and of the mesophyll size class. About 50% have entire margins, suggesting a subtropical climate. Later Paleocene floras suggest a slight cooling.

Lowland Malaysian genera are prominent in some Alaskan strata of early Eocene age, and the floras include *Alangium, Alnus, Anamirta, Barringtonia* (a tropical Pacific beach plant), *Cananga, Clerodendrum, Kandelia* (a mangrove), Lauraceae (five species), *Luvunga, Limacia, Melanorrhoea, Meliosma,* Menispermaceae (five paleotropical genera), *Myristica,* the dipterocarp *Parashorea, Phytocrene, Platycarya, Pyrenacantha, Sageretia, Saurauia, Stemonurus,* and *Tetracentron*. None are exclusively neotropical. The leaves are mostly broad-leaved evergreen. The most similar modern analogs of these floras occur in tropical or near-tropical frost-free climates with a mean annual temperature of ca. 18° C (warm mesothermal). J. A. Wolfe (1972, p. 212) applied the name Paratropical Rain Forest to the distinctive lowland vegetation. It apparently had two (rarely three) tree stories, abundant and diverse woody lianas, buttressed trees, and leaves of notophyll to mesophyll size classes and with drip-tips. Approximately 60–75% of the species had entire-margined leaves.

According to the summary by M. C. McKenna (1983, p. 469), most of the terranes south of the Denali Fault in Alaska were formed at sites far to the south and have been accreted to the North American Plate. By this view, fossil floras of Malaysian affinities on these terranes do not tell about ancient Alaska but about some other place (M. C. McKenna 1983). Indeed, the Eocene Beringian land bridge was primarily occupied by a broad-leaved deciduous forest (J. A. Wolfe 1985, cited in B. H. Tiffney 1985b) and may not have been a major route of migration for megathermal elements. A comparable situation exists with the Miocene Mint Canyon flora of southern California and the Carmel flora near Monterey, California, both of which have been displaced 200–300 km from the south (D. I. Axelrod 1986), and late Cretaceous floras from western California and British Columbia that were transported at least 2000 km (N. O. Frederiksen 1987). Paleomagnetic data cited in R. A. Spicer et al. (1987), however, suggest that the Yakutat terrane containing the plant fossils was located

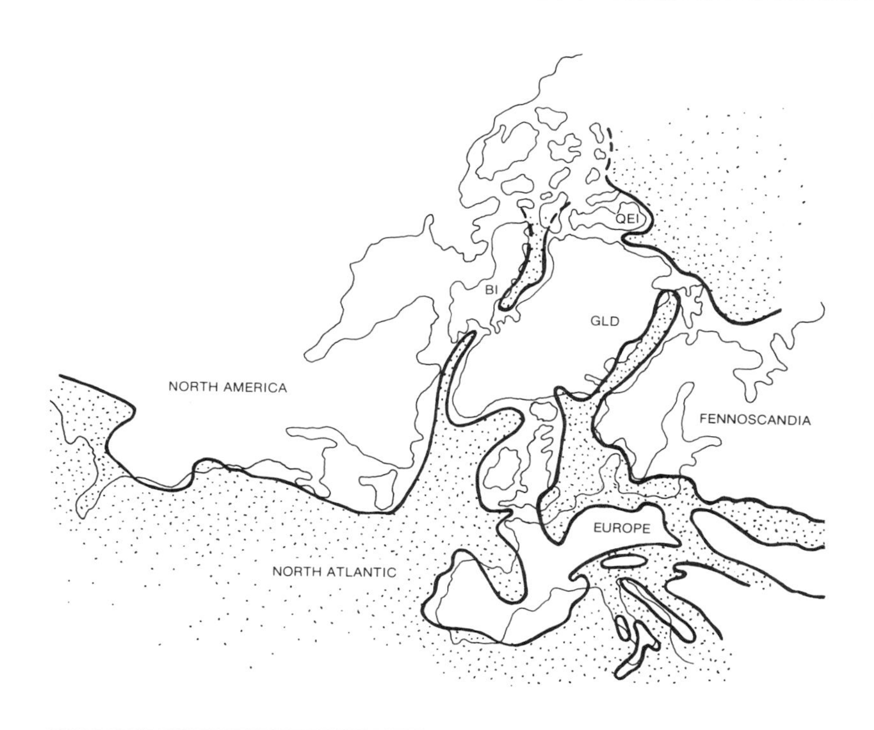

FIGURE 3.4. Paleogeography of eastern North America in the early Eocene. Stippled area = sea, heavy lines = paleocoastlines, light lines = present coastlines, BI = Baffin Island, GLD = Greenland, QEI = Queen Elizabeth Islands. [From B. H. Tiffney 1985b. Reprinted with permission from Journal of the Arnold Arboretum.]

off what is now central British Columbia and southeastern Alaska (as shown in J. A. Wolfe 1985).

From farther east, between latitudes 65° and 81° N, early Tertiary floras have been described from Ellesmere Island (R. L. Christie and G. E. Rouse 1976; S. Manum 1962), then located at about latitude 65°–70° N (J. A. Wolfe, pers. comm., 1989; 75.9° N fide M. C. McKenna 1983, p. 463), the Mackenzie River Delta (F. L. Staplin 1976; G. Norris 1982), Peel River (G. E. Rouse and S. K. Srivastava 1972), and eastern Greenland (B. E. Koch 1963; S. Manum 1962). The plants include *Abies, Glyptostrobus, Larix, Picea, Pinus, Metasequoia, Taxodium, Alnus, Betula, Castanea, Cercidiphyllum, Fagus, Liriodendron, Nyssa, Platanus, Populus, Pterocarya, Quercus, Ulmus,* and *Zelkova.* Surface marine water temperature bordering east Greenland was about 17° C in the late Cretaceous, and about 15° C bordering Ellesmere Island during the middle Eocene.

Land connection between North America and Asia existed for much of the early Tertiary via the Bering land bridge (D. M. Hopkins 1967), and perhaps until later in the Tertiary (J. T. Parrish 1987, p. 61). The latitude of the bridge, however, was ca. 75° N, some 7° farther north than the present Bering platform. It was also intermittent. Unfavorable climate and marine transgression formed a barrier in the Paleocene and middle Eocene, and the bridge was available for migration in the early and the late Eocene. Some filtering of Old World tropical to subtropical species must have occurred.

To the northeast, North America was connected to Europe by the now sundered landmass of Euramerica (fig. 3.4), which existed at the end of the Paleocene but was beginning to fragment by the early Eocene. Two routes for migration existed. There was a southern one along the Greenland-Scotland Ridge (the Thulean route, ca. 45°–50° N paleolatitude), disrupted in the early Eocene, but possibly providing land surfaces from North America east to Iceland until the Miocene. There was also a more northern route (the DeGeer route, ca. 10°–15° farther north, but still at a lower paleolatitude than Beringia), resulting from compression between Spitsbergen (Svalbard) and Greenland during the Eocene and from volcanic activities over the Iceland hotspot (M. C. McKenna 1983; B. H. Tiffney 1985b; P. A. Ziegler 1988). Migrants along this route were mostly deciduous elements through the early Oligocene, at which time the Greenland-Norwegian Sea connected with the Arctic Ocean. Prevailing climatic conditions of the two land bridges, rather than physical barriers, limited free exchange of floral elements between North America and Europe or Asia during the Tertiary. The early Eocene was the warmest interval of the Tertiary, and this was a time of rapid migration from Europe to North America over the North Atlantic land bridge (N. O. Frederiksen 1988).

Late Eocene to Late Miocene (35–10 Ma)

Analyses of Deep Sea Drilling Project cores record a significant drop in temperature in North America and elsewhere during the mid-Tertiary. The drop was particularly notable for winter minimum temperatures near the end of the Eocene, ca. 33 Ma. The lower temperature had an influence on the flora.

In the southeastern United States, the period from the middle Eocene to the early Oligocene is marked by two changes in vegetation. Fossil pollen floras record an increase in an oaklike form *(Quercoidites inamoenus)* and, more significantly, a decline in overall diversity (N. O. Frederiksen 1988). Climates fluctuated during this period (J. A. Wolfe 1978; J. A. Wolfe and R. Z. Poore 1982), but a significant cooling at the Upper Eocene–Lower Oligocene boundary (ca. 33–34 Ma) is evident.

There are several floras near the Eocene-Oligocene boundary at midlatitudes (40°–45° N) in western North America. These do not all preserve clear evidence of late Eocene cooling, because some are from localities with an insufficient number of closely spaced, sequential floras for critical comparison.

The Florissant flora of central Colorado (39° N latitude, ca. 34 Ma; H. D. MacGinitie 1953) is assigned to the latest Eocene by J. A. Wolfe and W. Wehr (1987, p. 7) and to the early Oligocene by D. I. Axelrod and P. H. Raven (1985, p. 23; S. L. Wing 1987), predating the period of climatic deterioration. It preserves an extensive flora of *Abies, Picea, Pinus, Chamaecyparis, Acer, Ailanthus, Amelanchier, Bursera, Cardiospermum, Carpinus, Carya, Castanea, Cedrela, Celastris, Celtis, Cercis, Cercocarpus, Colubrina, Crataegus, Daphne, Eugenia, Hydrangea, Lindera, Mahonia, Persea, Philadelphus, Platanus, Populus, Prosopis, Ptelea, Quercus, Ribes, Salix, Sambucus, Sapindus, Sassafras, Tilia, Trichilia, Ulmus, Vitis, Zelkova, Ziziphus,* and others. The basin of deposition was at an elevation of about 900–1000 m, with volcanic highlands to 2700 m to the immediate west. The climate at the lower elevations was warm temperate (average annual temperature 24° C), grading into cooler climates in the adjacent highlands. Annual rainfall was an estimated 525 mm. Mesic deciduous hardwoods grew along the streams and lakes, while more open vegetation occurred on the adjacent slopes. Affinities of the flora are with the temperate broad-leaved deciduous vegetation extending across into Asia and Europe north of latitude 45° N and southward where higher elevations existed, and with more subtropical vegetation farther to the south.

The Ruby Basin flora in southwestern Montana (latitude 45° N [H. F. Becker 1961]) is considered to be of middle Oligocene age (30.8–29.2 Ma [S. L. Wing 1987, p. 763]; it is approximately equivalent in age to the Florissant [J. A. Wolfe, pers. comm., 1989]) and includes many of the same elements as the Florissant flora, in addition to *Pseudotsuga, Glyptostrobus, Metasequoia, Alnus, Betula, Cornus, Fagus, Fraxinus, Myrica, Nyssa,* and others.

A last example of Eocene-Oligocene midlatitude western floras is the Copper Basin flora (ca. 40 Ma) of northeastern Nevada (ca. 42° N latitude). It contains a lake and streamside community of *Alnus, Acer, Amelanchier, Crataegus, Mahonia, Prunus,* and *Salix,* and a conifer–deciduous hardwood forest of *Sequoia, Chamaecyparis, Pseudotsuga, Cephalotaxus, Acer, Aesculus, Prunus, Sassafras,* and *Ulmus.* Montane conifers in the flora include *Abies, Picea, Pinus,* and *Tsuga.* The flora grew at an elevation of 1200 m; the climate was cool temperate (mean annual temperature ca. 11° C), and annual rainfall was between 1200 and 1500 mm.

The sequence of floras from western North America suggests the following general elevational zonation: lowlands (below ca. 300 m)—warm temperate, broad-leaved evergreen forest; 300–1000 m—temperate, mixed deciduous hardwood forest; 1000–1300 m—cool temperate, conifer–deciduous hardwood forest; and above 1300 m—cold temperate montane conifer forest with few deciduous hardwoods. Floras characteristic of the late Tertiary were in the cooler uplands during the Eocene, and they migrated into the lowland basins with the cooler climate later in the Cenozoic (D. I. Axelrod 1966).

In the Pacific Northwest (latitude 45° N), mean annual temperature is estimated to have been 12°–13° C, and in Alaska (latitude 60° N) at about 10°–11° C (J. A. Wolfe 1978).

Palynofloras from the MacKenzie Delta region of the Northwest Territories (latitude 69° N [J. C. Ritchie 1984, chap. 5]) preserve a clear record of lowering temperatures in the late Eocene. The Richards formation in this area encompasses the Eocene-Oligocene boundary. The middle Upper Eocene part of the Richards formation (G. Norris 1982) contains pollen of *Metasequoia, Picea, Pinus, Sequoia, Tsuga, Alnus, Betula, Castanea, Pterocarya, Quercus, Tilia,* and *Ulmus.* By the early Oligocene, diversity decreased sharply, and several thermophilous taxa—*Sequoia, Castanea, Quercus, Tilia,* and *Ulmus*—disappeared from this formation, only to reappear in the late Oligocene. A number of the North

American components of the Richards flora presently have their northern pollen rain limits in the Great Lakes region, reflecting the warmer climates of the late Oligocene.

A pollen flora from central British Columbia (K. M. Piel 1971) of late Oligocene age reflects a warming trend, as does the later portion of the Richards formation (fig. 3.2). The central British Columbia flora contains *Alnus, Carya, Engelhardia* type, *Juglans, Liquidambar, Pterocarya, Quercus,* and *Ulmus-Zelkova.*

The overall effect of the Eocene/Oligocene event was a southward shift of elements of the megathermal tropical vegetation, expansion of the high montane mixed coniferous forest at high latitudes and at high elevations further south, and at midlatitudes expansion of mesothermal broad-leaved temperate deciduous vegetation that would reach its maximum extent in the middle and late Miocene. The Rocky Mountains underwent intensive deformation and uplift in the Paleocene and, by the end of the Eocene, provided local highlands of significant relief (S. L. Wing 1987). The Sierra Nevada was not yet uplifted, and the volcanic pile representing the Cascade Ranges started to accumulate in the late Eocene. Their mean elevation was 700 m by the early Miocene.

Mid-Tertiary floras are rare in northeastern North America where Paleozoic and Mesozoic strata predominate. A notable exception is the Brandon lignite flora from Vermont that is of Oligocene to possibly early Miocene age (latitude 43°50' N [B. H. Tiffney 1985b; B. H. Tiffney and E. S. Barghoorn 1976; A. Traverse 1955]). The flora includes *Glyptostrobus, Pinus, Alangium, Carya, Castanea, Cyrilla, Engelhardia* type, Ericaceae, *Fagus, Gordonia, Ilex, Illicium, Juglans, Jussiaea, Liquidambar, Magnolia, Manilkara, Minusops, Morus, Nyssa, Parthenocissus, Planera, Pterocarya, Quercus, Rhamnus, Rhus, Symplocos, Tilia, Ulmus,* and *Vitis.* A warm temperate climate is inferred from this flora, in contrast to more subtropical conditions to the south. Affinities are with the broad belt of mesophyllous vegetation extending across Europe, Asia, and North America during the mid-Tertiary.

In western North America, late Oligocene plants are known from the Creede flora (27.2 Ma) of southwestern Colorado (D. I. Axelrod 1987). The flora is of interest because it provides insight into local communities of drier aspect within a region generally characterized by more humid vegetation. The flora apparently was deposited in an ecotone between a mixed conifer forest and a pinyon-juniper woodland scrub at 1200–1400 m elevation. Annual precipitation was an estimated 460–635 mm, and mean annual temperature was less than 2.5° C. J. A. Wolfe and H. E. Schorn (1989) revised this flora and state that it included *Abies, Juniperus, Picea, Pinus, Berberis, Cercis, Cercocarpus, Holodiscus, Mahonia, Populus, Prunus, Ribes, Salix, Sorbus,* and others. Affinities are mostly with the southern Rocky Mountain region, but a few elements of the flora have relationships to plants from temperate eastern Asia, eastern North America, and the Sierra Madre Occidental of Mexico. Just as the Eocene Copper Basin flora of Nevada contained cool temperate elements in the uplands that would spread during the late Tertiary period of lowering temperatures, the Oligocene Creede flora of southwestern Colorado contained elements that were able to thrive in the drier habitats that prevailed in the area during Pliocene and later times.

The final separation of continents in the Southern Hemisphere profoundly influenced the climate of the Northern Hemisphere. Sustained glaciations are evident in East Antarctica (middle Miocene). Australia separated further from Antarctica, and the Drake Passage between Antarctica and South America opened (middle Oligocene). As a result, cold water flowed northward into the southern oceans and circulated into equatorial latitudes, in turn strengthening high pressure systems and the drier climates associated with them in North America.

Drier climates and colder winters initiated the decline of tropical elements from the North American flora. The expansion of the mesothermal broad-leaved temperate forests occurred at midlatitudes. The early appearance of taxa adapting to these drier habitats was particularly evident in southwestern North America (D. I. Axelrod and P. H. Raven 1985). As the Rocky Mountains uplifted, coniferous forests characteristic of higher elevations expanded. The area of the southern Rocky Mountains and the Sierra Madre Occidental was an important center for the evolution of the Madro-Tertiary geoflora (D. I. Axelrod 1958) and for pines (D. I. Axelrod 1986b). Pollen of the Asteraceae first appeared in abundance at the Oligo-Miocene boundary (ca. 25 Ma), although some argue for an older origin (B. L. Turner 1977).

During the mid-Tertiary, a significant lowering of sea level was evident at about 30 Ma. Major volcanic activity occurred in western North America, the Andes, and the Philippines. The Mississippi Embayment, a remnant of the epeiric sea, was rapidly retreating southward. Temperate conditions continued to prevail across the North Atlantic and Beringia. Tectonic events, however, caused the further disruption of the North Atlantic land bridge. The cooling that occurred near the end of the Eocene (C. Pomerol and I. Premoli-Silva 1986) is reflected in the terrestrial vegetation of the Northern Hemisphere.

Late Miocene through Pliocene (10–2 Ma)

All through this period, a general cooling occurred. The geographic extent of the temperate deciduous hardwood forests declined from their former extensive ranges across the medium and high latitudes. Extensive grasslands appeared during this time, and at high latitudes and elevations coniferous forests continued to expand. Polar ice formed in the Arctic, and tundra elements appeared. The Sierra-Cascades reached substantial heights, and a rain shadow developed in the Great Basin. There was additional uplift in the Rocky Mountains and, at their southern end, the further development of a sclerophyllous element occurred.

The change in vegetation near the Miocene-Pliocene boundary is documented in the east by palynofloras from eastern Massachusetts (N. O. Frederiksen 1984). A middle Miocene assemblage from Martha's Vineyard represents a rich, warm temperate flora of *Abies, Picea, Pinus, Podocarpus, Tsuga, Alnus, Betula, Carya, Castanea, Fagus, Ilex, Liquidambar, Nyssa, Quercus,* and *Ulmus-Zelkova.* A Pliocene flora from the same area is depauperate and cool temperate in aspect. A similar change is evident in Miocene-Pliocene plant and animal communities across the midlatitudes of the Northern Hemisphere.

In the area that is now the plains states and provinces, colder winter temperatures were accompanied by reduced summer rains. The middle Miocene Kilgore flora from Nebraska (H. D. MacGinitie 1962) contains a valley element (*Carya, Liquidambar, Nyssa, Platanus, Populus, Tilia*), with open pine-oak grassy woodlands on the uplands. Other plants included *Acer, Celtis, Fraxinus, Prunus, Pterocarya, Ulmus,* and a more southern element of *Cedrela, Cordia, Diospyros,* and *Meliosma.* The composition suggests reduced rainfall but no well-developed prairie. Associated fossil faunas suggest frost-free climate. The affinities of this flora are with modern taxa growing east of the Rocky Mountains. Only six species of this flora are found also in Miocene floras of the Columbia Plateau, indicating that by the middle Miocene the cordillera was an effective barrier to biotic interchange between eastern and western North America.

Late Miocene to early Pliocene floras from Kansas, Nebraska, and Colorado (R. W. Chaney and M. K. Elias 1936; M. K. Elias 1942; J. R. Thomasson 1979, 1987) are not sufficiently diverse or extensive to reconstruct the vegetation in detail. The floras and associated faunas indicate the presence of savannas or savanna parklands, and temperatures that seldom, if ever, dropped to 0° C. There is no evidence to confirm the presence of treeless grasslands, semiarid conditions, and extremes of temperature that are currently characteristic of the region (J. R. Thomasson 1979).

Miocene pollen floras from the eastern foothills of the Rocky Mountains reflect an impoverished flora of *Artemisia, Sarcobatus, Ephedra, Eriogonum* (steppe, halophytic vegetation), *Salix,* Betulaceae, and riparian plants, and *Abies, Juniperus,* and *Pinus* on mountain slopes (E. B. Leopold and M. F. Denton 1987). D. I. Axelrod's (1985) summary of the grassland biome depicts a trend of gradually decreasing rainfall beginning about 16 Ma. E. B. Leopold and M. F. Denton (1987) noted that the Columbia Plateaus west of the Continental Divide consistently maintained deciduous hardwood and montane coniferous forests, both rich in woody genera. Evidence from the Great Plains east of the Rocky Mountains clearly indicates that deciduous forests with grassland elements existed in the valleys. Floristically these Great Plains floras bear little relation to those west of the Rocky Mountains.

Middle to late Miocene floras from western North America are extensive, particularly in the vicinity of the Columbia Plateau of central Oregon and adjacent regions (R. W. Chaney 1959; R. W. Chaney and D. I. Axelrod 1959). The western American, eastern Asian, and western European floras, as well as the few floras from mid- and eastern North America, record an extensive, temperate, mixed deciduous hardwood forest across the middle and high latitudes. This is the much debated Arcto-Tertiary geoflora (R. W. Chaney 1967; J. A. Wolfe 1972).

The Succor Creek flora (A. T. Cross and R. E. Taggart 1982; A. Graham 1965; R. E. Taggart and A. T. Cross 1980) of southeastern Oregon is typical of middle to late Miocene assemblages in the vicinity of the Columbia Plateau. It includes *Abies, Ginkgo, Glyptostrobus, Picea, Pinus, Tsuga, Acer, Ailanthus, Alnus, Amelanchier, Betula, Carya, Castanea, Cedrela, Celtis, Crataegus, Diospyros, Fagus, Fraxinus, Ilex, Juglans, Liquidambar, Magnolia, Mahonia, Nyssa, Ostrya, Platanus, Populus, Pterocarya, Quercus, Salix, Sassafras, Tilia, Ulmus,* and others (figs. 3.5, 3.6). Other Miocene floras east of the Cascade–Sierra Nevada ranges, and extending to British Columbia, add *Chamaecyparis, Metasequoia, Sequoia, Taxodium, Cercidiphyllum, Cinnamomum, Rosa,* and *Sorbus* (D. I. Axelrod 1986).

In the late Miocene, frost-sensitive plants such as *Cedrela* disappeared from the northern Rocky Mountains. The generic constituents of the flora on the Columbia Plateaus did not take on a modern character until at least late Pliocene (Blancan) or possibly Pleistocene times. Grassland and steppe elements (e.g., *Artemisia, Sarcobatus*) present in the region throughout the Miocene were numerically unimportant during that period. Data from the Snake River Plain indicate that grasses became sporadically abundant in the Pliocene 10 Ma after grasslands presumably developed in the Great Plains (E. B. Leopold and M. F. Denton 1987).

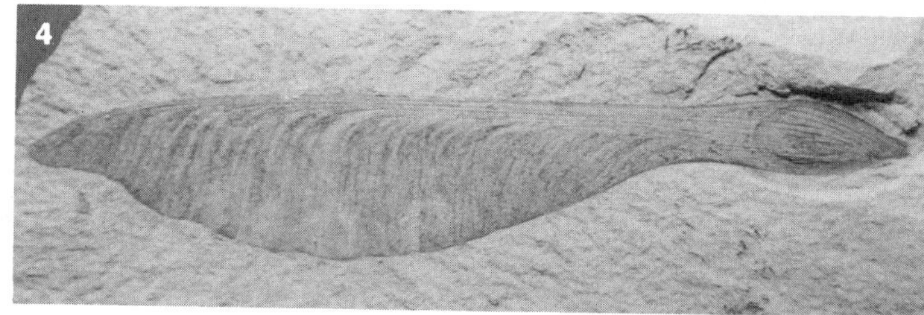

FIGURE 3.5(A). Representative specimens of plant macrofossils from the Succor Creek flora, southeastern Oregon, middle to late Miocene age. 1—*Sassafras columbiana,* 2—*Ulmus speciosa,* 3—*Fraxinus coulteri,* 4—*Acer bendirei* (=*A. chaneyi*).

In the southern Rocky Mountains, further development of the drier sclerophyllous vegetation was favored by a combination of lower minimum winter temperatures, reduced summer rainfall, and the effects of slope exposure, edaphic conditions, volcanism, and tectonic uplift. A tendency toward sclerophyllous vegetation is evident in the late Oligocene Creede flora of south central Colorado (D. I. Axelrod 1987; D. I. Axelrod and P. H. Raven 1985). The transition between the sclerophyllous vegetation to the south and the broad-leaved deciduous vegetation to the north was in central Nevada during the Miocene. Evergreen coniferous forests were prominent at high latitudes, and at high elevations further south.

With the late middle Miocene cooling, the west Antarctic Ice Sheet developed, and polar ice first appeared in the Arctic. Precursors to future tundra vegetation were evolving, and other elements coalesced into early versions of the boreal coniferous forest. Elements of the broad-leaved deciduous forest reached their maximum southern expansion in eastern Mexico.

By the beginning of the late Miocene (7–5 Ma), many Asian, neotropical, and paleotropical elements had disappeared from eastern North America. Nevertheless, the Brandywine flora of Maryland, questionably of late Miocene age, contained *Alangium, Pterocarya, Trapa,* and an *Ilex* similar to the Asian *I. cornuta* (L. McCartan et al. 1990).

In the midcontinental rain shadow to the east of the Rocky Mountains, reduced rainfall and lowered minimum winter temperatures further restricted arborescent vegetation to valley habitats. This favored continued

FIGURE 3.5(B). 5—*Quercus consimilis,* 6—*Betula thor,* 7—*Cedrela pteraformis* (with modern *Cedrela* seed), 8—*Glyptostrobus oregonensis,* 9—*Cedrela pteraformis,* 10—*Hydrangea bendirei.* [Photographs by author.]

development of herbaceous vegetation, but true prairie did not develop extensively until Quaternary time. Extensive grasslands probably began at the Miocene-Pliocene transition (7–5 Ma). This was the driest part of the Tertiary. Forests and woodlands were restricted, and grasses and forbs rapidly radiated (D. I. Axelrod 1985).

A middle Miocene flora from Carson Pass in the central Sierra Nevada suggests uplift of about 2300 m since that time (D. I. Axelrod 1986). In the Pliocene, the Cascade-Sierra Nevada and the Coast Ranges reached sufficient heights to create an effective rain shadow over the Basin and Range Province, resulting in a trend from mesic and summer-wet to xeric and summer-dry conditions. This trend strengthened in late Pliocene and Quaternary times. Sclerophyllous vegetation, containing elements adapted to local arid habitats since the early Tertiary, coalesced and spread across the drier low-elevation regions of southwestern North America.

Modern grassland and steppe vegetation in the Columbia Plateau region became widespread during late Pliocene (ca. 4.5 Ma) to Quaternary times (E. B. Leopold and M. F. Denton 1987). Elements of these vegetation types included *Sarcobatus* and other Chenopodiaceae, *Artemisia* and other Asteraceae, and grasses. Faunas were dominated by browsers and grazers.

As temperatures continued to decrease and rainfall became more seasonal, high elevation and high latitude coniferous evergreen forests expanded. This occurred at the expense of the broad-leaved deciduous forests,

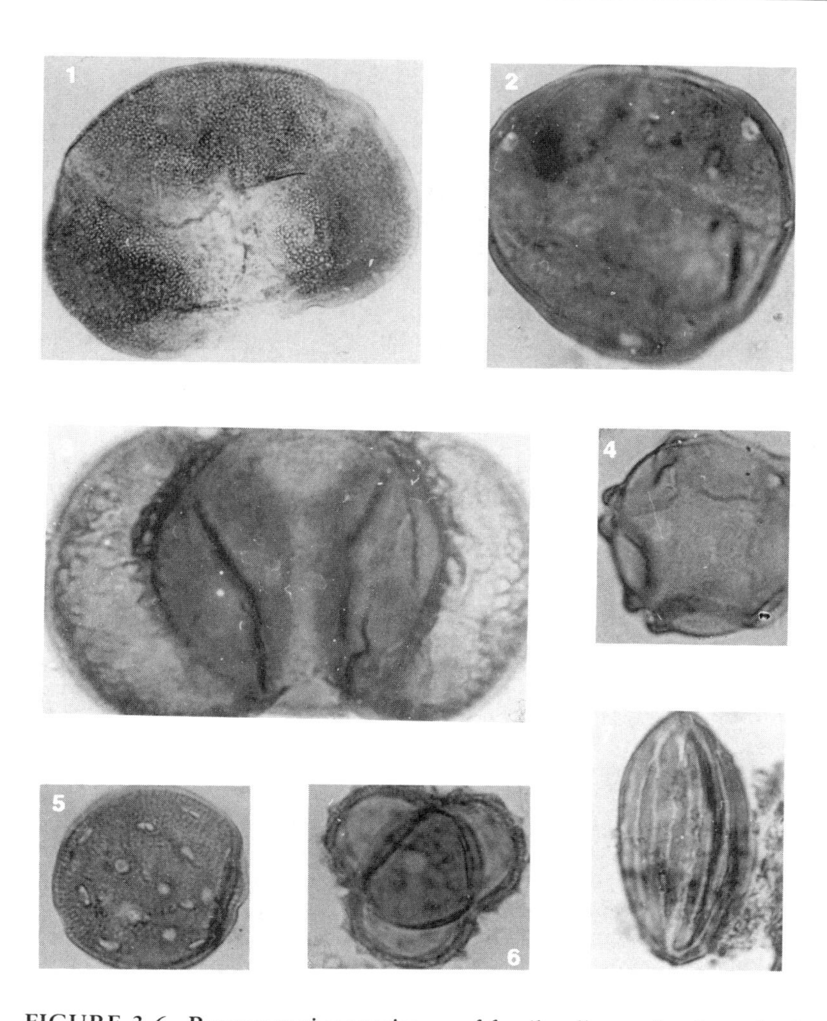

FIGURE 3.6. Representative specimens of fossil pollen grains from the Succor Creek flora, southeastern Oregon, middle to late Miocene Age. 1—*Picea*, 2—*Carya*, 3—*Pinus*, 4—*Alnus*, 5—Chenopodiaceae/ Amaranthaceae, 6—*Ambrosia*, 7—*Ephedra*. [Photographs by author.]

which were eliminated from many areas of western North America during the Pliocene.

The Pliocene Lava Camp site in Alaska borders on the Arctic Ocean (D. M. Hopkins et al. 1971) and contains *Picea* and *Pinus* as the dominants, along with *Abies*, *Larix* or *Pseudotsuga*, *Tsuga*, and Cupressaceae-Taxodiaceae. Deciduous gymnosperms *(Ginkgo, Glyptostrobus, Metasequoia)* and angiosperms *(Pterocarya)*, now restricted to East Asia, disappeared in the Pliocene. A narrow fringing zone of adapted tundra elements likely persisted along the expanding Arctic Ice Sheet from its initial appearance in the late Miocene, and perhaps locally along the North Slope region of Alaska, but no true tundra was evident.

The Pliocene vegetation of Alaska was largely coniferous forest. The climate was apparently cooler than that of the late Miocene. The coniferous forest disappeared from the Bering Sea area in the late Pliocene. It was replaced by an herbaceous and shrubby vegetation apparently dominated by Cyperaceae, Poaceae, Salicaceae, and Rosaceae. Typical tundra plants have not been recorded from this vegetation (J. A. Wolfe 1972). Miocene-Pliocene mosses from the Beaufort formation of Meighen Island near latitude 80° N (M. Kuc 1974) are typical of the boreal forest and not characteristic of arctic tundra.

The appearance of widespread prairie vegetation in midcontinental North America, plus sclerophyllous and coniferous vegetation in western and northern North America, was the first major disruption of the broad-leaved deciduous forest, the Arcto-Tertiary geoflora, that had extended across temperate latitudes of the Northern Hemisphere since late Eocene times. The deciduous forests of the Columbia Plateaus and those of eastern North America had been isolated geographically since at least the early Miocene by the montane coniferous forest and steppes of the Rocky Mountains and adjacent plains (E. B. Leopold and M. F. Denton 1987).

Prairie, sclerophyllous, and coniferous communities in western and southwestern North America spread during the Pliocene, and elements of the broad-leaved deciduous forest and associated fauna (particularly amphibians) that had extended into eastern Mexico during the middle to late Miocene became isolated in climatically comparable zones along the eastern escarpment of the Mexican Plateau at elevations of 1000–2000 m. In the Quaternary, versions of the modern broad-leaved deciduous forest persisted in the southeastern United States and in eastern Mexico. Thus the continuity of temperate elements of the biota, established during the middle to late Miocene cooling (A. Graham 1973b), was disrupted during Pliocene and later times, resulting in the present floristic and faunal relationship between eastern North America and eastern Mexico (A. Graham 1973; F. Miranda and A. J. Sharp 1950).

It is well known that the broad-leaved deciduous forests of eastern North America and eastern Asia are floristically related (A. Graham 1972, 1972b; A. Gray 1840, 1846; Li H. L. 1952; G. Davidse et al. 1983; B. H. Tiffney 1985; for patterns derived from molecular [isozyme] data, see M. T. Hoey and C. R. Parks 1991). This relationship was first recorded in 1750 by Linnaeus in the dissertation of his student Halenius (A. Graham 1966). It results from the maximum extension of the temperate deciduous forest in the mid-Tertiary and its disruption in western North America during the Pliocene and in western Europe during the Quaternary. The final events in the modernization of the North American flora were the climatic changes, anthropogenic influences, and vegetation responses during Pleistocene and Holocene times (see P. A. Delcourt and H. R. Delcourt, chap. 4).

Acknowledgments

I gratefully acknowledge the comments and helpful suggestions made by Daniel Axelrod, David Dilcher, Norman Frederiksen, and Jack Wolfe.

4. Paleoclimates, Paleovegetation, and Paleofloras during the Late Quaternary

Paul A. Delcourt
Hazel R. Delcourt

In the Quaternary period, which encompassed the last 2 million years (D. Q. Bowen 1985), Earth experienced an intensification of a trend toward climatic cooling and increased climatic variability that began in the late Tertiary (see A. Graham, chap. 3) and culminated in a series of at least 20 glacial-interglacial cycles. The earth at this time had a landmass configuration that made climatic systems at middle and high latitudes increasingly sensitive to seasonal variations in incoming solar radiation (A. L. Berger 1978; J. Imbrie and K. P. Imbrie 1979; N. G. Pisias and T. C. Moore Jr. 1981). Increases in both magnitude and frequency of climatic oscillations and profound environmental changes alternately driven by glacial and interglacial regimes resulted in the development of modern floristic and vegetational regions.

The origin of extant vascular plant genera and of many species occurred largely prior to the Quaternary. The Quaternary is characterized more by changes in the distributions of plant taxa and in the organization of plant communities (C. W. Barnosky 1987; L. E. Heusser and J. E. King 1988) than by the evolution of genera and species. One important event of the late Quaternary that affected subsequent biological diversity was the series of megafaunal extinctions that coincided with the spread of PaleoIndians between 12,000 and 10,000 years Before Present (yr B.P.) (P. S. Martin and R. G. Klein 1984). Prehistoric Amerinds influenced the distributions of certain plant species as well as the composition of plant communities, particularly during the last 5000 years, during which aboriginal populations became sedentary and grew native and introduced plants for food (R. J. Hebda and R. W. Mathewes 1984; P. A. Delcourt et al. 1986; H. R. Delcourt 1987; J. H. McAndrews 1988).

In this chapter, we summarize current views regarding the effects of climatic change on the vegetation and flora of North America during the Quaternary. We focus primarily on the past 20,000 years. This time interval represents late Pleistocene full-glacial conditions (20,000–15,000 yr B.P.), late-glacial climatic amelioration from 15,000 to 10,000 yr B.P., and Holocene interglacial conditions of the last 10,000 years. We review the broad patterns and timing of climatic, vegetational, and floristic changes that have occurred in North America and Greenland during this last glacial-interglacial cycle.

Climatic Change

Tectonic Framework

The closure of the Isthmus of Panama by tectonic activity in the late Pliocene epoch 3 million years ago (Ma) produced a continuous land bridge between North America and South America, prevented exchange of

Contribution Number 54, Center for Quaternary Studies of the Southeastern United States, University of Tennessee, Knoxville.

equatorial water between the Pacific and Atlantic oceans, and resulted in the strengthening of the Gulf Stream (W. A. Berggren and C. D. Hollister 1974). Moisture from storms that tracked northward along the warm Gulf Stream fed the accumulation of glacial ice in Greenland and northeastern Canada. Glacial tills in the midcontinental region of North America, dated from 2.4 to 2.2 Ma, are direct evidence of the expansion of major ice sheets (J. Boellstorff 1978).

Between 3 and 2.5 Ma, mountain glaciers developed at high latitudes of the Northern Hemisphere. During the last 2 million years, they expanded in suitable high elevation sites at middle latitudes concurrently with the uplift of the western Cordillera, particularly along the western Coast Ranges and Sierra Nevada (P. W. Birkeland et al. 1976; J. J. Clague 1989).

Because of these changes in configuration of continents and oceans, the Quaternary climates at middle and high latitudes became increasingly sensitive to seasonal variations in incoming solar radiation, and the onset of glacial-interglacial cycles began. Major papers by G. H. Denton and T. J. Hughes (1983) and by W. S. Broecker and G. H. Denton (1989) summarize the role of long-term changes in solar radiation in both initiating and amplifying climatic interactions among glaciers, oceans, and the atmosphere.

Milankovitch Cycles

J. Chappell (1978) provided a balanced summary of the many published theories that attempt to account for cyclic climatic changes and repeated episodes of continental glaciation during the Quaternary. Milutin Milankovitch [Milankovič] (1941) hypothesized that variations in the earth's orbit around the sun result in cyclic variation in the amount of solar radiation received by the earth, producing a "solar pacemaker" (J. D. Hays et al. 1976), which ultimately led to the climatic changes associated with the glacial-interglacial cycles (A. L. Berger 1978). Over each 100,000-year Milankovitch cycle, the amount of solar radiation received by the earth varies by 3.5%; the variation is attributed to changes in the eccentricity of the earth's elliptical path around the sun.

Two other Milankovitch cycles influence the seasonal contrast in radiation received at different latitudes. A 41,000-year cycle results from changes in the tilt of the earth's rotational axis, which in turn affects the seasonal contrasts between summer and winter. A 21,000-year cycle results from the precession of the equinoxes, in which the earth's axis of rotation moves slowly along a circular path. High latitudes above 65° N and 65° S, where continental ice sheets form and decay, are primarily affected by the 41,000-year cycle. The 21,000-year cycle amplifies climatic variations between latitudes 65° N and 65° S. Together, these two cycles are responsible for the glacial advances and retreats that are superimposed upon the 100,000-year glacial-interglacial cycles (J. Imbrie and K. P. Imbrie 1979).

Isotope curves developed from measurements of the ratio of ^{18}O to ^{16}O in the fossil shells of marine plankton reflect long-term oscillations in the volume of glacial ice and indicate that the 100,000-year Milankovitch cycle has become dominant in driving climatic change during the last 900,000 years (N. G. Pisias and T. C. Moore Jr. 1981). Over the course of each 100,000-year cycle (fig. 4.1A), glacial ice builds up slowly on the continents, with the glacial mode typically lasting for about 90,000 years (W. S. Broecker and J. van Donk 1970; W. A. Watts 1988). Glacial conditions then terminate rapidly as glaciers melt in response to increased solar radiation and maximum seasonal contrast in middle latitudes, conditions associated with the combined peaks of both the 21,000-year and the 100,000-year Milankovitch cycles (A. L. Berger 1978; J. Imbrie and K. P. Imbrie 1979).

The rapid termination of each glacial interval is triggered when the time of highest solar radiation coincides with the maximum seasonal range in solar radiation (fig. 4.1B; A. L. Berger 1978; J. E. Kutzbach 1987; W. A. Watts 1988). Increased summer temperatures melt glacial ice on land. Glacial meltwater returns to the oceans, forming a freshwater cap over denser, saline ocean water. Colder winter temperatures freeze upper marine waters, creating a layer of pack ice that in turn cuts off the supply of moisture evaporating from oceans to the atmosphere, further reducing the snow supply to disintegrating continental glaciers (W. F. Ruddiman and A. McIntyre 1981). The subsequent interglacial mode lasts for only 10,000 to 15,000 years. The overall shape of the glacial-interglacial curve of global ice volume (fig. 4.1B), as determined by $^{18}O/^{16}O$ ratios ($\delta^{18}O$), is asymmetrical and sawtoothed, not smoothly sinusoidal as are the fluctuations of solar radiation attributed to individual Milankovitch cycles (W. S. Broecker and J. van Donk 1970).

The Full-Glacial Climate

Proxy evidence of climatic change recorded on land (S. C. Porter 1983; H. E. Wright Jr. 1983; R. B. Morrison 1991) and in the oceans (CLIMAP Project 1981; W. F. Ruddiman and H. E. Wright Jr. 1987) provides constraints to guide computer simulations of full-glacial climate 18,000 yr B.P. (J. E. Kutzbach 1987; J. E. Kutzbach and H. E. Wright Jr. 1985; J. E. Kutzbach and P. J. Guetter 1986; COHMAP 1988; P. J. Guetter and

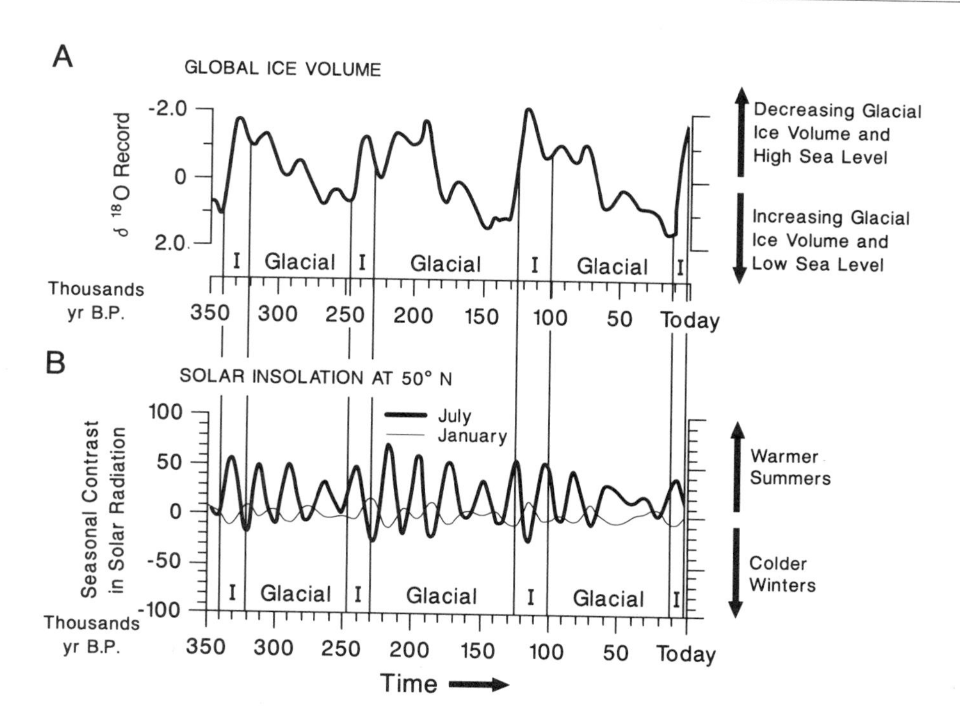

FIGURE 4.1. Paleoclimatic records of (A) global ice volume of glaciers inferred from the δ^{18}O record of deep-sea sediments and (B) seasonal contrast between winter and summer for incoming solar radiation (insolation) during the last three glacial-interglacial cycles of the past 350,000 years. Interglacial peaks (I) in the oxygen-isotopic record (δ^{18}O) represent times of minimal glacial-ice volume and of high sea level. The gradual drop in oxygen-isotope values during glacial intervals reflects expanding glaciers on land and dropping sea levels. The interglacial peaks of warmth in the oxygen-isotope record occur at approximately 100,000-year intervals: at 330,000, 235,000, 120,000, and 9,000 years ago. These times of interglacials typically coincide with maximum levels of solar radiation received by Earth and maximum seasonal contrast between hot summers and cold winters. [Primary data from A. L. Berger 1978; modified from W. A. Watts 1988, with permission from Kluwer Academic Publishers.]

J. E. Kutzbach 1990). This section is a synthesis of the results from computer simulations and available paleoclimatic proxy data.

During full-glacial conditions 18,000 years ago, three complexes of continental ice sheets dominated the northern half of North America, Greenland, and northern Europe (delineated by stippled areas of fig. 4.2). The North American complex included the western Cordilleran and eastern Laurentide ice sheets. The accumulation of water on land as glacial ice resulted in sea levels of full-glacial oceans being 100–120 m below modern levels. Lowered sea levels exposed extensive areas of coastal plains, particularly in Beringia (the land bridge that connected eastern Siberia and Alaska [D. M. Hopkins et al. 1982]), southern North America, southern Europe, and North Africa (J. E. Kutzbach and H. E. Wright Jr. 1985). The diagonal pattern on figure 4.2 identifies the winter extent of sea pack ice north of Beringia in the Arctic Ocean, and south to about 42° N in the North Atlantic Ocean (CLIMAP Project 1981).

Between latitudes 30° N and 60° N, fossil data and computer simulations indicate that full-glacial winters on land were an average of 6° C colder, and summers about 2° C cooler than today. In winter the polar jet stream split in the North Pacific Ocean (fig. 4.2). A northern branch of this polar jet stream passed along the northern perimeter of the Cordilleran and Laurentide ice sheets in northern Canada. The southern branch crossed the American Southwest at about latitude 30° N and flowed over the Southern High Plains and the middle latitudes of unglaciated southeastern North America. The southern jet stream would have funneled storms along this route and would have caused substantially cooler temperatures and increased winter precipitation across the Southwest and from about latitude 33° N to the glacial margin (about latitude 40°

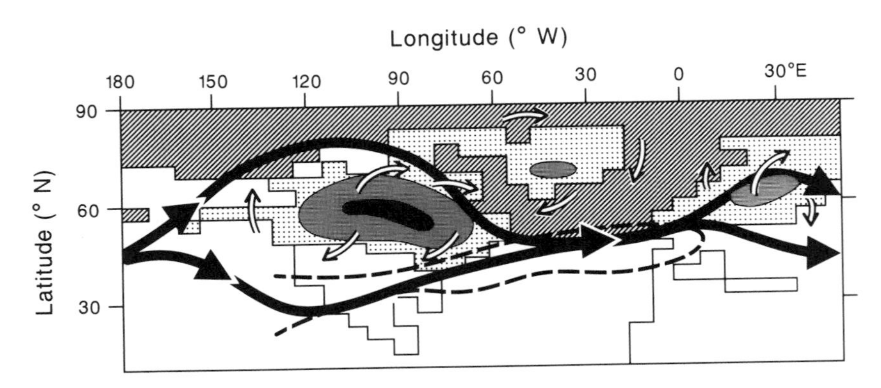

FIGURE 4.2. Full-glacial computer simulation of atmospheric circulation patterns in the Northern Hemisphere modeled for January of 18,000 years ago: polar jet stream (black arrows), anticyclonic winds (open arrows). Location (stippled pattern) and highest elevations (contoured regions with shaded patterns) are shown for North American (Cordilleran plus Laurentide), Greenland, and European ice sheets. The dashed lines along the southern branch of the jet stream delineate the latitudinal band of increased precipitation. Extent of sea pack ice is indicated by the diagonal pattern in the Arctic and North Atlantic oceans. [Modified from J. E. Kutzbach and H. E. Wright Jr. 1985, with permission from Pergamon Press.]

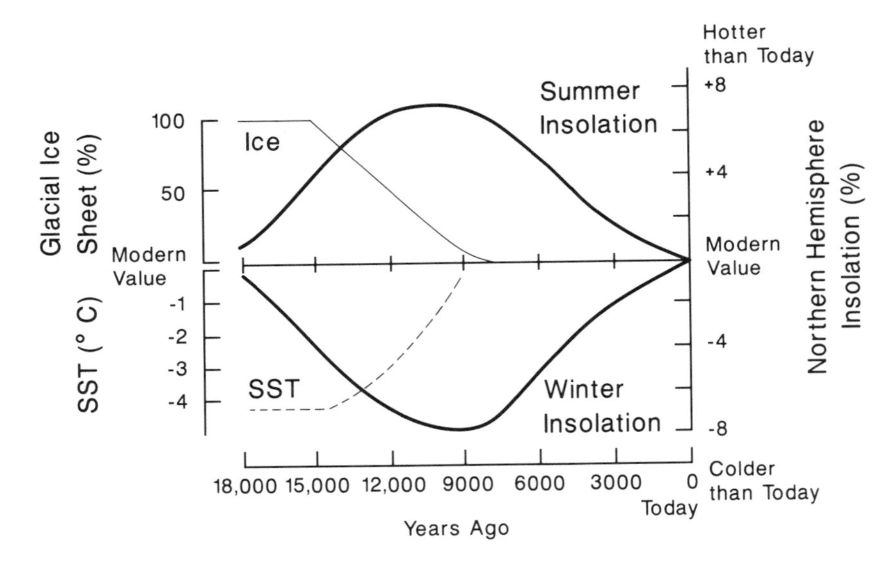

FIGURE 4.3. Summary of global climatic and environmental changes from peak times of maximum continental glaciation (100% area of ice sheets) and cool ocean waters (mean sea-surface temperatures [SST] 4°C cooler than modern values) 18,000 to 15,000 years ago, the late-glacial interval (15,000 to 10,000 years ago), and through the last 10,000 years of the present Holocene interglacial. Relative to modern climatic conditions, maximum seasonal contrast between winter and summer values of solar insolation occurred between 12,000 and 9000 years ago during the late-glacial/early Holocene transition. [Modified from P. J. Bartlein 1988, after J. E. Kutzbach and P. J. Guetter 1986, with permission from Kluwer Academic Publishers and the American Meteorological Society.]

N) in the Southeast (fig. 4.2; J. E. Kutzbach and H. E. Wright Jr. 1985; J. E. Kutzbach and P. J. Guetter 1986).

In western North America, this "pluvial" climatic regime led to a 10-meter rise in the groundwater table and the expansion of large, permanent pluvial lakes in which the water level rose as much as 335 m above modern levels (G. I. Smith and F. A. Street-Perrott 1983; L. Benson and R. S. Thompson 1987). Computer simulations of full-glacial climate in North America also specify an anticyclonic circulation of winds radiating outward from the ice sheets that would have brought cold, dry climatic conditions to the unglaciated area of the Pacific Northwest, and cool but very dry conditions to Beringia (fig. 4.2; J. E. Kutzbach and H. E. Wright Jr. 1985).

The Late-Glacial and Holocene Intervals

Computer simulations of climatic conditions (J. E. Kutzbach and P. J. Guetter 1986) show that during the portion of the late-glacial interval from 15,000 to 12,000 yr B.P., the two full-glacial branches of the jet stream (fig. 4.2) shifted in location and merged along one route as climate moderated (J. E. Kutzbach 1987). Between 12,000 and 9000 yr B.P., the route of this polar jet stream extended from latitude 52° N over the western mountain ranges of British Columbia and tracked southeastward across the midcontinent to about latitude 45° N along the southern Laurentide glacial margin. In the last 6000 years, the position of the jet stream has shifted seasonally in location, in winter occupying the midlatitudinal route from British Columbia to the northeastern United States (comparable with the path from 12,000 to 9000 yr B.P.), and in summer occupying a northerly route across Alaska and northern Canada from about 65° N to 70° N.

Changing environmental conditions through the past 18,000 years include a global decrease in extent of glacial ice, a postglacial increase of about 4° C in the global mean values for annual sea-surface temperature (SST), and seasonal changes in solar radiation received by the earth during summer and winter (fig. 4.3). Reconstructions based on Milankovitch cycle calculations indicate that, for the Northern Hemisphere, in the interval between 12,000 and 9000 yr B.P., there was maximum seasonal contrast in temperatures, with warmer summers and cooler winters (A. L. Berger 1978; J. E. Kutzbach 1987; P. J. Bartlein 1988). The Pleistocene-Holocene boundary at 10,000 yr B.P. consequently was a time of fundamental changeover in environmental and climatic conditions from a glacial to an interglacial mode (fig. 4.3). Holocene interglacial climates have been characterized by relatively warm conditions, by decreasing amounts of incoming solar radiation during summer, and by increasing solar radiation levels during the winter season, with an overall decrease in seasonal contrast (fig. 4.3; J. E. Kutzbach 1987).

Biotic Changes during the Last Glacial-Interglacial Cycle

Eastern North America

Various authors have synthesized the late Quaternary literature of plant fossil data for the eastern United States and Canada (J. C. Bernabo and T. Webb III 1977; P. J. H. Richard 1977; H. E. Wright Jr. 1981; M. B. Davis 1976, 1981, 1983; W. A. Watts 1983; V. M. Bryant Jr. and R. G. Holloway 1985; R. B. Davis and G. L. Jacobson Jr. 1985; P. A. Delcourt and H. R. Delcourt 1987, 1987b; G. L. Jacobson Jr. et al. 1987; J. C. Ritchie 1987; T. Webb III et al. 1987; T. Webb III 1988; J. V. Matthews Jr. et al. 1989; H. R. Delcourt and P. A. Delcourt 1991), and for northeastern Canada and Greenland (B. Fredskild 1973, 1985; P. J. H. Richard 1977; J. T. Andrews 1985; H. F. Lamb and M. E. Edwards 1988; J. V. Matthews Jr. et al. 1989). This regional review, based substantially on examples documented in P. A. Delcourt and H. R. Delcourt (1987, 1987b) and in H. R. Delcourt and P. A. Delcourt (1991), illustrates the response of temperate and boreal plant species and communities to climatic and environmental changes over the last glacial-interglacial cycle in eastern North America (from 25° N to 60° N and 50° W to 100° W).

During the peak of the last continental glaciation, between 20,000 and 18,000 yr B.P., the Laurentide Ice Sheet extended south to about latitude 40° N across the Great Lakes region (fig. 4.4). The Laurentide Ice Sheet influenced the full-glacial climate, locking the frigid Arctic air mass north of the great ice dome. The prevailing westerlies of the Pacific air mass thus dominated the region south of the glacial ice to 33° N. The southern branch of the full-glacial jet stream (fig. 4.2) extended across the middle latitudes as the Polar Frontal Zone that marked the climatic boundary between prevalence of the Pacific air mass and the Maritime Tropical air mass to the south (fig. 4.4). In a narrow, periglacial zone along the southern and eastern margin of the ice sheet, tundra communities were present only in localized areas of severe permafrost environments. These occurred in glacial reentrants such as in southeastern Iowa (R. G. Baker et al. 1986), at middle to high elevations within the central Appalachian Mountains, and on the coastal plain of the Northeast.

Boreal forest dominated in the belt south of the Laurentide ice to approximately latitude 34° N, extending westward across the Great Plains. Elements of decidu-

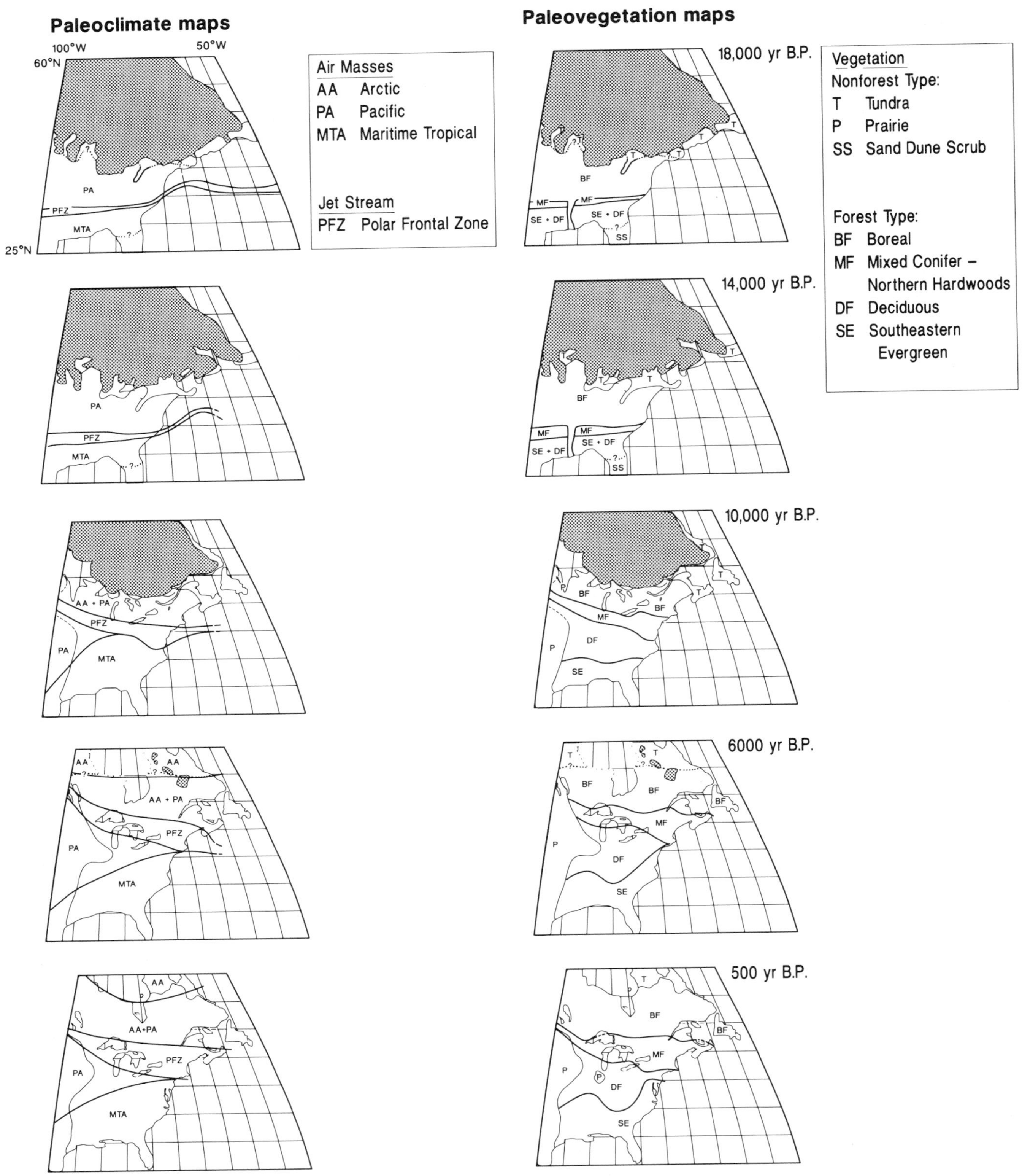

Paleoclimate maps

Paleovegetation maps

Air Masses
AA Arctic
PA Pacific
MTA Maritime Tropical

Jet Stream
PFZ Polar Frontal Zone

Vegetation
Nonforest Type:
T Tundra
P Prairie
SS Sand Dune Scrub

Forest Type:
BF Boreal
MF Mixed Conifer – Northern Hardwoods
DF Deciduous
SE Southeastern Evergreen

18,000 yr B.P.

14,000 yr B.P.

10,000 yr B.P.

6000 yr B.P.

500 yr B.P.

FIGURE 4.4. Glacial-interglacial changes in past climate, areal extent of glacial ice (shaded areas), and vegetation mapped across eastern North America for the last 18,000 years. The maps reflect full-glacial times (18,000 yr B.P.), the late-glacial interval (14,000 yr B.P.), and the present Holocene interglacial (10,000 yr B.P., 6000 yr B.P., and 500 yr B.P.). Paleoclimatic maps identify climatic regions dominated by Arctic, Pacific, and Maritime Tropical air masses. The typical path of the Polar Jet Stream corresponds with the Polar Frontal Zone. The paleovegetation maps are based on paleoecological reconstructions from 162 radiocarbon-dated plant-fossil sites. [After P. A. and H. R. Delcourt 1987, with permission from Plenum Publishing Corporation, Elsevier Science Publishers, and Springer-Verlag.]

ous forest in the southern region of the full-glacial boreal forest indicate that the boreal climate was less extreme than that of today (P. A. Delcourt et al. 1980). Based on quantitative comparison of fossil and modern pollen assemblages, midwestern plant communities dated from 28,000 until 15,000 yr B.P. were very similar in both forest composition and structure with their modern counterparts in the boreal forest biome (R. G. Baker et al. 1989). Prairie elements may have occurred in the understory of this boreal forest, but extensive prairie tracts were not present (H. E. Wright Jr. 1981). Between latitudes 34° N and 33° N, a narrow transition zone marked the ecotone between more northern boreal and more southern temperate communities.

South of latitude 33° N, across the southern Atlantic and Gulf coastal plains, floristic elements of temperate deciduous forest occurred with plant taxa characteristic today of southeastern evergreen forests (fig. 4.4; P. A. Delcourt and H. R. Delcourt 1987). In the Southeast, average temperatures during the full-glacial interval were probably similar to modern values; late Pleistocene evidence documents that sea-surface temperatures 18,000 yr B.P. in the northern Gulf of Mexico were less than 2° C cooler than those of today (CLIMAP Project 1981). This slight oceanic cooling decreased summer rates of evaporation from marine waters and substantially reduced the precipitation supplied by tropical storms and hurricanes across the interior of the Southeast (W. M. Wendland 1977; P. A. Delcourt and H. R. Delcourt 1984).

Sand dune scrub occupied the broad, exposed coastal plain of central and southern Florida. Tropical plant species were eliminated from southernmost Florida. Dry climate and a decline in water table by up to 20 m, resulting from lowered sea level (W. A. Watts 1983), may have been responsible for these xeric vegetation patterns across peninsular Florida.

In general, major vegetation patterns remained relatively consistent from 20,000 to 15,000 yr B.P. (P. A. Delcourt and H. R. Delcourt 1984). Between 15,000 and 10,000 yr B.P., late-glacial climatic warming and the northward retreat of the ice sheet resulted in widespread changes in climate and vegetation. Vegetation regions and biomes today correlate with distinctive climatic regions, delineated by mean positions of air mass boundaries (R. A. Bryson 1966; R. A. Bryson and F. K. Hare 1974b). This principle can be extended back in time to the analysis of former vegetation patterns (fig. 4.3; R. A. Bryson and W. M. Wendland 1967; P. A. Delcourt and H. R. Delcourt 1983, 1984). By 10,000 yr B.P., boreal forest spread throughout the deglaciated region immediately south of the Laurentide Ice Sheet, through the boreal climate zone where the Arctic air mass dominated in winter and the Pacific air mass was important in the summer. Portions of Nova Scotia, Newfoundland, and Labrador, however, were occupied by tundra (J. V. Matthews Jr. et al. 1989).

Across the Great Lakes and New England regions, the latitudinal belt of mixed conifer–northern hardwoods forest corresponded in distribution with the zone influenced by the average location of the polar jet stream, that is, the broad Polar Frontal Zone representing the prominent climatic boundary between boreal and temperate climatic regions. Deciduous forest spread northward with the areal expansion of the Maritime Tropical air mass. The southeastern evergreen forest remained confined to the Gulf and Atlantic coastal plains and in Florida it replaced the sand dune scrub. Prairie vegetation had developed by 10,000 yr B.P. in the central and southern Great Plains, where the relatively dry Pacific air mass dominated throughout the year.

By 6000 yr B.P. (fig. 4.4), remnants of continental ice sheets had largely retreated north of latitude 60° N. Tundra communities were widespread north of latitude 55° N both to the east and to the west of Hudson Bay, within the climatic region dominated exclusively by the Arctic air mass. Boreal forest occupied the zone between latitudes 50° N and 55° N, where Arctic and Pacific air masses prevailed in winter and summer seasons, respectively. The Polar Frontal Zone and mixed conifer–northern hardwoods forest occupied a wide band across the Great Lakes and New England regions.

A wedge of the warm, dry Pacific air mass and prairie vegetation extended eastward into Illinois and beyond as the prairie peninsula reached its easternmost extent during the middle Holocene (the Hypsithermal Interval, generally dated between 9000 and 4000 yr B.P. [H. E. Wright Jr. 1976]). With further northward influence of the warm, moist Maritime Tropical air mass, deciduous forest communities became widespread across the eastern United States, and southeastern evergreen forests spread northward along the Atlantic Coastal Plain and the Mississippi Alluvial Valley (P. D. Royall et al. 1991). With the rise in sea level and the increase in sea-surface temperatures to modern levels by 5000 yr B.P., the southern portion of the Florida peninsula was invaded by subtropical plant species.

Vegetation patterns for presettlement times, 500 yr B.P. (fig. 4.4), differ from those of the middle Holocene interval. During the last 3000 years, global climatic cooling resulted in a southward displacement of the ecotone between tundra and boreal forest across northeastern Canada (S. Payette and R. Gagnon 1985; S. Payette et al. 1989; J. V. Matthews Jr. et al. 1989). Similarly, the ecotone between boreal and mixed conifer–northern hardwoods forest was displaced southward across the Great Lakes region and New England (J. V. Matthews Jr. et al. 1989). A reduction in seasonal contrast in temperature within the late Holocene (fig. 4.3) diminished the frequency and intensity of summer droughts, reduced the occurrence of wildfire, favored invasion of grasslands by trees, and resulted in westward retraction of the prairie peninsula and its replacement by eastern forests from Minnesota to Illinois (fig. 4.4; E. C. Grimm 1983; G. L. Jacobson Jr. et al. 1987).

Changes in vegetation resulting from late Quaternary climatic changes can be mapped along boundaries between biomes that coincide with the thresholds of physiological tolerance of many dominants of the biomes (B. F. Chabot and H. A. Mooney 1985). It is widely recognized, however, that the responses of plant taxa to climatic change are individualistic (M. B. Davis 1981, 1983; T. Webb III 1988; P. A. Delcourt and H. R. Delcourt 1987, 1987b; H. R. Delcourt and P. A. Delcourt 1991). Figure 4.5 illustrates distribution patterns for the last 18,000 years for three major eastern North American tree taxa: spruce *(Picea)*, oak *(Quercus)*, and both the northern (N) and southern (S) groups of pine *(Pinus)*. Procedures for producing these paleo-population maps (contoured maps with inferred population abundance are expressed as percent forest composition) are given in P. A. Delcourt and H. R. Delcourt (1987, 1987b). Differences among these taxa are evident in the mean rates of migration, in the migration routes taken northward following glacial retreat, and in the changing locations of major population centers through time.

Spruces—including white spruce *(Picea glauca)*, black spruce *(P. mariana)*, and red spruce *(P. rubens)*—were dominant within boreal forests 18,000 yr B.P. (fig. 4.5). A full-glacial population center was located west of the Appalachian Mountains. As the Laurentide Ice Sheet retreated (P. F. Karrow and P. E. Calkin 1985; P. F. Karrow and S. Occhietti 1989), spruce developed its highest populations at the margins of proglacial lakes, i.e., lakes formed by ice-melt near the southern margin of the Laurentide glacier. By 10,000 yr B.P., outlying populations were persisting south of the main range of spruce at middle to high elevations in the central and southern Appalachian Mountains. Between 10,000 and 6000 yr B.P., the primary population centers of spruce shifted north to between 47° and 55° N. It reached its northernmost limit by 4000 yr B.P. as a major component of the boreal forest through central and eastern Canada (fig. 4.4). In the late Holocene, spruce populations generally shifted southward. Overall, the average late Quaternary rate for migration of spruce populations was 14.1 km per century (P. A. Delcourt and H. R. Delcourt 1987).

Eastern oak species were an important component of the eastern deciduous and southeastern evergreen forests south of 33° N latitude during full-glacial times (figs. 4.4, 4.5). Minor populations occurred northward within the boreal forest. By the early Holocene, oak species had invaded throughout the boreal forest, advanced to the tundra and glacial ice limits, and migrated as far north as latitude 51° N in central Canada. As early as 10,000 yr B.P., oak species were dominant or subdominant in southeastern evergreen forest as well as eastern deciduous forest that had expanded areally north to latitude 42° N. Overall, the mean rate of late Quaternary migration of oak populations was 12.6 km per century (P. A. Delcourt and H. R. Delcourt 1987).

Eastern North American pines are differentiated into two groups (W. B. Critchfield 1984; D. I. Axelrod 1986b) that today are either distributed primarily in the Appalachian Mountains and northward, or are generally restricted to the southeastern coastal plain. During the full-glacial interval, northern pines (including jack pine [*Pinus banksiana*], red pine [*P. resinosa*], and eastern white pine [*P. strobus*]) were dominant north of 34° N, within boreal forests located in the rain shadow east of the Appalachian Mountains on the central Atlantic Coastal Plain (fig. 4.5). Southern pine species were restricted to latitudes south of 33° N.

By 14,000 yr B.P., populations of jack and/or red pine that had survived the full-glacial period west of the Appalachians became extinct there. By 10,000 yr B.P., distributions of northern and southern groups of pines were more widely separated geographically than before. Population centers of northern pines had shifted northwest across the Great Lakes region. Southern pines had become important subdominants of the canopy on the southeastern Gulf Coastal Plain.

By 6000 yr B.P., northern pines became structural dominants within the mixed conifer–northern hardwoods forest and the southern half of the boreal forest (J. Terasmae and T. W. Anderson 1970; G. L. Jacobson Jr. 1979). Southern pines spread northward along the Atlantic Coastal Plain to southern New England. In the late Holocene, northern pines became widespread throughout the eastern boreal forest but maintained a primary population center in central Canada. Southern pines became dominants of the southeastern evergreen forest of the southern Atlantic and Gulf coastal plains (figs. 4.4, 4.5). During the late Quaternary, northern pines spread northward at a mean rate of 13.5 km per century; southern pines advanced northward along the Atlantic Coastal Plain at an average rate of 8.1 km per century (P. A. Delcourt and H. R. Delcourt 1987).

The difference between past forest communities and those of today can be measured by using an ordination technique called Detrended Correspondence Analysis (DCA) (fig. 4.6; P. A. Delcourt and H. R. Delcourt 1983, 1987, 1987b). This technique allows the identification

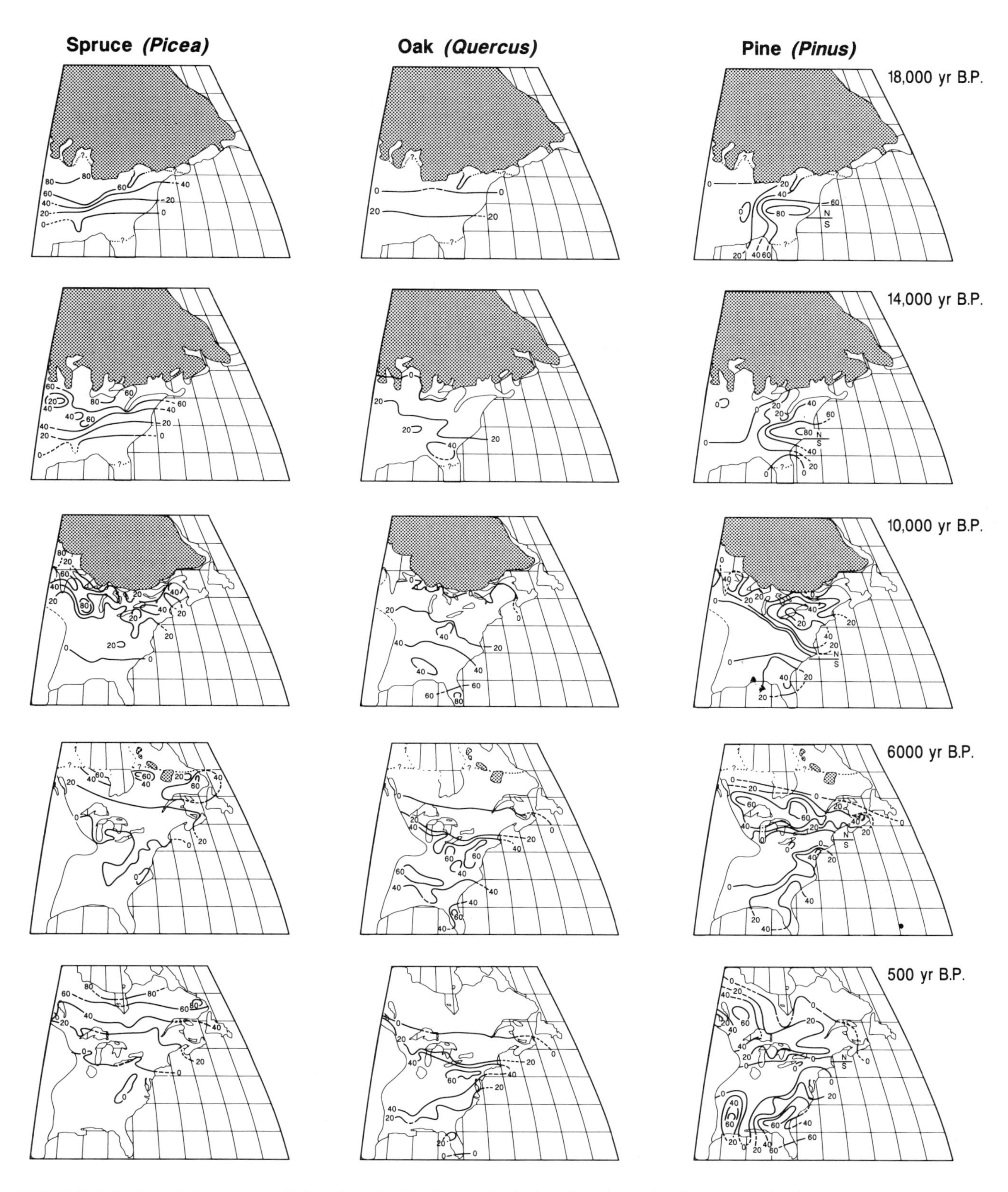

FIGURE 4.5. Population maps of changing distribution and relative abundance (with contoured values of % forest composition) for spruce *(Picea)*, oak *(Quercus)*, and pine *(Pinus)* trees during the past 18,000 years. In the time series of pine population maps, the letters N and S identify the mutual distributional border between northern and southern groups of pine species. Forest populations are not mapped across nonforested regions of tundra, prairie, or sand dune scrub. Shaded areas indicate location of the Laurentide Ice Sheet. [After P. A. and H. R. Delcourt 1987, with permission from Springer-Verlag.]

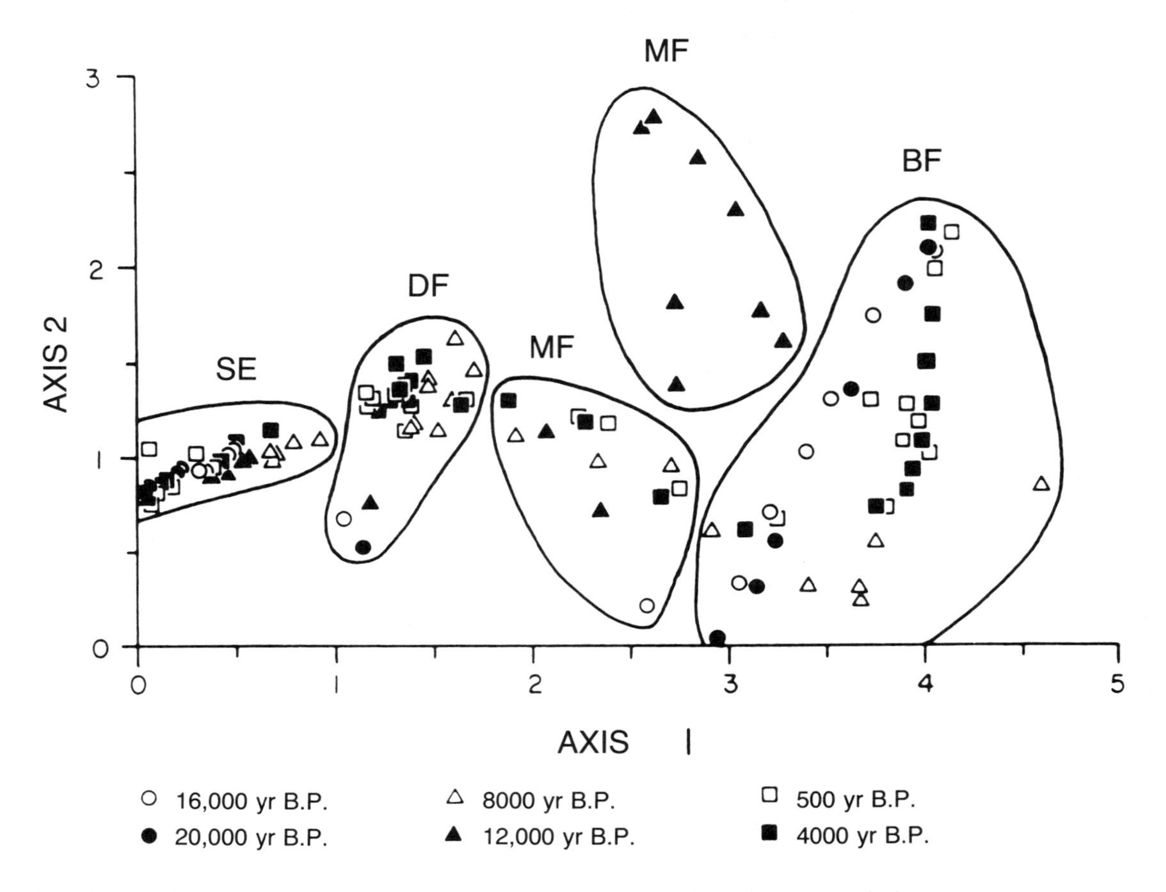

FIGURE 4.6. Late-Quaternary forests in eastern North America, characterized along a transect at 85° W longitude from 27° 30′ N to 50° N latitude. All forest samples for the last 20,000 years are plotted on the first two axes of the DCA (Detrended Correspondence Analysis) ordination. Symbol codes designate the age of the samples of reconstructed forest composition. Clusters denote the ordination space represented by the Southeastern Evergreen Forest (SE), Deciduous Forest (DF), Mixed Conifer–Northern Hardwoods Forest (MF, lower cluster with modern analogues; upper, late-glacial cluster without modern analogues), and Boreal Forest (BF). [Modified from P. A. and H. R. Delcourt 1987, with permission from Springer-Verlag.]

of plant communities that may have persisted through long periods of time as well as those that have been ephemeral. We studied the record of the compositional changes in plant communities that have occurred during the past 20,000 years along a south-to-north transect, following longitude 85° W from the Gulf of Mexico to the Laurentide glacier. The glacier retreated northward to Hudson Bay in the middle Holocene (fig. 4.7).

Several kinds of changes in plant communities occurred during the late Quaternary in eastern North America (fig. 4.6). The warm temperate vegetation of the southeastern coastal plains has remained relatively constant in its floristic composition through the last glacial-interglacial cycle. The eastern deciduous forest, however, which was restricted to a few small refuge areas during the late Pleistocene, emerged as a major vegetation type only in the Holocene. Today's mixed conifer–northern hardwoods forest has close analogues throughout the Holocene, but in late-glacial times it formed ephemeral communities unlike any in existence today. Since the last glacial maximum, boreal forest communities have exhibited a composition heterogeneous both in space and time. Close modern analogues exist for full-glacial boreal forest communities, but there are only poor modern analogues for the boreal forest

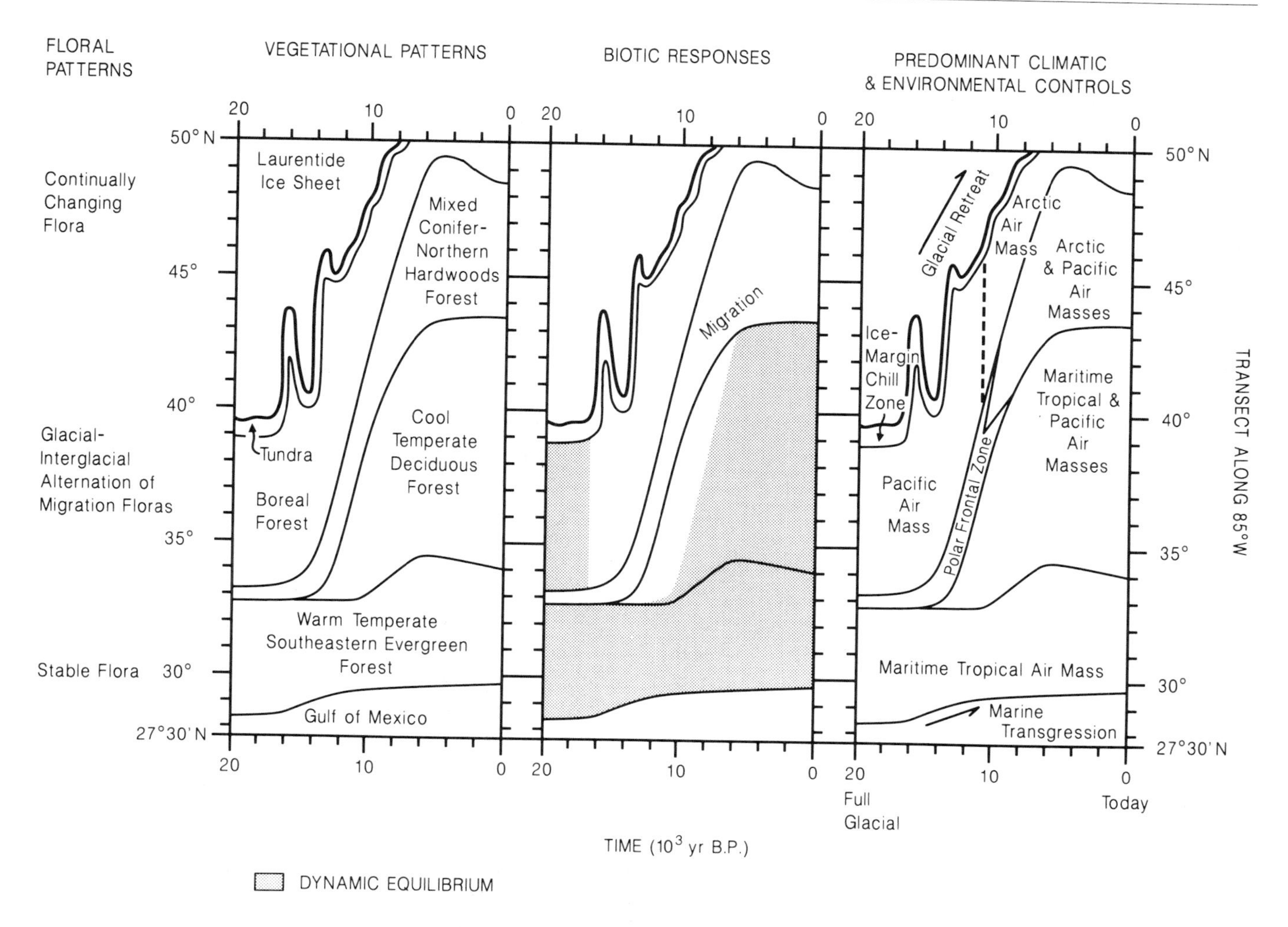

FIGURE 4.7. Glacial-interglacial model depicting changes in floral patterns, vegetational patterns, inferred biotic responses, and predominant climatic and environmental conditions across eastern North America (south-to-north transect of 85° W longitude) during the past 20,000 years. The shaded pattern on the panel for biotic responses identifies the areas and time intervals for which the vegetation attains dynamic equilibrium. On the far right panel of predominant climatic and environmental controls (environmental forcing functions), the vertical dashed line at 11,000 yr B.P. reflects the timing for renewal of southward flow of the Arctic air mass with late-glacial retreat and removal of the Laurentide Ice Sheet as a significant topographic barrier. [Modified from P. A. and H. R. Delcourt 1983, with permission from Quaternary Research.]

communities of the late-glacial/early Holocene transition (P. A. Delcourt and H. R. Delcourt 1987).

From these ordination results, further conclusions can be drawn concerning the degree of vegetational and floristic stability across eastern North America during the past 20,000 years (P. A. Delcourt and H. R. Delcourt 1983, 1987, 1987b). Over this time interval, no major relocations of warm temperate taxa have occurred in the region of the southern coastal plains, and the vegetation there has been in dynamic equilibrium (fig. 4.7). Between 34° N and 43° N (fig. 4.7), the vegetation has shifted from one state of dynamic equilibrium (boreal forest and boreal flora) in full-glacial times to another one (deciduous forest and corresponding temperate flora) in interglacial times, following a late-glacial and early Holocene transition period during which boreal taxa became locally extinct and temperate deciduous taxa immigrated. At higher latitudes (fig. 4.7), successive waves of invasion following deglaciation resulted in continual disequilibrium as tundra and then boreal forest communities became established through the late-glacial and early Holocene intervals. A trend toward equilibrium in species composition occurred only during the last several thousand years of the late Holocene.

Eastern Canadian Arctic and Greenland

During the full-glacial interval, the northernmost glacial margin of the Laurentide Ice Sheet extended from Labrador to the eastern margin of Baffin Island, and westward at about latitude 74° N through the Canadian Arctic Archipelago to Banks Island (J. T. Andrews 1987). Farther north, in the High Arctic zone of the Queen Elizabeth Islands and the corridor between Ellesmere Island and northern Greenland, the Quaternary history is not well known within the area occupied by local ice caps and unglaciated nunataks. During the retreat of the glacial ice from 12,000 until 9000 yr B.P., a rise in sea level and subsequent isostatic uplift of the ocean floor resulted in the exposure of land across formerly glaciated areas of the eastern Canadian Arctic and along the coastal perimeter of the Greenland icecap (references in J. T. Andrews 1985, 1987).

Concomitant with substantial retreat of glacial ice on the Ungava Plateau, the Canadian Archipelago, and Greenland in the early Holocene (10,000–9000 yr B.P.), a pioneer stage of arctic tundra developed on the newly exposed coastal zone (P. J. H. Richard 1977; B. Fredskild 1973, 1985; S. K. Short et al. 1985; H. F. Lamb and M. E. Edwards 1988; J. V. Matthews Jr. et al. 1989). By 10,000 yr B.P. at about latitude 60° N, in coastal zones of northern Quebec and southern Greenland, a High Arctic herb tundra, a sparse cover of arctic or arctic-alpine species of herbs (such as *Oxyria digyna, Koenigia islandica,* and *Saxifraga oppositifolia*), ferns, and club-mosses initially colonized the severe periglacial environments of unstable fell-fields and frost-churned soils. This herb tundra community was replaced by a heath tundra between 9100 and 8500 years ago with the invasion of heath dwarf shrubs such as crowberry *(Empetrum hermaphroditum)* and bilberry *(Vaccinium uliginosum)*. Along both the western and eastern coastlines of Greenland, similar communities developed between 9600 and about 8000 yr B.P. In northwestern Greenland (latitude 75° N) comparable stages of pioneer herb tundra and heath dwarf shrub tundra were established between 8600 and 7700 yr B.P. and 7700 and 6600 yr B.P., respectively (B. Fredskild 1985).

During the middle Holocene interval, shrub populations of shrub willow and dwarf birch (fig. 4.8) invaded northward into Greenland along two dispersal routes, forming communities of Low Arctic shrub tundra. Long-distance dispersal of seeds, probably from northwestern Europe, facilitated colonization by shrub willow *(Salix)*, possibly as early as 9400 yr B.P. in southern Greenland (fig. 4.8A). By 8000 yr B.P. in eastern Greenland, populations of three shrub willows *(Salix arctica, S. herbacea,* and *S. glauca)* and one dwarf birch *(Betula nana)* were successfully established (B. Fredskild 1973, 1985). The maritime corridor of northeastern North America provided a second route for plant migration to southern and western Greenland (J. V. Matthews Jr. et al. 1989). Shrub willows colonized between 8900 and 6700 yr B.P. These were followed by successive invasions of dwarf birches, first *Betula nana* (8000–6500 yr B.P.), then *B. glandulosa* (5700–3800 yr B.P.), as well as populations of juniper *(Juniperus communis)* (7000–6000 yr B.P.) (fig. 4.8B, C; B. Fredskild 1985).

The development of subarctic vegetation in southernmost Greenland coincided with the arrival of tree birch *(Betula pubescens)* between 3800 and 3600 yr B.P., an example of "sweepstakes dispersal" from European seed sources. Tundra species typically reached their northernmost distributional limits between 9000 and 5000 yr B.P., a time of relatively high seasonal contrast and of peak warm, dry climatic conditions.

Two stepwise changes in climate occurred in the last half of the Holocene interglacial. Both climatic shifts involved increases in precipitation and markedly cooler temperatures, with the first climatic change occurring in the middle Holocene (between 5000 and 3500 yr B.P.) and the second change occurring in the late Holocene (ca. 2000 yr B.P.). Pronounced late Holocene cooling resulted in reduced populations of juniper across coastal Greenland. This juniper decline and reduced amounts of total pollen influx observed at many sites have been interpreted as a progressive replacement of flowering plants by lichens and bog mosses within the arctic tundra vegetation (B. Fredskild 1985).

Western North America

Literature for late Quaternary changes in floras and vegetation has been summarized for western North America (V. M. Bryant Jr. and R. G. Holloway 1985; R. S. Thompson 1988), for the American Southwest (R. G. Baker 1983; W. G. Spaulding et al. 1983; P. V. Wells 1983; B. F. Jacobs et al. 1985; T. R. Van Devender et al. 1987; J. L. Betancourt et al. 1990), for the northwestern United States and western Canada (M. Tsukada 1982; R. G. Baker 1983; C. J. Heusser 1983;

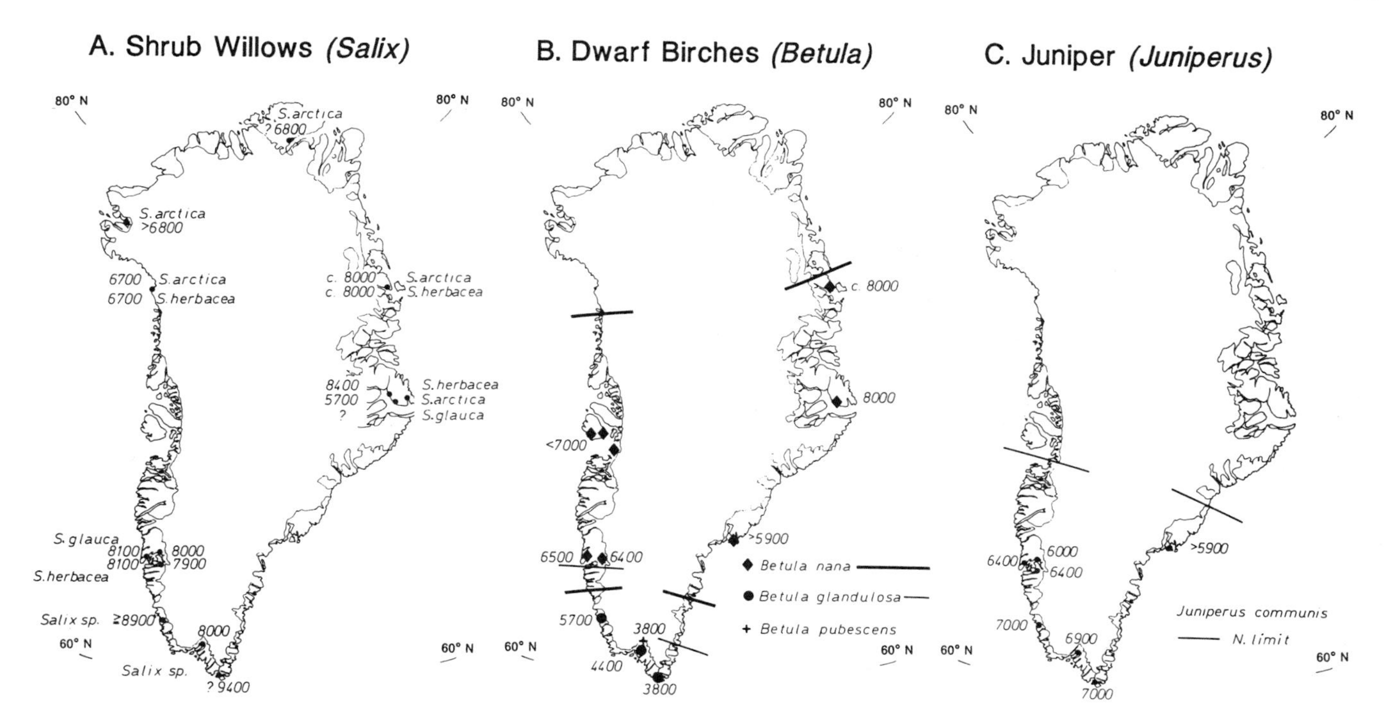

FIGURE 4.8. Holocene immigration for populations of (A) shrub willows *(Salix arctica, S. herbacea,* and *S. glauca),* (B) dwarf birches *(Betula nana, B. glandulosa,* and *B. pubescens),* and (C) juniper *(Juniperus communis)* along eastern and western coastal corridors of Greenland. The numbers on each map correspond with the first time (in yr B.P.) for postglacial arrival of the taxon's populations as documented by radiocarbon dating of plant-fossil evidence. [Modified from B. Fredskild 1985, with permission from Routledge.]

J. C. Ritchie 1980, 1987; P. J. Mehringer 1985; C. W. Barnosky et al. 1987; J. V. Matthews Jr. et al. 1989; J. M. Beiswenger 1991), and for northwestern Canada and Alaska (J. V. Matthews Jr. 1974; D. M. Hopkins et al. 1982; T. A. Ager 1983; C. J. Heusser 1983, 1983b; J. C. Ritchie 1984, 1987; C. W. Barnosky et al. 1987; H. F. Lamb and M. E. Edwards 1988; J. V. Matthews et al. 1989).

The American Southwest

During the last full-glacial interval, the Aleutian low pressure system in the North Pacific Ocean intensified, and both the zone of prevailing westerly winds and one branch of the polar jet stream shifted southward (fig. 4.2). This resulted in regional cooling of 7° to 8° C and a persistent Pacific frontal track of winter storms that brought moisture across the American Southwest (G. R. Brakenridge 1978; J. E. Kutzbach and H. E. Wright Jr. 1985; J. E. Kutzbach and P. J. Guetter 1986). Cooling at high elevations produced an elevational lowering of periglacial environments (T. L. Péwé 1983) and regional snowlines. Elevational limits for growth of mountain glaciers were lowered by as much as 1000 m (S. C. Porter et al. 1983). As far south as northern Arizona, the timberline was lowered by 800 to 1000 m (G. R. Brakenridge 1978).

Across the unglaciated portions of southwestern North America, at lower elevations on valley floors and lower montane slopes of closed basins, the cool, moist "pluvial" climate led to the expansion of large freshwater "pluvial" lakes (fig. 4.9; see L. Brouillet and R. D. Whetstone, chap. 1). The pluvial lakes produced a lake effect on regional climates, locally enhancing precipitation and reducing seasonal temperature contrasts (L. Benson and R. S. Thompson 1987). Four factors—increased precipitation, reduced evaporation, high groundwater tables, and increased soil-moisture levels—favored overland water flow. Drainage networks of perennial streams expanded, increasing habitat

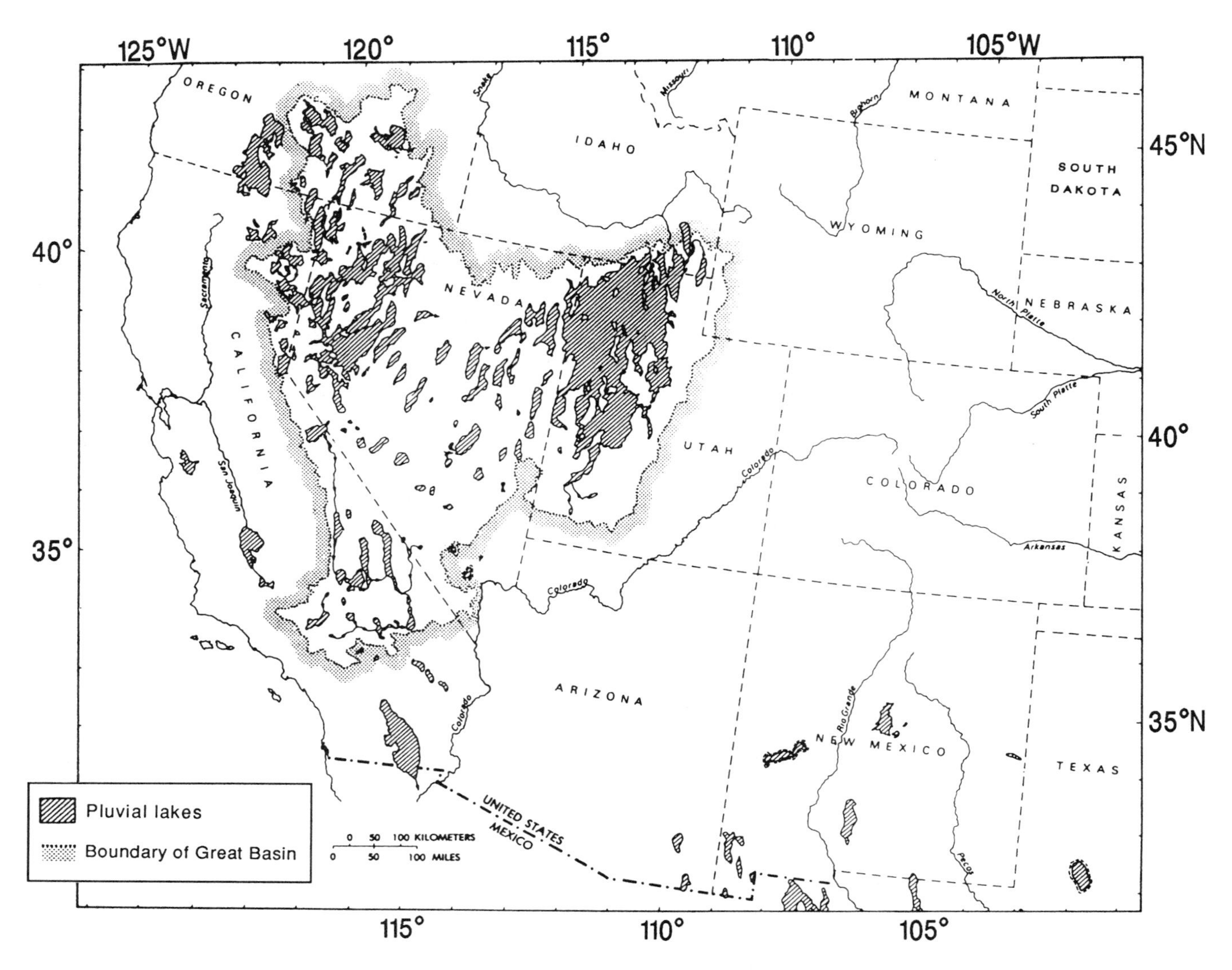

FIGURE 4.9. Late Pleistocene distribution of pluvial lakes that expanded in area between 25,000 and 10,000 yr B.P. within the western United States. The Great Basin Physiographic Province (shown with a dotted boundary) contains nearly all of these major pluvial-lake basins. [Modified from G. I. Smith and F. A. Street-Perrott 1983, with permission from University of Minnesota Press.]

available for riparian plant communities (G. I. Smith and F. A. Street-Perrott 1983).

Under this full-glacial climatic regime, woody perennials encountered a wide spectrum of habitats through a broad elevational range. Equable full-glacial conditions permitted range extensions of forest plants along their lower elevational and southern distributional limits, where they overlapped the ranges of, but did not necessarily replace, desert scrub taxa (T. R. Van Devender et al. 1987). The comparison of modern ranges of individual species with their Pleistocene fossil occurrences (W. G. Spaulding et al. 1983) indicates a general elevational shift downward in their distributions of 200–1200 m (fig. 4.10). This is especially apparent along the lower elevational limits of many tree species and the upper limits of desert shrub taxa.

This full-glacial reassortment of plant species did not simply produce a "telescoping" of biomes with the same communities of coevolved species. Instead, new combinations of species competed for space, light, and nutrients (K. Cole 1985). Full-glacial pluvial climates favored the expansion of alpine tundra and steppe

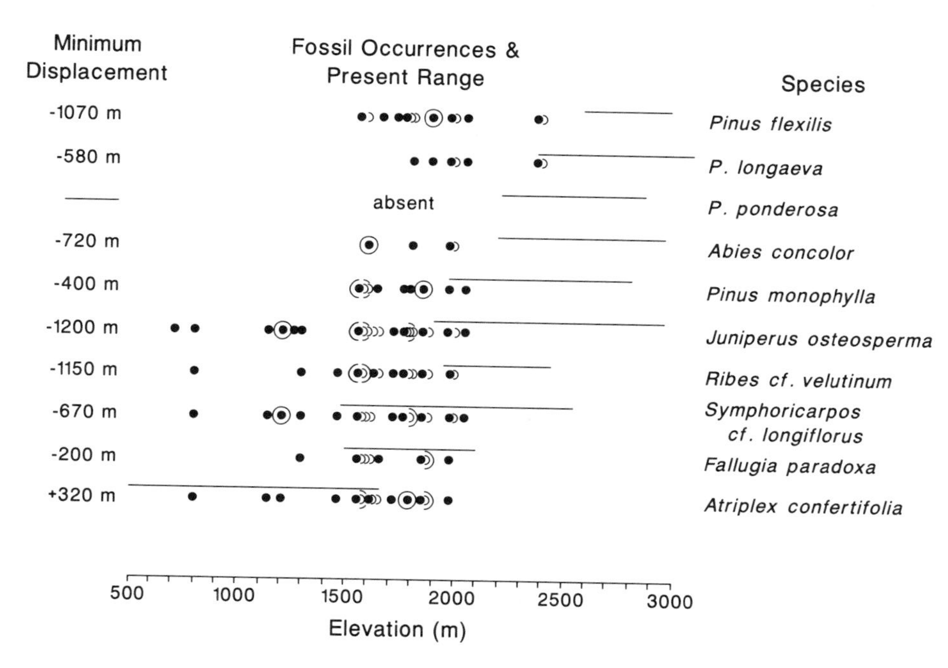

FIGURE 4.10. Glacial and interglacial shifts in elevational ranges of selected trees and shrubs on calcareous substrates in southern Nevada. Horizontal lines represent present elevational ranges and dots represent fossil occurrences prior to 11,000 yr B.P. Circled dots represent collections from several sites with the same elevation. Values for minimum displacement are conservative estimates of elevational adjustments in the ranges of the taxa over the last glacial-interglacial cycle. [Modified from W. G. Spaulding et al. 1983, with permission from University of Minnesota Press.]

communities at high elevations, of subalpine forest communities at intermediate elevations, and of woodlands dominated by trees such as juniper (*Juniperus* spp.) across lower montane slopes (fig. 4.11). Full-glacial biotic communities were bracketed between highland mountain glaciers and lowland pluvial lakes.

Woodland communities rather than desert communities prevailed during most of the Quaternary. Many species that evolved as early as 8 to 5 Ma in the late Miocene (D. I. Axelrod 1979) and that are constituents of the present floras of the Chihuahuan, Sonoran, Mojave, and Great Basin deserts (M. G. Barbour and N. L. Christensen, chap. 5) survived as understory plants or in disturbed openings of glacial-age woodlands. In the late-glacial period, with a shift northward of the polar jet stream, the climates changed from a glacial (pluvial) mode to an interglacial (interpluvial) mode. Warmer temperatures, diminished moisture supply, and increased evaporation resulted in pronounced drying and shrinking of pluvial lakes between 14,000 and 10,000 yr B.P. (fig. 4.12).

By 8000 yr B.P., climate-induced reshuffling of plant species in the late Pleistocene and early Holocene intervals produced new desert scrub and desert communities at low elevations that displaced woodlands and invaded exposed playa flats (former pluvial lakes; fig. 4.12). Forest communities persisted at intermediate elevations and expanded to higher mountain summits (K. Cole 1985).

During the early and middle Holocene, the winter peak in precipitation favored one suite of taxa characteristic of the Mojave and Great Basin deserts (T. R. Van Devender et al. 1987). The early Holocene development of summer monsoons with a new summer peak in moisture, the result of the seasonal shift of the Maritime Tropical air mass expanding northwestward from the Gulf of Mexico, favored a second suite of desert plants characteristic of the modern Chihuahuan and Sonoran regions. Migration of desert taxa continued to enrich modern desert scrub and desert grassland communities until stabilization of the floras was attained at about 4000 yr B.P. (W. G. Spaulding et al.

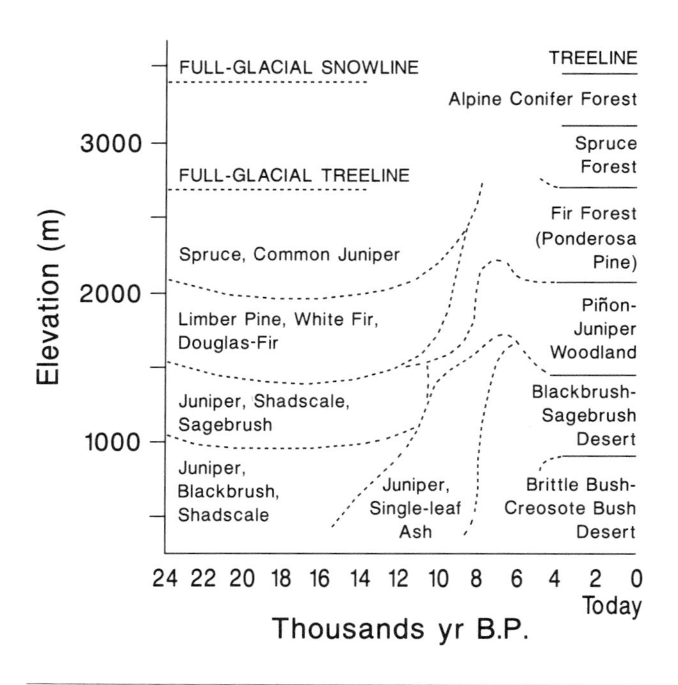

FIGURE 4.11. Elevational zonation of major plant communities in northern Arizona during the last 24,000 years. Plant-fossil data are from radiocarbon-dated pack-rat midden deposits from the Grand Canyon. Position of full-glacial treeline is inferred from fossil-pollen evidence in the White Mountains, and full-glacial snowline is inferred from distribution of glacial deposits on the San Francisco peaks. [Modified from K. Cole 1985, with permission from University of Chicago Press.]

1983; T. R. Van Devender et al. 1987; R. S. Thompson 1988).

In the coastal zone of western California, in southern Arizona, and in western Mexico, the interglacial (interpluvial) climate was characterized by moderate temperatures, cool wet winters and hot dry summers (Mediterranean-type climate), and seasonal droughtiness. Fires increased in frequency, particularly during the Holocene interglacial peak in warmth and aridity. Sclerophyllous scrub vegetation, including chaparral, became more widespread during the middle Holocene Hypsithermal interval from about 8000 to 4000 yr B.P. (D. I. Axelrod 1989).

Chaparral taxa evolved over the past 50 million years in response to the generation of suitable montane habitat, and their populations expanded in response to major tectonic uplift in the late Pliocene and Quaternary. While chaparral plants were not the evolutionary product of Mediterranean-type climates established during the Quaternary interglacials, their predominance on the landscape has been affected by glacial-interglacial cycles of climatic change, which have alternately suppressed and then augmented wildfire disturbance regimes (D. I. Axelrod 1989). These chaparral taxa have presumably assembled into Mediterranean scrub communities during interpluvial periods and disassembled during pluvial periods. Earlier throughout the Tertiary period and during the pluvial glacial ages of the Quaternary, the chaparral taxa persisted as understory scrub within woodland communities.

Northwestern United States and Southwestern Canada

During full-glacial times, the Cordilleran Ice Sheet extended from southern Alaska through British Columbia to its southernmost limit in northern Washington, Idaho, and Montana (D. B. Booth 1987; J. J. Clague 1989). It flowed westward across a 50-km stretch of Pacific Coastal Plain exposed by a lowering of sea level by at least 100 m (C. W. Barnosky et al. 1987). Small pockets of unglaciated land (J. J. Clague 1989), such as in the Queen Charlotte Islands in coastal British Columbia, served as glacial-age, floristically diverse refugia for herbs, willow, Sitka spruce *(Picea sitchensis)*, and lodgepole pine *(Pinus contorta)* (B. G. Warner et al. 1982; J. V. Matthews Jr. et al. 1989).

Along the eastern base of the Rocky Mountains, glacial ice flowing eastward from the Cordilleran Ice Sheet merged with the western and southwestern limits of the Laurentide Ice Sheet, resulting in continuous late Wisconsinan ice contact of these two ice sheets from west central Alberta, across northeastern British Columbia, to the southeastern corner of the Yukon (N. W. Rutter 1984). Thus during full-glacial time 20,000 to 18,000 years ago, a complex of extensive continental glaciers stretched across North America from the eastern coast of the Pacific Ocean to the northwestern margin of the Atlantic Ocean (J. T. Andrews 1987; D. B. Booth 1987; J. J. Clague 1989).

During peak glaciation 18,000 yr B.P., this continen-

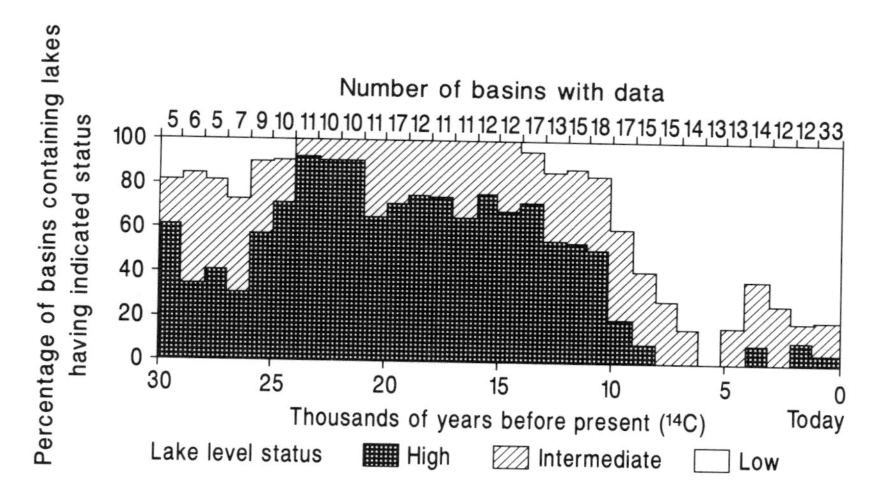

FIGURE 4.12. Relative percentages of late Quaternary pluvial lakes in the Western United States, with lake stage in water-level fluctuation (low, intermediate, or high levels) plotted against time (yr B.P.) for the past 30,000 years. The greatest proportion of pluvial lakes achieved high lake stages during the full-glacial and late-glacial intervals between 25,000 and 10,000 yr B.P. During the last 10,000 years of this Holocene interglacial interval, most glacial-age pluvial lakes have dropped to relatively low water stages within areally restricted, permanent saline lakes or represent ephemeral playa lakes. [Modified from G. I. Smith and F. A. Street-Perrott 1983, with permission from University of Minnesota Press.]

tal mass of glacial ice produced a climatic regime of "glacial anticyclone" (fig. 4.2), generating strong, cold winds that radiated in a clockwise direction off the southern Cordilleran and Laurentide ice margins and swept toward the west and southwest (J. E. Kutzbach and H. E. Wright Jr. 1985). These strong easterly winds maintained severely cold and dry conditions through most of the unglaciated northwestern United States. Rigorous permafrost environments with patterned ground, ice wedges, and cryoplanation terraces developed across Montana, Wyoming, Idaho, and the eastern two-thirds of Washington (T. L. Péwé 1983; J. M. Beiswenger 1991).

Westerly winds sweeping across the Pacific Ocean brought cool, moist conditions only to the maritime area west of the Coast Ranges, the first orographic barrier encountered. The eastward transport of oceanic moisture by westerly winds was effectively stopped by the combination of the prevailing, periglacial easterly winds and by the coastal mountain ranges, reinforcing the rain shadow farther east. Because it is directly related to distance from moisture source, alpine glaciation was most prominent in the Pacific maritime zone of the Coast Ranges and Cascade Range. Formation of mountain glaciers was progressively more limited farther to the east in the Rocky Mountains (S. C. Porter et al. 1983).

C. J. Heusser (1983) and C. W. Barnosky et al. (1987) reviewed the patterns of vegetation in the Pacific Northwest relative to the last advance and retreat of the Cordilleran Ice Sheet during the last 20,000 years. South of the glacial limit in western Washington 20,000–18,000 yr B.P. (fig. 4.13), cool, humid conditions favored the growth of alpine glaciers in the Olympic Mountains of the Coast Ranges and inland in the Cascade Range. On the exposed Pacific Coastal Plain between 20,000 and 16,800 yr B.P., the vegetation was a subalpine parkland with a mixture of lowland and montane species, including Sitka spruce *(Picea sitchensis)*, western white pine *(Pinus monticola)*, lodgepole pine *(P. contorta)*, mountain hemlock *(Tsuga mertensiana)*, and western hemlock *(T. heterophylla)* (Bogachiel Valley, fig. 4.13; C. J. Heusser 1983). In the rain shadow east of the Olympic Mountains (the first Coast Range), the lowland vegetation of the Puget Trough was composed of a mosaic of grass-sedge-wormwood *(Artemisia)* tundra and parkland with trees of lodgepole pine and probably Engelmann spruce *(Picea engelmannii)* (Davis Lake and Battle Ground Lake, fig. 4.13; C. W. Barnosky et al. 1987).

In the Columbia Basin, east of the glacier-mantled

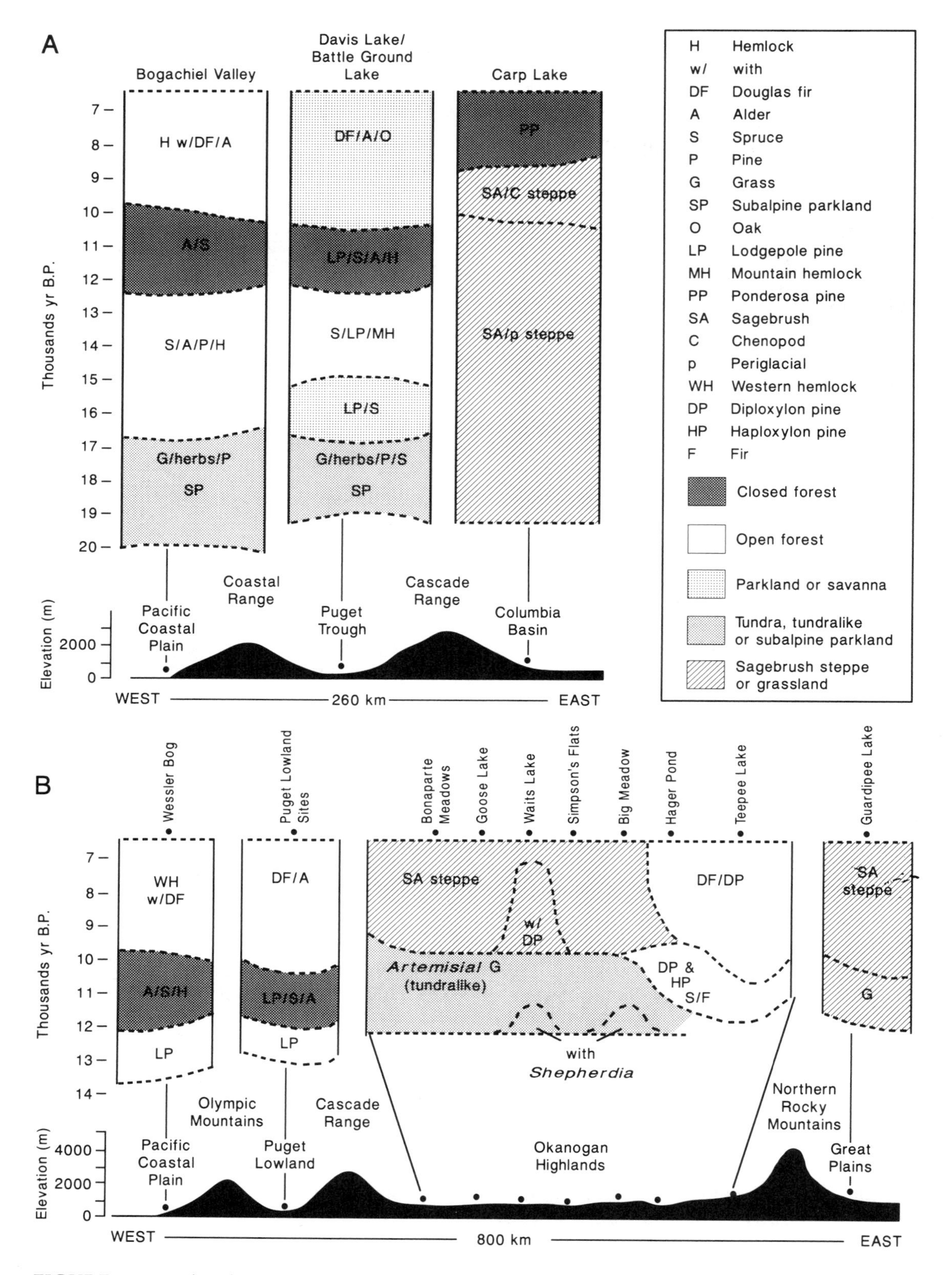

FIGURE 4.13. Glacial-interglacial changes in vegetation. (A) In unglaciated terrain across western Washington. The vegetation reconstructions from 20,000 to 7000 yr B.P. are based on paleoecological evidence preserved within sites along a west-to-east transect from the Pacific Coastal Plain to the Columbia Basin. (B) Postglacial vegetation reconstructed from 14,000 to 7000 yr B.P. from deglaciated plant-fossil sites, situated along the southern terminus of the Cordilleran Ice Sheet from the Pacific Coastal Plain of northwestern Washington east to the Great Plains of northern Montana. [Modified from C. W. Barnosky et al. 1987, with permission from the author.]

Cascade Range, a sparse vegetation of periglacial steppe persisted from 20,000 to about 10,000 yr B.P. Under the influence of cold, extremely dry, easterly winds, this lowland steppe vegetation consisted of grasses, sagebrush *(Artemisia)*, and alpine herbs (Carp Lake, fig. 4.13). A forest zone may have been compressed elevationally between the mountain glaciers and the lowland steppe, surviving at intermediate elevations along the eastern slope of the Cascade Range, or, alternatively, alpine tundra at its lower limit may have merged with steppe at its upper limit (C. W. Barnosky et al. 1987). Either situation would have allowed the combination of alpine and steppe floras within this important periglacial steppe vegetation (fig. 4.13) that extended from central Washington east to Wyoming (P. J. Mehringer Jr. 1985; C. W. Barnosky et al. 1987; J. M. Beiswenger 1991).

Between 17,000 and 15,000 yr B.P., increased seasonal contrast in temperatures and the warming of summers were first expressed by retreat of alpine glaciers toward mountain summits (S. C. Porter et al. 1983) and the shift from subalpine parkland toward open forest on the lowlands of western Washington (fig. 4.13A; C. J. Heusser 1983). Late-glacial forests diversified from about 16,000 until 12,500 yr B.P. Major expansions of tree populations, particularly of Sitka spruce and red alder *(Alnus rubra)*, occurred on the Pacific Coastal Plain. Inland in the Puget Trough, lodgepole pine colonized deglaciated terrain (C. W. Barnosky et al. 1987).

Substantial melting and retreat of the Cordilleran Ice Sheet occurred between 12,500 and 10,000 yr B.P. in the Pacific Northwest. This reduced glacial mass diminished both the glacial source for and the strength of anticyclonic easterly winds. The enhanced prevailing westerly winds from the Pacific Ocean resulted in widespread warming of temperate postglacial climates. Closed forests of red alder, spruce, lodgepole pine, and hemlocks developed west of the Cascade Range. Populations of temperate trees, such as grand fir *(Abies grandis)*, western hemlock, and mountain hemlock, shifted elevationally from lowlands into nearby montane habitats, and these trees also migrated northward into suitable maritime areas of temperate, humid conditions (fig. 4.13B; C. W. Barnosky et al. 1987).

At this time of transition between glacial and interglacial climatic regimes, the western slopes of the northern Rocky Mountains served as an increasingly effective orographic barrier, screening moisture from the now-prevailing westerly winds. By about 10,500 yr B.P. in western Montana (Tepee Lake, fig. 4.13B), the periglacial tundra was replaced as open coniferous forests were established, dominated by a variety of pines that probably included western white pine *(Pinus monticola)*, white bark pine *(P. albicaulis)*, and lodgepole pine *(P. contorta)*, as well as by subalpine fir *(Abies bifolia)* and spruce *(Picea engelmannii* and possibly white spruce [*P. glauca*]) (R. N. Mack et al. 1983). Tundralike grasslands or sagebrush steppes persisted within dry rainshadow regions directly east of the Cascade Range and in the Great Plains east of the northern Rocky Mountains (fig. 4.13B; R. G. Baker 1983; C. W. Barnosky et al. 1987).

During the Holocene epoch, warmest and driest summers (fig. 4.3) occurred between 10,000 and 7000 yr B.P., accentuating drought stress during the growing season, especially for plants on well-drained or xeric sites (C. W. Barnosky et al. 1987). In response, early Holocene forests developed more open canopies. Mesophytic conifers, such as spruce and hemlock, became less important in these forests, whereas more drought-tolerant species, such as Douglas-fir *(Pseudotsuga menziesii)* and red alder, became dominant (R. G. Baker 1983; C. W. Barnosky et al. 1987).

Northwestern Canada and Alaska

During the maximum extent of continental glaciation, unglaciated regions with cold, dry climates and sparse treeless vegetation occurred across Beringia to the Mackenzie River Delta in the Northwest Territories (south to 60° N along the margins of the Cordilleran and Laurentide ice sheets) (D. M. Hopkins et al. 1982; J. C. Ritchie 1984; J. V. Matthews Jr. et al. 1989). In this region, full-glacial fossil-pollen assemblages of wormwood *(Artemisia)*, grasses, and sedges have low total values of pollen influx. This has been interpreted to indicate a landscape mosaic that included many types of herbaceous tundra, interspersed by periglacial areas of polar desert communities and open ground maintained by freeze-thaw processes (C. E. Schweger 1982; J. V. Matthews Jr. 1982; T. A. Ager 1983; J. C. Ritchie 1984, 1987; C. W. Barnosky et al. 1987; H. F. Lamb and M. E. Edwards 1988). This interpretation differs from earlier proposals of either a continuous "arctic steppe" herb tundra (J. V. Matthews Jr. 1976; R. D. Guthrie 1984) or of barren polar desert or fell-field (L. C. Cwynar and J. C. Ritchie 1980; J. C. Ritchie and L. C. Cwynar 1982). Low icecaps occupied the Alaskan Brooks Range and the Ahklun Mountains (T. D. Hamilton and R. M. Thorson 1983).

The Cordilleran Ice Sheet served as a major barrier to the northward dispersal of boreal and temperate plants during full-glacial times. At the maximum extent of Pleistocene glaciation, 20,000 to 18,000 years ago, relatively few summits of the Coast Ranges, MacKenzie Mountains, and Rocky Mountains rose above the Cordilleran glacial ice surface. Generally above 2500 m elevation, these isolated montane summits provided sites for alpine glaciers and periglacial "nunataks," serving as possible refugia for polar desert and periglacial steppe communities (fig. 1.12 in J. J. Clague 1989).

During the late-glacial interval the southern margin of the Cordilleran glacier rapidly retreated northward after 14,000 yr B.P. It retracted to the United States–Canada border by 11,000 yr B.P. (D. B. Booth 1987). Between about 13,500 and 11,000 yr B.P., deglaciated terrain and proglacial lakes formed an ice-free corridor between the retreating margins of the Cordilleran and Laurentide ice sheets (N. W. Rutter 1984), creating a continuous corridor for plant and animal migration across the western interior of Canada (J. C. Ritchie 1980, 1987; J. V. Matthews Jr. et al. 1989). This midcontinental route was used by both eastern and western plant species between 13,500 and 9000 yr B.P.

Boreal trees, such as white spruce *(Picea glauca),* black spruce *(P. mariana),* and jack pine *(Pinus banksiana),* migrated northward along this route from refugia in the western Great Lakes region (J. C. Ritchie and G. M. MacDonald 1986; J. C. Ritchie 1987). Western populations of lodgepole pine expanded from a full-glacial refuge in the unglaciated portions of the Pacific Northwest and initiated their northward migration through this corridor more than 12,200 years ago (G. M. MacDonald and L. C. Cwynar 1985; L. C. Cwynar and G. M. MacDonald 1987).

Cordilleran ice retreated across the Pacific Coastal Plain between 13,500 and 9500 yr B.P. The plain was then repeatedly uplifted by isostatic rebound and submerged by rising sea levels (J. J. Clague 1989). The Aleutian Islands were deglaciated between 12,000 and 10,000 yr B.P., and the Gulf of Alaska was ice-free by 10,000 yr B.P. The present coastline of western British Columbia and southern Alaska became free of glacial ice between 13,500 and 9500 yr B.P., providing a western maritime route for plant invasions (T. D. Hamilton and R. M. Thorson 1983; J. J. Clague 1989).

In unglaciated regions of eastern Beringia, i.e., Alaska, the northern Yukon, and Northwest Territories, full-glacial climatic conditions provided a longitudinal gradient of cool, mesic conditions in the west and much colder and drier conditions nearer the continental ice sheets. This environmental gradient was reflected in the change from mesic tundra meadows of grasses, sedge, *Artemisia,* and shrub willow *(Salix)* in western Alaska to more xeric, discontinuous herb tundra and polar desert communities in northwestern Canada (fig. 4.14; T. A. Ager 1983; J. C. Ritchie 1984, 1987).

By 14,000 yr B.P., climate ameliorated across Beringia. Summer temperatures rose concomitant with the increased seasonality of solar radiation. Coastal portions of the Bering land bridge were inundated as sea levels rose (C. W. Barnosky et al. 1987). Increased availability of water from melting glacier margins, and closer proximity to coastal waters, combined to produce more precipitation. This late-glacial shift toward warmer, wetter summers favored the expansion of formerly small, scattered populations of dwarf birch and many heath species (family Ericaceae). The formerly extensive herb tundra was replaced at about 14,000 yr B.P. by birch–heath shrub tundra through virtually all of the Alaskan mainland (T. A. Ager 1983; C. W. Barnosky et al. 1987). Maritime meadows of herb tundra, dominated by grasses and sedges, were established on deglaciated terrain. These have persisted for the last 10,000 years on the chain of the Aleutian Islands, which were geographically isolated by Holocene sea-level rise.

The opening of the ice-free corridor in the interior of western Canada between 13,500 and 11,000 yr B.P. provided the opportunity for a sequence of rapid northward immigrations of plant taxa between the southern Alberta corridor portal and the northern portal of the unglaciated Yukon and Northwest Territories. Other species migrated southward from the Beringian refugium onto terrain recently exposed by retreating glaciers, many tracking along the eastern edge of the Rocky Mountain foothills southward for thousands of kilometers (J. C. Ritchie 1984; J. V. Matthews Jr. et al. 1989). By 10,500 yr B.P., herb tundra communities were invaded by shrubs such as dwarf birch and heaths.

Passage of a fundamental bioclimatic threshold 10,000 years ago shifted the competitive balance from dominance of late Pleistocene herbaceous plants to arboreal plants in the early Holocene. J. C. Ritchie et al. (1983) presented environmental and paleobotanical evidence to suggest the existence of a regional maximum in summer temperatures at high latitudes in Beringia 10,000 years ago, a climatic phenomenon related to maximum seasonal contrast that was a consequence of Milankovitch cycles (fig. 4.3). Between 10,000 and 9000 yr B.P., large shrubs and pioneer trees became established, including willow, juniper *(Juniperus communis),* balsam poplar *(Populus balsamifera),* and white and black spruce. The modern landscape mosaic of herb and shrub tundra and woodland was established by 9000 yr B.P. in the far northwest of Canada (J. C. Ritchie 1984).

Populations of trees established in the Yukon in the early Holocene provided the seed source for dispersal into central and western Alaska along the westward-flowing network of the Yukon River and its tributaries (fig. 4.14). The forests of central Alaska, containing white and black spruce, white birch *(Betula papyrifera),* balsam poplar, and trembling (quaking) aspen *(Populus tremuloides),* were established between 10,000 and 8000

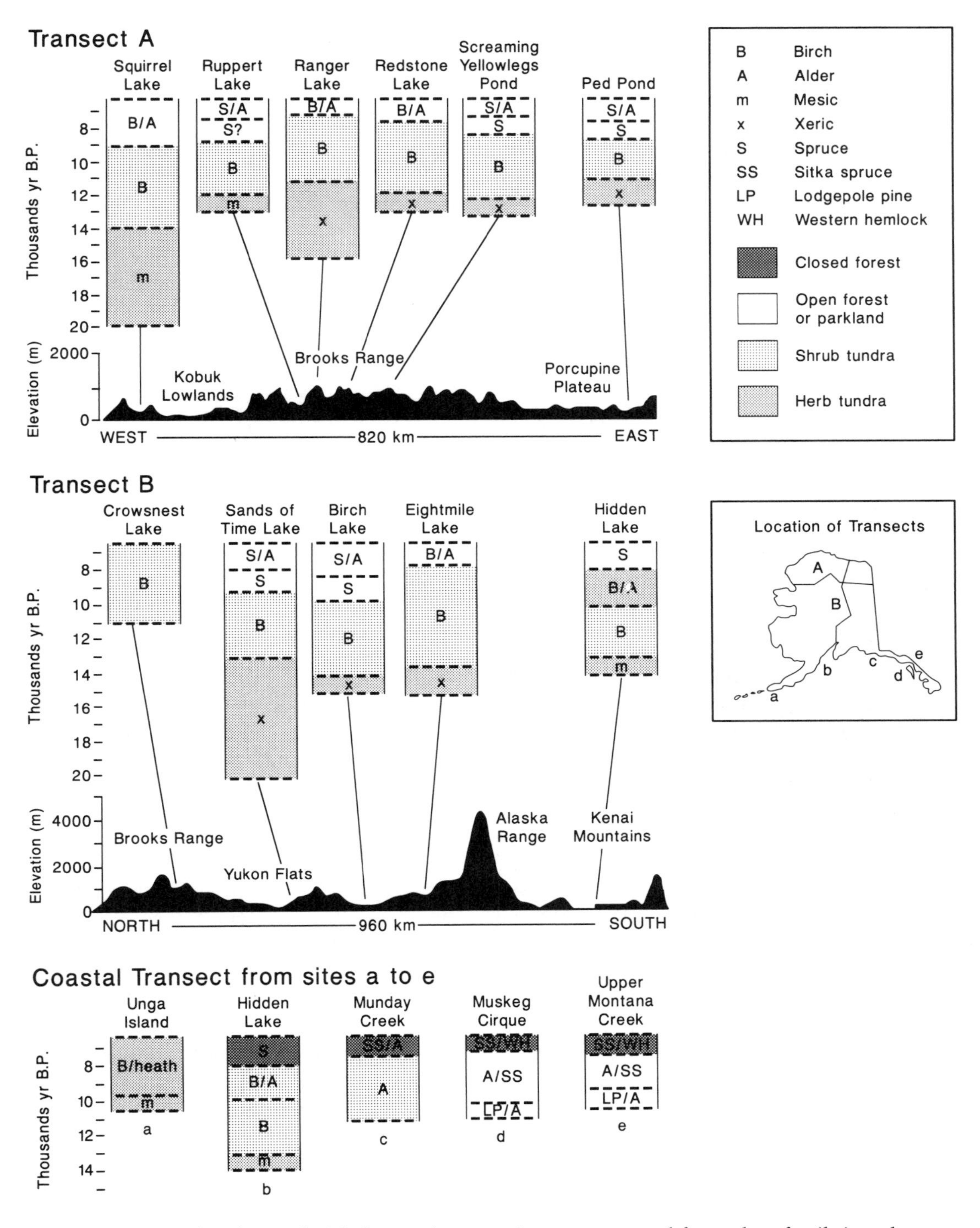

FIGURE 4.14. Glacial-interglacial changes in vegetation reconstructed from plant-fossil sites along three transects across Alaska. [Redrawn from C. W. Barnosky et al. 1987, with permission from the author.]

yr B.P. (fig. 4.14). By about 5000 yr B.P., they were present along their current western limit (C. J. Heusser 1983; T. A. Ager 1983).

The deglaciated coast of western British Columbia and southern Alaska provided the second principal route for postglacial plant migration. By 11,000 yr B.P., lodgepole pine and red alder colonized disturbed sites, forming an open parkland as far north as the panhandle of southeastern Alaska. Coastal coniferous forests were established across northwestern British Columbia and southeastern Alaska. The entry of Sitka spruce occurred between 10,500 and 8500 yr B.P. Successive invasions of western and mountain hemlock occurred in this region between about 8500 and 2700 yr B.P. Coastal forests dominated by Sitka spruce and mountain hemlock reached their modern western limits in south central Alaska within the last 4000 years (T. A. Ager 1983; C. W. Barnosky et al. 1987). This relatively late northwestward extension of Pacific coastal forests has been attributed to a late Holocene increase in the frequency of storms and the supply of moisture they brought (C. J. Heusser 1983b; C. W. Barnosky et al. 1987).

R. J. Hebda and R. W. Mathewes (1984) documented the postglacial advance of western redcedar *(Thuja plicata)*, an endemic species that is today an important component of the coniferous forests of the cool, moist Pacific coastal slope. Increased percentages of pollen grains similar to those of western redcedar in the fossil record indicate that it initially colonized the Puget Trough of western Washington as early as 10,000 years ago. Populations of western redcedar advanced through the deglaciated Fraser Lowlands of southwestern British Columbia between 8000 and 6500 yr B.P., achieving their present northern range in northwestern British Columbia by about 6000 years ago. A maritime climatic shift toward increased moisture favored a major increase in population size of western redcedar from 5000 to 2000 yr B.P. This late Holocene expansion of cedar resulted in the relatively recent codominance in Pacific coastal forests of western redcedar and western hemlock, the previous Holocene dominant.

Quaternary Floras

The modern flora of North America has developed as the constituent plant taxa were evolving at different times and in response to different kinds of selection pressures (B. H. Tiffney 1985, 1985b). The development of the flora during the last 3 million years of the late Tertiary and Quaternary reflects contributions from three groups of floristic elements: (1) "relict floras" that evolved much earlier in the Mesozoic or early Cenozoic and that persist as relatively old, unchanging relicts of past floras; (2) "orthoselection floras" that include a suite of species changing in response to a consistent, long-term environmental trend; and (3) "migration floras" that exhibit substantial shifts in distributional ranges as they migrate across relatively large distances, tracking major climatic changes between alternating glacial and interglacial regimes (V. P. Grichuk 1984).

The relatively ancient or relict floras comprise taxa that have survived the late Cenozoic onset of increasing environmental and climatic oscillations of ever-greater magnitude (E. B. Leopold and H. D. MacGinitie 1972; D. Q. Bowen 1985), and that have persisted to the present day within refugial regions such as the Gulf Coastal Plain (P. S. Martin and B. E. Harrell 1957; E. L. Little Jr. 1971b; A. Graham 1972b, 1973b; A. J. Sharp 1972, 1972b; see A. Graham, chap. 3). In part, these relict floras represent plants with closely related vicariant species, species pairs with disjunct populations in widely separated regions of the world today (H.-L. Li 1972). Elements of these floras have been displaced by the gradual plate-tectonic movements of continents, and their distributional ranges have been fragmented by mountain building, climatic change, and loss of migrational corridors such as the Beringian and North Atlantic land bridges (J. A. Wolfe 1972; B. H. Tiffney 1985, 1985b; D. I. Axelrod 1986, 1986b).

Two Quaternary examples illustrate orthoselection floras (V. P. Grichuk 1984). These are floras that have responded to long-term trends in climate and in regional uplift of terrain. At high latitudes during the late Pliocene and early Pleistocene, long-term climatic cooling shaped the patterns of taxonomic differentiation and favored the latitudinal expansion of tundra and taiga communities across the Arctic (D. I. Axelrod 1986b). This late Cenozoic, climate-driven shift in both composition and dynamics of tundra ecosystems provoked the in situ evolution of plant and animal taxa (H. Hara 1972; J. A. Wolfe 1972; "phyletic ecosystem evolution" sensu J. V. Matthews Jr. 1974).

At middle latitudes in the American West, the combination of climatic cooling and continued late Cenozoic mountain uplift, particularly within the Coast Ranges, the Cascade Range, the Sierra Nevada, and the southern Rocky Mountains, intensified the orographic influence of the rainshadow effect within and immediately east of the north-south trending ranges of the Cordillera. This long-term cooling, decreased summer precipitation, and increased continentality of montane sites both impoverished the modern Cordilleran woody flora and enhanced the elevational differentiation of alpine tundra and herbaceous steppe floras, as it did also with assemblages of subalpine arboreal conifers (E. B. Leopold and H. D. MacGinitie 1972; D. I. Axelrod and P. H. Raven 1985; D. I. Axelrod 1986, 1986b, 1988; C. W. Barnosky 1987).

During the Quaternary, migration floras moved over hundreds to thousands of kilometers, each species tracking the shifts in climatic regions according to its particular ecological requirements (V. P. Grichuk 1984). Alternating glacial and interglacial climatic regimes triggered the differential responses of species, expressed in the location and areal extent of their distributions, in their demography and competitive abilities, and in the spectrum of their habitats (M. B. Davis 1976, 1981, 1986; P. A. Delcourt and H. R. Delcourt 1987).

W. B. Critchfield (1984) used data from plant fossil and contemporary genetics to develop a model of response of the North American temperate and boreal conifers to glacial-interglacial cycles. He suggested that, as each episode of continental glaciation was initiated, the combination of areal expansion of glacial ice masses and shifting climatic zones would displace some conifer populations, resulting in both contraction and fragmentation of their ranges. During times of maximum continental glaciation, the fragmented conifer populations may undergo genetic differentiation within isolated refugia. With the onset of interglacial conditions, migration and expansion of distributional ranges would reestablish genetic exchanges between populations within species and among closely related species that were formerly isolated.

The model by W. B. Critchfield (1984) identifies four principal kinds of Pleistocene climatic impacts on the genetic structure of conifer populations and on patterns of floristic change: (1) reduced genetic variation, (2) extinction of taxa, (3) geographic and temporal redistribution of genetic variation, and (4) increased genetic variation in a taxon. The first of these population-level responses to Quaternary environmental changes is the long-term loss of genetic variability. This may be reflected in reduced "ecological amplitude" and progressive restriction in the spectrum of habitats that plants occupy (R. O. Kapp 1977).

In extreme cases, extinction may result, a second type of population response. For example, in the late-glacial interval 11,000 years ago, a large-coned morphotype of white spruce *(Picea glauca)* became extinct in the southeastern United States (P. A. Delcourt et al. 1980; W. A. Watts 1980b; P. D. Royall et al. 1991). W. B. Critchfield (1984) suggested that this now-extinct fossil spruce may have represented a distinct species rather than an ecotype of white spruce.

Populations may also respond to change by redistribution of genetic variation. During the Quaternary, this may have occurred in two ways within North American conifers (W. B. Critchfield 1984). First, with the onset of glacial climates and more severe environments, genetic impoverishment might have resulted from the early-glacial fragmentation and local elimination of populations, with the genetic pool residing in transitory, migratory races. Second, geographic races that were isolated in glacial-age refugia might have experienced relatively infrequent and brief flushes of genetic exchange during interglacial intervals. Within the current distributional range of some trees, such as lodgepole pine *(Pinus contorta)*, geographic gradients in genetic structure from central to marginal populations may reflect interglacial paths of migration that have resulted from multiple and successive events of long-distance dispersal and subsequent establishment of outlier populations (G. M. MacDonald and L. C. Cwynar 1985; L. C. Cwynar and G. M. MacDonald 1987).

The last type of Pleistocene climatic impact on the coniferous tree flora is that of enhanced genetic variation. Opportunities were generated for hybridization and introgression as interglacial expansion in distributional ranges removed intervening barriers and facilitated genetic exchange among populations both within species and among closely related species groups (W. B. Critchfield 1984).

Through the last 900,000 years of the Quaternary, a period of 90,000 years of each 100,000-year glacial-interglacial cycle has been characterized by gradual cooling (N. G. Pisias and T. C. Moore Jr. 1981), and the onset of each 10,000-year interglacial interval has been marked by rapid climatic warming and increased seasonality of climates (fig. 4.15; J. Imbrie and K. P. Imbrie 1979). Because climates and environments have not been constant during the Quaternary, we might anticipate that any evolution that has occurred within the North American flora during the past several million years has not been gradualistic. Rather, we hypothesize that bursts or pulses of macroevolution (S. M. Stanley 1979) would have occurred every 100,000 years, modulated by the climatic and environmental changes triggered by Milankovitch cycles and coinciding with the transition from glacial to interglacial climatic regimes (fig. 4.1).

The relatively short time span from peak glacial to peak interglacial conditions is the time of greatest environmental instability during a glacial-interglacial cycle (fig. 4.3), and it is characterized both by the greatest magnitude and the greatest rate of change of global temperature (fig. 4.15). We postulate that a pulse of extinctions would result during times of rapid change from glacial to interglacial conditions, because the thresholds of temperature tolerance of plant species would be exceeded during these periods of high seasonal contrast. Extinctions would trigger a subsequent wave of speciation as other taxa diversify and fill the recently vacated niches. Minor readjustments would continue into the interglacial interval, so that eventually the rate of speciation would offset the rate of ex-

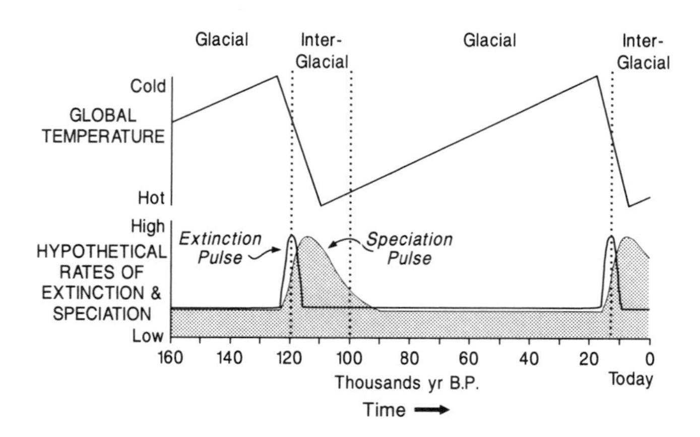

FIGURE 4.15. Macroevolutionary model showing hypothetical rates of extinction and speciation during a 100,000-year glacial-interglacial cycle.

tinction. During the last stages of each interglacial interval and into the next developing glacial interval, evolution would be gradualistic (fig. 4.15).

During glacial intervals North American plant populations would be fragmented in two ways: (1) mountain and continental glaciers expand to occupy formerly vegetated areas and (2) climatic cooling and shifts in climatic zones result in displacement of migration floras in unglaciated regions (R. A. Bryson and W. M. Wendland 1967; P. A. Delcourt and H. R. Delcourt 1984; J. E. Kutzbach and H. E. Wright Jr. 1985). For example, in the American West, full-glacial conditions provoke elevational adjustments in species ranges and their dispersal along montane corridors (W. G. Spaulding et al. 1983; P. V. Wells 1983; K. Cole 1985; C. W. Barnosky et al. 1987; T. R. Van Devender et al. 1987; R. S. Thompson 1988). Likewise, in middle latitudes of the unglaciated Southeast, a widespread latitudinal displacement in species distributions occurs during glacial times (M. B. Davis 1981, 1983, 1986; W. A. Watts 1983; P. A. Delcourt and H. R. Delcourt 1987; T. Webb III 1988). Thus glacial-age conditions may isolate populations within separate refugia, facilitating genetic drift and interpopulation heterogeneity.

During each glacial-interglacial transition, floras would be affected in several ways. Heightened seasonal contrast would produce unstable and retreating glaciers, resulting in availability of newly deglaciated territories for plant colonization. Increased seasonality of climate would also release some plant species whose populations were formerly constrained by some limiting bioclimatic threshold factor. Therefore individualistic plant migrations and the intermingling of floras would be facilitated (W. A. Watts 1988). The extinction of some species would occur because of geographic restriction coupled with a loss of genetic variability. In contrast, opportunities for genetic divergence leading to new ecotypes, subspecies, and species would increase for other plants, as selection intensified because of environmental change and competition accentuated among ephemeral assemblages of migrating species.

In the region of the unglaciated southeastern United States, the results of both types of processes can be seen in the development of the Holocene interglacial flora. With the onset of the present interglacial, climatic warming and increased seasonality caused the decline of formerly widespread populations of eastern spruce species along their southern periphery. Meanwhile, the slow rate of melting of the Laurentide Ice Sheet prevented the rapid spread of spruce northward and resulted in an environmental bottleneck that caused the collapse of former population centers of spruce, the extirpation of fragmented populations along the southern periphery of their range, and the extinction of one morphotype or species of white spruce (J. C. Bernabo and T. Webb III 1977; P. A. Delcourt et al. 1980; W. B. Critchfield 1984; P. A. Delcourt and H. R. Delcourt 1987).

In the central and southern Appalachian Mountains, populations of arctic-alpine and boreal taxa were progressively restricted during the early Holocene to high mountain peaks. Interglacial range restrictions may have resulted in speciation of narrowly endemic plants such as *Geum peckii* (P. S. White 1984b), as well as of certain boreal and cool temperate trees including red spruce *(Picea rubens)*, Fraser fir *(Abies fraseri)*, and Carolina hemlock *(Tsuga caroliniana)* that are today restricted in distributional range to the central and southern Appalachian region (E. L. Little Jr. 1971b).

PART III CONTEMPORARY VEGETATION AND PHYTOGEOGRAPHY

5. Vegetation

Michael G. Barbour
Norman L. Christensen

Covering nearly 60° of latitude and 145° of longitude, the North American continent extends through a striking diversity of climates and supports a rich array of vegetation types. Most of the world's major plant formations are represented. Tundra, boreal forest, montane conifer forest, temperate deciduous forest, prairie, steppe, woodlands of deciduous and evergreen trees, semiarid desert scrub, Mediterranean scrub, and tidal marshes occupy significant portions of the continent. Prominently missing are extremely arid deserts and several tropical grassland, savanna, and forest types.

We have been assisted in our review by two recent volumes on North American vegetation: B. F. Chabot and H. A. Mooney (1985) and M. G. Barbour and W. D. Billings (1988). We have also drawn heavily on the plant geography text by R. Daubenmire (1978). Less formal summaries prepared by M. G. Barbour, J. H. Burk, and W. D. Pitts (1987) and J. L. Vankat (1979) were also useful. We have surveyed regional summaries and research papers but have tried to limit references cited to modern summaries, older classics, and a few research articles that support important concepts about a particular vegetation type. Therefore, the review we present is not exhaustive, nor does it include aspects on the history of vegetation mapping in North America. Furthermore, we have concentrated on the current distribution of natural vegetation, rather than on human-modified landscapes or on movements of floras or vegetation through geologic time (but see A. Graham, chap. 3 and P. A. Delcourt and H. R. Delcourt, chap. 4).

The themes we emphasize in this chapter are: (1) relationships between vegetation and regional climate, (2) environmental gradients and episodic stresses that create vegetation gradients over space and time, and (3) the nature of ecotones that exist between all major vegetation types. We begin our survey with tundra vegetation—at the northern limits of plant life—and work clockwise through the continent, to boreal forest, eastern deciduous forest, grassland, desert, woodland, western conifer forest, and coastal vegetation. A map of the major vegetation types of North America north of Mexico is given in figure 5.1. Because of the map's large scale, it does not show many ecotones, mosaics, or local subtypes of vegetation. Every regional specialist who examines our map will likely find some detail to be in error. We alert the reader to the fact that figure 5.1 represents an oversimplification of nature, and we apologize for necessary local distortions.

We dedicate this chapter to John W. Harshberger, who wrote the earliest summary of North America vegetation in 1911.

Arctic Ecosystems

Overview

The Arctic, that region north of the climatic limit of boreal forest, is vegetated by a diverse assemblage of plant communities. L. C. Bliss (1981) distinguished the Low Arctic region, dominated by tundras, from the High

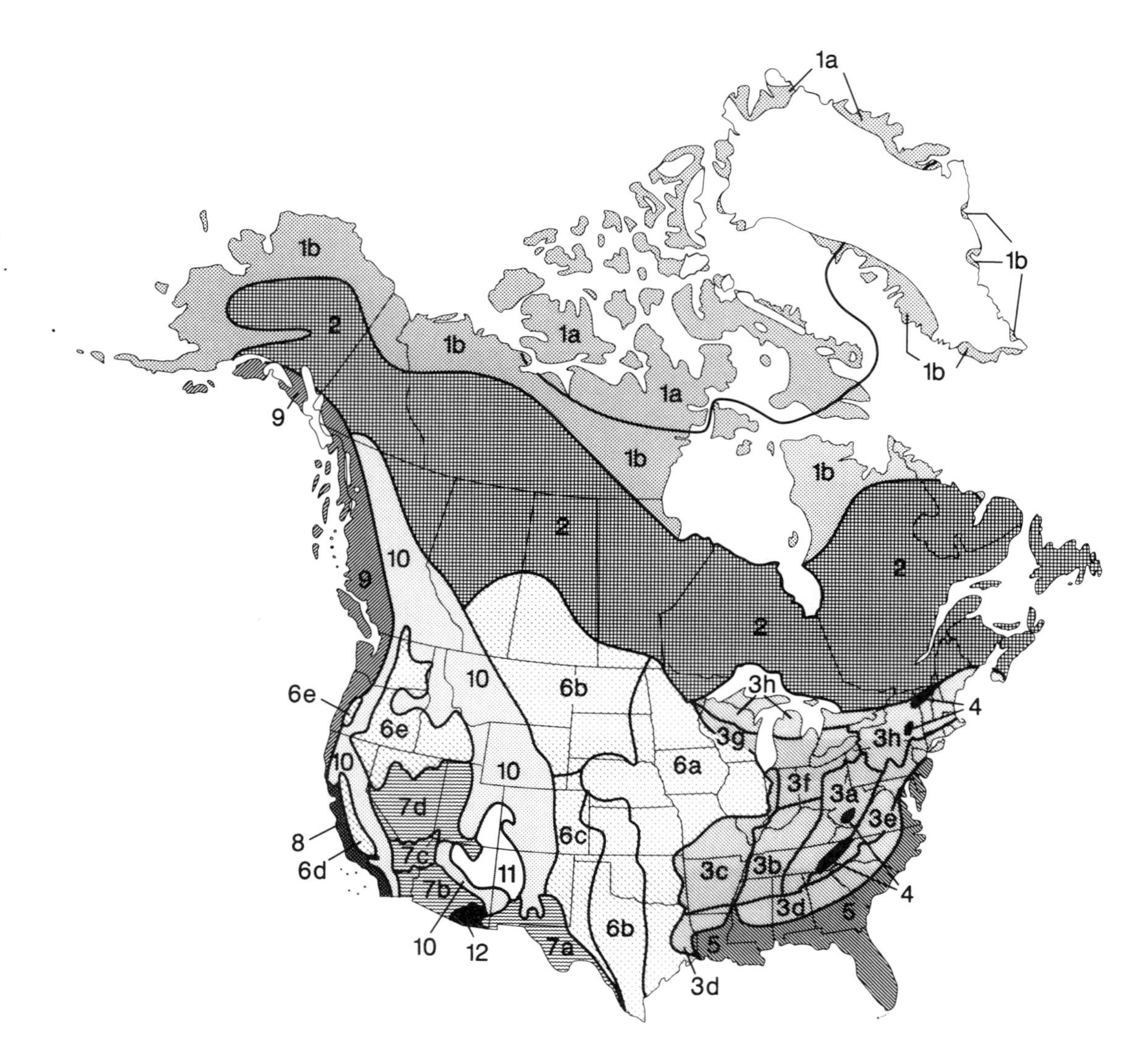

FIGURE 5.1. Vegetation of North America. Vegetation types that occupy less than 1% of the North American landmass are not shown. Most ecotones, mosaics, and local subtypes are not indicated. 1a, High Arctic tundra and polar desert. 1b, Low Arctic tundra. 2, boreal forest or taiga. 3a, mixed mesophytic deciduous forest. 3b, western mixed mesophytic forest. 3c, oak and hickory association. 3d, oak, hickory, and pine association. 3e, oak and chestnut association. 3f, beech and maple association. 3g, maple and basswood association. 3h, northern hardwoods association. 4, high elevation Appalachians. 5, southeastern Coastal Plain. 6a, tallgrass prairie, including the prairie peninsula and oak savanna ecotone with the eastern deciduous forest. 6b, mixedgrass prairie, including aspen parkland ecotone with the boreal forest. 6c, shortgrass prairie. 6d, central California grassland. 6e, intermountain grassland or shrub steppe, including Palouse and Columbia Basin areas. 7a, Chihuahuan warm desert scrub. 7b, Sonoran warm desert scrub. 7c, Mojave warm desert scrub. 7d, Intermountain, Great Basin, or cold desert scrub. 8, Mediterranean woodlands and scrublands. 9, Pacific Coast conifer forest. 10, western montane conifer forests, including Rocky Mountains, Coast Ranges, Cascade Range, Sierra Nevada, Transverse Range, Peninsular Range, and scattered peaks of the intermountain region. 11, mosaic of desert grassland, pinyon-juniper woodland, petran chaparral, and Madrean woodlands and scrublands. 12, mosaic of desert grassland, warm desert scrub, and Madrean woodlands and scrublands. [Adapted from map developed by W. D. Billings, in M. G. Barbour and W. D. Billings (1988), with permission from Cambridge University Press.]

FIGURE 5.2. Graminoid-moss tundra near Barrow, Alaska. [Courtesy of W. D. Billings.]

Arctic region in which polar deserts are prevalent. We use the term "tundra" to refer to a continuum of treeless plant communities varying in stature from shrub heaths (1–5 m high) to low sedges, grasses, and cryptogams (fig. 5.2). Plant cover is virtually 100% in tundras, but generally less than 50% (often less than 5%) in polar deserts (L. C. Bliss 1988).

Low Arctic landscapes occur as far south as 55° N latitude in eastern Canada, but generally north of 60° N latitude in western North America. This southern limit coincides roughly with the southern extent of continuous soil permafrost (L. C. Bliss 1988). Other factors regulating the tundra-taiga ecotone include length of the growing season (generally fewer than 90 days) and patterns of disturbance (especially fire) (W. C. Oechel and W. T. Lawrence 1985). High Arctic ecosystems are largely confined to the islands of Canada's Northwest Territory and northern Greenland. Suitable sites for plant growth at the northern limit of High Arctic polar desert appear to be restricted by water-saturated soils in the spring and water deficits in the summer (L. C. Bliss 1988). Together, arctic ecosystems cover over 4 million km^2, or approximately 19% of North America north of Mexico.

Compared to temperate and boreal ecosystems, arctic ecosystems are quite young. During Pliocene times (6–3 million years before the present, M.Y.B.P.), nearly all of the present Arctic was covered by coniferous forest. Tundra may have occurred on the most northern Canadian islands. The majority of arctic species are thought to have evolved from ancestors in montane central Asia, with fewer having been derived from Rocky Mountain species (W. D. Billings 1974; L. C. Bliss 1988). At the onset of the Pleistocene the circumpolar arctic flora probably included over 1500 species, but successive glacial advances and accompanying climate change have reduced this number to fewer than 1100 (D. Löve and A. Löve 1974), of which about 700 occur in North America.

Human utilization and impacts on arctic landscapes have been modest. The native Inuit and Eskimos have made only limited use of tundra plants. More recently, local areas have been affected by grazing of livestock. Exploitation of mineral resources, especially oil, presents a potential for alteration of some arctic landscapes.

As might be expected, arctic plants are well adapted to grow at low temperatures. Optimal temperatures for shoot growth are generally between 10° and 20° C, and many species can maintain net photosynthesis at temperatures near 0° C. G. R. Shaver and W. D. Billings (1977) found that root growth in *Eriophorum angustifolium* was optimal at 1°–4° C. In general, plant height is correlated with the height above ground at which summer temperatures are warmest (L. C. Bliss 1988). Furthermore, shrub, cushion, and mat growth forms create a significant boundary layer in which air temperatures may be 5°–10° above ambient. Arctic species are also adapted to take advantage of the abbreviated growing season. Shoot growth in most vascular plants begins immediately after snowmelt, utilizing photosynthetic reserves from the previous year (F. S. Chapin III and G. R. Shaver 1985). Buds are "preformed" at the end of each growing season to facilitate rapid deployment of shoots and roots at the beginning of the next growing season.

In polar desert regions, short growing seasons and water deficits limit annual net primary production to only 5–30 g/m^2 (L. C. Bliss 1988). In Low Arctic wet tundra ecosystems, net primary production is 150–200 g/m^2/yr. Low atmospheric nutrient input, soil anoxia, and temperature-limited rates of mineralization result in widespread deficiencies of nitrogen and phosphorus in these habitats (F. S. Chapin III and G. R. Shaver 1985). In well-drained Low Arctic shrub tundras, primary production is 250–400 g/m^2/yr.

Regional Variation

Total plant cover and stature, as well as relative abundance and diversity of vascular plants, generally decrease with increasing latitude. Diversity and relative abundance of lichens and bryophytes, however, are highest in High Arctic ecosystems. Within a particular region, vegetation composition and structure may vary considerably along microclimatic and hydrologic gradients, and in relation to patterns of snowmelt and disturbance.

Tall shrub tundra occurs on well-drained, nutrient-rich sites in the Low Arctic. These areas are most common on river flood plains or other areas with a deep active layer (L. C. Bliss 1988). Shrubs in genera such as *Alnus, Betula,* and *Salix* grow to a stature of 1–5 m on such sites.

Low shrub tundra (shrub stature 40–60 cm) ecosystems are common on rolling upland sites across the southern Low Arctic. Diminutive individuals of the same *Alnus, Betula,* and *Salix* spp. found in high shrub tundra occur with *Arctostaphylos alpina, A. rubra, Empetrum nigrum, Ledum palustre, Rubus chamaemorus, Vaccinium uliginosum,* and *V. vitis-idaea,* in varying combinations. In some locations *Carex* spp. and a diverse array of lichens and mosses are especially abundant. The relative importance of these taxa often varies in relation to seasonal patterns of snow accumulation and melt (L. C. Bliss 1988). It is believed that changes in climate and fire frequency during the past two centuries have caused replacement of forest tundra by low shrub tundra in many areas near the arctic/boreal ecotone (I. G. W. Corns 1974; J. C. Ritchie 1977, 1984).

Dwarf shrub heath tundras occur as small patches (usually less than 1 ha) on well-drained soils on flood-plain and upland areas. Low stature (10–20 cm) shrubs belonging to the Ericaceae, Empetraceae, and Diapensiaceae dominate. The patchy distribution of these ecosystems is related to local heavy accumulation (greater than 20–30 cm) of winter snow (L. C. Bliss 1988). Thus, the relative importance of this vegetation type in various arctic landscapes varies with topographic patterns that regulate snow distribution. Cotton grass–dwarf shrub heath tundras are most common in upland areas between the mountains and wet coastal plain in Alaska and the Yukon. The presence of prominent tussocks of *Eriophorum vaginatum* (tussock grass) in a heath matrix distinguishes this community from other heath tundras. Considerably less diverse dwarf shrub heath tundras may occur in local mesic sites associated with late snowmelt in the High Arctic (L. C. Bliss 1988; L. C. Bliss and J. Svoboda 1984).

Graminoid-moss tundras dominate poorly drained areas across the Low Arctic. Sedges, such as *Carex* and *Eriophorum* spp., and grasses, including *Alopecurus, Arctagrostis, Arctophila,* and *Dupontia,* dominate these wetlands along with a diverse assemblage of bryophytes. Local variation in relative abundance of graminoid-moss tundra species often occurs as a consequence of small-scale variations in topography and depth of the active layer that influence local soil hydrology. Graminoid-moss tundras are common in High Arctic landscapes where drainage is impeded (L. C. Bliss and J. Svoboda 1984).

L. C. Bliss (1988) distinguished three arctic semidesert communities: (1) Cushion plant-cryptogam communities are dominated by cushionlike or mat-forming herbs, such as *Draba corymbosa, Dryas integrifolia,* and *Saxifraga* spp., and lichens and mosses. Scattered graminoids are virtually always present. This vegetation is common on warm, dry sites with a relatively long growing season. (2) Cryptogam-herb vegetation, in which lichens and mosses account for 50–80% cover, occurs on moister sites. (3) Graminoid steppe, dominated by *Alopecurus alpinus* and *Luzula confusa,* has a distribution that appears to be related to slate-derived soils on some Canadian islands (L. C. Bliss and J. Svoboda 1984).

Polar desert ecosystems include herb barrens and snowflush communities and occur on upland plateaus above 200 m. Plant cover in herb barrens varies with

FIGURE 5.3. Forest dominated by *Picea mariana* surrounding a sedge bog in southern Ontario.

the availability of soil; in rocky areas cover may be less than 5%. Common vascular plants include *Draba corymbosa, D. subcapitata, Minuartia rubella, Papaver radicatum, Puccinellia angustata,* and *Saxifraga oppositifolia.* Many of these areas may have been vegetated only since the Little Ice Age (400–130 yr B.P.). Snow-flush communities occur where snow meltwater is present all summer. These sites vary considerably in species composition and are characterized by greater plant cover and greater species diversity than are adjacent barrens.

Boreal Forests

Overview

Boreal forest or taiga is typified by relatively dense stands of evergreen coniferous trees of modest stature and comparatively low species diversity (fig. 5.3). The boreal forest formation is actually composed of a complex of plant communities that include bogs and meadows in addition to forest stands. This formation extends as a continuous band across North America, coinciding with H. Walter's (1979) Zonobiome VIII. Boreal forest reaches its southernmost extent in southeastern Canada and the Great Lakes states, and extends northwest into central Alaska. It commonly spans over 10° latitude and dominates 28% of the North America landmass north of Mexico, making it the most extensive formation in North America.

The northern transition from boreal forest to tundra is probably related to short growing seasons (fewer than 90 days) and permafrost conditions unfavorable to tree growth (D. L. Elliott and S. K. Short 1979; R. A. Black and L. C. Bliss 1978, 1980; D. L. Elliott-Fisk 1983). In the western Canadian Rocky Mountains, boreal forest intergrades and shares many species with montane coniferous ecosystems. In southern Manitoba and western Ontario, taiga borders on prairie. Across the Great Lake states and southern Canada, the boreal forest borders on mixed deciduous forest.

Conifers *(Abies, Larix, Picea,* and *Pinus)* comprise the bulk of taiga biomass. Most ecosystems, however, also include a variety of deciduous tree, shrub (especially Ericaceae), and herb species. In general, species diversity within individual communities increases with the length of the growing season, favorable drainage, and increasing soil fertility (D. L. Elliott-Fisk 1988).

The modern boreal flora is generally thought to be derived largely from the Arcto-Tertiary geoflora (J. A. Larsen 1980). Nearly all of the extant tree genera were present in late Tertiary times. Most of the region now occupied by taiga was glaciated repeatedly during Quaternary times, when boreal taxa were found as far south as the Carolinas and the lower Mississippi Embayment

(P. A. Delcourt et al. 1980; W. A. Watts 1980). Boreal forest is currently distributed across terrain that was covered by continental glaciers just 14,000 years ago. These landscapes have supported boreal forests for only 3000–6000 years (J. C. Ritchie and G. A. Yarranton 1978). D. L. Elliott-Fisk (1988) attributed the relative impoverishment of the boreal flora in part to Quaternary disturbance and the youth of the landscape.

Human impacts on boreal forest landscapes include timber and mining activities, alteration of fire events, and hydrologic modification. In more northern regions, destruction of permafrost accompanying human land use has resulted in landscape degradation. Nevertheless, extensive vigorous tracts of boreal forest remain and provide a major renewable timber resource. Because of the limited economic value of boreal wetlands, many of these landscapes have remained virtually unaffected by human activities (D. L. Elliott-Fisk 1988).

Boreal plants must be able to tolerate extremes of environment. Summers are cool and winters are extremely cold. Annual precipitation varies between 300 and 900 mm, with most falling during the summer. On average, the precipitation/evapotranspiration ratio (P/E) is greater than 1.0, but high summer evapotranspiration, coupled with porous soils, can result in local water deficits during the growing season. Nevertheless, water stress generally is not a limiting factor in boreal forests (W. C. Oechel and W. T. Lawrence 1985). H. Walter et al. (1975) found that the distribution of boreal forest was most strongly correlated with growing season length (number of days with a mean temperature greater than 10° C). At the taiga-tundra ecotone, the growing season is less than 90 days, whereas the growing season is greater than 120 days at the transition to temperate ecosystems.

Local distribution of boreal communities is heavily influenced by topography and soils. Glacial features such as kettle holes, eskers, moraines, and outwash plains create considerable topographic and soil variation at nearly all spatial scales. Spodosols (podzols), i.e., soils with an infertile, leached A horizon and a well-defined B horizon high in iron and aluminum, dominate upland areas. On glacial tills and outwash plains, clay layers may impede drainage and greatly influence forest composition (M. R. Roberts and N. L. Christensen 1988). Histosols (organic soils) are prevalent in wetlands.

In northern areas, permanently frozen ground or permafrost significantly influences vegetation distribution. The depth of thaw determines the volume of soil available for root growth and resource exploitation. Vegetation changes that affect gain and loss of soil heat can result in considerable changes in permafrost distribution (W. C. Oechel and W. T. Lawrence 1985).

Many boreal forest species are able to carry out net photosynthesis over a range of temperatures from less than 0° C to greater than 40° C (T. Vowinckel et al. 1975). Such wide temperature tolerance is due in part to acclimation during the growing season (W. C. Oechel and W. T. Lawrence 1985). Rates of photosynthesis are generally greater for deciduous taxa than for evergreen ones. Evergreen species, however, are generally able to photosynthesize at lower temperatures and thus experience a longer growing season.

Net primary production (NPP) on upland sites varies from 480 $g/m^2/yr$ on sterile sandy soils to 1280 $g/m^2/yr$ on mesic fertile sites (J. S. Olson 1971). In black spruce muskegs, NPP may be less than 130 $g/m^2/yr$ (K. Van Cleve et al. 1983). Nutrients, particularly nitrogen and phosphorus, are limited in most boreal ecosystems. This may be in part responsible for the prevalence of evergreen taxa (F. S. Chapin III 1980). Nearly all boreal forest vascular plants possess mycorrhizae that aid in nutrient uptake (D. Malloch and B. Malloch 1981).

Regional Variation

J. A. Larsen (1980) divided the North American taiga into seven regional zones based on physiography and variations in the relative importance of dominant species and associations. The Alaskan region (zone 1) is dominated by *Picea glauca* on mesic sites and *P. mariana* on poorly drained sites. Abundant fire and floodplain disturbance result in a complex mosaic of successional types with abundant *Alnus crispa, Populus balsamifera, Betula papyrifera,* and *Salix* spp. The Cordilleran region (zone 2) is influenced by the diverse topography of the northern Rocky Mountains. *Picea mariana*–dominated forests are found here primarily on north-facing slopes. The interior region (zone 3) extends from the Cordilleran foothills to the edge of the Canadian Shield. In addition to *Picea* spp., *Pinus contorta* var. *latifolia* and *Pinus banksiana* are important. On the Canadian Shield (zone 4), *Picea mariana* forests characterize the north and grade into a mixture of forest types to the south. *Abies balsamea* becomes increasingly important here. J. S. Rowe (1977) identified 18 different boreal associations in J. A. Larsen's Gaspé-Maritime region (zone 5). *Abies balsamea* is often dominant, and pure stands of *Picea mariana* are rare. The Labrador-Ungava region (zone 6) is characterized by vegetation continua associated with moisture and geologic gradients. Species of *Abies balsamea, Betula papyrifera, Larix laricina, Picea* spp., *Pinus banksiana, Populus balsamifera, P. tremuloides,* and *Thuja occidentalis* are important. Along the northern boreal boundary (zone 7), *Picea mariana* is dominant, with *P. glauca* occurring on well-drained sites and *Larix laricina* in wet areas.

D. L. Elliott-Fisk (1988) classified upland ecosystems into closed forest, lichen woodland, and tundra-forest ecotone. Spruce-feathermoss forests are the most widespread closed forest type. Either *Picea glauca* or *P. mariana* may dominate the canopy of such forests, with a nearly continuous bryophyte stratum of *Hylocomium splendens* and *Pleurozium schreberi.* The flora of understory vascular plants is most diverse beneath *Picea glauca–Abies balsamea*–dominated canopies. Lichen woodlands are characterized by scattered individuals of *Picea glauca* or *P. mariana* and an understory covered by the lichens *Stereocaulon paschale* or *Cladonia stellaris* (K. A. Kershaw 1977). These woodlands are considered by some to be successional following fire (J. C. Ritchie 1962; K. A. Kershaw 1977) and by others to be stable climaxes (J. A. Larsen 1980; D. L. Elliott-Fisk 1983).

The tundra-forest ecotone is composed of scattered *Picea mariana,* sometimes occurring in clumps, within a surrounding matrix of tundra. This ecotone is relatively narrow in the east and west, but it is 200–300 km wide in central Canada (D. L. Elliott-Fisk 1988). Although the general location of this ecotone is certainly related to climate (D. Löve 1970), tree distribution in this ecotone is also influenced by fire (R. A. Black and L. C. Bliss 1978; D. L. Elliott-Fisk 1983) and hydrology (D. L. Elliott-Fisk 1988).

Shrublands dominated by *Alnus crispa, Betula glandulosa,* and *Salix* spp. are common throughout the taiga. Near the northern treeline, such communities may be stable and more properly classified as tundra, but in other areas they are often successional to forest (D. L. Elliott-Fisk 1988).

Wetlands constitute a major part of the boreal landscape. Ombrotrophic bogs—peatlands that receive nutrient inputs only from the atmosphere—are dominated by *Sphagnum* and a diverse assemblage of Ericaceous shrubs. Herbs in the Asteraceae, Cyperaceae, Orchidaceae, and Poaceae are also characteristic of such wetlands (J. A. Larsen 1982). *Picea mariana* and *Larix laricina* are the most common tree species in such bogs. Fen-type wetlands are considerably less acidic and more nutrient enriched. They share many tree species in common with bogs, but little overlap occurs in their herbaceous flora (J. A. Larsen 1982).

Eastern Deciduous Forests

Overview

The eastern deciduous forest province contains a diverse array of forests dominated by winter-deciduous trees (figs. 5.4, 5.5). The closed canopies of these forests reach to a height of over 25 m, with a complex understory. The broad northern transition of deciduous forest to taiga begins at about 45° N latitude. The western grassland–deciduous forest ecotone is often abrupt, but it meanders between 90° W and 100° W longitude. The eastern deciduous forest occupies approximately 11% of the continent and straddles H. Walter's (1979) Zonobiomes V (warm temperate maritime) and VI (temperate with a cold winter—nemoral). In the southeastern coastal plain, coniferous and broad-leaved evergreen trees become more common. This region is discussed separately.

The northern transition from deciduous to boreal forest is thought to be a consequence of decreasing growing season length. Growing seasons must be sufficiently long to allow deciduous species to produce enough photosynthate to amortize the cost of new leaves each year (D. J. Hicks and B. F. Chabot 1985). The infertile soils typical of boreal climes probably also favor evergreen conifers relative to deciduous trees (F. S. Chapin III 1980). In general, deciduous forest gives way to grassland as water deficits increase to the west. In Minnesota, where cooler temperatures result in lower evapotranspiration, about 60 cm of rain is sufficient to support deciduous forest. In Texas and Oklahoma, however, deciduous forests require 90–100 cm of rain because of higher temperatures, greater evapotranspiration, and longer growing seasons. Local variations in hydrology and soil characteristics also affect water availability and the location of this ecotone (E. L. Braun 1950). Finally, animal grazing and fire have clearly influenced the distribution of deciduous forest relative to grassland (A. M. Greller 1988).

During the Tertiary, the progenitors of modern deciduous forest taxa were distributed across what are now boreal regions of North America (J. A. Wolfe 1978). The eastern deciduous forest is thought to have become restricted to its current range during the Quaternary. Repeated Pleistocene glaciations resulted in extensive readjustment of deciduous forest ecotones. For example, during full-glacial periods, deciduous forest species were confined to favorable sites along the Gulf region (P. A. Delcourt and H. R. Delcourt 1977, 1980). Considerable variation existed among deciduous forest species with respect to rates and routes of migration during the last glacial retreat, beginning 14,000 yr B.P. (M. B. Davis 1981, 1983). Indeed, it appears that migrations of some species are still continuing (M. B. Davis 1987).

Very little of the eastern deciduous forest has escaped human impact (J. Bakeless 1961; M. Williams 1989). Native Americans may have influenced forest structure in many areas by burning the understory vegetation to improve browse for game (S. J. Pyne 1982); the extent of such burning is a matter of debate (E. W.

FIGURE 5.4. Mixed mesophytic forest dominated by *Liriodendron tulipifera, Tilia,* and *Magnolia* in the southern Blue Ridge Mountains of North Carolina.

B. Russell 1983). With European colonization, the frequency and intensity of fires increased in many locations. Most arable land was cleared for agriculture, and wooded areas were selectively cut and used for livestock grazing (W. Cronon 1983; A. E. Cowdrey 1983).

Widespread land abandonment in many regions over the past century has created a landscape mosaic of successional fields (for a historical account see N. L. Christensen 1989). Abandoned fields are initially invaded by a predictable sequence of herbs, among which the Asteraceae and Poaceae are prominent. Patterns of tree invasion vary by region. In the northeast, successional forests are dominated by a mixture of hardwoods. In the west and middle Atlantic region, *Juniperus virginiana* is prominent, and in the southeast, *Pinus echinata, P. taeda,* and *P. virginiana* are widespread in old fields. These successional forests have become important economic resources in many areas.

Introduction of exotic species has had a very significant impact on the North American deciduous forest. For example, *Ailanthus altissima,* a native of southern China, is a widespread invader of disturbed areas. Honeysuckle *(Lonicera japonica)* and kudzu *(Pueraria lobata)* are other examples of introduced pernicious weeds (R. L. Stuckey and T. M. Barkley, chap. 8). Introduced plant diseases such as the chestnut blight and Dutch elm disease have greatly altered forest structure and composition throughout this region (E. L. Braun 1950).

Within the deciduous forest region are strong latitudinal and longitudinal climatic gradients. In the south, mean annual temperatures approach 19° C, rainfall may exceed 1400 mm, and the frost-free growing season often exceeds 280 days. In the mixed mesophytic forest region near the southern Appalachians, annual rainfall averages more than 2000 mm. Net primary production

FIGURE 5.5. Forests dominated by oak-hickory in the Piedmont region of North Carolina.

is highest in this region, exceeding 2500 g/m^2/yr on fertile soils (D. L. DeAngelis et al. 1981). In general, rainfall declines to the north and west across this province. For example, at the deciduous forest–grassland boundary in Minnesota, mean annual temperature is 8°–9° C, annual precipitation is 600–700 mm, and the frost-free growing season is generally less than 150 days. On dry, infertile sites, net primary production is only 1200–1500 g/m^2/yr (D. L. DeAngelis et al. 1981).

Soils vary considerably with respect to climatic gradients and the character of parent rocks. In general, the soils (Inceptisols, Spodosols, and Alfisols) derived from glacial tills and outwash in the northern portion of this region are considerably less fertile than the ancient, residual soils (mostly Ultisols) to the south (D. Steila 1976; see also D. Steila, chap. 2).

Regional Variation

The eastern deciduous forest biome can be divided into eight regions or associations as indicated in figure 5.1, based on variations in species composition and physiography (E. L. Braun 1950; A. M. Greller 1988). The mixed mesophytic forest represents the center of diversity for deciduous forest vegetation (fig. 5.4). The forest canopy includes a diverse array of taxa including *Fagus grandifolia, Liriodendron tulipifera, Acer saccharum,* and various species of *Tilia, Aesculus, Quercus, Fraxinus,* and *Carya. Castanea dentata* was a prominent member of this community prior to the chestnut blight epidemic. These forests also possess very diverse herb and shrub floras.

S. A. Cain (1944) and E. L. Braun (1950) proposed that parts of this region served as refugia for this association during Pleistocene glaciations and that the relative lack of disturbance accounted for their high diversity. We now know, however, that most deciduous forest taxa were displaced hundreds of kilometers farther south during the last ice age (M. B. Davis 1983), and there is no evidence that they did survive in the areas where they are so richly represented now. Reliable rainfall and fertile soils may better explain the complexity of these communities (P. A. Delcourt and H. R. Delcourt 1987). *Quercus* and *Carya* spp. increase in dominance, and overall species richness declines in the western mesophytic region, probably in response to decreasing rainfall.

The oak-hickory association extends westward to the

grassland ecotone (fig. 5.5). In northern areas these forests may be savannalike, dominated by *Quercus macrocarpa*. Farther south, *Quercus stellata, Q. macrocarpa,* and *Carya* spp. dominate. J. L. Vankat (1979) combined E. L. Braun's oak-pine-hickory association with the oak-hickory. He argued that the prevalence of pines *(Pinus echinata, P. taeda,* and *P. virginiana)* in this region, which includes the southeastern Piedmont, is largely related to patterns of human-caused disturbance and old-field succession. *Quercus alba, Q. rubra, Q. velutina,* and *Carya* spp. dominate relatively undisturbed sites in this area (H. J. Oosting 1942).

Because of the chestnut blight, the composition of the oak-chestnut association has changed considerably (E. W. B. Russell 1987). Presettlement forests were dominated by *Castanea dentata, Quercus prinus, Q. rubra,* and *Carya* spp. In the understory *Rhododendron* spp. were often abundant. Loss of chestnut from the canopy has resulted in increased dominance of *Carya* and *Quercus* spp. (C. Keever 1953; J. F. McCormick and R. B. Platt 1980; S. L. Stephenson 1986).

The beech-maple association predominates on glaciated terrain in Ohio, western Indiana, and southern Michigan. *Fagus grandifolia* and *Acer saccharum* are canopy dominants, often growing with lesser numbers of *Fraxinus* and *Ulmus* spp. *Fagus grandifolia* is replaced by *Tilia americana* west of Lake Michigan in the maple-basswood association. Throughout this region *Quercus* and *Carya* spp. are important on drier sites, while *Fraxinus, Ulmus,* and *Acer rubrum* are important on the wettest soils (A. M. Greller 1988).

The northern hardwoods association is transitional to boreal forest. This association consists of a mosaic of community types largely distributed in relation to soil characteristics and patterns of disturbance. On the most fertile soils in the absence of fire or other disturbance, mixed conifer–hardwood forest, dominated by *Acer saccharum, Fagus grandifolia, Tilia americana,* and *Tsuga canadensis,* occur. On sandy, less fertile soils *Pinus strobus* and *P. resinosa* are canopy dominants, and on the driest, least fertile soils *Pinus banksiana* forms dense, even-aged stands. This latter community type is maintained by frequent, often intense wildfires (H. L. Hansen et al. 1974). See R. S. Rogers (1980, 1981) for a description of the understory vegetation of these forests. Toward the northern portion of this transition, boreal species such as *Picea glauca* (in the west) and *Picea rubens* (in the east) become more important. *Picea mariana* and *Larix laricina* dominate boggy areas, which may be abundant on this landscape. Widespread cutting and burning in many areas have resulted in extensive stands of successional taxa such as *Betula, Populus tremuloides, P. grandidentata,* and *Prunus pensylvanica*.

High Elevation Appalachian Ecosystems

Overview

Higher elevations of the Appalachian Mountains support a unique assemblage of subalpine and alpine vegetation types distributed in an ecological "island archipelago" from New England to southwestern North Carolina. The upper elevational limit for deciduous forest is approximately 760 m in the Green Mountains of Vermont and White Mountains of New Hampshire. The transition to subalpine vegetation rises to 1280 m in the Catskill Mountains of New York. To the south the highest elevations of the Appalachian complex are below the subalpine transition until one reaches the Allegheny Mountains of West Virginia. Farther south in Tennessee and North Carolina, the Blue Ridge and Unaka mountains (including the Great Smoky Mountains) rise well above the 1400–1500 m elevational limit of deciduous forest.

Spruce-fir forests

As one nears the elevational limit of deciduous forest, red spruce *(Picea rubens)* increases in importance and forms dense, relatively even-aged stands above this transition. At higher elevations *Abies* increases in importance and may form pure stands if elevations are sufficient. Other commonly associated species include *Betula allegheniensis, B. lenta, B. papyrifera* (north), *Sorbus americana, Vaccinium* spp., and *Viburnum* spp. Within each region a variety of subtypes have been identified based on subordinate vegetation (D. L. Crandall 1958; E. L. Core 1966). H. J. Oosting and W. D. Billings (1951) argued that, despite the north-to-south elevational shift in the spruce-fir zone, the climatic conditions were generally similar, i.e., generally harsh winters and short frost-free growing seasons (less than 120 days). They noted, however, that temperature extremes and snowfall were less, and humidity and growing season precipitation were greater, in the southern Appalachians than in the mountains of New England. Soils in all areas are usually relatively infertile spodosols, but organic horizons are generally thicker at northern sites (H. J. Oosting and W. D. Billings 1951).

Clear floristic differences exist between spruce-fir forests in the southern Appalachians and those from West Virginia northward. In the south the fir species is the endemic *Abies fraseri,* whereas *Abies balsamea* dominates to the north. H. J. Oosting and W. D. Billings (1951) identified a number of bryophyte, herb, and shrub taxa that were also unique to each of these areas. Palynological evidence suggests that many species now unique to northern areas were present in the southern Appalachians during the last ice age. Thus these floristic variations probably reflect climatic conditions unique to each of these regions rather than evolutionary change due solely to geographic isolation.

Natural disturbance is important in Appalachian spruce-fir landscapes. Fungal pathogens and insects such as the spruce budworm cause high tree mortality in some places and account for the relative abundance of successional species such as *Betula* spp., *Prunus pensylvanica,* and *Sorbus americana.* At higher elevations wind and ice damage may create "waves" of tree mortality, producing a mosaic of different age trees (D. G. Sprugel 1976).

Human impacts have greatly altered the structure of Appalachian spruce-fir forests. Logging and human-set fires resulted in the complete loss of many southern forests early in this century (C. F. Korstian 1937). More recently, introduction of an exotic insect, the balsam woolly aedelgid, has resulted in rapid decline of *Abies fraseri* throughout its range. It is generally agreed that throughout the Appalachians, spruce-fir ecosystems have been hard hit by acid rain and other atmospheric pollutants (National Research Council 1989).

Alpine Tundra

True timberlines, i.e., elevations above which trees cannot establish and grow, occur only on the summits of the White Mountains of New Hampshire. Although the elevation is only 1900 m, the environment here is among the harshest of any alpine habitat (W. D. Billings 1988). These ecosystems have a high (70%) floristic affinity to arctic ecosystems even though they are isolated from the Arctic by over 1000 km (L. C. Bliss 1963; W. D. Billings 1988).

In the most exposed snow-free areas *Diapensia lapponica* dominates with *Juncus trifidis.* Sedge meadows dominated by *Carex bigelowii* are common in snowy and foggy locations. Other important species include *Arenaria groenlandica, Ledum groenlandicum, Vaccinium vitis-idaea,* and *V. uliginosum.*

Grass and Heath Balds

Grass and heath balds represent the only treeless upland vegetation types in the southern Appalachians. Unique soil and historical factors appear to be much more important than elevation in the distribution of these ecosystems.

Grass balds are most common on mountaintops and ridges between 1700 and 1800 m (D. M. Brown 1941; A. F. Mark 1958). Lower elevation grass-dominated areas were likely created and maintained by human disturbance in the past two centuries (P. J. Gersmehl 1971). Dominant plants in high elevation balds include *Danthonia compressa, Potentilla canadensis,* and *Rumex acetosella.* Mosses such as *Polytrichum commune* and *P. juniperinum* are especially common in some locations (A. F. Mark 1959).

A. F. Mark (1958) argued that high elevation balds were created simultaneously as spruce-fir forests were displaced from mountaintops by climatic warming 8000–5000 years ago. Subsequent cooling resulted in retreat of deciduous forest species. Poor seed dispersal and competition from the already established herbs, however, have prevented reinvasion by spruce and fir. Shallow soils, fire, and animal grazing may also play roles in exclusion of spruce and fir.

Heath balds are closed-canopy shrub communities dominated by *Kalmia latifolia, Leiophyllum lyonii, Rhododendron catawbiense,* and *R. maximum* (R. H. Whittaker 1979). They are characteristic of ridges with shallow soils between 1400 and 2000 m elevation. S. A. Cain (1930) suggested that such shrublands were produced and maintained by disturbances such as fire or landslide. Whittaker argued that they were "topographic climaxes" dominated by plants adapted to shallow infertile soils on steep slopes.

Southeastern Coastal Plain Ecosystems

Overview

The southeastern coastal plain is composed of alluvial and marine sediments deposited since Mesozoic times and considerably reworked during the Pleistocene by fluvial and coastal geomorphic processes (G. E. Murray 1961). It covers 3% of the North American landscape. This province is often classified as the southern mixed hardwood region of the deciduous forest, based on the assumption that successional change in the absence of disturbance such as fire would result in a climax forest dominated by a mixture of deciduous and evergreen angiosperms and evergreen gymnosperms (E. L. Braun 1950; E. Quarterman and C. Keever 1962). N. L. Christensen (1988) argued that the unique soils, topography, climate, and disturbance regime of the coastal plain make succession to such vegetation unlikely in many locations. Instead, upland sites are normally vegetated by pine-dominated forests, and low sites support a diverse array of paludal and alluvial wetlands (figs. 5.6, 5.7).

Vegetation composition of the coastal plain varied considerably during the Pleistocene. During full-glacial periods the drop in sea level exposed vast areas of the Continental Shelf for colonization by terrestrial plants. Peninsular Florida, for example, was 600 km wide at the peak of the last ice age. During several of the past interglacial periods, sea level may have been 85 m higher

FIGURE 5.6. Fire-maintained pine flatwoods dominated by *Pinus elliottii* and *Serenoa repens* in central Florida.

than at present (G. E. Murray 1961; T. M. Cronin et al. 1981).

The advent of Native American people on the coastal plain appears to coincide with the close of the last ice age (C. Hudson 1976). Intensive land use by Indians was greatest on river flood plains, near the coast, and around swamp complexes. Indian-set fires probably had a major impact on the entire landscape (A. E. Cowdrey 1983; N. L. Christensen 1988).

European colonization of continental North America began in the coastal plain. Flood plains and many upland areas were cleared for agriculture. The abundant pine forests spawned a timber and naval stores industry that continues to the present (A. E. Cowdrey 1983). Like their Native American predecessors, European colonists were quick to understand the need for frequent fires to maintain many coastal plain communities (S. J. Pyne 1982).

Upland Pine Forests

We divide the upland pine forests into three general types: northern pine barrens, xeric sand communities, and mesic pine communities. Northern pine barrens are confined to the coastal plain north of Delaware Bay. Pine-oak forests dominated by *Pinus rigida*, *Quercus stellata*, and *Q. marilandica* dominate the most favorable sites. The relative importance of oak species increases as fire frequency decreases (R. T. T. Forman and R. E. Boerner 1981). On coarse-textured soils and with increasing fire frequency, tall-stature (greater than 10 m) *Pinus rigida* trees form a closed canopy with an understory of *Gaylussacia baccata*, *Quercus* spp., and *Vaccinium* spp. Frequent fires maintain the dwarf pine plains, dominated by shrubby ecotypes of *Pinus rigida*, *Quercus ilicifolia*, and *Q. marilandica*. These communities have floristic affinities with the southern Appa-

FIGURE 5.7. Cypress *(Taxodium distichum* var. *imbricarium)* forest growing in a shallow lake on the South Carolina coastal plain.

lachians (e.g., *Comptonia peregrina, Kalmia latifolia,* and *Quercus ilicifolia*) as well as the southeastern coastal plain (e.g., *Clethra alnifolia, Gaylussacia dumosa, G. frondosa,* and *Ilex glabra*).

Well-drained coarse sands provide xeric habitats on much of the southern coastal plain despite abundant rainfall. The xeric sand communities that occupy such sites include sandhill pine forest and sand pine scrub. Sandhill pine forest occurs throughout this region and is dominated by a broken overstory of *Pinus palustris.* Understory trees include *Diospyros virginiana, Nyssa sylvatica* var. *sylvatica,* and *Quercus laevis. Aristida stricta* often dominates the herb layer, although on the most sterile sands, lichens and mosses provide the greatest ground cover. Frequent surface fires (every 3–6 years) maintain the open structure of these forests and are essential to the reproduction of *Pinus palustris* (N. L. Christensen 1988). In peninsular Florida where fires occur at less frequent intervals (every 30–50 years), sand pine scrub is the predominant vegetation. Here, *Pinus clausa* is the dominant canopy species, and a diverse array of shrubs and low palms dominate the understory. In the absence of fire, these forests slowly succeed to xeric oak woods, or "hammocks," dominated by *Quercus virginiana* (R. L. Myers 1985).

Mesic pine communities include so-called "flatwoods" and "savannas." Flatwoods are generally dominated by an even-aged closed canopy of *Pinus palustris, P. taeda, P. serotina,* and/or *P. elliottii* (fig. 5.6). The relative frequency of these pine species is dependent on variations in site fertility, drainage, and fire history (C. D. Monk 1968). Understory vegetation in these forests often includes abundant shrub species such as *Ilex glabra, Myrica* spp., and *Serenoa repens.* Pine savannas occur most frequently at transitions between xeric pine communities and wetlands. Depending on hydrologic conditions and fire frequency, these savannas are characterized by a very open canopy of *Pinus palustris, P. taeda,* or *P. serotina.* The understory herb layer includes a very diverse assemblage of herbs and forbs, all of which are adapted to frequent surface fires. Abundant insectivorous plants (species of *Sarracenia, Pinguicula, Drosera,* and *Dionaea muscipula*) testify to the low availability of nutrients in savanna soils (N. L. Christensen 1988).

Upland Hardwood Forests

In the absence of fire, broadleaf forests do indeed dominate some coastal plain sites. On sandy soils in Florida and across the Gulf, oak hammocks with an overstory of *Quercus virginiana* occur. These communities often have a shrubby understory that includes *Serenoa repens, Ilex, Quercus,* and *Lyonia.* On more mesic sites, particularly on fertile soils derived from limestone or fine-textured sediments, forests dominated by a mixture of evergreen and deciduous trees are prevalent. In Florida and along the Gulf, *Magnolia grandiflora* and *Fagus grandifolia* are prominent overstory trees, along with *Quercus* spp., *Liquidambar styraciflua,* and *Pinus taeda.* In the Carolinas such forests are dominated by *Carya* spp., *Fagus grandifolia, Quercus alba,* and *Q. laurifolia* (E. Quarterman and C. Keever 1962).

Alluvial Wetlands

Subdued topography and complex drainage patterns result in an abundance of wetlands on coastal plains landscapes. Wetlands occupy over 10^7 ha, or 15% of the total land surface of the southeastern coastal plain, and roughly half of these ecosystems are alluvial or associated with streams and rivers. Vegetation varies in relation to frequency of flooding, which affects such ecologically important factors as soil aeration and water and nutrient availability. R. T. Huffman and S. W. Forsythe (1981) proposed a zonal classification of alluvial wetland vegetation based on inundation frequency.

Zone I plant communities occupy permanent water courses and impounded areas. Where water flow is rapid, submerged aquatic plants with streamlined leaves, such as *Alternanthera philoxeroides,* are common. Where flow is sluggish, emergent broad-leaved taxa such as *Pontederia* and *Sagittaria* spp. and floating aquatics such as *Azolla, Lemna,* and *Spirodela* are common.

Zone II, river swamp forests, occupy river flood plains that are inundated for much of the year but may be exposed during the growing season. *Taxodium distichum* var. *distichum* is the most typical tree in this zone. *Taxodium distichum* var. *imbricarium* (= *T. ascendens*) is more common on sandy substrates and in impounded areas such as oxbow lakes (fig. 5.7). *Chamaecyparis thyoides* and various species of *Nyssa* codominate in many locations. In less frequently inundated portions of this zone, *Acer rubrum, Liquidambar styraciflua, Quercus laurifolia,* and *Ulmus americana* are important.

Zone III, or lower hardwood swamp forest, soils are saturated 40–50% of the year. Plants in this zone must tolerate inundation in the early part of the growing season and water deficits in the late summer. Nutrient subsidization from alluvial sediments contributes to the high fertility and diversity of these forests. Dominant trees include *Quercus lyrata* and *Carya aquatica. Cornus foemina, Gleditsia aquatica, Ilex verticillata,* and *Itea virginica* are common understory associates.

Zone IV forests occur in backwater and flat areas that are inundated only 20–30% of the year, and very infrequently during the growing season. *Quercus laurifolia, Q. phellos, Fraxinus pennsylvanica,* and *Ulmus americana* typify canopy trees in this zone; *Carpinus caroliniana, Ilex decidua, Crataegus* spp., and *Sabal minor* are common understory associates.

Zone V, transition to upland, comprises the highest locations of the active flood plain and includes levees, higher terraces and flats, and sand ridges and dunes. Soils here are generally saturated less than 15% of the year. Dominant trees include *Quercus michauxii, Q. falcata,* and *Q. nigra* on fine-textured soils, and *Quercus virginiana* on sandy sites. Understory trees and shrubs include *Ilex opaca, Asimina triloba, Lindera benzoin, Serenoa repens,* and *Sabal palmetto.*

Paludal Wetlands

Because input of water to paludal (nonalluvial) wetlands is primarily from rainfall or groundwater flow, plant production is often nutrient limited. Graminoid-dominated wetlands or wet "prairies" are maintained by frequent disturbance such as fire. Dominant genera in these habitats include *Carex, Cladium, Juncus, Muhlenbergia, Panicum, Rhynchospora,* and *Scirpus.*

Poor drainage in flat interstream areas results in peat accumulation and bog development over thousands of hectares. Pocosins, or shrub bog vegetation, are characteristic of the centers of such bogs (N. L. Christensen et al. 1981; R. R. Sharitz and J. W. Gibbons 1982). These peatlands typically have a dense impenetrable cover of deciduous and evergreen shrubs with scattered emergent *Pinus serotina.* Shrub diversity may be quite high (20 species/ha) with *Cyrilla racemiflora, Ilex glabra, Lyonia lucida,* and *Zenobia pulverulenta* being most abundant. In the most nutrient-limited areas, shrub height may be less than 1 m. Toward the bog margins on thinner peats, nutrient availability and tree and shrub height increase, with little change in species composition. Fires occur in these ecosystems at intervals of 30–50 years.

Bay forests are common at the margins of such bog complexes and on shallow peats. They also occur in shallow depressions such as lime sinks. These forests are dominated by broad-leaved evergreen trees such as *Gordonia lasianthus*, *Magnolia virginiana*, and *Persea borbonia*. Understory species include many of the shrubs found in pocosins. Atlantic white cedar swamp forests, dominated by *Chamaecyparis thyoides*, occur in similar habitats but are maintained by fires at 80–100 year intervals.

Cypress heads (also called "domes" or wet hammocks) occur in poorly drained depressions throughout the coastal plain. *Taxodium distichum* var. *imbricarium* (= *T. ascendens*) often forms a monotypic, relatively closed canopy. These areas are generally inundated more frequently than depressions that support bay forest. Compared to Zone I swamp forest, these areas are profoundly nutrient limited. S. Brown (1981) found that annual inputs of phosphorus to cypress heads were only one ten-thousandth that of floodplain cypress swamps.

Tropical Hardwood Hammocks

Within the contiguous United States, only at the very southern tip of Florida does the vegetation have a tropical aspect. The climate in this area corresponds to H. Walter's (1979) Zonobiome II, tropical with summer rains (A. M. Greller 1980). The vast majority of this area is vegetated by wetlands of various kinds, or upland pine forest. Tropical hardwood hammocks, however, occur where soils are well drained and fire is infrequent (R. M. Harper 1911; F. C. Craighead Sr. 1971; D. Wade et al. 1980).

Tropical hardwood hammocks vary in size from less than 0.2 ha to over 50 ha. Although abundant at one time, only a few hundred of these "islands" have survived human land development.

Broad-leaved evergreen taxa with definite tropical affinities include *Bursera simaruba*, *Ficus* spp., *Metopium toxiferum*, and *Swietenia mahogani*. Temperate species that are important in other coastal plain communities, such as *Quercus virginiana*, often occur as scattered individuals in these forests. Lianas and vines, as well as epiphytes in the Bromeliaceae and Orchidaceae, give these forests a particularly tropical appearance (F. C. Craighead Sr. 1971).

Grasslands

Overview

Grassland vegetation is dominated by herbaceous plants that form 1–2 canopy layers (fig. 5.8). Woody plants are restricted to local areas of distinctive topography, soil, or protection from fire. "Prairie" denotes relatively humid grasslands of nearly continuous cover, dominated by tallgrass species. "Plains," "high plains," or "steppe" denote more arid, open vegetation dominated by shortgrasses. The grasses are largely perennial, and they may be either rhizomatous sod-formers or nonrhizomatous bunchgrasses.

Grasslands occur in H. Walter's (1979) Zonobiome VII: regions of temperate, arid macroclimates too xeric for closed forest and too mesic and cold for desert scrub. Specifically excluded from this section are meadows within forested areas, tundra, freshwater marsh, tidal saltwater marsh, and tropical savannas.

Poaceae (along with Cyperaceae) dominate the biomass, but forbs (in particular members of the Asteraceae and Fabaceae) account for most of the species. P. G. Risser (1985) estimated that North American grasslands contain 7500 species. Ecotype differentiation among wide-ranging taxa is a common theme (C. McMillan 1959; J. A. Quinn and R. T. Ward 1969). Species richness increases with topographic diversity, precipitation, and distance from human disturbance in the past century (R. T. Coupland 1979). Few taxa are endemic to grasslands, however, and D. I. Axelrod (1985) used this fact to support his contention that grasslands are geologically recent, dating back only 7–5 million years. In addition, large areas of grassland were displaced during Pleistocene glacial advances. J. A. Young et al. (1976) have suggested that the intermountain grassland is especially fragile and easily modified by human disturbance because of its evolutionary youth.

In pristine times, grassland may have covered as much as 25% of the landmass north of Mexico (P. L. Sims 1988), but our own estimate is 21%. It is the second-largest formation in North America. Human activities, however, have reduced the extent of natural grasslands and modified their composition. Modifications resulted from removal of native grazing animals, introduction of domesticated livestock, clearing of land for agriculture, modification of the natural fire regime, and the purposeful or accidental introduction of aggressive weeds from other continents. All six major grasslands of North America were originally dominated by perennial grasses and forbs, but human activities have led to an increase in cover by such other growth forms as annuals, subshrubs, shrubs, and trees.

Although each regional grassland is floristically rich, a given local patch of grassland is dominated by only 4–5 grass species (P. L. Sims 1988). Important growth forms include: (1) perennial sod-forming (rhizomatous or stoloniferous) species; (2) nonrhizomatous or short-rhizomatous perennial bunchgrass species; (3) cool-season C_3 species active in fall, winter, and spring; (4)

FIGURE 5.8. Mixedgrass prairie in central Minnesota.

warm-season C_4 species active in summer; and (5) annuals (J. R. Carpenter 1940; P. G. Risser et al. 1981; P. L. Sims et al. 1978). These growth forms are not mutually exclusive: annuals and perennials can be either cool-season or warm-season types, and certain species can be rhizomatous in some habitats but nonrhizomatous in others.

Bunchgrasses tend to dominate the desert, intermountain, and California regions. Cool-season grasses tend to dominate north of 45° N, in regions with a July minimum cooler than 8° C (P. G. Risser 1985; J. A. Teeri and L. G. Stowe 1976), and also historically in California. The several growth forms appear to exhibit metabolic, phenologic, and morphologic behaviors that adapt them to such stresses as drought, high temperatures, nutrient limitations, grazing, and fire (P. G. Risser 1985). Bunchgrasses exhibit less resistance to intense grazing than do sod-forming grasses, and R. N. Mack and J. N. Thompson (1982) hypothesized that they evolved with lower grazing pressure than did sod-formers.

C. O. Saure (1958), among others, argued that grasslands are so diverse and so often owe their existence to edaphic or fire-related circumstances that there is no "grassland climate." In North America, grasslands occur where annual precipitation is 250–625 mm, a marked dry season of several weeks to several months in length occurs, the P/E ratio is 0.2–1.0, the annual moisture deficit is greater than 300 mm (N. L. Stephenson 1990), the growing season is accompanied by high temperatures and wind, and great variability in weather exists from year to year (P. G. Risser 1985; P. L. Sims et al. 1978). Soils tend to be heavy and deep. Natural fires from dry lightning strikes can be expected with a periodicity of 5–10+ years. Beyond this, climates do vary widely, from continental, to Mediterranean, to dry subtropical (P. L. Sims 1988). Annual net productivity, as a consequence, varies from a high of 500–1000 g/m^2 in the tallgrass prairie to 50 g/m^2 in the most arid regions.

Regional Grasslands

The central grasslands consist of three major grassland types. The tallgrass prairie, also known as the true prairie (J. E. Weaver and T. J. Fitzpatrick 1934), extends from southern Manitoba to Texas (D. D. Diamond and F. E. Smeins 1985). It is dominated by *Andropogon gerardii, Schizachyrium scoparium,* and *Sorghastrum nutans.* Common associates, often more locally restricted, include *Panicum virgatum, Sporobolus asper, S. heterolepis,* and *Stipa spartea.* These grasses are a mixture of sod-formers and bunchgrasses, with foliage 40–100+ cm tall and inflorescences up to 200+ cm tall.

Most of the tallgrass prairie is now in cultivation. Along its eastern edge, at approximately 95° W longitude, the true prairie forms a broad ecotone with two major forest types. In the north, from Saskatchewan and Manitoba to Minnesota, the ecotone with the boreal forest is an aspen parkland 75–175 km wide. *Populus tremuloides* and *P. balsamifera* are favored over grasses, apparently because of the decreasing amount of precipitation in the spring months of April through June, which has a negative effect on the reproduction of dominant grasses (J. Looman 1983).

Farther south the ecotone with the deciduous forest is an oak savanna 50–100 km wide. This ecotone also projects for a total of some 250,000 km^2 east across Illinois, to Indiana, Ohio, and Kentucky, in what has been called the prairie peninsula (E. N. Transeau 1935). Common trees in the savanna include *Acer negundo, Carya texana, Quercus macrocarpa, Q. marilandica, Q. stellata, Q. velutina,* and *Ulmus americana* (J. E. Weaver and T. J. Fitzpatrick 1934; R. C. Anderson and L. E. Brown 1986). The factors that historically have been thought to preclude tree invasion along this ecotone include high fire frequency, absence of ectomycorrhizal fungi in prairie soil (R. W. Goss 1960; D. White 1941), and drought episodes.

Consecutively farther west of the tallgrass prairie are the mixedgrass and shortgrass steppes or high plains. The mixedgrass prairie (fig. 5.8) is floristically the most complex of the central grasslands, because it is an ecotone/tension zone between the tallgrass and shortgrass prairies, thus containing taxa characteristic of both (J. S. Singh et al. 1983). As annual climate becomes relatively humid or arid, the species of tallgrass or shortgrass prairies are alternately favored (A. W. Küchler 1972). Dominants of the shortgrass prairie include *Agropyron smithii, Bouteloua gracilis, Buchloë dactyloides, Hilaria jamesii, Koeleria cristata, Muhlenbergia torreyi, Sporobolus cryptandus,* and *Stipa comata.* Warm-season grasses predominate.

The shortgrass prairie extends southward like a finger into Mexico, mainly at 1100–2500 m elevation along the flank of the Sierra Madre, from the state of Chihuahua at the United States/Mexico border to 22° N latitude in the state of Jalisco (J. Rzedowski 1978). The grasses are 20–70 cm tall. Generally, less than 50% of the ground is covered by vegetation. Dominants include *Bouteloua curtipendula, B. gracilis,* and *B. hirsuta.* Overall, the flora has a pronounced Mexican cast that includes endemic species in *Andropogon, Aristida, Erioneuron, Heteropogon, Hilaria,* and *Lycurus.* Human disturbance has led to invasion of *Acacia schaffneri, Juniperus monosperma, Opuntia* spp., *Prosopis velutina, Quercus chihuahuensis, Q. cordifolia, Q. emoryi,* and *Yucca decipiens* in parts of this central Mexican zacatal (J. Rzedowski 1978; see also S. Archer et al. [1988] for a model of *Prosopis glandulosa* invasion into southern Texas grassland).

The warm desert grassland covers plateaus greater than 1000 m elevation at the edge of the Sonoran and Chihuahuan deserts in western Texas, southern New Mexico, southeastern Arizona, and northeastern Sonora. An outlier is a shrub steppe in southeastern Utah (N. E. West 1988). In pristine times the desert grassland was dominated by short bunchgrasses such as *Bouteloua eriopoda, Hilaria belangeri, H. jamesii, H. mutica, Muhlenbergia porteri,* and *Oryzopsis hymenoides.* Within the past century, desert shrubs (mainly *Larrea divaricata* subsp. *tridentata, Flourensia cernua, Prosopis glandulosa,* and *P. velutina*) have expanded in cover, sometimes twentyfold, and have become dominants (L. C. Buffington and C. H. Herbel 1965; R. R. Humphrey and L. A. Mehrhoff 1958; J. Rzedowski 1978).

The intermountain grassland (often called a steppe or shrub steppe) extends from western Wyoming through northwestern Utah, southern Idaho, northern Nevada, and northeastern California, and into the Columbia Basin of eastern Oregon and the Palouse area of southeastern Washington. An outlier is the prairie of the Willamette Valley of Oregon and of coastal northern California, which was dominated by *Danthonia californica* in pristine times (J. F. Franklin and C. T. Dyrness 1973; H. F. Heady et al. 1988). *Artemisia* shrubs (*A. arbuscula, A. rigida, A. tridentata*) are typically associated with the cool-season bunchgrasses *Aristida longiseta, Elymus lanceolatus* (= *Agropyron dasystachyum*), *E. cinereus, Festuca idahoensis, Koeleria cristata, Pascopyrum smithii* (= *Agropyron smithii*), *Poa fendleri, P. nevadensis, P. sandbergii, Pseudoroegneria spicata* (= *Agropyron spicatum*), *Sitanion hystrix, Sporobolus airoides,* and *Stipa comata* (A. Cronquist et al. 1972+, vol. 1; J. F. Franklin and C. T. Dyrness 1973; P. L. Sims 1988; E. W. Tisdale 1986; N. E. West 1988). This grassland has been severely affected by overgrazing, burning, and the introduction of aggressive annual forbs and grasses, especially *Bromus tectorum.*

Finally the central California grassland uniquely evolved in a Mediterranean-type climate, and consequently many of its pristine dominants were endemic. This grassland has been completely modified in the past 150 years, and few relict areas remain. From limited information, W. J. Barry (1972) and H. F. Heady (1988) estimated that the original dominants at elevations below 100 m were the cool-season bunchgrasses *Stipa pulchra* and *S. cernua.* On gentle slopes just above valley floors, *Stipa* was joined by *Elymus glaucus, Festuca californica, Melica californica, M. torreyana, Muhlenbergia rigens,* and *Sitanion hystrix.* Bunchgrasses contributed 50% cover, with native annual grasses and an-

FIGURE 5.9. Sonoran Desert near Tucson, Arizona, dominated by *Carnegiea gigantea*.

nual and perennial forbs seasonally filling in the spaces between bunches. The grassland covered 13–14% of California's area, occupying much of the Sacramento Valley, the northern half of the San Joaquin Valley, and coastal valleys from Monterey to San Diego. Today, most of the land originally dominated by bunchgrass vegetation has been converted either to agriculture and urban use, or to an annual grassland dominated by Eurasian species of *Avena, Bromus, Hordeum,* and *Lolium* (R. M. Love 1975).

Desert Scrub

Overview

Deserts occupy 5% of the North American landmass north of Mexico. They occur in the southwestern part of North America at elevations below 1700 m, where annual precipitation is less than 400 mm and highly variable from year to year. On a global scale, "true deserts" are areas receiving less precipitation than 120 mm/yr (A. Shmida 1985). North American deserts receive more precipitation and qualify only as "semiarid." They are dominated by desert scrub vegetation consisting of evergreen or drought-deciduous shrubs and subshrubs that provide an interrupted cover (typically 5–25%) and have low annual net productivity (less than 100 g/m^2 [S. R. Szarek 1979]; figs. 5.9, 5.10). Associated growth forms are diverse and include phreatophytic trees (roots penetrate to ground water), stem succulents, bunchgrasses, ephemerals, and the ecophysiological C_3/C_4/CAM photosynthetic syndromes (F. S. and C. D. Crosswhite 1984; P. R. Kemp 1983; F. Shreve 1942).

Desert scrub in the intermountain region (fig. 5.11)

FIGURE 5.10. Mojave Desert in southern California.

has developed under a regime of the majority of annual precipitation falling as snow, which is unique in North American deserts. This intermountain (cold) desert falls within H. Walter's (1979) Zonobiome VII, with the temperate grasslands, whereas the warm deserts fall within his Zonobiome III, subtropical arid. Annual water budget deficits are 600–900 mm in the cold desert, but 2000–3000 mm in the warm deserts (N. L. Stephenson 1990). In the arid American West, seasonality of precipitation changes from a summer peak in the east to a winter peak in the west. Winter and summer rainfall events differ in intensity and associated temperatures. Consequently, it is not surprising that annual and perennial flora and vegetation vary with longitude (J. A. MacMahon 1979, 1988; J. A. MacMahon and F. H. Wagner 1985; F. Shreve 1942).

The easternmost and largest desert is the Chihuahuan. It has a total area of 450,000 km^2, is situated at an elevation of 1100–1500 m, and receives a mean annual precipitation of 150–400 mm, 60–80% of which falls in summer. Its soils are often rich in limestone or gypsum and often have endemic plants (J. A. MacMahon 1988). This desert includes western Texas, southern New Mexico, and portions of Chihuahua, Coahuila, Durango, Zacatecas, and San Luis Potosí, south to 22° N latitude (J. A. MacMahon and F. H. Wagner 1985; P. S. Nobel 1985).

West of the Chihuahuan Desert is the Sonoran Desert (fig. 5.9). Its low elevation (below 600 m) and southern position make it the warmest of the North American deserts. Fewer than 1% of annual hours experience temperatures lower than 0° C (M. G. Barbour, J. H. Burk, and W. D. Pitts 1987). Annual precipitation is 50–300 mm, more or less evenly divided between winter and summer. This desert includes southwestern Arizona, southeastern California, most of Baja California, and northwestern Sonora, for a total area of 300,000 km^2 (J. A. MacMahon and F. H. Wagner 1985; P. S. Nobel 1985). It has a rich and bizarre array of growth forms. The Sonoran and Chihuahuan deserts nearly meet in eastern Arizona, being separated by only 250 km of desert grassland. The two deserts exhibit high floristic similarity; more than 50% of their genera are shared (J. A. MacMahon and F. H. Wagner 1985). Overgrazing, climatic change, and fire suppression have allowed the vegetation of both deserts to expand into adjacent grassland.

North of the Sonoran Desert is the Mojave, smallest of the four deserts (140,000 km^2 [J. A. MacMahon 1988]; fig. 5.10). It includes portions of southern Cal-

FIGURE 5.11. Intermountain desert scrub dominated by *Artemisia* spp., in Yellowstone National Park, Wyoming.

ifornia, southern Nevada, southwestern Utah, and northwestern Arizona. Its position and varied topography (from below sea level to 1200 m elevation) make it a "bridge" between the warm deserts and the cold intermountain desert. Annual precipitation is 40–275 mm, most falling in winter, some as snow. It is the driest of North American deserts and has vegetation with the least cover. Frost occurs 2–5% of yearly hours (M. G. Barbour, J. H. Burk, and W. D. Pitts 1987). Despite the Mojave's ecotonal position, 25% of its total flora is endemic. Even more striking, 80% of its annual flora is endemic (J. A. MacMahon 1988).

The intermountain, or Great Basin, desert extends eastward from central Nevada through central Utah and southwestern Colorado to northern New Mexico and Arizona (fig. 5.11). Elevations are 1200–1600 m. Precipitation (100–300 mm/yr) has a winter peak, and most falls as snow (M. M. Caldwell 1985). Frost occurs in 5–20% of yearly hours (M. G. Barbour, J. H. Burk, and W. D. Pitts 1987). Succulent species are uncommon. The few *Opuntia* spp. that occur in the Great Basin are obviously frost-tolerant; P. S. Nobel (1985) has documented their degree of frost-tolerance. The portion of this arid area that extends north of 42° N latitude is more properly defined as a shrub steppe (N. E. West 1988); consequently we have described it in the grassland section as "intermountain grassland."

Two major landforms in all four deserts are bajadas (gentle alluvial slopes with coarse soil) and playas (basins with no external drainage). Bajadas support the regional, zonal community type, the biomass of which is typically dominated by C_3 and CAM (Crassulacean Acid Metabolism) syndromes (J. P. Syvertson et al. 1976). Playas support azonal halophytic plant communities, rich in C_4 taxa (J. P. Syvertson et al. 1976). Bajada and playa communities often exhibit sharp boundaries not completely explained by gradients of soil chemistry and structure (D. H. Gates et al. 1956; J. E. Mitchell et al. 1966). Other desert habitats include arroyos (washes or small canyons created by intermittent streams), which support riparian vegetation, sand dunes (accumulated from the dry beds of Pleistocene lakes), which support a highly endemic flora (J. E. Bowers 1984), and oases, which in the Sonoran Desert support striking groves of *Washingtonia filifera* (R. J. Vogl and L. T. McHargue 1966).

Some desert taxa are geologically old, and it appears that desert environments have existed since the Paleozoic (A. Shmida 1985). The prevailing opinion, however, is that modern North America deserts—and their floras—are geologically recent (D. I. Axelrod 1950, 1979).

Regional Deserts

Chihuahuan Desert vegetation can be classified into half a dozen major communities. The most common one is dominated by *Larrea divaricata* subsp. *tridentata*, associated with *Acacia neovernicosa, Agave lecheguilla, Dalea formosa, Ephedra* spp., *Flourensia cernua, Fouquieria splendens, Jatropha dioica, Koeberlinia spinosa, Krameria* spp., *Opuntia phaeacantha*, and a rich collection of *Yucca* spp. (J. A. MacMahon 1988). Density of individual perennial plants may often exceed 4000/ha, and their cover may often exceed 20%. Leaf succulents *(Agave, Dasylirion, Yucca)* dominate the visual aspect of Chihuahuan vegetation. The southern extension of the Chihuahuan Desert in Mexico deserves more attention by researchers (J. Rzedowski 1978).

Sonoran Desert vegetation is characterized by an arboreal element (fig. 5.9). The diversity of growth forms reaches its peak in Arizona uplands and central Baja California. Overstory elements include columnar cacti (e.g., *Carnegiea gigantea* and *Pachycereus pringlei*), or trees such as *Cercidium floridum, C. microphyllum, C. praecox, Olneya tesota, Pachycormus discolor*, and *Yucca valida*, the inflorescences of *Agave deserti* and *A. shawii*, and the bizarre boojum, *Fouquieria columnaris* (R. R. Humphrey 1974). Tall shrubs include *Acacia greggii, Fouquieria splendens, Krameria grayi, K. parvifolia, Larrea divaricata* subsp. *tridentata, Lycium andersonii*, and *Simmondsia chinensis*. Subshrubs and small cacti include *Encelia farinosa* and several species of *Ambrosia, Jatropha*, and *Opuntia*. The composition of the herbaceous element depends on the season. Winter annuals flower in April–May, summer annuals flower in August–September. Few species are capable of doing both. Summer and winter annuals differ in morphology, photosynthetic pathway, phenology, and length of life (J. H. Burk 1988; J. R. Ehleringer 1985; T. W. Mulroy and P. W. Rundel 1977). Total plant cover for all four canopy layers can reach 25–50%.

The Sonoran Desert is rich in Cactaceae. Ecological literature is extensive for saguaro, *Carnegiea gigantea*, which visually characterizes the Sonoran landscape. It exhibits the classic CAM syndrome of traits—shallow roots, slow growth rate, high water-use efficiency, and capability of tolerating relatively high tissue temperatures (greater than 60° C in some cases [P. S. Nobel 1985]). In addition, the saguaro has a life span greater than 200 years. Its young plants are dependent on "nurse plants" for microenvironmental protection. Present population structure suggests senescence, perhaps because of grazing pressure by rodents or larger vertebrates (W. A. Niering et al. 1963, but see J. Vandermeer 1980).

In contrast to the Chihuahuan and Sonoran deserts, the Mojave has a low, 1–2 canopy layer and a rather monotonous aspect (fig. 5.10). Perennial plant density is typically less than 2000 individuals/ha, and cover is less than 10%. The bajadas are thoroughly dominated by *Larrea divaricata* subsp. *tridentata*, associated with *Ambrosia dumosa, Grayia spinosa, Lycium* spp., *Krameria parvifolia*, and *Yucca schidigera* (N. H. Holmgren 1972; F. C. Vasek and M. G. Barbour 1988). The Joshua-tree, *Y. brevifolia*, is more characteristic of the upper elevation limits of the desert than it is of the regional bajadas. Columnar succulents are absent, but smaller conspicuous cacti include *Echinocactus polycephalus, Ferocactus cylindraceus* (= *F. acanthodes*), and *Opuntia* spp. In southern Nevada, the warm desert and cold desert plant communities meet and are differentiated sharply along subtle environmental gradients (J. C. Beatley 1975; W. H. Rickard and J. C. Beatley 1965).

Mojave Desert vegetation is remarkably sensitive to soil compaction (R. C. Stebbins 1974). Several studies have documented the process and timing of secondary succession following disturbance by off-road vehicles, military maneuvers, pipeline burial, powerline construction, and even agricultural clearing (D. E. Carpenter et al. 1986). Succession leads through a seral stage of short-lived subshrubs (species of *Acamptopappus, Gutierrezia, Haplopappus, Hymenoclea, Leucelene, Salazaria, Sphaeralcea, Thamnosma*) and ultimately back to the original dominants within 65–130(–500+) years.

T. J. Mabry et al. (1977) have summarized the biology of *Larrea divaricata* subsp. *tridentata*. This species dominates all three warm deserts. *Larrea* is not genetically homogeneous over this large range, but it exists as a distinct chromosome race and ecotype in each desert (M. G. Barbour 1969; T. W. Yang 1970). It can spread vegetatively, forming clones estimated to be several thousand years old (F. C. Vasek 1980).

Intermountain desert scrub has low species and growth-form richness (fig. 5.11). In general, *Artemisia* spp. constitute more than 70% of the cover and more than 90% of the biomass (N. E. West 1988). Species and subspecies of *Artemisia arbuscula, A. longiloba, A. nova*, and *A. tridentata* are segregated on gradients of moisture and elevation. Associated woody perennials include *Chrysothamnus nauseosus, C. viscidiflorus, Ephedra viridis, Grayia spinosa, Purshia tridentata*, and *Tetradymia glabrata*. Clearing, overgrazing, burning, and introduction of alien plants have led to significant modifications in the vegetation. Noxious or poisonous annuals, such as *Bromus tectorum, Ranunculus testiculatus, Elymus caput-medusae, Halogeton glomeratus, Onopordum acanthium, Salsola ibirica, S. kali*, and *Sisymbrium altissimum*, have become common (A. Cronquist et al. 1972+, vol. 1; N. E. West 1988; J. A. Young et al. 1988).

Alkaline areas, warmer and drier than sagebrush-dominated areas, support the C_4 shrub *Atriplex confertifolia,* associated with *Artemisia spinescens,* several other *Atriplex* spp., and a variety of halophytic or glycophytic taxa (W. D. Billings 1949; A. Cronquist et al. 1972+, vol. 1).

Playas are dominated at their least saline edge by *Sarcobatus vermiculatus.* More saline zones support *Allenrolfea occidentalis* and *Distichlis spicata* var. *stricta.* Other common halophytes include *Ceratoides* (= *Eurotia) lanata, Kochia americana, Salicornia utahensis, Suaeda fruticosa,* and *S. torreyana.* Many of these taxa occur on saline soils in the warm deserts and grasslands as well (A. Cronquist et al. 1972+, vol. 1).

Mediterranean and Madrean Scrublands and Woodlands

Overview

In North America, Mediterranean and Madrean areas replace each other along an axis from central California to central Mexico. The northwest portion of the axis is Mediterranean; the southeast portion is Madrean. They share a common flora and vegetation but experience different climates.

Mediterranean ecosystems are found in five locations throughout the world (M. L. Cody and H. A. Mooney 1978): the Mediterranean region, the Cape region of South Africa, south and southwest Australia, central Chile, and southern Oregon to northern Baja California. The regions all lie between 40° and 32° N or S latitude, occupy west or southwest edges of continents, receive 275–900 mm annual precipitation (more than two-thirds of which falls in winter), experience less than 3% of annual hours with frost, and have hot, dry summers. The vegetations in all five Mediterranean regions share many traits at a superficial level, but M. G. Barbour and R. A. Minnich (1990) summarized many important differences in both abiotic environment and vegetation. In North America, Mediterranean vegetation includes—from the most mesic to the most xeric—mixed evergreen forest, oak woodland and savanna, grassland, and several types of scrub, e.g., northern coastal scrub, chaparral, southern coastal scrub (figs. 5.12, 5.13).

Closely related Madrean vegetation extends from central Arizona to southern New Mexico, southwestern Texas, and south along the Sierra Madre Occidental to southern Durango, at about 22° N latitude (D. E. Brown 1982, 1982b; H. A. Mooney and P. C. Miller 1985; C. P. Pase and D. E. Brown 1982, 1982b; J. Rzedowski 1978). Mexico is a center of oak speciation (ca. 150 species), and oak woodlands account for 5–6% of Mexico's land area (R. Daubenmire 1978; J. Rzedowski 1978). In North America north of Mexico, however, Mediterranean and Madrean vegetations dominate only 1% of the land area.

In western North America, Mediterranean vegetation evolved from the Madro-Tertiary geoflora, which had differentiated by late Miocene (A. Graham, chap. 3). As aridity expanded in the southwestern United States, the Madrean geoflora moved north and west from central Mexico, supplanting a retreating Arcto-Tertiary geoflora. By 15–14 million years ago, a Mediterranean climate existed throughout much of low elevation California (D. I. Axelrod 1958, 1975). Evergreen elements dominate Madrean vegetation. In contrast, predominantly deciduous scrub and woodland vegetation ("petran") occurs in more northerly regions (north of 33° N latitude) in the colder areas along the east and west slopes of the Rocky Mountains. This vegetation is discussed later, at the end of the section on Intermountain Region Montane Forests.

Mixed Evergreen Forest

Mixed evergreen forest has a rather dense, species-rich overstory of hardwood and needle-leaf evergreen trees. The shrub and herb strata are depauparate. W. S. Cooper (1922) included it in his "broad sclerophyll forest formation." It is an ecotone, generally at 600–1500 m elevation, between the lower oak woodland and the higher montane conifer forest. The average annual precipitation of 800–900 mm is at the wet extreme of the Mediterranean type climate (M. G. Barbour 1988; J. Rzedowski 1978; up to 1700 mm in Oregon, according to J. F. Franklin and C. T. Dyrness 1973). This forest occurs in interior valleys of Oregon south of 43° N (J. F. Franklin and C. T. Dyrness 1973), is well represented in the Siskiyou Mountains (R. H. Whittaker 1960), extends through the Coast, Transverse, and Peninsular ranges of California, and parallels oak woodland distribution in Mexico along the middle elevation slopes of the Sierra Madre Occidental in Mexico (J. Rzedowski 1978).

In Oregon and California, *Quercus chrysolepis* is a unifying taxon for this forest. Common associates include *Acer macrophyllum, Arbutus menziesii, Calocedrus decurrens, Chrysolepis chrysophylla, Lithocarpus densiflorus, Pinus coulteri, P. ponderosa, Pseudotsuga menziesii, P. macrocarpa, Quercus kelloggii,* and *Umbellularia californica* (M. G. Barbour 1988; J. O. Sawyer et al. 1988). Many of these genera, but only a few of the species, continue into the mixed evergreen forest of central Mexico (J. Rzedowski 1978). Pines are an important element there, in particular *Pinus cooperi, P. durangensis, P. engelmannii, P. leiophylla, P. lumholtzii,* and *Pinus ponderosa* var. *arizonica* (D. E. Brown 1982).

FIGURE 5.12. Transition from California grasslands to oak woodland dominated by *Quercus douglasii* and *Q. wislizenii,* in the foothills of the central Sierra Nevada.

Oak Woodland

Oak woodland generally lies below mixed evergreen forest, but floristic elements of the two may interdigitate, cooccur, or occupy different slope aspects. Oak woodland experiences a significantly drier and warmer climate than that of mixed evergreen forest; mean annual precipitation is about 600 mm (M. G. Barbour 1988; D. E. Brown 1982b).

Oak woodland in Oregon and California is characteristically two-layered, with an overstory canopy 5–15 m tall and 30–80% closed, which is composed of evergreen and deciduous trees (fig. 5.12). The most characteristic taxa are *Quercus agrifolia, Q. douglasii, Q. engelmannii, Q. garryana, Q. kelloggii, Q. lobata,* and *Q. wislizenii,* associated with *Aesculus californica, Juglans californica,* and *Pinus sabiniana* (M. G. Barbour 1988; R. Daubenmire 1978; J. F. Franklin and C. T. Dyrness 1973; J. R. Griffin 1988). The understory is dominated by species characteristic of adjacent grassland, although there may be scattered shrubs (*Arctostaphylos* spp., *Heteromeles arbutifolia, Rhus diversiloba, Symphoricarpos albus*). Oak woodland typically exists within a complex mosaic of grassland, savanna, woodland, and chaparral, which all share the same macroclimate but differ in fire frequency, soil texture, soil depth, and slope aspect. Regeneration of the deciduous oaks *Q. douglasii, Q. engelmannii,* and *Q. lobata* has been poor for the past century, apparently because of high populations of seed predators, compaction of soil by cattle, and competition with introduced annual grasses (for example, see M. T. Borchert et al. 1989).

Madrean oak woodland in Arizona, New Mexico, Texas, and Mexico (encinal, bosque de encino) is more variable in height, has a more continuous shrub layer less than 2 m tall, and exhibits a richer diversity of vines and epiphytes than does California oak woodland. More than one-third of annual precipitation falls in summer—three times as much as in California (H. A. Mooney and P. C. Miller 1985). The vegetation occupies slopes at 900–2100 m elevation. The most characteristic oaks in Arizona and New Mexico are *Quercus arizonica, Q. emoryi, Q. gambelii, Q. grisea, Q. hypoleucoides, Q. oblongifolia,* and *Q. rugosa,* and they are associated with *Pinus cembroides, P. edulis, P. leiophylla, P. ponderosa,* and *P. strobiformis* (R. K. Peet 1988).

In Mexico, oak of varying degrees of deciduousness often cooccur; among them are white oaks *Quercus arizonica,* the closely related forms *Q. santaclarensis, Q. grisea, Q. chihuahuensis,* and *Q. chuchiuchupensis* (=

FIGURE 5.13. California chaparral dominated by *Adenostoma fasciculatum, Arctostaphylos* spp., and *Ceanothus* spp. in the San Raphael Mountains in California. Southern coastal sage scrub dominates the slopes in the distance.

Q. toumeyi), and black oaks *Q. albocincta* and *Q. durifolia* (D. E. Brown 1982b; A. S. Leopold 1950). D. E. Goldberg (1982) has decribed how evergreen and deciduous Madrean oak elements segregate along microenvironmental gradients of soil nutrient level, soil pH, and abundance of seed predators. Mexican oak woodlands have been degraded by cutting, overgrazing, and erosion, and they require serious attention by conservationists (J. Rzedowski 1978).

Chaparral

Chaparral is a dense, one-layered scrub, 1–3 m tall, composed of rigidly branched, sclerophyllous, C_3 shrubs with small leaves and extensive root systems (although the root:shoot ratio is less than one). Herbs are infrequent. Fires cycle every 25–75 years, and most woody species are capable of stump sprouting. Ground cover is close to 100% and the leaf area index = 2. Net annual productivity is 400–800 g/m^2 (T. L. Hanes 1988; J. E. Keeley and S. C. Keeley 1988; H. A. Mooney and P. C. Miller 1985). The scrub vegetation called "chaparral" may be subdivided into three different types: Californian chaparral, Madrean (or interior) chaparral, and southern coastal scrub (fig. 5.13).

Chaparral often occurs on steep slopes with coarse soil at 400–800 m elevation on coast-facing slopes, at 900–1800 m elevation for interior locations, and at 1700–2400 m elevation in Mexico (C. P. Pase and D. E. Brown 1982). Californian chaparral occurs from Baja California to southern Oregon; Madrean chaparral extends from northern Mexico to northwestern Arizona, north central New Mexico, and western Texas. Madrean and Californian chaparral are both considered to be climax types, despite the frequency of fire (C. P. Pase and D. E. Brown 1982).

Many species range throughout Californian and Madrean chaparral, but each area has its separate dominants. Californian chaparral is characterized by *Adenostoma fasciculatum, Arctostaphylos* spp., *Ceanothus* spp., *Heteromeles arbutifolia, Garrya* spp., *Quercus dumosa, Rhamnus californica, R. crocea,* and *Rhus* spp. (T. L. Hanes 1988). Chaparral in southern California and northern Baja California is more diverse in growth forms and woody species than the chaparral in northern California (H. A. Mooney and D. J. Parsons 1973). Madrean chaparral north of the international border is dominated by *Quercus turbinella,* associated with *Arctostaphylos pringlei, Ceanothus greggii, Cercocarpus betuloides, Garrya* spp., *Rhamnus* spp., *Rhus* spp., and several other scrub oaks.

In Mexico, *Quercus intricata* often dominates this chaparral, associated with *Q. pringlei* and *Q. pungens.* This Madrean chaparral is floristically and ecologically closely related to Californian chaparral, despite the fact that the climate is not Mediterranean. Madrean areas experience significant amounts of summer as well as winter rain. Consequently, J. L. Vankat (1989) suggested that "if chaparral is climatically determined, it is associated with a warm temperate climate that has seasonal drought, the timing and duration of which may not be critical."

A variation of chaparral, called southern coastal scrub, dominates slopes below chaparral in southern California and in Baja California (fig. 5.13). The adjective "coastal" is a misnomer, but it is too well accepted in the literature to replace. This scrub does occur on ocean-facing slopes, but it also covers desert-facing slopes some distance inland. The vegetation has a lower, more open canopy than chaparral, with a lower leaf area index of 1.3. The foliage is soft, pubescent, gray, and drought-deciduous, rather than hard, glabrous, and evergreen. Net annual productivity is half that of chaparral. Dominant taxa include *Artemisia californica, Encelia californica, Lotus scoparius, Eriogonum fasciculatum, Malosma laurina, Salvia* spp., *Rhamnus ilicifolia, Rhus* spp., *Toxicodendron* spp., and *Yucca whipplei.* Succulent species increase to the south, in Baja California. Coastal scrub is sometimes seral to chaparral, sometimes climax; as a consequence, typically narrow ecotones between the two may correspond either to persistent microenvironmental differences or to disturbance boundaries. The ecology, dynamics, and distribution of coastal sage scrub have most recently been described by J. T. Gray and W. H. Schlesinger (1981), H. A. Mooney (1988), and W. E. Westman (1981, 1983).

Pacific Coast Coniferous Forest

Overview

This forest has been called the most luxuriant, most productive vegetation in the world (R. Daubenmire 1978; J. F. Franklin and C. T. Dyrness 1973). It is dominated by a rich diversity of massive, long-lived tree species, underlain by equally rich canopies of shrubs, herbs, and cryptogams. It occupies a coastal, low elevation strip (0–1000 m) that extends from 36° N latitude in Monterey County, California, to 61° N latitude at Cook Inlet, Alaska, an area of about 3% of the North American landmass.

Climate is maritime, with narrow diurnal 6°–10° C (daytime maximum minus nighttime minimum) and seasonal 7°–17° C (mean of warmest month minus mean of coldest) thermoperiods. Hard frosts and persistent snow are uncommon. Studies in growth chambers with seedling conifer elements of this forest, such as *Sequoia sempervirens,* show that optimum growth occurs at zero thermoperiod (H. Hellmers 1966). Annual precipitation is 800–3000 mm, 80% or more of which falls 1 October–1 April (that is, in the cool season). Summers are relatively dry, but they are moderated by cloud cover and fog drip.

Net annual productivity is 1500–2500 g/m^2, and standing aboveground biomass is 80,000–90,000 (–230,000) g/m^2. Both ranges exceed the values for any other forest (J. F. Franklin and C. T. Dyrness 1973). Furthermore, the productivity is high despite an average age of overstory dominants at 400–1200 years. Hardwoods are minor elements in this forest, generally restricted to specialized or seral habitats, and they have been declining in diversity and abundance for at least the past 1.5 million years. Conifers account for more volume over hardwoods by 1000:1 for several reasons: moderate temperatures permit net photosynthesis outside the normal growing season (as little as 30% of annual net photosynthesis may occur during the growing season); conifer needles are tolerant of the moderately high water stresses that develop during summer (south of 51° N); conifers have a high nutrient-use efficiency, and time periods between disturbance—on the order of several centuries—suit the life cycles of populations with massive, long-lived individuals (J. F. Franklin and C. T. Dyrness 1973; J. P. Lassoie et al. 1985; N. L. Stephenson 1990; R. H. Waring and J. F. Franklin 1979).

Regional Vegetation

R. Daubenmire (1978) has broken this coastal strip, which he calls the *Tsuga heterophylla* province, into three areas and forests. The northernmost forest, from Cook Inlet to the southern tip of Alaska, has a relatively simple overstory. *Tsuga heterophylla* is the dominant species, extending from low elevations to the subalpine zone. Minor species include *Chamaecyparis nootkatensis, Picea sitchensis,* and *Tsuga mertensiana. Picea* is seral. Scattered shrubs of *Cornus canadensis, Rubus pedatus,* and *Vaccinium* spp. overlay a continuous carpet of mosses. *Sphagnum* moss is capable of invading this climax forest, raising the water table, and setting in motion a retrogressive succession back to bog and wet meadow (W. H. Drury Jr. 1956).

A central section, much richer in tree species, extends from the southern Alaska–British Columbia border to the Oregon-California border. *Tsuga heterophylla* is

FIGURE 5.14. Old-growth Pacific coast coniferous forest, dominated by large *Tsuga heterophylla* and *Pseudotsuga menziesii.* A human figure (in white/light colors) appears in the lower middle foreground, highlighted in shade. Cathedral Grove Provincial Park, British Columbia.

dominant, and important associates include *Picea sitchensis* (along the most coastward strip), *Pseudotsuga menziesii, Abies amabilis,* and *Thuja plicata* (fig. 5.14). *Taxus brevifolia* is a common subcanopy tree, and in the ground layer are many understory shrubs, ferns, herbs, bryophytes, and lichens (J. F. Franklin and C. T. Dyrness 1973; V. J. Krajina 1965). Similar vegetation extends inland, on west-facing slopes, to the central Rocky Mountains as a peninsula created by storm tracks that extend as far southeast as Yellowstone National Park and as far northeast as Jasper National Park (44–53° N latitude [R. K. Peet 1988]). On the most mesic sites—sandwiched between lower elevation *Pseudotsuga* forest and higher elevation *Picea* subalpine forest—*Tsuga heterophylla* dominates luxuriant forests, with *Thuja plicata* and *Abies grandis* as characteristic associates. Basal areas are 100–200 m^2/ha, no different from those of western Washington and Oregon; ericaceous shrubs, however, are less common while herb diversity is higher (R. Daubenmire 1978).

The successional nature of these forests has been difficult to discern because of the long-lived nature of most species, including seral species. For example, *Pseudotsuga menziesii* is a common seral tree in the region that can maintain populations or individuals for 400–1000 years while slowly being replaced by *Tsuga heterophylla* (J. F. Franklin and C. T. Dyrness 1973). *Tsuga heterophylla* is understood to be the climatic climax species, because of all major western forest species it is the most moisture-demanding, the least heat-tolerant, the least drought-tolerant, and among the most shade-tolerant (W. K. Smith 1985; D. Minore 1979).

A variant of the *Tsuga heterophylla* forest occupies the immediate coast, generally within a few kilometers of the ocean except where river valleys and level terrain (such as the west side of the Olympic Peninsula) offer more area below 600 m elevation. *Acer macrophyllum* and *Picea sitchensis* are more abundant, the nurse log phenomenon of tree sapling establishment is very evident, the overstory canopy is less completely closed (especially where *Thuja plicata* and *Picea sitchensis* are common), individual trees are more massive, and the understory is either more dense (especially *Gaultheria shallon* and *Vaccinium* spp.) or more open due to elk grazing. Vegetation in Olympic National Park epitomizes this variant, often referred to as a "temperate rainforest."

The southern part of this Pacific Coast forest, from the California-Oregon border to the Big Sur coast at 36° N, is dominated by *Sequoia sempervirens. Abies grandis, Lithocarpus densiflorus, Pseudotsuga menziesii,* and *Tsuga heterophylla* may be associated, but they are seral (R. Daubenmire 1978; P. J. Zinke 1988). Coast redwood generally occurs within 35 km of the coast and below 1000 m elevation. It dominates a more or less continuous forest zone in northernmost California, but much of its range is characterized by discontinuous groves. Since approximately 1850, when commercial logging of redwood began, 90% of old-growth forest acreage has been removed by clear-cutting (P. Hyde and F. Leydet 1969).

Sequoia occupies both riparian floodplains and slopes. It is uniquely tolerant of fire and silt deposition following flood because of its ability to form adventitious roots and shoots. The time interval between fire or flood disturbances appears to be 30–60 years. Associated species are not as tolerant of disturbance. E. C. Stone and R. B. Vasey (1968) hypothesized that redwood is seral and will not maintain itself without such disturbance, but S. D. Viers Jr. (1982) accumulated data for some stands suggesting that disturbance is not always essential.

Western Montane Coniferous Forests

Overview

Coniferous forests clothe the slopes of the Rocky Mountains and their extensions into Mexico (Sierra Madre Oriental and Sierra Madre Occidental), the Sierra Nevada–Cascade axis, the Coast Range of British Columbia, Washington, Oregon, and northern California, the Transverse and Peninsular ranges of southern California and Baja California, and scattered ranges and northern plateaus of the intermountain region (figs. 5.14, 5.15). They extend from 65° N to 19° N latitude and from 100° W to 140° W longitude. North of Mexico, these forests dominate 7% of the landmass.

Floristically, this enormous expanse is not homogeneous. For example, montane areas of north and south extremes of the Cascade–Sierra Nevada axis exhibit a Sorensen community similarity coefficient of only 30 (0 = total dissimilarity, 100 = identity [D. W. Taylor 1977]). The region does have a long, common paleobotanical history through the Tertiary (D. I. Axelrod and P. H. Raven 1985; D. I. Axelrod 1976), and some species have wide distribution limits through much of the area (though they may have climax status in one region and seral status in another, or they may be differentiated into regional subspecific taxa). Among the trees, only *Populus tremuloides* is found throughout the area; *Pinus ponderosa* and *Pseudotsuga menziesii* occur through most of the area (H. A. Fowells 1965).

Zonation of forest types along elevation gradients is a common theme recognized since C. H. Merriam (1898), and the forests characteristic of each elevational zone share many physiognomic similarities from one mountain chain to another mountain chain. Low elevation forests tend to be rather open savannas or woodlands, intermingled with species from grasslands, broad-leaved woodlands, deserts, or chaparral. Frequent fires seem essential to maintain the open, two-layered structure, and these forests have been modified by overgrazing and fire suppression policies of the past century. Midmontane forests are typically rich in overstory and understory woody species. Upper montane and subalpine forests are denser, simpler, and experience deep, long-lasting snowpacks.

Local zonation is affected by aspect, exposure, soil depth, distance from the ocean, and latitude, as well as by elevation; consequently zonation is not consistent from place to place, even within a floristically homogeneous region. In addition, western montane forests are characterized by frequent disturbance—fire, wind, insects, disease, browsing, avalanche, landslide, weather extremes, volcanism, and logging—and as a result a mosaic of communities exists within a given elevational zone (R. K. Peet 1988). J. R. Habeck (1988) has also pointed out that the vegetation is still recovering from glacial retreat and the Hypsithermal interval, and he concluded that any modern description only succeeds in capturing "a moment in an ever-changing phenomenon."

Climate throughout most of the region, except the Madrean, has a pronounced winter peak in precipitation. Annual precipitation typically ranges from 600 mm at low elevations *(Pinus ponderosa)* to more than 2000 mm at high elevations (spruce-fir), and annual productivity ranges from 600 to 1100 g/m^2 (R. K. Peet 1988). The pinyon-juniper lower fringe, where present, receives as little as 250 mm of precipitation and exhibits significantly lower productivity (D. E. Brown 1982; A. Cronquist et al. 1972+, vol. 1; J. F. Franklin and C. T. Dyrness 1973; N. E. West 1988). Fire cycles may be as long as 200–400 years in subalpine spruce-fir and as short as 5–12 years at low elevations (R. K. Peet 1988).

FIGURE 5.15. Mixed conifer forest of the central Sierra Nevada. The fire-scarred tree in the center is *Sequoiadendron giganteum*, the giant sequoia.

Rocky Mountain Forests

Detailed reviews of these forests have been written by R. Daubenmire (1943) and R. K. Peet (1988). Each divides the Rocky Mountains into somewhat different regions and zones; for consistency, we shall use Peet's more recent treatment. He divided the range into four areas: far north (approximately 53°–65° N latitude), north (45°–53°), south (35°–45°), and Madrean (19°–35°), the last being largely a simplification for convenience, given the high degree of variation within it (J. Rzedowski 1978).

Far northern Rocky Mountain forests are relatively simple, and they broadly intergrade with surrounding lowland boreal forest. The common montane tree species, *Abies bifolia, Picea engelmannii,* and *Pinus contorta,* are all capable of hybridizing with their boreal counterparts, *A. balsamea, Picea glauca,* and *Pinus banksiana,* respectively. Many understory herbs—e.g., species of *Clintonia, Cornus, Linnaea, Pyrola, Sorbus*—are shared between montane and boreal forests (R. Daubenmire 1978). *Picea glauca* dominates lower and midmontane slopes, and *A. bifolia* dominates the subalpine. Timberline is at 1400 m elevation. Toward the northern end of the Rocky Mountains, the range of *P. glauca* becomes constricted to a narrowing band at low elevations, and the range of *Picea engelmannii* increases in the subalpine (P. Achuff 1989). This brief summary does not do justice to the complexities of vegetation within British Columbia. A fine vegetation map, at a scale of 1:2 million, identifies a dozen montane biogeoclimatic zones; we have mentioned but a few (British Columbia Ministry of Forests 1988).

Vegetation of the northern Rocky Mountains, as mentioned earlier, has a Cascadian aspect, due to penetration by westerly storm tracks. J. R. Habeck (1988) recently reviewed this region in detail. Low elevations, 800–1500 m, support a *Juniperus scopulorum–Pinus ponderosa* woodland intermixed with *Cercocarpus ledifolius* and *Purshia tridentata* shrubs. A low elevation ecotype of *Pinus flexilis* is also present. The montane zone, up to 1800 m, is a productive forest rich in tree species such as *Abies grandis, Pseudotsuga menziesii, Thuja plicata,* and *Tsuga heterophylla.* Above, in the subalpine, *Abies bifolia* and *Picea engelmannii* dominate mesic sites, while *Pinus albicaulis* and *Larix lyallii* occur on more extreme sites. The deciduous *Larix* is unique in the northern Rocky Mountains. It is able to establish stands above evergreen conifers, in what would otherwise be tundra, and it creates a suitable microenvironment for montane herbs to follow upslope with it (S. F. Arno and J. R. Habeck 1972). Timberline in the northern Rocky Mountains lies at 2700 m. Timberline rises to the south at a rate of +100 m per degree latitude (R. K. Peet 1988).

Zonation in the southern Rocky Mountains proceeds from a pinyon-juniper woodland through a *Pinus ponderosa* parkland, a *Pseudotsuga menziesii* forest (with *Abies concolor* and *Picea pungens*), and a *Pinus contorta* forest to a dense *Abies bifolia–Picea engelmannii* subalpine forest (with an open *Pinus aristata* woodland on exposed sites). Although *Pinus contorta* is largely a seral species throughout western forests, at its southern limit it is capable of climax status. In the Rocky Mountains it is typically serotinous (cones remain closed at maturity), but the degree of serotiny is known to vary with stand history (P. S. Muir and J. E. Lotan 1985).

The Madrean portions of the Rocky Mountains are complex in their flora and topography, being less continuous. Zonation typically proceeds from a pinyon-juniper woodland (north) or an oak-pine mixed evergreen forest (south) through a *Pinus ponderosa* parkland, and a dense *Pseudotsuga menziesii* forest (associated with *Abies concolor* and *Pinus flexilis,* among several other taxa with more Madrean affinities) to a subalpine forest of *Picea* and *Abies.* Ponderosa pine parkland has 50–75% tree cover, 8–40 m tall, overtopping a well-developed grass understory (C. P. Pase and D. E. Brown 1982). *Pinus ponderosa* changes from var. *scopulorum* in the north to var. *arizonica* in the south. In Arizona and New Mexico, *Picea engelmanii* and *Abies bifolia* "var. *arizonica*" dominate mesic sites in the subalpine, while *Pinus aristata* and *P. flexilis* dominate dry sites (C. P. Pase and D. E. Brown 1982b).

Pacific Northwest Montane Forests

In the Coast Range, and along the west face of the northern Cascade Range, montane vegetation above the temperate rainforest proceeds through an *Abies amabilis* zone to a *Tsuga mertensiana* zone and an *Abies lasiocarpa* subalpine zone (J. F. Franklin and C. T. Dyrness 1973). The *Abies amabilis* forest receives a modest 1–3-m-deep snowpack and annual precipitation greater than 2000 mm. From Mt. Rainier south, its overstory, including *Abies procera, Pinus monticola,* and *Pseudotsuga menziesii,* overlies rich ericaceous shrub, herb, and moss canopies. To the north, *Tsuga heterophylla* is a common associate. The higher *Tsuga mertensiana* forest receives heavy snowpacks up to 7.5 m deep, and its lower limit appears to coincide with freezing elevation in winter (E. B. Peterson 1969). Associates include *Abies amabilis* and *Chamaecyparis nootkatensis.*

Eastern slopes of the Cascade Range, and mountains to the east, exhibit a complex zonation that proceeds from a *Juniperus occidentalis–Artemisia* savanna through a *Pinus ponderosa* forest (sometimes with *Pseudotsuga menziesii*) to a mesic *Abies lasiocarpa* or *Tsuga mertensiana* forest above 1500 m elevation.

Extensions of the Coast and Cascade ranges south of 43° N latitude and into northern California experience a Mediterranean precipitation regime. As a consequence, they have a Sierran zonation of species and are included in the next section.

Montane Forests of Alta and Baja California

These forests have recently been summarized by M. G. Barbour (1988), P. W. Rundel et al. (1988), J. O. Sawyer and D. A. Thornburg (1988), and R. F. Thorne (1988). The region may be subdivided into a northern portion (southern Cascade, Siskiyou Mountains, Klamath Mountains, north Coast Range of California, Sierra Nevada) and a southern portion (Transverse and Peninsular ranges, extending south to the San Pedro Mártir of Baja California).

In the northern portion, typical zonation proceeds upward from a mixed evergreen forest through a relatively open mixed conifer forest, to a denser *Abies magnifica* forest, and to an open mixed subalpine woodland.

The mixed conifer forest has received considerable research and sylvicultural attention because of its commercial value. The overstory exhibits shared or shifting dominance by five conifer species: *Abies lowiana* (= *A. concolor* var. *lowiana*), *Calocedrus decurrens, Pinus lambertiana, P. ponderosa,* and *Pseudotsuga menziesii.* At the lower edge *Pinus ponderosa* predominates; at the upper edge *Abies lowiana* predominates. (This white fir is the western counterpart of the Rocky Mountain *A. concolor.* It is capable of hybridizing with *A. grandis* of the Pacific Northwest [J. F. Franklin and C. T. Dyrness 1973].) Within this zone are 75 groves of *Sequoiadendron giganteum,* the most massive trees in the world (fig. 5.15). Their habitat restrictions, community dynamics, geologic record, and problematic future have been described by D. I. Axelrod (1986c) and P. W. Rundel (1971, 1972).

All forests in this lower montane zone are tolerant of, and dependent on, frequent ground fires to maintain their openness and mixed dominance (B. M. Kilgore 1973; J. L. Vankat and J. Major 1978). Fire suppression and postlogging management have seriously modified these forests; more than half the pristine old-growth acreage has been cut (R. J. Laacke and J. F. Fiske 1983). Another threat is air pollution damage, once restricted to southern California, but now documented for the southern Sierra Nevada as well (W. T. Williams et al. 1977).

The upper montane zone is dominated thoroughly by *Abies magnifica* on mesic sites and by *Pinus contorta* or *Populus tremuloides* on wet sites. *Abies magnifica* forests receive the highest snowpacks of any Californian vegetation, 2.5–4 m deep and lasting 200 days. The lower limit of the forest appears to correspond with average elevation of freezing temperature in winter (M. G. Barbour et al. 1991). They are floristically simple forests—often monospecific in the overstory and with few understory shrubs and herbs—but the trees are impressive in size, with mature ones 30–45 m tall and 1.5 m dbh (diameter at breast height) (M. G. Barbour and R. A. Woodward 1985). *Abies magnifica* forms hybrid swarms with *A. procera* of the Pacific Northwest. *Pinus contorta* here is the nonserotinous var. *murrayana* (H. A. Fowells 1965).

The mixed subalpine woodland, in contrast to the *Abies* forest, has only 5–40% canopy cover provided by trees 10–15 m tall. Unlike the subalpine zone of the Rocky Mountains and the Pacific Northwest, *Picea* and *Abies* are absent. Dominants include *Pinus albicaulis, P. balfouriana, P. contorta* var. *murrayana, P. flexilis, P. monticola,* and *Tsuga mertensiana.* We have adopted the "mixed subalpine" name applied by the Society of American Foresters (F. H. Eyre 1980) to emphasize the pattern of shared or shifting dominance by these six taxa. Timberline increases in elevation from 2900 m in the southern Cascades to 3400 m at the southern limit of the Sierra Nevada.

In southern California and Baja California, the mixed conifer forest becomes simplified to a yellow pine forest, with mixtures of *Pinus jeffreyi* and *P. ponderosa* that depend on elevation and aridity. In Baja California mountains, fire suppression has never been practiced. There, *P. ponderosa* is absent, and the lower montane forest is dominated by *P. jeffreyi* and *P. quadrifolia* (R. A. Minnich 1987; M.-F. Passini et al. 1989). Above this is an open upper montane forest with *Abies concolor, Calocedrus decurrens, Pinus lambertiana,* and *P. jeffreyi. Juniperus occidentalis* var. *australis* occurs on rock outcrops throughout this zone, as it does to the north. The subalpine zone, 2500–3000+ m in elevation, supports either relatively dense *Pinus contorta* forest or open *P. flexilis* woodland (R. F. Thorne 1988). Southern mountains are not tall enough to exhibit a climatic timberline.

The eastern, desert-facing slopes of Californian mountain ranges fall steeply, have shallower soils, and support less continuous forests with less well-defined zones. The flora has a pronounced desert aspect.

Intermountain Region Montane Forests

Between the Cascade-Sierra axis and the Rocky Mountains, and between 45° N and 37° N latitude, are mountain chains isolated like islands amidst a sea of shrub steppe and desert scrub. Because of their relative aridity, and because of the accidental vagaries of plant migrations during the Pleistocene, zonation and species richness on these peaks are not as complex as on the Rockies or the Sierra Nevada. These montane forests have been well described by W. D. Billings (1951), A. Cronquist et al. (1972+, vol. 1), and F. C. Vasek and R. F. Thorne (1988).

Elevations between 1500 and 2500 m have a pinyon-juniper woodland, one of the most extensive vegetation types in southwestern North America, covering 170,000 km^2 (D. E. Brown 1982; N. E. West 1988). The woodland is composed of trees 3–7 m tall with rounded crowns that generally do not touch each other. They overlie an understory of cold desert shrubs, C_3 bunchgrasses, and forbs with varying cover. Modern woodlands are denser than those of 150 years ago, perhaps because grass cover in pristine woodlands was able to carry fire more frequently than the overgrazed modern cover. The most widely ranging components are *Juniperus osteosperma*, generally at lower elevations, and *Pinus monophylla*, generally at higher elevations, but A. Cronquist et al. (1972+, vol. 1) and N. E. West (1988) reported several other species of pine and juniper that may be more important locally. Not discussed here are other *Pinus-Juniperus* woodlands along the eastern flank of the Rocky Mountains and into the Madrean region of Mexico (R. K. Peet 1988; J. Rzedowski 1978).

Above pinyon-juniper is an *Abies concolor* montane forest, followed by a subalpine woodland (approximately 3000 m elevation) dominated by *Pinus flexilis* and/or *P. longaeva*. The latter species (the Great Basin bristlecone pine) contains individuals that are approaching 5000 years, older than those of any other taxon. On some very arid intermountain ranges, the *Abies* zone is absent and pinyon pines run up the entire slope, stopping only when reaching the shade of bris-

FIGURE 5.16. Salt marsh dominated by *Spartina alterniflora,* near Wilmington, North Carolina.

tlecone pines. On some relatively mesic ranges, the montane zone includes *Pinus contorta, P. ponderosa,* and *Pseudotsuga menziesii,* and the subalpine zone includes *Abies lasiocarpa* and *Picea engelmannii.*

A deciduous scrub may, in places, either replace the pinyon-juniper woodland or lie just above it as a dry margin to montane forest. This scrub has been called petran chaparral or Great Basin montane scrubland. It differs from typical Californian or interior chaparral by species composition and predominance of deciduous elements (D. E. Brown 1982b). *Quercus gambelii* (in the broad sense) is dominant, associated with *Amelanchier alnifolia, A. utahensis, Arctostaphylos patula, Ceanothus fendleri, C. velutinus, Cercocarpus montanus, C. ledifolius, Cowania mexicana, Symphoricarpos* spp., and *Prunus virginiana,* among others. The scrub is 1–6 m tall and ranges from dense to open; its physiognomy and herbaceous composition have been degraded by overgrazing throughout its distribution.

Tidal Wetlands

Overview

Salt marshes are coastal meadowlands subject to periodic flooding by the sea (fig. 5.16). They are typically restricted to shorelines with low-energy waves, such as estuaries or areas behind barrier islands or spits, but otherwise they can occur on stable, subsiding, or rising coasts (V. J. Chapman 1976). The vegetation is usually a single, nearly closed layer of perennial herbs. The flora is rather simple. For example, tidal marshes along the Atlantic and Gulf coasts of North America combined support only 347 taxa of vascular plants, those along the Pacific Coast from the tip of Baja California to Point Barrow have 78 species, and the marshes of Hudson Bay contain 28 species (B. L. Haines and E. L. Dunn 1985).

Perhaps as a consequence of the simplicity of the flora, several species have enormous ranges that take them through nearly all of North America *(Atriplex patula, Distichlis spicata, Salicornia virginica, Spartina alterniflora, Triglochin maritima).* Poaceae, Juncaceae, and Chenopodiaceae dominate the temperate zone salt marshes, but subtropical and arctic marshes are more diverse at the family level (L. C. Bliss 1988; W. A. Glooschenko 1980; R. F. Thorne 1954). Subtropical coastlines in North America, generally south of 29° N latitude, tend to be dominated by a group of woody taxa collectively called mangroves. Mangrove swamps are not salt marshes in the physiognomic sense defined above, but for completeness they are included in this section.

Environmental stresses on salt marsh vegetation include flooding—with attendant mechanical disturbance and anaerobic conditions—and salinity. As the land slopes upward from the sea, these two stresses decline in intensity. As a consequence, vegetation is divided into two or more zones. The most typical zones are low marsh and high marsh. Low marsh extends approximately from mean sea level to mean high water and is flooded once or twice a day, whereas high marsh is irregularly flooded once a month or less, only at highest tides or during storms. The ecotone between the two is characteristically abrupt, and its location has been ascribed variously to physical factors and competition (M. D. Bertness and A. M. Ellison 1987; B. E. Mahall and R. B. Park 1976).

The low marsh tends to be a narrow fringe, only 1–30 m wide, whereas the high marsh may be hundreds or even thousands of meters wide (J. P. Stout 1984). With farther distance inland or inward along river mouths, salinity declines and complex, diverse, brackish marshes become dominant. These are not discussed in this section. Areal estimates for tidal salt marshes along the coast of the United States have been compiled, but not for Canada. Our estimate, largely based on a map in B. L. Haines and E. L. Dunn (1985) and our own calculations, is that these marshes occupy 1% of the North America landmass.

Net annual productivity varies with dominant species and latitude, but overall for North America it is 300–2000 $g/m^2/yr$ (M. Josselyn 1983; D. M. Seliskar and J. L. Gallagher 1983; J. P. Stout 1984), the highest values being for the Gulf Coast, which has a year-long growing season.

Regional Salt Marsh Vegetation

Arctic salt marshes are scattered, and their short plants (3–5 cm tall) exhibit only 15–25% cover. No estimates of their area or productivity seem to exist in the literature. Their extent is limited by annual ice scour, lack of fine sediment, limited tidal amplitude, low salinity of coastal waters, and a short growing season (L. C. Bliss 1988). The low marsh is dominated by *Puccinellia phryganodes*, and the high marsh is a mixture of *Atriplex patula, Calamagrostis deschampsioides, Carex bipartita, C. maritima, C. neglecta, C. ramenskii, C. salina, C. subspathacea, C. ursina, Chrysanthemum arcticum, Cochlearia officinalis, Dupontia fisheri, Festuca rubra, Glaux maritima, Plantago maritima, Potentilla egedii, Ranunculus cymbalaria, Salicornia europaea, Scirpus maritima, Senecio congestus, Stellaria crassifolia, S. humifusa, Suaeda maritima,* and *Triglochin maritima* (L. C. Bliss 1988; V. J. Chapman 1976; W. A. Glooschenko 1980, 1980b; W. A. Glooschenko et al. 1988; K. A. Kershaw 1976; K. B. Macdonald and M. G. Barbour 1974).

The Atlantic Coast low marsh, from the Gulf of St. Lawrence to southern Florida, is dominated by cordgrass, *Spartina alterniflora* (fig. 5.16). It is a dense, usually monospecific community. Away from the tide line, tall cordgrass (1–2 m) gives way to sterile, short cordgrass (less than 1 m). Transplant and genetic studies have demonstrated that these are ecophenes (not genetically fixed) rather than ecotypes (C. M. Anderson and M. Treshow 1980; I. Valiela et al. 1978). The upper marsh is dominated by *Spartina patens,* associated most commonly with *Distichlis spicata, Juncus gerardii, Limonium carolinianum, Salicornia virginica, Suaeda linearis,* and *Triglochin maritima* (M. D. Bertness and A. M. Ellison 1987; V. J. Chapman 1976; W. A. Glooschenko 1980; B. L. Haines and E. L. Dunn 1985; R. J. Reimold 1977). This corresponds to the "northern cordgrass prairie" of A. W. Küchler (1964).

Along the United States portion of the Gulf of Mexico, cordgrass continues to dominate the low marsh. The high marsh is thoroughly dominated by *Juncus roemerianus,* but associates include *Batis maritima, Borrichia frutescens, Carex* spp., *Distichlis spicata, Fimbristylis castanea, Limonium carolinianum, Salicornia virginica, Scirpus* spp., *Sesuvium portulacastrum, Spartina patens,* and *Suaeda linearis* (R. H. Chabreck 1972; V. J. Chapman 1976; L. N. Eleuterius 1980; J. P. Stout 1984; R. F. Thorne 1954). This is the "southern cordgrass prairie" of A. W. Küchler (1964). *Juncus roemerianus* is the most abundant salt marsh species on the Gulf Coast; it also extends east and north to Virginia (where it replaces *J. gerardii*) and inland to brackish marshes. It exists in tall (1–2 m) and sterile short (less than 1 m) forms, which occupy different zones, just as cordgrass does.

South of 29° N latitude, on both sides of the Florida peninsula, mangroves assume increasing dominance (A. F. Johnson and M. G. Barbour 1990). *Rhizophora mangle* is the most seaward species, followed by a zone of *Avicennia germinans* (= *A. nitida*), *Laguncularia racemosa,* and *Conocarpus erecta.* The *Avicennia-Laguncularia* zone includes the salt marsh species *Batis maritima, Salicornia virginica, Sesuvium portulacastrum,* and *Suaeda linearis.* Sand, marl, or limestone substrates here also support *Monanthochloë littoralis* and *Sporobolus virginicus* (R. F. Thorne 1954).

The Pacific Coast salt marsh progresses northward through subtropical, temperate, and cool temperate vegetation regions (K. B. Macdonald and M. G. Barbour 1974). A number of subtropical high marsh plants also occur as elements of strand vegetation here and

FIGURE 5.17. Beach and dune vegetation near Cape Sabal, Florida. [Courtesy of L. Gunderson.]

along the Gulf of Mexico (M. G. Barbour et al. 1987; A. F. Johnson 1977, 1982; A. F. Johnson and M. G. Barbour 1990; P. Moreno-Casasola 1988).

From central Baja California to northern California, the temperate low marsh is dominated by *Spartina foliosa* and the high marsh is dominated by *Salicornia virginica* (associated with its parasite *Cuscuta salina*), *Distichlis spicata*, *Frankenia grandifolia*, *Grindelia stricta*, *Jaumea carnosa*, *Limonium californicum*, *Suaeda californica*, and *Triglochin maritima* (M. Josselyn 1983; K. B. Macdonald and M. G. Barbour 1974; B. E. Mahall and R. B. Park 1976; L. Neuenschwander et al. 1979; J. B. Zedler 1982). From Oregon through western Alaska, *Puccinellia* spp. replace cordgrass in an open low marsh, and the high marsh is characterized by *Carex lyngbyei*, which also extends into brackish areas. Associated taxa include *Deschampsia caespitosa*, *Distichlis spicata*, *Glaux maritima*, *Grindelia stricta*, *Jaumea carnosa*, *Juncus* spp., *Plantago maritima*, *Potentilla egedii*, *Salicornia virginica* and its parasite *Cuscuta salina*, *Stellaria humifusa*, and *Triglochin maritima* (J. F. Franklin and C. T. Dyrness 1973; W. A. Glooschenko 1980; K. B. Macdonald and M. G. Barbour 1974; D. M. Seliskar and J. L. Gallagher 1983).

Beach and Frontal Dune Vegetation

Overview

"Beach" is that strip of sandy substrate adjacent to open coast that extends from mean tide line to the top of the frontal dune—or in the absence of a frontal dune, to the farthest inland reach of storm waves (fig. 5.17). It is characterized by a maritime climate, high exposure to salt spray (oceanic coasts) and sand blast, and a shifting, sandy substrate with low water-holding capacity and low organic matter content, and it is subject to episodic overwash. Reviews of entire coastlines (G. J. Breckon and M. G. Barbour 1974; M. G. Barbour et al. 1985; M. G. Barbour, M. Rejmanek, et al. 1987; P. Moreno-Casasola 1988) indicate that many beach taxa have wide latitudinal and longitudinal distributions, the limits of which relate more to substrate texture and chemistry, wave energy, and frequency of disturbance than to zonal climate.

The dominant growth forms are (1) rhizomatous perennial grass or (2) prostrate, succulent, perennial forb (M. G. Barbour 1992). Woodiness, sclerophylly, and annuality are uncommon features (M. G. Barbour 1992). Plant cover tends to be low (5–20%), and any given beach is characterized by only a dozen or fewer taxa. As in the salt marsh, species tend to be zoned with distance back from tide line: some taxa characteristically dominate the leading edge of vegetation and others dominate more inland strips. Species richness, growth-form diversity, and plant cover all increase with distance inland. The frontal dune may be densely covered by grasses or forbs—indeed, its existence is due to the sand-stilling nature of vegetation. To our knowledge, productivity estimates have not been made for beach vegetation in North America; judging from biomass estimates of Pacific and Gulf Coast strands made by M. G. Barbour and R. H. Robichaux (1976) and F. W. Judd et al. (1977), they may approximate those of steppe and desert. Strand vegetation occupies much less than 1% of the North American landmass, and the land it occupies is mobile. Beaches can build seaward during calm seasons and be eroded during storm seasons.

Barrier beaches on islands may also migrate over time when erosion and building occur simultaneously, but on different parts of the island. Unfortunately, the aesthetic qualities of beachfront land have attracted road and home construction, leading to a degradation of the beach habitat along most coastlines (e.g., R. Dolan et al. 1973; Corps of Engineers 1973). Revegetation, groin construction, and beach "nourishment" (additions of dredged sand) are some techniques that have been used for strand reclamation, with a record of only limited success.

To the lee of the frontal dune, or beyond the reach of storm waves, a variety of habitats and a mixture of communities exist: low swales with marsh plants, where deflation has brought down the soil surface close to ground water; stable dunes with woody vegetation, where plant cover is more nearly continuous and soils are colored by significant amounts of organic matter; and transition zones with an admixture of typically inland taxa. An elegant study by J. S. Clark (1986) demonstrated that because the disturbance frequency is so high, these communities are not in equilibrium with their present physical environments. The vegetation of these more inland habitats are too complex for us to summarize in this brief continental review, so we limit our discussion to beach and frontal dune ("strand").

Regional Vegetation

Beaches facing the Arctic Ocean are few and narrow, and they have a rather abrupt transition to tundra. Their dominant taxa have a circumarctic flavor, and they extend southward on both Atlantic and Pacific coasts for various distances: *Angelica lucida, Cochlearia officinalis* subsp. *groenlandica, Elymus mollis* (= *E. arenarius* subsp. *mollis* and *Lymus mollis*), *Festuca rubra, Honkenya peploides, Lathyrus japonicus, Ligusticum scoticum, Matricaria ambigua, Mertensia maritima,* and *Senecio pseudo-arnica* (L. C. Bliss 1988; G. J. Breckon and M. G. Barbour 1974; K. B. Macdonald and M. G. Barbour 1974). The grasses *Elymus* and *Festuca* are dominants.

Pacific Coast beaches may be divided into 5–7 ecofloristic zones, and some of these correspond to climatic zones in the Köppen system (G. J. Breckon and M. G. Barbour 1974; M. G. Barbour et al. 1975). Flora with arctic and subarctic ranges, such as *Elymus mollis,* characterize beaches to 54° N latitude, but a temperate ecofloristic zone occupies the region 54°–34° N. Common taxa here include *Abronia latifolia, Ambrosia chamissonis, Atriplex leucophylla, Calystegia soldanella, Lathyrus littoralis,* and *Poa douglasii.* Introduced plants, *Ammophila arenaria* (which has largely displaced *Elymus mollis*), *Cakile maritima, C. edentula,* and *Mesembryanthemum chilense,* also exist along this portion of the coast. South of 37° N latitude, grasses no longer dominate the foredune, and they are replaced with prostrate forbs such as *Abronia maritima* (M. G. Barbour and A. F. Johnson 1988). Beaches in the region 34°–24° N latitude are arid, and their flora has a desert aspect. South of 24° N latitude, such tropical species as *Ipomoea stolonifera, Scaevola plumieri, Sesuvium verrucosum,* and *Sporobolus virginicus* are dominants (A. F. Johnson 1977, 1982, 1985).

Along the United States shore of the Gulf of Mexico, 73 common taxa form three vegetation regions (M. G. Barbour, M. Rejmanek, et al. 1987). *Uniola paniculata* dominates everywhere except Louisiana, where it is replaced by a dune ecotype of *Spartina patens. Spartina patens* ranges extensively on the eastern seaboard, from 50° N latitude in Canada to 16° N latitude in Central America. In the northern part of its range, it occurs exclusively in salt marshes; in the south, exclusively on dunes; but in the middle of its range, distinct ecotypes occur in dune, marsh, and swale habitats (J. A. Silander 1979; J. A. Silander and J. Antonovics 1979).

Other important local or regionwide Gulf strand plants include *Cenchrus incertus, Croton punctatus, Ipomoea stolonifera, Iva imbricata, Oenothera humifusa, Panicum amarum, Schizachyrium maritimum,* and *Sesuvium portulacastrum.* Many of these taxa continue south along the Mexican portion of the Gulf (I. Espejel 1986; P. Moreno-Casasola 1988; J. D. Sauer 1967). C_4 taxa constitute more than 40% of all plant cover in the subtropical Gulf, in contrast to less than 25% along temperate North American shores (M. G. Barbour et al. 1985). A special calcareous shell substrate—which occurs in southern Florida, the Florida Keys, and parts of Mexico—creates narrow beaches that support a unique community that deserves more study (M. G.

Barbour, M. Rejmanek, et al. 1987; A. F. Johnson and M. G. Barbour 1990; P. Moreno-Casasola 1988).

We have not found a thorough regional review of Atlantic Coast strand vegetation. There appear to be three ecofloristic zones along that coast. Several Gulf taxa continue around the Florida peninsula and up the Atlantic Coast to approximately the Virginia–North Carolina border (approximately 37° N latitude), and their presence defines a southern zone (fig. 5.17). Barrier islands are characteristic of this region, just as they are of the Gulf Coast (R. Dolan et al. 1980; J. G. Ehrenfeld 1990; S. P. Leatherman 1982; G. F. Levy 1990; C. A. McCaffrey and R. D. Dueser 1990). *Uniola paniculata* is the beach dominant, with *Andropogon virginicus, Cenchrus tribuloides, Croton punctatus, Diodia teres, Erigeron canadensis, Euphorbia* (= *Chamaesyce*) *polygonifolia, Heterotheca subaxillaris, Hydrocotyle bonariensis, Iva imbricata, Oenothera laciniata, Panicum amarum, Physalis maritima,* and *Triplasis purpurea* among the important associates (J. M. Barry 1980; N. L. Christensen 1988; P. J. Godfrey and M. M. Godfrey 1976; R. Stalter 1974). Ecotypic differences between Gulf and Atlantic populations of *Uniola* have been described by E. D. Seneca (1972).

The mid-Atlantic zone is dominated by *Ammophila breviligulata,* most commonly associated with *Andropogon virginicus, Arenaria peploides, Artemisia stelleriana, Cakile edentula, Euphorbia polygonifolia, Lathyrus japonicus, Polygonum glaucum,* and *Solidago sempervirens* (A. F. Johnson 1985b; W. E. Martin 1959; B. Robichaud and M. F. Buell 1973).

Moving west along the St. Lawrence River, one leaves a saline coast and enters the lacustrine beach system of the Great Lakes. The vegetation of sandy coasts here has been described in detail for only a few places (e.g., R. G. Morrison and G. A. Yarranton 1973, 1974; J. S. Olson 1958). Its character has been regionally summarized by M. A. Maun (1992). Beach dominants include the grasses *Ammophila breviligulata* and *Calamovilfa longifolia,* and the forbs *Cakile edentula, Corispermum hyssopifolium, Euphorbia polygonifolia,* and *Salsola kali.* All of these, save *Calamovilfa,* extend to the Atlantic Coast, and some evidence indicates that ecotypic differentiation between maritime and lacustrine habitats has occurred (R. S. Boyd and M. G. Barbour 1986; P. R. Baye 1990). Behind the foredune ridge is a grassland-heath scrub with such tallgrass prairie species as *Andropogon gerardii, A. scoparius, Sorghastrum nutans,* and *Stipa spartea.* The grasses are associated with *Arctostaphylos uva-ursi, Juniperus communis,* and *Prunus pumila* shrubs.

Returning to the maritime Atlantic Coast, one sees that arctic and subarctic elements are gradually added with increasing latitude, with dominance shifting from *Ammophila* to *Elymus* along the Newfoundland Coast (approximately 50°–52° N latitude). The Atlantic Coast to the north is characterized by a circumarctic element described at the beginning of this section. Beach vegetation of the Canadian maritimes has been described in detail by D. Thannheiser (1984) and G. Lamoureux and M. M. Grandtner (1977, 1978). A littoral fringe, dominated by *Cakile edentula,* is replaced farther inland by *Elymus mollis, Honkenya peploides,* and *Mertensia maritima* (the latter is especially common on shingle or coarse sand beaches). More inland yet, dunes are dominated by *Ammophila breviligulata, Festuca rubra,* and *Lathyrus japonicus.*

Conclusion

Our tour of North American vegetation began with the Arctic, and it has ended there. We can conclude for virtually every vegetation type that the depth of our understanding does not match the scale of its distributional range. In some areas, alpha-level descriptions of modern and past vegetation still remain to be accumulated and published. In other regions, such important ecosystem attributes as productivity and nutrient cycling are essentially unknown. And in most cases, we have very rudimentary knowledge about the autecology of more than one or two dominant species: details about their demography, ecophysiology, response to human-mediated disturbance, and biotic relationships with associated plants and animals await discovery. Compared to some parts of the world, such as Europe, our mastery of vegetation is very superficial.

The passage of time will not guarantee better understanding, because the extent of natural vegetation annually grows smaller. Vast areas of some types of western and northern vegetation still remain as wilderness, but many of our deciduous forests, grasslands, deserts, woodlands, montane coniferous forests, beaches, and marshes are endangered. Historical accounts that document what we had 100–200 years ago (J. Bakeless 1961; M. Williams 1989) are, in turns, exhilarating and depressing as a sense of discovery turns into a sense of loss.

6. Phytogeography

Robert F. Thorne

Because of the basically temperate nature of the flora of the United States of America and Canada, the floristic richness of this huge area is considerably less than that of Mexico and several of the other American nations to our south. According to my system of classification, North America north of Mexico has indigenous representatives of 211 flowering plant families. This number can be compared with the number of indigenous angiosperm families known to occur in Mexico (236), Central America including Mexico (245), and South America (273). Other than Mexico, the richest flowering-plant family representations in the New World are found in Peru (227), Brazil (218), Costa Rica (214), Panama (206), West Indies (194), and Bolivia (190).

Comparable figures for regions of the Old World are available for Asia excluding Malesia (289 families), China (260), Africa south of the Sahara (247), Malesia (248), Australia (224), and Madagascar (196). Except for China, most comparable temperate areas of the Old World are considerably less rich in indigenous angiosperm families than North America north of Mexico: the former Soviet Union (153); Europe including Mediterranean Africa (152); islands in the Atlantic Ocean, e.g., Macaronesia (Canary Islands, Madeira, and Azores), Bermuda, and St. Helena (124); and New Zealand (111). These and other indigenous angiosperm floras of the world are listed in table 6.1.

Within North America (fig. 6.1), the family representation diminishes with increasing distance from the tropics, as might be expected in a basically tropically oriented class. The richest American flowering plant family floras are found in Florida (187) and Texas (177), again as might be anticipated from their southern latitudes. Somewhat less rich are the more temperate states of Georgia (171), Louisiana (164), California (140), and New York (140). Much more depauperate are the floras of such boreal areas as Manitoba (106), Newfoundland (86), Alaska (75), Labrador (67), and Greenland (52).

Floristic Regions of North America

An excellent discussion of the floristic regions and provinces of North America north of Mexico is included in *Floristic Regions of the World,* by Armen L. Takhtajan (1986) in collaboration with Arthur Cronquist. Their treatment divides North America north of Mexico into two floristic kingdoms, the Holarctic and Neotropical, the former with two subkingdoms and four

The author of this essay uses his own classification system for the angiosperms. His family and subfamily circumscriptions differ somewhat from those of Cronquist, whose scheme is used in the taxonomic treatments of the Flora of North America. The numbers of families and subfamilies attributed to the various geographic or chorionomic regions differ slightly from what those numbers would be if Cronquist's scheme were used in this essay; the ratios among these numbers, however, would remain about the same, e.g., there would still be about twice as many families of flowering plants in California as in Alaska.—Eds.

TABLE 6.1 Putatively Indigenous Angiosperm Families and Additional Subfamilies of the World

	Dicots	*Monocots*	*Totals*	*Total Families & Additional Subfamilies*
Asia *(excl. Malesia)*				
Families	227	62	289	
Add. Subfamilies	125	31	156	445
South America *(incl. Trinidad)*				
Families	217	56	273	
Add. Subfamilies	103	36	139	412
Central America *(incl. Mexico)*				
Families	198	47	245	
Add. Subfamilies	95	35	130	375
Africa *(S of Sahara)*				
Families	192	55	247	
Add. Subfamilies	95	29	124	371
Malesia *(Malaya-Fiji)*				
Families	194	54	248	
Add. Subfamilies	81	29	110	358
Australia *(incl. Tasmania)*				
Families	164	60	224	
Add. Subfamilies	78	33	111	335
North America *(N of Mexico)*				
Families	167	43	210	
Add. Subfamilies	76	26	102	312
West Indies *(incl. Bahamas)*				
Families	159	37	194	
Add. Subfamilies	65	28	93	289
Madagascar & Comoros Families	154	42	196	
Add. Subfamilies	67	25	92	288
Pacific Basin *(excl. Hawaii)*				
Families	142	33	175	
Add. Subfamilies	60	24	84	259
New Caledonia & Loyalties				
Families	121	34	155	
Add. Subfamilies	46	16	62	217
Europe *(incl. Med. Africa)*				
Families	116	36	152	
Add. Subfamilies	48	14	62	214
Indian Ocean Islands				
Families	115	32	147	
Add. Subfamilies	49	19	68	215
Atlantic Islands				
Families	96	28	124	
Add. Subfamilies	32	8	40	164
New Zealand				
Families	88	23	111	
Add. Subfamilies	19	8	27	138
Hawaiian Islands				
Families	70	14	84	
Add. Subfamilies	15	5	20	104
Subantarctic Islands				
Families	35	10	45	
Add. Subfamilies	2	3	5	50
Antarctica				
Families	1	1	2	
Add. Subfamilies	0	0	0	2

regions with ten provinces, one of which, the Sonoran, has three (I have added one more) subprovinces represented north of the border (fig. 6.2). The Neotropical Kingdom is represented only by the West Indian Province of the Caribbean region, which includes the southern third of the Florida peninsula and the adjacent Florida Keys. These phytochorionomic units, or natural floristic units, are discussed briefly below as to their locations, general characteristics, floristic content, primary vegetation types, and degrees of endemism. A. L. Takhtajan's book presents more detailed information about each unit. The floristic regions presented below and on the map legend (fig. 6.2) are numbered in accord with Takhtajan. The missing numbers correspond to regions outside of North America north of Mexico.

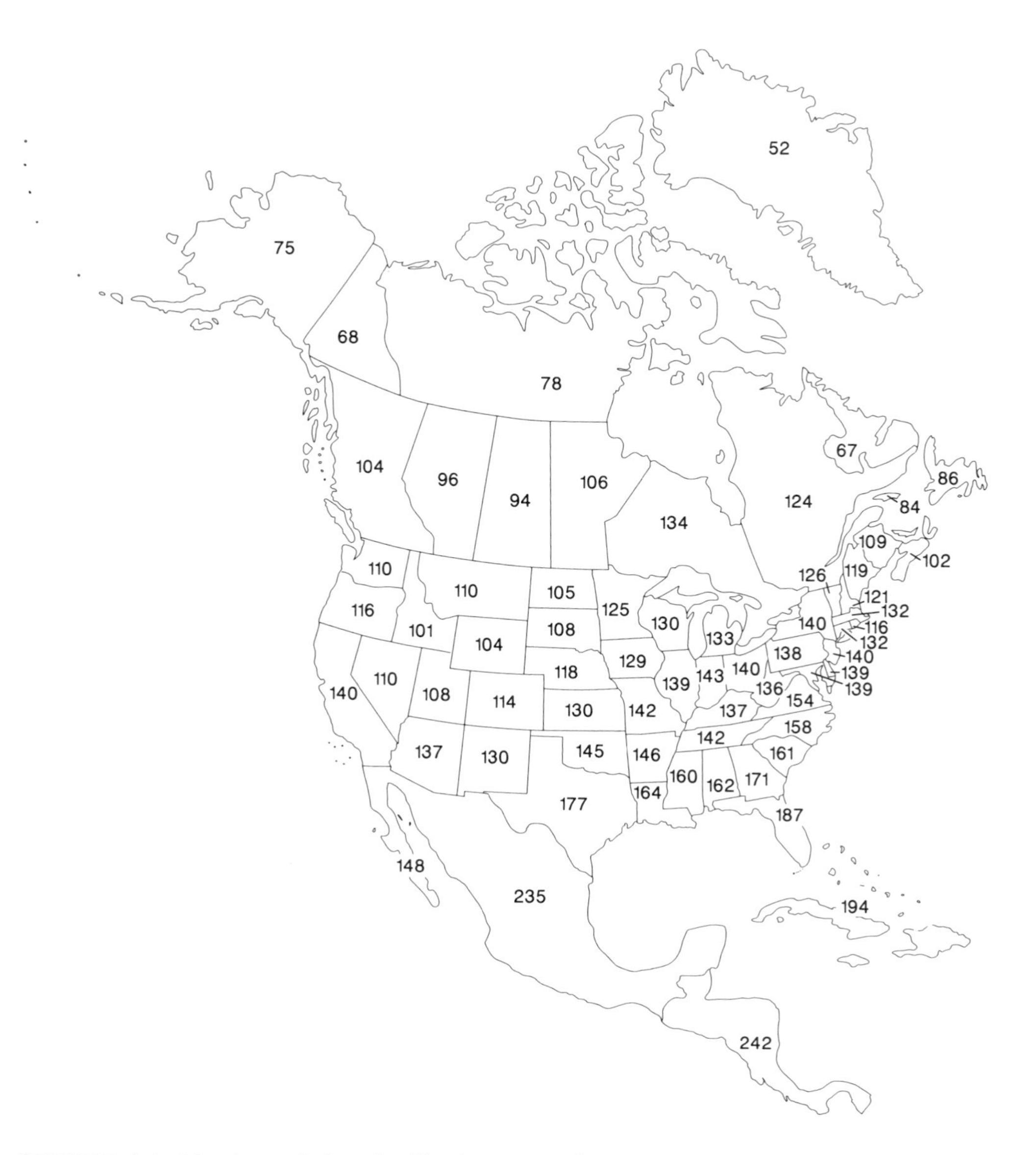

FIGURE 6.1. Numbers of plant families by state and province.

A. Holarctic Kingdom
(Boreal Subkingdom)

1. CIRCUMBOREAL REGION

The Boreal Subkingdom of the Holarctic Kingdom has three regions in North America. The first is the vast Circumboreal region, represented in North America by two provinces, the Arctic and Canadian provinces.

• 1a. Arctic Province

The Arctic Province in North America consists of all the treeless land of the far north, mostly north of the Arctic Circle, i.e., most of coastal Alaska (except the southeastern Panhandle) and northern coastal Canada, and all of the ice-free Canadian Archipelago and Greenland. The flora is depauperate, lacking endemic families and having very few genera largely restricted to circumarctic (and alpine) ranges, e.g., *Arctagrostis, Arctous, Braya, Diapensia, Dupontia, Loiseleuria,* and *Oxyria* (R. F. Thorne 1972). Many of the species are circumpolar, and strictly endemic species in each hemisphere are probably fewer than one hundred.

E. Hultén (1963) regarded 352 species as circumpo-

A. Holarctic Kingdom

(BOREAL SUBKINGDOM)

1. Circumboreal Region
 1a. Arctic Province
 1b. Canadian Province
3. North American Atlantic Region
 3a. Appalachian Province
 3b. Atlantic and Gulf Coastal Plain Province
 3c. North American Prairies Province
4. Rocky Mountain Region
 4a. Vancouverian Province
 4b. Rocky Mountain Province

(MADREAN OR SONORAN SUBKINGDOM)

9. Madrean Region
 9a. Great Basin Province
 9b. Californian Province
 9c. Sonoran Province
  9c.1. Mojavean Subprovince
  9c.2. Sonoran Subprovince
  9c.3. Chihuahuan Subprovince
  9c.4. Tamaulipan Subprovince
 9d. Mexican Highlands Province

B. Neotropical Kingdom

23. Caribbean Region
 23a. Central American Province
 23b. West Indian Province

FIGURE 6.2. [Regional schema adapted from A. L. Takhtajan 1986, with permission from University of California Press.]

Betula papyrifera. **Pen-and-ink drawing by Frederick Walpole (1861–1904). [Hunt Institute, Carnegie Mellon University, Pittsburgh, indefinite loan from Smithsonian Institution.]**

lar or nearly so, plus another 107 species with similar ranges but represented by different races or slightly different species on opposite sides of the Atlantic Ocean or the Bering Sea. The only complete flora of the Arctic region (N. Polunin 1959) lists 892 vascular plant species in 230 genera of 66 families (or 70 by my interpretation). N. Polunin's (1940) flora of the Canadian eastern Arctic consists of 297 vascular plant species in 38 families. The flora of Greenland (T. W. Böcher et al. 1968), not including adventive and introduced species, contains 496 vascular plant species in 61 families.

The vegetation consists of tundra grading northward into polar desert. Trees are absent, and the flora consists primarily of low shrubs of *Salix, Betula,* and Ericaceae, and of herbaceous perennials, especially in the Brassicaceae, Caryophyllaceae, Cyperaceae, and Poaceae. Species of *Dryas, Oxyria, Pedicularis, Potentilla,* and *Saxifraga* are characteristic. Annuals are virtually absent; lichens and bryophytes are abundant. Because of the extensive Pleistocene glacial denudation of most of this province, the present vegetation can be only a few thousand years old. The floristic elements, however, may be relatively ancient, having survived the glacial advances in refugia both north and south of the ice sheets. With the melting of the glaciers, the arctic species rapidly spread into their present far northern ranges.

- 1b. Canadian Province

South of the Arctic Province, the Canadian Province forms a wide band across Alaska and Canada north of the Rocky Mountains and North American Atlantic regions. It also includes parts of northern New England, Michigan, and Minnesota. Endemics are few, represented by perhaps two dozen species found primarily in the western and eastern ends of the province. Dominant are the extensive conifer forests in this heavily glaciated land of mostly low relief with many lakes, slow streams, and *Sphagnum* bogs. The flora is somewhat richer than the Arctic Province but, aside from the trees, the species are likewise mostly widespread. Many are circumboreal. In contrast to the distribution of species in the province, no families and perhaps only 50 genera are primarily circumboreal (and subalpine) in distribution. The dominant trees, *Picea glauca, Abies balsamea, Larix laricina, Picea mariana, Pinus banksiana, Betula papyrifera, Populus tremuloides,* and *P. balsamifera,* are distinct from but closely related to Eurasian species. Many of the shrubs, aquatics, and other herbaceous species are circumboreal or closely related to Eurasian species.

Because of the repeated Pleistocene glaciations, the vegetation and present flora of the Canadian Province, like that of the Arctic Province, are very recent, composed largely of immigrant elements from unglaciated territories to the south and west and from such ice-free refugia that might have offered asylum within the present provincial boundaries, as in Alaska-Yukon. Again there is no complete floristic treatment of the Canadian Province, but Frère Marie-Victorin's *Flore Laurentienne* (1935), largely in the Canadian Province, can serve as a partial example. This work includes 1568 species in the Laurentian area, but Marie-Victorin cited for the whole flora of Quebec, including the Arctic Province and some of the Appalachian Province, 1917 species in 554 genera.

To the north the coniferous forests become sparse and low (taiga) and grade into the arctic tundra, usually without a sharp Arctic treeline. To the south many species of the Canadian Province range into the Appalachian and Rocky Mountain provinces as components of spruce-fir forests. Thus the boundaries of the Canadian Province are nowhere sharp.

3. NORTH AMERICAN ATLANTIC REGION

The North American Atlantic region stretches across North America from the Atlantic Ocean to the Rocky Mountains and from south central Canada to the Gulf of Mexico. It includes three provinces, the Appalachian, the Atlantic and Gulf Coastal Plain, and the North American Prairies. The flora is relatively rich and characterized by considerable endemism. There are two monotypic families, Hydrastidaceae and Leitneriaceae, and perhaps 100 endemic or near-endemic genera, some with extremely limited ranges. Despite this endemism, the linkage between the floras of eastern North America and eastern Asia is well known and much researched. I have counted (R. F. Thorne 1972) at least 74 genera that are restricted to eastern North America and Asia (most only in eastern and southeastern Asia). Another 68 genera are found in eastern North America and Eurasia and also in western North America and/or the Mexican highlands. An additional 118 genera are wide-ranging in primarily temperate areas; of these, 62 are circum-North Temperate and 56 are represented also in the temperate regions of one or more of the southern continents.

Most of the conspicuous woody genera that are common to eastern North America and eastern Asia usually have species in one region represented closely related to species in the other. The fossil record indicates that in Tertiary times a widespread subtropical or warm temperate flora extended across the Northern Hemisphere, with migration frequent between the two continents. Because the eastern Asiatic flora is much richer and more archaic than the eastern North American flora, possibly there was greater movement from Asia to North America, though certainly North America suffered much more heavily in later Tertiary time from climatic and glacial catastrophes than did Asia.

• 3a. Appalachian Province

The Appalachian Province includes that part of upland eastern North America that was largely covered by deciduous forest in historic times. It stretches from southernmost Ontario and Quebec to central Georgia and Alabama, includes most of Arkansas and part of eastern Texas, and reaches west through the Ouachita Mountains, Ozark Plateau, eastern Iowa, and southeastern Minnesota. Excluded are the coastal plains, including the Mississippi Embayment, and the treeless prairies and plains. It is bounded on the north by the Canadian Province, on the east and south by the Atlantic and Gulf Coastal Plain Province, and on the west by the North American Prairie Province. As might be expected, much floristic intercourse exists between these adjacent provinces and the Appalachian Province.

No floristic treatment has been done just for this province, but much of its territory (plus some Canadian, Coastal Plain, and Prairie provinces as well) is included in the floras by M. L. Fernald (1950) and H. A. Gleason and A. Cronquist (1963). The former includes 4425 indigenous and 1098 adventive or naturalized species; the latter about 4600 species. Two states entirely within the Appalachian Province are Indiana, with 1838 indigenous and 302 introduced species (C. C. Deam 1940), and West Virginia, with 2155 total species (P. D. Strasbaugh and E. L. Core 1978). If the species of the rich flora of the southeastern portion of the Appalachian Province are added (an area excluded from the works of M. L. Fernald, and H. A. Gleason and A. Cronquist), the total vascular flora for the province might be about 5500 to 6000 species.

In addition to the Eurasian–eastern North American disjuncts of the region as discussed above, mention should be made of the strong floristic relationships between the southern Appalachians and western North America. C. E. Wood Jr. (1971) found that 65% of the approximately 557 genera of seed plants in the southern Appalachians extend to western North America, and 158 of these have one or more taxa with wide disjunctions between the Appalachians and the West.

During the repeated Pleistocene glaciations and coastal plain inundations, the flora of the Appalachian Province found refuge in the Appalachian Highlands on the east and south and in the Ozarks to the southwest. This resulted in some allopatric speciation in the two refuge areas. Specific endemism is very high in this province, with numerous further endemics shared with the Coastal Plain Province. The great deciduous forest that once covered the Appalachian Province has been largely destroyed, or at least greatly altered, by the immigrant Europeans who replaced the Amerinds. In the coves of the southern Appalachians, however, especially the Great Smoky Mountains, there still exists a rich, mixed mesophytic forest of large trees of a score or more species in such genera as *Acer, Aesculus, Betula, Carya, Fagus, Liriodendron, Magnolia, Prunus, Quercus,* and *Tilia.* Various plant communities exist within the deciduous forest, as the *Fagus grandifolia–Acer saccharum* com-

***Abies fraseri*. Pen-and-ink drawing by Leta Hughey (late nineteenth–early twentieth century). [U.S. Department of Agriculture Forest Service Collection, Hunt Institute, Carnegie Mellon University, Pittsburgh.]**

munity to the north, *Acer saccharum–Tilia americana* community to the northwest, and *Quercus-Carya* communities on the drier fringes to the east, south, and west.

In addition to the dominant deciduous trees of the province, especially to the north and in the Appalachian highlands, are a number of conifers: *Pinus strobus, P. resinosa, Tsuga canadensis, T. caroliniana,* and *Taxus canadensis*. In the highest elevations of the Appalachians a spruce-fir forest is dominant, with *Abies balsamea* and *Picea glauca* both being replaced to the south by the endemic *Abies fraseri* and *Picea rubens*. Unfortunately this forest is now under severe stress from acid rain and other aerial pollution, especially on Mt. Mitchell, at 2000 m the highest mountain in eastern North America.

Among the special areas of considerable specific and subspecific endemism in the Appalachian Province are the shale barrens of western Virginia and neighboring states, the granite flatrocks of the Georgia Piedmont and adjacent states, the Paleozoic sedimentaries of the uplifted Cumberland Plateau, and the cedar glades of central Tennessee, Kentucky, and Alabama. Some of the endemics of these special habitats have western affinities, and they presumably reached their present habitats during the warmer, drier, postglacial hypsithermal period.

• 3b. Atlantic and Gulf Coastal Plain Province

Lying to the east and south of the Appalachian Province, the Atlantic and Gulf Coastal Plain Province ranges from the southern tip of Nova Scotia to eastern Texas. From the much attenuated coastal strip of Cape Cod, the New England islands, and Long Island, the coastal plain widens in New Jersey to a broad plain through the Delmarva Peninsula, Virginia, the Carolinas, much of southern Georgia, and most of Florida. Along the Mississippi River is a deep embayment extending to the southern tip of Illinois, the Mississippi Embayment. Physiographically this province is mostly sharply demarked from the Piedmont of the Appalachian Province by the Fall Line, although the vegetation and flora overlap considerably with those of the neighboring province.

Endemism is high in the Coastal Plain Province, where the flora contains several hundred endemic species, several endemic genera, and one endemic family, the monotypic Leitneriaceae. Because of repeated Pleistocene inundations, it has a young flora recruited largely from the much more ancient Appalachian Province, and to a lesser extent from the West Indian Province to the south and the North American Prairies Province to the west. Although many of the floristic elements are apparently of very recent origin, many presumably archaic genera are represented, such as *Ceratiola, Croomia, Dionaea, Franklinia, Gordonia, Illicium, Leitneria, Rhapidophyllum, Sarracenia, Schisandra, Taxodium, Taxus,* and *Torreya,* in addition to the many more widespread Arcto-Tertiary (or Boreo-tropical) genera.

Although no floristic treatment for the Coastal Plain Province has been attempted, the richness of the flora is indicated by some local floras completed for various parts of the coastal plain, as southern New Jersey with 1373 vascular plant species (W. Stone 1912), Delmarva

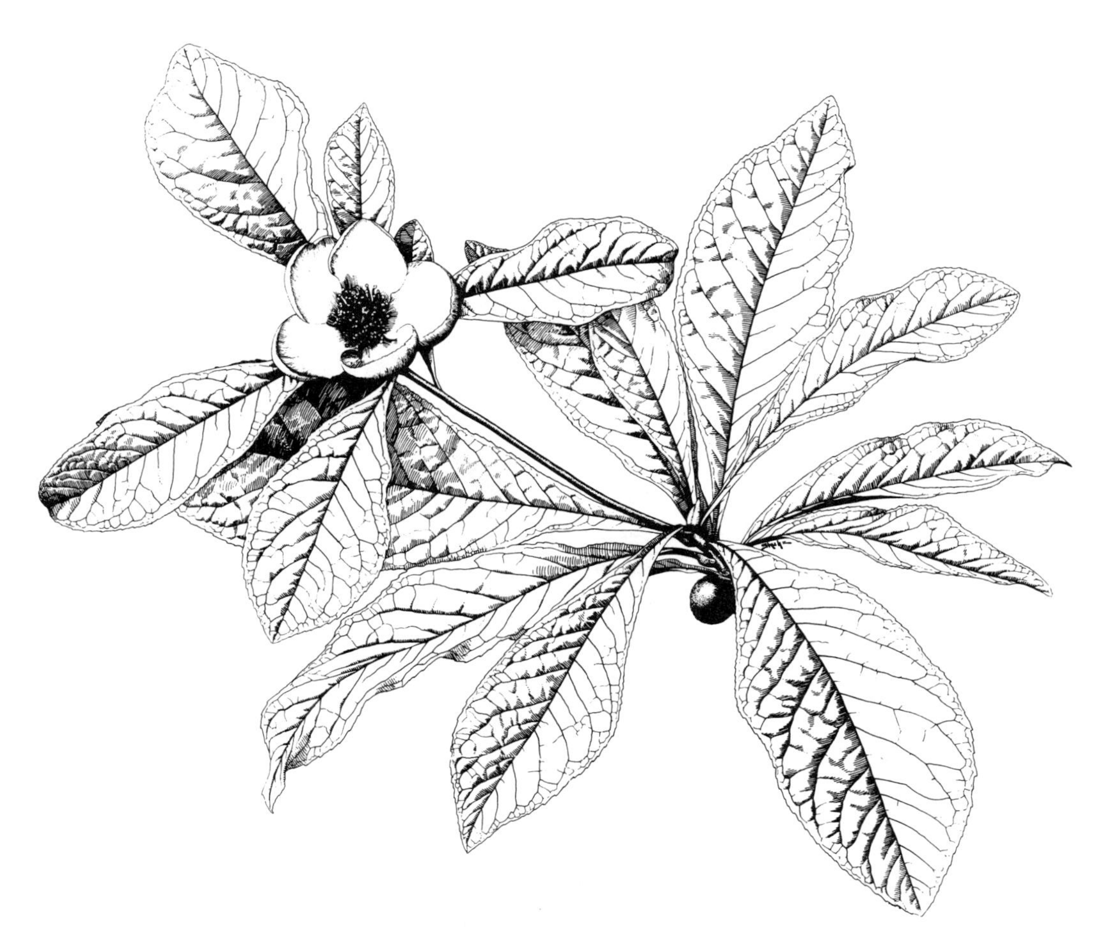

***Franklinia altamaha*. Pen-and-ink drawing by James M. Shull (1872–1948). [U.S. Department of Agriculture Forest Service Collection, Hunt Institute, Carnegie Mellon University, Pittsburgh.]**

Peninsula (mostly Atlantic Coastal Plain) with 2111 species (R. R. Tatnall 1946), southwestern Georgia on the Gulf Coastal Plain with 1750 species (R. F. Thorne 1954b), and central Florida with 2197 species (R. P. Wunderlin 1982).

The vegetation is conspicuously dominated by various species of *Pinus*, especially *P. rigida* to the north and *P. palustris* and *P. elliottii* to the south. The extensive pine forests are a response to repeated fires and to excessively or poorly drained, sandy soils. When recurring fire is eliminated, the vegetation rapidly becomes dominated by the economically less valuable hardwoods, such as species of *Quercus* and *Carya* on drier soils and *Fagus, Magnolia,* evergreen *Quercus, Persea,* and other genera on moister, richer soils. Extensive swampy areas, as characterized by the Dismal Swamp of Virginia and the Okefenokee Swamp of Georgia, are dominated by the two subspecies of *Taxodium distichum,* and species of *Nyssa, Fraxinus, Populus, Ulmus,* and other broad-leaved hardwoods where the flooding is seasonal.

Coastal marshes are dominated by *Juncus roemerianus* and various coarse, rhizomatous grasses, especially of the genus *Spartina*. Floristically very rich are the moist, sandy, acid pinelands with their abundance of species of Cyperaceae, Eriocaulaceae, Poaceae, Xyridaceae, terrestrial orchids, and insectivorous members of Droseraceae, Lentibulariaceae, and *Sarracenia*. Other habitats of special floristic significance are the bay-heads, with their broad-leaved evergreens of genera such as *Gordonia, Magnolia, Persea, Ilex, Cyrilla,* and *Cliftonia,* and the sand hills, especially of Georgia and Florida, with their numerous endemics.

Several areas of extremely high endemism on the Florida Coastal Plain seem to indicate refugia that were not flooded during the Pleistocene. Among these flo-

***Andropogon gerardii*. Pen-and-ink drawing by unknown artist. [Hitchcock-Chase Collection of Grass Drawings, Hunt Institute, Carnegie Mellon University, Pittsburgh, indefinite loan from Smithsonian Institution.]**

ristically significant areas are the Orange Island refugium of central peninsular Florida, the banks of the Apalachicola River, the Marianna Red Hills, and coastal flatwoods in the Florida panhandle. The Apalachicola and Marianna areas have had their floras much enhanced by Appalachian elements carried down the Chattahoochee River (R. F. Thorne 1949).

Many of the genera and species that are conspicuous on the southern coastal plain are also found in the highlands of Mexico and Guatemala. Among these genera are *Carpinus, Carya, Clethra, Fraxinus, Illicium, Liquidambar, Magnolia, Nyssa, Ostrya, Pinus, Platanus,* and *Quercus*.

• 3c. North American Prairies Province

The North American Prairies Province is a huge grassland province of prairies and plains, the latter rather equivalent to steppes, lying between the Appalachian deciduous forest on the east and the Rocky Mountains on the west, the Canadian coniferous forests on the north, and the arid semideserts to the southwest. Only at the Front Ranges of the Rocky Mountains are the boundaries of this province well defined. At least formerly, prairies reached eastward through the Midwest to Ohio or farther east (the "prairie peninsula"), intergrading on the uplands with the deciduous woodlands of the lowlands. This intergradation still obtains throughout Iowa and the eastern Dakotas, southward to northern Oklahoma. The virgin tallgrass prairie remaining today is largely restricted to a few botanic preserves of limited area. There remain, however, some extensive tracts of tallgrass prairie in more or less pristine condition in the Flint Hills of east central Kansas. Much of the region has been grazed, but it has not been severely mismanaged.

Northward the grasslands in Minnesota and central

Buchloë dactyloides. **Pen-and-ink drawing by Leta Hughey (late nineteenth–early twentieth century). Top: staminate plant; bottom: pistillate plant. [U.S. Department of Agriculture Forest Service Collection, Hunt Institute, Carnegie Mellon University, Pittsburgh.]**

Canada intergrade with *Populus tremuloides* copses on the border of the Canadian Province, and westward they grade into the coniferous forests of the Rocky Mountain Province. Southwestward some mesquite-grasslands occupy highlands in the Chihuahuan and Sonoran desert subprovinces.

Because of this extensive intergradation on most boundaries and the occupation of drier, open sites in adjacent provinces of members of the grassland flora, endemism is rather limited in the Prairies Province. There are no endemic families, few if any endemic genera, and perhaps fewer than 50 endemic species. Grasses dominate the province although showy perennial forbs, especially of the Asteraceae and Fabaceae, are conspicuous. To the east, the tallgrasses of the genera *Andropogon, Sorghastrum, Panicum,* and *Spartina* are dominant. In the more arid west, the short grasses *Buchloë dactyloides, Bouteloua gracilis,* and *Bouteloua hirsuta* are dominant. Between the two north-south strips of grassland are midgrasses such as *Bouteloua curtipendula.*

The mostly recent and adventive flora of this province is somewhat limited because of the general lack of forests and highlands. Including the Black Hills, South Dakota has a vascular plant flora of 1585 species (T. Van Bruggen 1976). The vascular plant flora of the Great Plains in the United States, one-fifth of the area of the conterminous United States of America, apparently consists of only 3067 taxa (Great Plains Flora Association 1977, 1986).

Much of the Prairies Province was glaciated during the Pleistocene glacial incursions; hence the recentness of the flora, especially in the northern portions. In the east, greater moisture and absence of fire have allowed deciduous forests to invade much of the former prairie peninsula of the Midwest. A few patches of grassland

Lomatium triternatum. Pen-and-ink drawing by unknown artist. [U.S. Department of Agriculture Forest Service Collection, Hunt Institute, Carnegie Mellon University, Pittsburgh.]

and the black prairie loam, among the richest soils in the world, indicate the former distribution of prairie. The grasslands, like the pine forests of the coastal plain, were to a large extent maintained by fire, formerly set by lightning and by the Amerinds. In western Iowa, fire, light grazing, or late harvests of wild hay are necessary to maintain prairie in good health. Otherwise, rhizomatous shrubs, weedy trees, or introduced perennial grasses take over.

Several areas within the Prairie Province are outliers of adjacent provinces, including especially the Black Hills of South Dakota, Pine Hills of Nebraska, and Edwards Plateau of Texas. The former two, though dominated by Cordilleran conifers, have many grassland elements, and the Black Hills have numerous boreal and eastern American elements as well. The Edwards Plateau has grassland and Chihuahuan and Tamaulipan desert elements, but also at least 20 endemic species of great interest, especially in the genera *Buddleja, Dasylirion, Forestiera, Styrax,* and *Yucca.*

4. ROCKY MOUNTAIN REGION

The high Rocky and Cascade-Sierran mountain systems of western North America, along with the Coastal Ranges south from Kodiak Island of Alaska to the San Francisco Bay Area of California, are included by A. L. Takhtajan (1986) in two provinces of the Rocky Mountain region, the inland Rocky Mountain Province and the coastal Vancouverian Province. This region has no strictly endemic vascular plant family but it has many endemic or near-endemic genera and numerous endemic species. Other than *Arnica, Castilleja, Erigeron,* and *Lomatium,* no genus seems to have its major center of diversity in this region; the widespread sedge genus *Carex* is represented by about 200 species.

The vegetation is dominantly coniferous forest with

the greatest diversity of conifers in the New World. Most coniferous species, other than *Pinus ponderosa, P. contorta,* and *Pseudotsuga menziesii,* are not dominant and widespread through both provinces. Deciduous trees form a minor part of the vegetation except for the numerous groves of quaking aspen, *Populus tremuloides.* Above timberline in the highest mountains is often a considerable area of alpine tundra with many arctic species, especially in the more northern alpine areas.

• 4a. Vancouverian Province

The Vancouverian Province extends as a coastal strip from Kodiak Island, Alaska, southward through the Alaskan panhandle and British Columbia, widening in Washington and Oregon to include the Cascade Mountains plus the Olympic Mountains and other Coastal Ranges, and in California stretching, according to A. L. Takhtajan (1986), through the Klamath and Coastal Ranges to the San Francisco Bay Area and through the Cascades and to the south end of the Sierra Nevada. I prefer this treatment of these areas to the more inclusive Californian Province of P. H. Raven and D. I. Axelrod (1978). On the other hand, I think the Warner Mountains of northeastern California are better included in the Madrean Great Basin Province rather than the Vancouverian Province. The mixture of species from both provinces makes placement of this latter range rather arbitrary.

As with most other floristic regions, there is no floristic treatment of the Vancouverian Province, and the authors, such as C. L. Hitchcock et al. (1955–1969) and C. L. Hitchcock and A. Cronquist (1973), of the regional treatments that most nearly cover the province have not presented a statistical summary for their works. Three local floras within the Vancouverian Province do give some useful data. According to J. A. Calder and R. L. Taylor (1968), the Queen Charlotte Islands of British Columbia have a vascular flora of 593 species and subspecific taxa in 277 genera and 69 families. W. C. Muenscher (1941) found in Whatcom County, the northwesternmost county of Washington, 1008 species in 422 genera and 91 families.

G. N. Jones (1936), for the Olympic Peninsula of Washington, listed 1015 species, with 138 species or 13% adventive, about one-third the flora of the state. He reported 20 species or infraspecific taxa as restricted to the Olympic Peninsula, but 140 species of the Cascade Mountains as missing from the Olympic Mountains. Two regional floras that cover more than the Vancouverian Province are that for British Columbia by R. L. Taylor and B. MacBryde (1977), with 3137 vascular plant species and infraspecific taxa, 2475 or 79% indigenous, in 744 genera and 131 families, and that for Oregon by M. E. Peck (1961), with 3370 species and 604 infraspecific taxa in 749 genera and 117 families. Surely, the flora of the Vancouverian Province must include at least 3000 to 3500 species of indigenous vascular plants.

As to endemism in the Vancouverian Province, C. V. Piper (1906) reported 158 species and 27 subspecies as restricted to the state of Washington, and G. L. Stebbins and J. Major (1965) indicate 406 endemic species for their Vancouverian areas in California and immediately adjacent Oregon. Thus, perhaps 500 to 600 species are restricted, or largely so, to this province. Among the more noteworthy endemics are such plants as *Sequoia sempervirens, Sequoiadendron giganteum, Vancouveria,* the insectivorous *Darlingtonia californica,* the mycophytic ericad genera *Allotropa, Hemitomes, Pityopus, Pleuricospora,* and *Sarcodes,* as well as *Oemlera, Whipplea,* and *Romanzoffia.*

The Klamath region is rich in locally restricted species, some of them apparently archaic relicts, others specialists on serpentine and peridotite substrata. In a study area of about 13 square kilometers around English Peak in the Marble Mountain Wilderness area of Siskiyou County, northwestern California, F. W. Oettinger (1975) found 452 indigenous and 29 introduced vascular plant species in 243 genera of 64 families. Twenty-four of the indigenous species are largely restricted to the Klamath Mountains region. In the nearby Trinity Alps, W. J. Ferlatte (1974) treated 571 species in 266 genera and 73 families. C. E. Wood Jr. (1971) pointed to this region, especially the Siskiyou Mountains, as the "richest conservatory of relict groups of plants in western North America." Perhaps the most interesting of the relicts is *Picea breweriana,* belonging to a Eurasian section. The flora of this region is transitional to the adjacent Californian region. Other areas of high endemism in the Vancouverian Province are the Olympic and Wenatchee ranges of Washington, the Columbia River Gorge, the southern Cascade Mountains in California, and the Sierra Nevada.

The vegetation of the Vancouverian Province, like the Rocky Mountain Province, is dominated by conifers, but broad-leaved trees are more conspicuous in the Vancouverian Province than in the Rocky Mountain ranges. Among the broad-leaved trees are *Acer macrophyllum, Alnus rubra, Cornus nuttallii, Fraxinus latifolia, Populus trichocarpa, Quercus* spp., *Arbutus menziesii,* and *Umbellularia californica.* The last two trees are evergreen as are many of the common shrubs, such as *Mahonia aquifolia (Berberis), Gaultheria shallon,* and *Rhododendron macrophyllum.*

Coastal slopes of British Columbia and Olympic Peninsula of Washington support a very wet and tall temperate rainforest, dominated by huge trees of *Picea sitchensis, Thuja plicata,* and *Tsuga heterophylla,* accompanied by smaller broad-leaved *Alnus rubra, Acer macrophyllum, Cornus nuttallii,* and *Arbutus men-*

***Picea sitchensis*. Pen-and-ink drawing by Frederick Walpole (1861–1904). [Hunt Institute, Carnegie Mellon University, Pittsburgh, indefinite loan from Smithsonian Institution.]**

ziesii. To the east and south, summer drought increases fire frequency and thus favors *Pseudotsuga menziesii.* An attenuated version of the temperate rainforest to the south is the redwood forest, flourishing on seaward slopes that are mostly bathed in fog year-round. The major tree of this forest, *Sequoia sempervirens,* is one of the tallest trees in the world. The species is constantly under attack by timbermen and is now, except in limited federal and state parks, often being replaced by the faster growing *Pseudotsuga menziesii,* favored economically by the timber interests.

As in the Rocky Mountains, the Vancouverian ranges, particularly the southern Cascades and Sierra Nevada, show striking vertical zonation in the forest cover. Above the Madrean chaparral and foothill woodland of the western Sierran slopes, the lowest montane zone is mostly an open *Pinus ponderosa* forest, grading upward into a mixed conifer forest of *Pinus jeffreyi, P. lambertiana, Abies lowiana,* and, locally, *Sequoiadendron giganteum,* the world's largest, and among the oldest, of living things.

Above the mixed conifer forest, *Abies magnifica* and *Pinus contorta* subsp. *murrayana* are dominant, often in rather pure stands. The subalpine forest above them, upward to timberline, is dominated by *Tsuga mertensiana* and several pines, *Pinus monticola, P. albicaulis, P. balfouriana,* and *P. flexilis.* Above timberline the alpine tundra harbors many arctic-alpine and circumboreal species as well as local endemics. The Vancouverian Province contains many wide-ranging species, as well as more local species, intruding from adjacent Boreal and Madrean provinces.

• 4b. Rocky Mountain Province

The Rocky Mountain Province includes the Rocky Mountains and associated ranges from northern British Columbia to the Blue and Wallowa ranges of central Oregon, the Uinta Mountains of northeastern Utah, and the south end of the Rockies in northern New Mexico. Because the Canadian parts of this province have been heavily glaciated, the flora there is very recent, relatively depauperate, and low in endemism. There is much intergradation with the Canadian Province northward, and many of the species are circumpolar or circumboreal, especially in the alpine areas.

Southward in the United States, where glaciation was merely local, the flora is much richer and the degree of endemism considerably higher. According to A. L. Takhtajan (1986), five genera are endemic to this province, and 10 more are shared with the Vancouverian Province. Takhtajan lists about 40 endemic species for such special areas in the province as the sagebrush lands

***Arbutus menziesii.* Pen-and-ink drawing by Frederick Walpole (1861–1904). [Hunt Institute, Carnegie Mellon University, Pittsburgh, indefinite loan from Smithsonian Institution.]**

of central Washington, the John Day Valley in north central Oregon, the Palouse country of southeastern Washington and adjacent Idaho, the Snake River Canyon between the mountains of northeastern Oregon and those of west central Idaho, and the sagebrush lands and seleniferous outcrops of southern Wyoming. He also mentions about 50 species as being quite or nearly confined to the southern Rocky Mountains.

Aside from the floristic elements fingering into the Rocky Mountains from adjacent provinces, several geographic elements are of considerable interest. W. A. Weber (1953, 1965, 1987), among others, has noted the following: an eastern deciduous forest element limited to mesic and cool places along the eastern edge of the uplift, circumpolar and boreal elements, relict Madro-Tertiary species, a southwestern desert-steppe element, and a particularly interesting element analogous to one in the Altai Mountains of central Asia in similar semi-arid, high elevation areas. Weeds, at least in the Colorado Rockies, are largely natives of southeastern Europe and Asia Minor.

Because many floras lack statistical summaries, it is difficult to obtain statistics on the size of the Rocky Mountain flora. J. M. Coulter and A. Nelson (1909) found 2733 species in the Central Rocky Mountains, and W. A. Weber and R. C. Wittmann (1992) located 3088 vascular plants (including infraspecific taxa) for the whole state of Colorado. P. A. Rydberg (1917) estimated about 4000 species for the Rocky Mountains. C. V. Piper and R. K. Beattie (1901) listed 663 species just for their flora of the Palouse region of eastern Washington and adjacent Idaho. Probably the province contains about 4000 to 4500 species.

The vegetation of the Rocky Mountain Province is conspicuously zoned vertically. Open stands of *Pinus ponderosa* form the lowest forest zone; they give way above to *Pseudotsuga menziesii,* which in turn in the uppermost forests is replaced by *Abies bifolia* and *Picea engelmannii.* Especially at middle elevations, *Populus tremuloides* and the fire-dependent pine, *Pinus contorta,* are often abundant, as is also *Picea pungens* southward. Scattered small trees of *Juniperus scopulorum* often are found in the drier foothills, and in southern Colorado and New Mexico scrubby stands of *Quercus gambelii* often occur below the forest.

(Madrean or Sonoran Subkingdom)

9. MADREAN REGION

The flora of the American Southwest and adjacent areas of northern and central Mexico is so distinct from the two other Holarctic subkingdoms, the Boreal and the Tethyan (or Mediterranean), that it deserves its high chorionomic rank. The flora is largely locally derived, and the region takes its name from the Sierra Madre Occidental of Mexico. This single Madrean region is represented in North America north of Mexico by three provinces, the Great Basin, Californian, and Sonoran, and the last by four subprovinces, Mojavean, Sonoran, Chihuahuan, and Tamaulipan.

The Madrean flora is exceedingly rich and distinctive. At least three families, Fouquieriaceae, Simmondsiaceae, and Setchellanthaceae, are endemic, and five other wider-ranging families, Crossosomataceae, Garryaceae, Lennoaceae, Limnanthaceae, and Stegnospermataceae, have their principal development within the region. Several larger families have their major centers of diversity in the region, e.g., Onagraceae, Polemoniaceae, and Hydrophyllaceae. Other higher taxa especially well developed there are Agavaceae, Alliaceae, Apiaceae, Arbuteae, Asteraceae, Boraginaceae, Brassicaceae, Cactaceae, Orcuttieae, Papaveraceae, Polygonaceae, Rhamnaceae, and Scrophulariaceae.

According to A. L. Takhtajan (1986), more than 250 genera and probably more than half of the species are endemic. Most of the genera are common to two or more of the Madrean provinces; likewise many of the species. Most of the Madrean taxa are adapted to arid-desert or Mediterranean-type climates and have a long history of such adaptation since the more humid, widespread climates of the late Cretaceous and early Tertiary.

• 9a. Great Basin Province

The Great Basin Province includes most of the Great Basin between the Rocky Mountains and Cascade–Sierra Nevada axis, including the Warner Mountains of California, the Snake River Plains of southern Idaho, and most of the Colorado Plateau, including the Uinta Basin but not the Uinta Mountains. Provincial boundaries are mostly well defined by the bases of the surrounding ranges. In Arizona the Mogollon Rim is a conspicuous boundary. The physiographic Great Basin consists of many north-south trending mountain ranges separated by often broad, alkaline basins without external drainage. Much of the area lies above 1300 meters and is subject to the continental climate of hot summers and cold winters. Most of the limited precipitation comes as snow during the winters.

The Great Basin flora has been developed largely since Miocene time and has been shaped by drought and by the cold winters, thus restricting migration from the warmer areas to the south. The higher ranges show a strong Rocky Mountain influence, and the lower elevations harbor wide-ranging species of arid lands from many of the adjacent provinces. Thus, endemism is relatively low with but few endemic genera and perhaps 25% specific endemism (A. L. Takhtajan 1986). The three largest genera in the province are *Astragalus* with 175 species, and *Eriogonum* and *Penstemon* with more than 100 species each. Because each genus has 50 to 60 species endemic to the province, the Great Basin can be considered their center of diversity. Other well-developed genera are *Cymopterus, Lomatium,* subgenus *Oreocarya* of *Cryptantha, Chrysothamnus, Erigeron, Phacelia, Castilleja,* and *Gilia.* All these genera taken together have about 275 endemic species. Many of these and local species of other genera in the province appear to be of rather recent origin. Edaphically distinct habitats such as tuffaceous slopes, hanging gardens on sandstones in the Colorado River drainage, sand dunes, dolomites, or soils rich in gypsum or heavy metals often have their own local species.

The flora of the Great Basin Province is included within the multivolume *Intermountain Flora* by A. Cronquist et al. (1972+). The boundaries of that flora are essentially the same as our defined province, except that it includes all of Utah. Unfortunately, the work is incomplete, and the authors have not yet produced a floristic analysis. Utah and Nevada comprise most of the Great Basin; hence, two recent floristic studies can give some approximation of the Great Basin flora. In *A Flora of Nevada,* J. T. Kartesz (1988) mentioned more than 3000 species of vascular plants, and B. J. Albee et al. (1988), in *Atlas of the Vascular Plants of Utah,* included 2438 mapped and 384 listed taxa for a total for Utah of 2822 vascular plants. Considering that both treatments range beyond the Great Basin and have numerous species in common, one might estimate that the Great Basin flora is somewhat less than 3000 species of vascular plants.

The vegetation is characterized by the wide-ranging Great Basin sagebrush, *Artemisia tridentata* and its close relatives, on better soils, and by members of the Chen-

***Artemisia tridentata*. Pen-and-ink drawing by Frederick Walpole (1861–1904). [Hunt Institute, Carnegie Mellon University, Pittsburgh, indefinite loan from Smithsonian Institution.]**

opodiaceae, especially species of *Atriplex,* on more alkaline soils. Before the severe overgrazing of much of these lands, many grasses accompanied the sagebrush. Above the sagebrush zone on the mesas is often found a woodland of scattered junipers or junipers and pinyons, *Juniperus* and *Pinus* species. Often an open *Pinus ponderosa* forest, or in the south a chaparral dominated by *Quercus gambelii,* occurs above this zone, where somewhat more moisture is available. At higher montane to alpine elevations, the flora is largely a Rocky Mountain flora with dominance by conifers of the genera *Pseudotsuga, Picea,* and *Abies,* and at timberline by species of *Pinus.* One of these is the bristlecone pine, *P. longaeva,* members of which are the world's oldest living vascular plants, attaining ages of nearly 5000 years. Broad-leaved tree species are rather few and include a few species of *Populus* and *Acer.* The alpine flora consists largely of wide-ranging genera but with few circumboreal arctic-alpine species.

• 9b. Californian Province

The floristically rich Californian Province occupies most of cismontane California and a portion of adjacent northwestern Baja California. As defined by A. L. Takhtajan (1986), it includes the great Central Valley, inner North Coast ranges, South Coast ranges, Transverse and Peninsular ranges of southern California, and foothills of the Cascade and Sierra Nevada ranges up to the ponderosa pine forests at about 1200 meters in the southern Sierra Nevada. This is a narrower definition than that of P. H. Raven and D. I. Axelrod (1978), but it has the advantage of conforming more exactly to the area subject to the Mediterranean-type climate of moist, cool winters and hot, dry summers.

The Californian Province, as defined by Raven and Axelrod, was calculated by them to contain 4452 species of vascular plants, with 2125 of those (47.7%) considered endemic to the province. As redefined here, the province may contain fewer than 4000 species with about half of them (50%) endemic. A few of the more than 50 California Province endemic or near-endemic genera are *Adenostoma, Bergerocactus, Carpenteria, Cneoridium, Dendromecon, Fremontodendron, Jepsonia, Lyonothamnus, Neostapfia, Odontostomum, Ornithostaphylos, Pickeringia,* and *Romneya.* Among the larger genera and higher taxa that have their main, or at least a principal, center of diversity within the Californian Province are the following: *Arctostaphylos*, Brodiaeinae of Alliaceae, *Calochortus, Caulanthus, Streptanthus, Ceanothus, Cryptantha* (subg. *Krynitzkia*), *Downingia, Dudleya,* Eritrichieae of Boraginaceae, Eriogonoideae of Polygonaceae, Gileae of Polemoniaceae, Hydrophyllaceae, Limnanthaceae, *Lotus* (subg. *Hosackia*), Madiinae of Asteraceae, *Mimulus,* Onagreae and Epilobieae of Onagraceae, Orcuttieae of Poaceae, and Eschscholzioideae and Platystemonoideae of Papaveraceae. Also, *Astragalus* and *Cupressus* have radiated strongly in the province.

In addition to numerous floristic elements intruding into the province from adjacent provinces, there is a strong link between the Mediterranean climatic areas of central Chile and California. Nearly 100 closely related species pairs or identical species are known. Mostly they are self-compatible annuals of open habitats with small disseminules. Their Californian distribution seems to be primary, and their dispersal apparently is recent, repeated, and by migrating birds, most probably by shore birds.

There is also a less obvious linkage with the Mediterranean region of southern Europe and north Africa, ignoring the obvious large contingent of annual grasses and weeds brought from the Mediterranean by early Europeans. *Styrax officinalis* and other taxa with Mediterranean affinities are discussed in more detail later in this chapter.

The vegetation of the Californian Province is varied and shows some elevational zonation. The Central Valley, which in the recent geologic past was a large inland sea, was in early historical time a vast prairie apparently dominated by bunchgrasses, species of *Stipa, Poa, Melica, Elymus, Aristida, Festuca,* and *Koeleria,* accompanied after moister winters by a lush display of showy forbs. Lower areas and shallow basins with an impervious substrate, filled with water in the winter and desiccated through evaporation in the spring and early summer, gave rise to a special vernal-pool or vernal-marsh flora of nearly 200 species (R. F. Holland 1976), many of them endemic to the province. From a total of 128 genera with species found in vernal pools, 16 have their Californian species essentially limited to this habitat (R. F. Thorne 1984). Other areas of the valley with high alkalinity have equally special saltbush scrub communities. Now most of these plant communities in the Central Valley have been obliterated by agricultural, commercial, and residential development.

Except for the riparian gallery-woodlands of *Platanus racemosa* and species of *Salix, Populus,* and *Quercus,* valley grassland was largely treeless. Around the margins above the valley floor, a grassy savanna with scattered trees of several species of *Quercus* merges upward into open foothill woodland of *Quercus douglasii, Pinus sabiniana,* and *Aesculus californica.* On drier foothill and lower montane slopes, sclerophyllous chaparral replaces the oak and oak-pine woodlands.

A mixed evergreen forest occurs in moister canyons and on shaded slopes of the Coast, Transverse, and Peninsular ranges, and includes the conifers *Pseudotsuga macrocarpa* and *Calocedrus decurrens,* broad-leaved evergreen species of *Quercus* and *Umbellularia californica,* and such deciduous species as *Acer macrophyllum* and *Fraxinus dipetala.* The higher montane and subalpine slopes of these ranges support forests similar to those found in the Sierra Nevada, including, with increasing elevation, forested zones dominated by *Pinus ponderosa* and *P. jeffreyi, Abies concolor* and *Pinus lambertiana, Pinus contorta* subsp. *murrayana,* and at timberline *Pinus flexilis,* often mixed with *P. contorta* subsp. *murrayana.* The few southern Californian peaks with timberline and krummholz have a very limited alpine fell-field cushion community, with a mixture of widespread arctic-alpine and local species.

Occupying drier slopes below the chaparral in interior valleys, and especially along the coast, a low sage scrub of soft (malacophyllous) shrubs of *Salvia, Artemisia, Eriogonum,* and other genera was formerly very extensive. Other special plant communities of the Californian Province include closed-cone coniferous woodlands dominated by various species of fire-oriented *Cupressus* and *Pinus*; open woodlands dominated by *Pinus coulteri, Juniperus californica,* or *Pinus torreyana*; serpentine woodland; maritime-desert scrub; and various other maritime communities of coastal bluffs, dunes, salinas, marshes, and submarine habitats.

• 9c. Sonoran Province

The arid North American Southwest, especially the Mojave, Sonoran, and Chihuahuan deserts, is mostly included within the Sonoran Province of the Madrean region. It covers most of northwestern Mexico and much of the southwestern United States from southern California and southern Nevada to southeastern Texas. North of the border are four readily distinguished subprovinces, the Mojavean, Sonoran, Chihuahuan, and Tamaulipan.

9c.1. *Mojavean Subprovince* Takhtajan prefers to treat the Mojave Desert as a district of the Sonoran Desert, but I consider it as a well-marked subprovince that is rather distinct in elevation, climate, and flora from the more tropical, lower Sonoran Desert. As defined here, this subprovince includes Death Valley and adjacent ranges bordering the Great Basin Province (the Inyo region of P. H. Raven and D. I. Axelrod 1978) south through the "high" desert of San Bernardino County, California, to the base of the Transverse ranges, including the Little San Bernardino and Eagle desert ranges separating this desert from the Colorado portion of the Sonoran Desert. To the east, the Mojave Desert extends through southern Nevada to the southwestern tip of Utah and south into northwestern Arizona. The borders with the Great Basin and Sonoran Desert are not well defined. The Mojave Desert is probably best characterized by *Yucca brevifolia*, the Joshua-tree, whose distribution is essentially that of the Mojave Desert.

According to Barry Prigge (pers. comm.), the Mojave Desert, defined approximately as above, possesses a total flora of 2085 taxa (including infraspecific entities), 186 of them naturalized. The 1878 species are included in 652 genera and 114 families. With some species counted in more than one group, the flora is apportioned among 1041 perennial herbs, 566 woody plants (trees, shrubs, and half-shrubs), 837 annuals, 81 biennials, and 150 others (parasites, vines, succulents, aquatics, etc.). Our floristic treatment of the higher ranges and Kelso Dunes of the eastern Mojave Desert in California (R. F. Thorne et al. 1981) listed 783 species with 717 of them indigenous.

P. H. Raven and D. I. Axelrod (1978) cited 108 species as being endemic to the Mojave Desert, including their Inyo region, though G. L. Stebbins and J. Major (1965) counted 138 endemics in 70 large- and intermediate-size genera and 78 relict genera for the California Inyo and Mojave regions.

The Mojave Desert is so diverse in topography, geology, and climate that I have listed (R. F. Thorne 1982) 21 plant communities and 9 subcommunities for the Mojavean Subprovince in habitats ranging upward in elevation from alkali sinks to montane white fir–pinyon woodland.

9c.2. *Sonoran Subprovince* North of the Mexican border, the Sonoran Desert ranges from the southern border of the Mojave Desert in California, southern Nevada, and southwestern Utah southward to include much of southern transmontane California, where it is called the Colorado Desert, and much of southwestern Arizona below 1050 meters to the southeastern corner of that state. Much more of the Sonoran Desert is found in the Mexican states of Baja California and Sonora. The total area of the desert north of the border in California and Arizona is about 105,420 km^2 or 56,571 mi^2 (F. Shreve 1951). It is lower and hotter than the Mojave Desert, with more summer rainfall and less frost in winter.

The flora, in consequence of this warmer climate, is more subtropical, largely Madro-Tertiary in origin, and contains many genera not found in the Mojavean Subprovince. In California alone, about 25 genera do not reach north to the Mojave Desert, among them *Ayenia, Bursera, Calliandra, Carnegiea, Castela, Colubrina, Fouquieria, Horsfordia, Koeberlinia, Olneya, Simmondsia,* and *Ziziphus*. There seems to be no available count of the species found in the Sonoran Desert north of the border, although P. H. Raven and D. I. Axelrod (1978) stated that about 100 species of the Colorado Desert are found nowhere else in California. F. Shreve and I. L. Wiggins (1964), in their valuable two-volume *Vegetation and Flora of the Sonoran Desert*, offered no statistical analysis of the flora. They did, however, treat 138 vascular plant families in the entire subprovince. A count in the two volumes gave 2683 species for the entire Sonoran Desert flora.

The degree of endemism in the Sonoran Desert north of the border has not been calculated, although G. L. Stebbins and J. Major (1965) have 27 endemics in the Colorado Desert of California in the 70 large- and intermediate-size genera, plus 39 relict genera. A. L. Takhtajan (1986) estimated endemism in the whole Sonoran Province as probably more than 25%.

Some of the plant communities of the Mojave Desert do not occur in the Sonoran Desert, among them the white fir–pinyon woodland, the Joshua-tree woodland, the blackbush scrub, the pigmy sagebrush scrub, and some of the alkali scrubs. On the other hand, such low desert woodlands (e.g., microphyll, arborescent cacti, desert oasis, and desert riparian) and stem-succulent (cactus) scrub are abundantly developed (R. F. Thorne 1982; D. E. Brown 1982c). In fact, the Sonoran Desert is best characterized by the microphyll woodlands, dominated by microphyllous trees of the Fabaceae and Rhamnaceae, by the arborescent cacti *Carnegiea, Pachycereus,* and *Stenocereus,* and by the great abundance of other succulent-stemmed cacti.

9c.3. *Chihuahuan Subprovince* The Chihuahuan Desert lies southeast of the Sonoran Desert, ranging north of the border from the southern edge of eastern Arizona to trans-Pecos Texas and southward well into Mexico, especially in Chihuahua and Coahuila. Because of its higher average elevation (mainly above 1070 m), the Chihuahuan Desert is cooler than the Sonoran Desert and subject in all parts to frost during the winter. Like the Sonoran Desert, a great variety of life forms are present, although, as in the two other deserts just discussed, great tracts are monotonous open stands of *Larrea divaricata* subsp. *tridentata*. Trees are found only along the drainageways, and the commonest tall plants are species of *Yucca* and cylindropuntias *(Opuntia)*. This desert is especially well endowed in gypsum-rich habitats, and A. L. Takhtajan (1986) listed nearly 50 species largely restricted to gypseous habitats there.

J. Henrickson (pers. comm.) has listed for the entire Chihuahuan Desert 3233 species of vascular plants with

***Larrea divaricata* subsp. *tridentata*. Pen-and-ink drawing by unknown artist. [U.S. Department of Agriculture Forest Service Collection, Hunt Institute, Carnegie Mellon University, Pittsburgh.]**

545 of them endemic, or 15.2% endemism. The relatively low level of endemism is probably due to the broad contacts with other drylands that allowed migration of plant taxa. In addition, the southwestern deserts expanded relatively recently in Holocene times (R. F. Thorne 1986). The isolation of the Baja California peninsula and the maritime nature of much of the Sonoran flora there presumably accounts for the greater endemism of the Sonoran Desert, estimated at 20–25%.

9c.4. *Tamaulipan Subprovince* The Tamaulipan Subprovince, because of the greater available moisture, is less desertlike than the other Sonoran subprovinces, and it is more of a thornscrub or mesquite-grassland. In southern Texas, the Tamaulipan Subprovince occupies some 8 million hectares (20 million acres) (D. S. Correll and M. C. Johnston 1970) and is known as the Rio Grande Plains or South Texas Plains. The rolling to level topography, ranging from sea level to 1000 feet, is dissected by streams and arroyos flowing into the Rio Grande or the Gulf of Mexico. The vegetation is largely open prairies, mostly with *Prosopis,* and brushlands with shrubby, often spiny, species of *Celtis, Ziziphus, Karwinskia, Porlieria, Aloysia, Condalia, Castela, Acacia, Leucophyllum,* and other genera with Mexican affinities. Much of the area is now rangeland.

This area of Texas and the coastal strip about Brownsville are of special interest to our study, for they accommodate representatives of genera and even families not found indigenous elsewhere in North America north of Mexico. Among these are *Phaulothamnus spinescens* of Achatocarpaceae, *Xylosma flexuosa* of Flacourtiaceae, *Helietta parvifolia* and *Esenbeckia berlandieri* of Rutaceae, *Urvillea ulmacea* and *Serjania brachycarpa* of Sapindaceae, to mention only a few. Along the south Texas coast, the mangroves, marine phanerogams, *Batis maritima, Sabal mexicana,* and other tropical species add greatly to the diversity of the flora.

B. Neotropical Kingdom

23. CARIBBEAN REGION

• 23b. West Indian Province

The only portion of North America north of Mexico that is included in the Neotropical Kingdom is the southern third of the Florida peninsula and the Florida Keys. This subtropical area is included in the West Indian Province of the Caribbean region. To botanists trained in temperate North America, this limited area of Florida is a different botanical world, dominated by West Indian genera, tropical families, palms, mangroves, epiphytes, showy tropical cultivated plants, and tropical weeds.

In addition to southern Florida and its Keys, the West Indian Province includes the Bermuda, Bahama, and Greater Antilles and Lesser Antilles archipelagoes. The province as a whole probably contains at least 200 endemic genera and perhaps 50% species endemism (A. L. Takhtajan 1986). There is one endemic family, Goetzeaceae, and the Floridian portion of the province is rather low in endemism because of the Pleistocene and post-Pleistocene immigration of the West Indian biota into Florida. R. W. Long and O. Lakela (1971) listed 1647 species for the three southernmost counties of Florida (Collier, Dade, and Monroe) with about 9% of the species endemic to Florida. They consider 61% of the flora to be tropical in relationship, with 91% of these elements being species occurring in the Caribbean area. Endemism in the tropical element is especially low and surely is very recent. The sandhill endemics, on the other hand, are non-Caribbean in origin, and mostly much older, many probably of Tertiary origin.

Nineteen indigenous families are wholly or largely restricted to southern Florida in the flora of North America north of Mexico: among them are the Avicenniaceae, Canellaceae, Combretaceae, Cymodoceaceae, Goodeniaceae, Meliaceae, Myrsinaceae, Myrtaceae, Olacaceae, Piperaceae, Rhizophoraceae, Surianaceae, Theophrastaceae, and Zamiaceae. Also 13 of the 17 naturalized families in Florida are similarly restricted (R. F. Thorne 1985). They are found primarily in the more tropical plant communities, such as the marine meadows, mangrove swamps, tropical beaches and dunes, and shore and tropical hammocks. This West Indian flora also includes 187 genera, with at least 128 more introduced and presumably naturalized. The West Indian element occurs much farther north than the three counties treated by R. W. Long and O. Lakela (1971) as tropical Florida. Many are found northward to Lake Okeechobee, and the tropical maritime species range well up the coasts to Cedar Keys and even to St. Marys River on the Atlantic Coast.

Origins and Relationships of the Present Flora

The present flora of North America north of Mexico, as mentioned earlier, contains indigenous representatives of 211 angiosperm families and 99 additional subfamilies from the world flora of 443 angiosperm families and 272 additional subfamilies as recognized in my current classification. Twenty-eight pteridophyte and gymnosperm families are also represented. This can be compared with the much richer, more tropical flora of Central America and Mexico, which has indigenous representatives of 245 angiosperm families and 130 additional subfamilies. The richest flora of higher taxa of angiosperms is possessed by Asia (excluding Malesia) with 289 indigenous families and 156 additional subfamilies represented. Table 6.1 allows comparison of the North American flora with that of other continental and oceanic chorological regions.

Only three families are restricted to North America north of Mexico: Hydrastidaceae, Limnanthaceae, and Leitneriaceae, each monogeneric. Five more families and four subfamilies, however, are shared with Mexico or extend farther into Central America and the West Indies: Crossosomataceae, Fouquieriaceae, Garryaceae, Simmondsiaceae, Stegnospermataceae, the papaveraceous Eschscholzioideae and Platyspermoideae, dracaenaceous Nolinoideae, and agavaceous Yuccoideae. The parasitic Lennoaceae reach slightly farther south into Colombia in northwestern South America. The relatively poor representation of endemic families and subfamilies in North America is a result of the long involvement of the continent with Eurasia and the more recent Pliocene linkup with South America through the Central American isthmus. North America thus lacks the isolation that results in the high family and generic endemism found in Australia, South America, Africa south of the Sahara, Madagascar, and New Caledonia.

North American Loss of Floristic Richness

The North American flora was much richer in the past. For example, the present flora of California contains indigenous species of 140 angiosperm families, but fossils record at least 40 more families no longer found in the state (R. F. Thorne 1986). Several families, such as Bombacaceae, Chloranthaceae, Dilleniaceae, Gunneraceae, Icacinaceae, Proteaceae, Sabiaceae, and Winteraceae, occur in the New World but are not now found north of Mexico. Others, such as Alangiaceae, Cercidiphyllaceae, Eupomatiaceae, and Trapaceae, no longer occur as natives in the Western Hemisphere. Many genera represented in Tertiary or Cretaceous strata of North America have also vanished as natives from the New World, e.g., *Caldesia, Cunninghamia, Cyclocarya, Engelhardia, Eucommia, Euodia, Exbucklandia, Glyptostrobus, Mastixia, Metasequoia, Nypa, Paliurus, Paulownia, Phellodendron, Platycarya, Pterocarya, Rhoiptelea, Sciadopitys,* and *Zelkova* (B. H. Tiffney 1985b, among others). This depauperization of our flora is well documented by J. A. Wolfe and T. Tanai (1987) for *Acer* and by S. R. Manchester (1987) for the Juglandaceae. An extensive review is provided by A. Graham (chap. 3).

Recent Additions to the North American Flora

The present subtropical West Indian flora of southern Florida must be a very recent addition, for much of the peninsula was under seawater until Pleistocene time. Some of the 19 indigenous families that are wholly or largely restricted to southern Florida are listed above in the discussion of the West Indian Province, along with the habitats in which most of them are found. Also mentioned above, 187 indigenous genera of West Indian origin and 128 naturalized genera are found now in southern Florida. The frequent hurricanes that sweep over southern Florida from the adjacent Bahamas, Cuba, and other West Indies can account for most of these Caribbean floristic elements now indigenous in southern Florida. The warm Gulf sea currents and migrating birds are presumably responsible for the other indigens, and the intentional or accidental activities of humankind for the more recent adventives.

Origins of the North American Flora

The origins of the North American flora are diverse. In addition to the numerous autochthonous elements, as in the Madro-Tertiary geoflora in the Southwest and the Arcto-Tertiary and Boreotropical geofloras throughout the continent, immigration from the other continents has enriched the flora.

MEXICAN RELATIONSHIPS

Another pattern of distribution concerns the plants of deserts and bordering semiarid areas. These are now concentrated south of the Mexican border, where warm deserts are larger than to the north. These plants seem clearly to have originated as the arid areas of North America expanded through the Tertiary—for the past 65 million years, and especially in the last 15 million years. They include western North American representatives of Achatocarpaceae, Agavaceae, Bignoniaceae, Buddlejaceae, Burseraceae, many Cactaceae, Cochlospermaceae, Dracaenaceae, Ebenaceae, Flacourtiaceae, Fouquieriaceae, Lennoaceae, Malpighiaceae, Martyniaceae, Menispermaceae, Nyctaginaceae, Passifloraceae, Rafflesiaceae, Sapotaceae, Simaroubaceae, and Sterculiaceae among others.

SOUTH AMERICAN RELATIONSHIPS

Other immigrants may well have come through Mexico and Central America but have originated in South America. The dominant plant of our southwestern deserts, *Larrea divaricata* subsp. *tridentata,* seems to be conspecific with the Patagonian population of *L. divaricata* subsp. *divaricata.* Likewise, *Frankenia salina* is disjunct between Chile and the Californias, as is *Koeberlinia spinosa* between Bolivia and our Southwest, and *Capparis atamisquea* between Argentina-Chile and Mexico-Arizona. Also with strong links to temperate South America are the southwestern species of *Menodora, Prosopis, Lycium,* and *Nicotiana.* All of these except *Prosopis* are represented also in South Africa. Lacking from South America now but with relatives in Africa are *Thamnosma* of the Rutaceae and *Selinocarpus* of the Nyctaginaceae, the former being found in Namibia, Somalia, and Arabia and the latter only in Somalia. A good many of the herbaceous species common to temperate South America and western North America seem to have emigrated southward, rather than northward from South America, as discussed earlier.

EURASIAN RELATIONSHIPS

Many of the circumboreal and circumarctic species found in Canada and Alaska, and some in higher mountains south of the Canadian border, whether originated in North America or Eurasia, obviously migrated readily and perhaps frequently between the two continents via the North Atlantic and Beringian land bridges when they were operative. Less commonly disseminules may have been blown across ice-bound straits or frozen seas. Where there has been evolutionary diversification, as in *Diapensia lapponica, Armeria maritima,* and *Scheuchzeria palustris,* it is sometimes possible to make out the probable land bridge used.

Some of the circumboreal species or genera are bipolar, presumably having been carried to Fuegia by migrating birds. Good examples of such are *Empetrum, Hippuris,* and *Primula.* The Old World distribution of other species offers the clue to the origins of our representatives. Thus our American species of *Echinopanax, Lysichiton, Paeonia,* and *Phyllospadix,* found only in the West, have most likely migrated over Beringia. The boreal muscatel, *Adoxa moschatellina,* is closely related to two recently described genera in China, and it may well have originated there. It could have reached North America over either land bridge.

Two archaic genera of the southeast, *Illicium* and *Schisandra,* have numerous relatives in eastern and southeastern Asia and surely are of Asiatic origin. Common or dominant western American species, such as *Artemisia tridentata, Ceratoides lanatum,* and *Datisca glomerata,* likewise have rather certain Asiatic origin. Species of *Atriplex, Kochia, Mahonia, Monolepis, Prunus* subg. *Amygdalus, Stipa,* and *Suaeda* have strong links with the arid areas of central Asia. They presumably immigrated via Beringia.

MEDITERRANEAN RELATIONSHIPS

There is a considerable Mediterranean element in western North America, which includes such taxa as *Fagonia, Peganum mexicanum, Oligomeris linifolia, Plantago ovata, Senecio mohavensis,* and *Styrax officinalis.* Other southwestern genera with strong relationships to the Mediterranean region, including North Africa, are *Antirrhinum, Cupressus, Juniperus, Pistacia,* and possibly *Astragalus, Lupinus,* and *Trifolium.*

NATURALIZED FLORA

In any discussion of the North American flora, one cannot disregard the past and present heavy addition of those species brought into the continent by humankind, either by accident or by intention (see R. L. Stuckey and T. M. Barkley, chap. 8). The Californian flora of perhaps 5900 indigenous species has been enriched by more than 1000 naturalized plants brought in by immigrants, from the Spanish conquistadores to our present peripatetic citizens. These naturalized taxa may comprise one-sixth or more of our present North American flora, which has been estimated to consist of perhaps 18,000 species north of the Mexican border.

PART IV HUMANKIND AND THE FLORA

7. Taxonomic Botany and Floristics

James L. Reveal
James S. Pringle

In the Beginning

The modern history of systematic botany and floristics in North America began when the first Europeans landed on these shores and began to collect objects of curiosity. It is imperative to use the term "modern," for long before colonization of the New World by Europeans, the Native Americans, who had arrived millennia earlier, had developed their own systems of classification, means of identification, and associated nomenclature. Unlike that of their European counterparts, their knowledge was transferred by the spoken rather than the printed word and was mostly lost as their civilizations fell to the invaders. To a great degree, it was not until the twentieth century that Native Americans were recognized as knowledgeable about their plants. By then, European thought dominated botany, and the Native American's botanical understanding was passed on only in an occasional native name retained in a Latinized form.

It was not until Columbus's second voyage, in 1493, that New World plants and animals were taken across the Atlantic. For the European scientific community, the unfamiliar specimens were a source both of great intellectual curiosity and of philosophical concern. The curiosities were clearly different from their Old World counterparts, and in some instances they were entirely novel. The likes and near-likes could be associated, but the distinctly different were philosophically troublesome.

The Spanish of the early sixteenth century were the first to describe the flora of the New World. Gonzalo Fernández de Oviedo y Valdes (1478–1557) visited several of the Caribbean islands and portions of Central America, trying to fit the tropical vegetation he observed into a classification scheme that recognized only six species of trees with persistent green leaves. Oviedo had become acquainted with native New World plants of equal or even greater value than those introduced to the New World by the Spanish, and he urged their use. He was ignored.

Nicolas Bautista Monardes (1493–1578) never saw the New World. His interests were the new medicines and new remedies he felt certain existed. He classified plants according to their medicinal properties, and for the American ones he often retained the native names. He accepted treatments recommended by the Amerinds, but as a firm believer in the Doctrine of Signatures, he occasionally modified them.

The missionary Jose d'Acosta (1539–1600) spent 20 years in Peru, returning to Spain in 1588 to publish various works on the New World. He urged scholars to regard the majority of living things in the New World as unique and not to assign them established European names. He described numerous native economic and medicinal plants and commented on the diversity of potatoes, tomatoes, and chili peppers he had found in the market; he also mentioned cacao and coca.

During this period, intellectual thought often was dominated by religious dogma. Scholarly investigations in the natural sciences began primarily in northwestern Europe. The first naturalists often had to flee the upheavals of the Protestant Reformation and, as a result, many traveled widely and learned from others. In this way, a more unified system of classification and nomenclature began to develop.

Herbals, those great tomes illustrated with woodcuts, were the primary botanical publications of the age. At first they were little more than restatements of Dioscorides or other classical authors, but as the herbals were developed over the next two centuries, new species and remedies were incorporated, including the wonders of the New World. Of equal importance was the development of botanic gardens, first established in Pisa in 1543. These soon became centers of scientific importance because not only could plants of faraway places be seen, but their medicinal properties could be determined also.

The herbalists of continental Europe were emulated by Englishmen such as William Turner (1510–1568), "Father of English Botany," and John Gerard (1545–1612) of London. Later John Tradescant (d. 1638) and his son, John Tradescant the Younger (1608–1662), set up a garden at Lambeth and, with royal patronage, established (1629) a natural history museum known as "Tradescant's Ark." To increase the holdings of the museum, the younger Tradescant repeatedly traveled to Virginia, beginning in 1637, to collect plants and animals for display. In 1621, Magdalen College granted Oxford University 5 acres to establish the Oxford Physic Garden, and although it did not reach prominence until after the Stuart Restoration of 1660, it became the one garden that received hundreds of Virginia plant species prior to 1700.

When the first English colonists arrived on the East Coast, they carried with them their English herbals. Although a few native American plants, such as potato, tobacco, maize, and Jerusalem artichoke, were accounted for in Gerard's *Herball or Generall Historie of Plantes* (London, 1597), the truly valuable American medicinal plants were either still unknown or considered inferior to European plants. Four decades later, in his *Theatrum Botanicum* (London, 1640), John Parkinson (1567–1650) accounted for some 3800 different kinds of plants, including many from eastern Canada, New England, and Virginia.

The French were equally active in the exploitation of New World plants. Samuel de Champlain (1567–1635), who was also an apothecary, made several voyages to North America and took special interest in the settlement of the St. Lawrence Valley. In 1610, he established a garden at Port-Royal (now Annapolis Royal in Nova Scotia) where he grew several American and European plants. He also sent seeds and living specimens to Jean Robin (1550–1629) in Paris.

The illustrated *Canadensium Plantarum Historia* (Paris, 1635), written by the Paris physician Jacques Philippe Cornut (1606–1651), accounted for 38 species from temperate eastern North America. It was this work that Linnaeus consulted a century later to understand better the plants of this region. Several French colonists contributed material to botanists working in Paris at the Jardin du Roy. Notable among these were two "Médecins du Roy," Michel Sarrazin (1659–1735), correspondent of Tournefort, and Jean-François Gaultier (d. 1756), correspondent of Duhamel de Monceau, who published *Traité des Arbres et Arbustes...* (2 vols., Paris, 1755). Gaultier received the active support of Interim Governor Barrin de la Galissonière, who encouraged explorers and officers to send specimens and seeds through him and Gaultier to Paris. Gaspard Bauhin (1560–1624) listed several North American plants in his *Pinax Theatri Botanici...* (Basel, 1623), the first attempt to summarize the already confusing array of names and synonymy.

The New England colonists learned from the Amerind the values of some plants, but little of scientific or medical value was recorded. Thomas Hariot, a member of Sir Walter Raleigh's company that founded the colony at Roanoke in 1585, wrote a florid description of the region (*A Brief and True Report of the New Found Land of Virginia...*, London, 1588) to encourage others to settle in the New World. Captain John Smith (*A True Relation of...Virginia...*, London, 1608) championed the region around Chesapeake Bay in equally glowing terms, describing the great forests, the abundance of wildlife, and the beauty of the land. While these publications were of limited scientific value, the maps and illustrations that accompanied them did much to demonstrate the bounty that awaited anyone willing to take the risk of settling the land.

To Europeans, the temperate New World was an opportunity awaiting exploitation. The sixteenth and early seventeenth centuries were a period of conquest, not scientific exploration. The first 150 years of exploitation of the New World set the scene for changes in Europe. The wealth so generated helped create a class of people that had free time, education, and an interest in the sciences. It also provided the means whereby individuals of talent, but not of wealth, could find the means to study the natural and physical world.

The philosophy of science that greeted the discovery of the New World was one hardened by centuries of total faith in the teachings of the ancient past. The immediate challenge the New World presented was that it called into question the accepted norm. As Francis

Bacon wrote in *Novum Organum* (London, 1620), knowledge "must begin anew" from its very foundation. New means of evaluating human observations had to be devised to organize knowledge and to explain what exists.

Bacon realized that no one person could know or explain all, and as with the arts, which he deemed to be equal to the sciences, no one person could be fully accomplished in every aspect. He argued for corporate undertakings involving both those with the means to support the arts and sciences and those with the ability to be artists and scientists. Only by working jointly, and over generations, could humankind hope to comprehend all that exists in the world.

For the sciences, cooperation was slow to establish itself. The first science club was formed by Wadham College, Oxford, in 1648. A decade later the Oxford Club was founded with Robert Boyle, Robert Hooke, and John Evelyn among its members. By 1660 the concept of a Royal Society had taken shape, and soon thereafter formal scientific bodies were established in both England and France.

The societies provided a meeting place to witness experiments, hear papers, and generally discuss ideas of interest. They also became focal points where others, outside Europe, could send contributions in the form of specimens or objects of curiosity that were housed and maintained by the membership, or as manuscripts and letters reporting on observations. Through publications, patronage, and especially correspondence, the royal societies encouraged people to contribute. For a few individuals scattered along the Atlantic Coast of North America, this was a needed avenue to inform others of the curiosities of their new land.

By 1783, numerous men from North America had been elected to fellowship, and many hundreds more were correspondents. The societies not only provided a venue for science to be addressed, they actively promoted it. Governors, heads of major companies, and large landowners were elected to membership and then were encouraged to support individuals who could make observations on the natural and physical world. Tracts on natural products were written; plants, animals, and minerals were collected and described; the weather was recorded; and the movements of comets and planets were noted.

Still, important voices were heard outside the societies. John Josselyn (1608–1675) spent several years in New England, and in his *New-England's Rarieties Discovered...* (London, 1672), he described numerous new species of plants. Although Josselyn's drawings were crude, Linnaeus cited this work in 1753. This and Josselyn's later *Account of Two Voyages to New England...* (London, 1674) were the most complete summary of the North American flora for more than a century.

The 20-year period following the publication of the second part of *Plantarum Historiae Universalis Oxoniensis* (Oxford, 1680) by Robert Morison (1620–1683), was one of dramatic change in botanical thought in Europe. Nehemiah Grew (1641–1712) published his *Anatomy of Plants...* (London, 1682), and John Ray (1627–1705) summarized the known flora of the world in his three-volume *Historia Plantarum...* (London, 1686–1704), in which he applied the philosophy he had outlined in his *Methodus Plantarum Nova...* (London, 1682). John Ray and two competitors, Leonard Plukenet (1642–1706) and James Petiver (1663–1718), all members of the Royal Society as well as London's informal Temple Coffee House Botany Club, sought to obtain and study the new plants coming into England from throughout the world. In France, Joseph Pitton de Tournefort (1656–1708) toiled to organize the equally overwhelming wealth of new material arriving there, producing his three-volume work *Eléments de Botanique...* (Paris, 1694).

Early knowledge of the temperate eastern North American flora came primarily from Virginia. Foremost in providing specimens and new observations was John Banister (1650–1692). During his years at Oxford, Banister had attended Morison's lectures at the Botanic Garden. To complete his *Plantarum Historiae Universalis Oxoniensis,* Morison persuaded Bishop Henry Compton to send Banister as a minister to the James River area. Banister sailed to Virginia in 1678, where he began his descriptions, observations, and drawings of plants and animals. After Morison's accidental death in 1683, Jacob Bobart the Younger, with the help of William Sherard and Samuel Dale, completed the third part of Morison's *Historia* (Oxford, 1699), including over 400 Virginia species. Bobart shared Banister's catalog and plant specimens with John Ray, who published them in the second and third volumes of his *Historia Plantarum.* Plukenet also reproduced Banister's plant drawings in his *Phytographia* (7 parts, London, 1691–1705).

A wealth of information had been supplied by Banister and three collectors in Maryland: Reverend Hugh Jones (1671–1702), William Vernon (d. 1711?), and Dr. David Krieg (d. 1710), a correspondent of Petiver. This allowed Plukenet, in the last three parts of his *Phytographia* (1696–1705), and Ray, in the final volume of *Historia Plantarum,* to account for nearly a thousand species of temperate North American plants.

Petiver encouraged many of his other North American correspondents to send him seeds and specimens.

Robert Steevens (flourished 1700), Edmund Bohun (fl. 1699–1703), and Robert Ellis (fl. 1700–1704), all of Charles Town and its environs, routinely sent Carolina plants. Surgeons and captains of ships also collected specimens for Petiver along the shores from Hudson Bay to the Florida Keys. Petiver soon had many plants in cultivation, describing and illustrating them in his publications.

As the seventeenth century passed into the eighteenth, the natural sciences divided into a series of more specialized fields, botany diverging from zoology, and taxonomy from medicine. Many talented individuals became interested in understanding plant products and set to work at the task. Yet, the volume of novelties coming from India, southern Africa, Java, China, Japan, Turkey, and the American colonies was overwhelming the available resources. The value of Bauhin's *Pinax* was rapidly declining, so much so that at Oxford, William Sherard (1659–1728) began to draft a new edition. He was forced, however, to ask John Jacob Dillenius (1687–1747) to assist him as the volume of new names became too great. Bauhin had accounted for some 6000 names; Ray, in 1704, treated some 18,000.

To summarize all of this nomenclature and render from it a sensible taxonomy required specimens from many sources and the cooperation of others. For Sherard and Dillenius, that cooperation was minimal. Tournefort and Sebastien Vaillant (1669–1722) in Paris and Paul Hermann (1646–1695), Herman Boerhaave (1658–1736), and others at Leiden all sent them specimens. Their greatest problem, however, was a fellow Englishman, Sir Hans Sloane (1660–1753).

Sloane had collected mainly in Jamaica from 1687 until 1689, amassing a large cabinet of curiosities that he began to study on his return to London. Sloane's marriage to a wealthy widow provided him with the income and social status to rise to a position of power as Physician to the Queen and secretary of the Royal Society. He outlived most of his contemporaries, and in time he acquired the libraries, correspondence, and collections of Plukenet, Petiver, and a host of others. The result was a collection so large that the British Museum was established to hold it.

The herbaria Sloane obtained were critical to the efforts of Sherard and Dillenius, but these herbaria were so unorganized, and Sloane had delayed access to them for so long, that even after Sherard's death in 1728, Dillenius was still unable to study them. Furthermore, William Sherard's brother James insisted that Dillenius divert his efforts to *Hortus Elthamensis...* (London, 1732), to the neglect of William's intended new Pinax. That manuscript was never finished, and the constant proliferation of names, together with the lack of a single source for all names, caused chaos for all who sought to identify and classify plants.

An Outline of Rationales

The modern taxonomic era began in 1753 with the publication by Carl Linnaeus (1707–1778) of a two-volume work entitled *Species Plantarum...* (Stockholm). The 20-year period leading to that publication was a time of intensive study for the young Swedish naturalist, isolated in a remote land far from the libraries, botanical gardens, and collections so vital to Dillenius. To expand his horizons, Linnaeus traveled to Holland in 1735 to obtain his doctorate and to study plants. His primary effort was to gain an appreciation for and a knowledge of the multitude of new plants being discovered in far-off lands.

The major innovation for which Linnaeus is credited today is the consistent use of binomial nomenclature. Yet the set of principles he formulated while yet a young man is of greater importance to taxonomy. His *Fundamenta Botanica...* (Amsterdam, 1736), with its 365 aphorisms, was an outline of the rationales he used to compose his *Systema Naturae...* (ed. 1, Leiden, 1735). These rationales matured over the following 15 years and formed the basis of *Philosophia Botanica...* (ed. 1, Stockholm, 1750). The concepts outlined by Linnaeus have come down to the present as some of the traditions and fundamental precepts in biological taxonomy.

At the time, the classification, identification, and naming of plants were, at best, nationalistic and personal, at worst, chaotic. Classification schemes were designed only to retrieve information, not to express relationships; identifications were based on regional features, not diagnostic characters; and nomenclature was a personal expression of opinion as to position, location, and features as compared to similar plants. Before Tournefort had developed his concept of the genus, plants often were grouped together by location, presumed affinities, or assumed virtues rather than by the possession of similar morphologic features.

Whereas taxonomic concepts and principles had been stated by previous authors, none seemed to carry sufficient weight to overcome national boundaries. Consequently, before Linnaeus, what was practiced by the English was not necessarily followed by the French or Spanish. Some authors, such as Ray and Plukenet, wrote descriptions based on individual specimens; Tournefort wrote his to fit an abstracted concept; but none had what might be regarded as a modern species concept. It was this innovation that Linnaeus contributed.

The modifications Linnaeus made to his aphorisms from 1736 to 1750 were mandated by experience gained from their use. Linnaeus's polynomials were diagnostic, each headed by a consistently used and uniformly applied generic name. The genera were arranged into orders and classes that reflected the system of classification Linnaeus had devised for the major groups of plants he accepted: algae, fungi, bryophytes, ferns, grasses, and the remaining herbs or trees. The Linnaean scheme made it possible to identify readily a given plant without an accompanying illustration. The idea that a given kind of plant should have but a single name was novel; but, of course, as other members of the same genus were found, the diagnostic features, and therefore the phrase name, had to be changed. To Linnaeus and his early followers, it was not the name that was consistent but the biological entity represented by the name.

In retrospect, Linnaeus's classification scheme was artificial in that it was based largely on the number and nature of reproductive structures; it was, however, practical. There was no evident concept of grouping organisms together that were related; that notion was to come later in the century. The value of Linnaean principles, however, was immediately obvious to many in Holland, and the name of Linnaeus preceded him as he toured France and England, visiting Bernard de Jussieu (1699–1776) in Paris, Dillenius at Oxford, and Philip Miller (1691–1771) at Chelsea. Boerhaave, Jan Fredrik Gronovius (1690–1762), and Adriaan van Royen (1704–1779) championed Linnaeus's work in Holland, as did Peter Collinson (1694–1768) in England. Through these individuals, Linnaeus's ideas were soon widely adopted, and the published works of the young Swede were dispersed widely, even to the American colonies.

The Clerk from Gloucester County

Botanical explorations along the Atlantic Coast of North America became limited as interest in this flora diminished with the introduction starting in 1700 of the diverse Chinese flora. Correspondents of Petiver continued to supply the London apothecary with plants from the Carolinas and Florida. Surprisingly rich collections of Hudson Bay plants were collected by Richard Tilden (fl. 1700–1707) in 1700, and by Dr. John Smart, a ship's surgeon, in 1708. Hannah Williams, perhaps the first woman collector in temperate North America, who sent mainly butterflies to Petiver, encouraged Joseph Lord (1672–1748) to collect. Lord became the foremost collector in Carolina. From 1701 to 1710, Petiver accounted for many of Lord's new species. John Lawson (d. 1711), a surveyor, also collected plants for Petiver, traveling far to the west and into many unexplored areas.

Mark Catesby came to Williamsburg, Virginia, in 1712 and collected throughout much of the area. The specimens were misplaced, however, and unstudied until 1981, although Catesby's collections in the Sherardian Herbarium at Oxford had been studied by Frederick Pursh, who based some new species on his collections. In 1719, when Catesby returned to England, Samuel Dale (1659–1739) remarked to Sherard that of the 70 specimens he had examined, at least half represented new species. When Catesby came back to the New World in 1722, he resided at Charles Town and traveled widely over the next four years, going as far south as northern Florida and the Bahama Islands. He was able to amass a huge collection of plants, animals, fossils, and other items of interest. He also skillfully sketched and painted many of the plants and animals he had seen in their natural habitats. When Linnaeus came to Oxford in August 1736, he probably examined Catesby's specimens, for they were the foundation for Catesby's *Natural History of Carolina, Florida and the Bahama Islands...* (2 vols, London, 1730–1747). Even so, Catesby had a lesser impact on Linnaeus than did his successor, John Clayton (1693–1773).

Clayton arrived in Virginia around 1720 and took up residence near Williamsburg where he served as the clerk of Gloucester County while maintaining a productive plantation and a lively correspondence with fellow botanical enthusiasts. Catesby and Clayton knew one another, and after Catesby returned to England, Clayton began to botanize actively, sending to Catesby at Oxford a large shipment of dried specimens collected in 1734. Catesby was ill-prepared to identify the specimens, and reluctantly he sent this collection, and another received in 1735, to Gronovius in Holland, who was equally unable to cope with the problem.

Unaware of their fate, Clayton kept sending specimens to England. Catesby kept a few of the dried specimens for the herbarium at Oxford, and Collinson went through the seeds and retained for himself what he wished, but the bulk of the material continued on to Holland. Gronovius soon asked the young Linnaeus, then at the estate of George Clifford (1685–1760) at Hartekamp, to examine those specimens Gronovius was unable to place or to confirm those he recognized to be new.

Meanwhile, Clayton drafted a catalog of the plants of Virginia, following Ray's method, and sent it to Catesby, asking that he find a publisher. The manuscript went to Gronovius who, with the assistance of Linnaeus, completely revised the text according to the Linnaean method and published the work in 1739 as the first part of *Flora Virginica...* (2 vols., Leiden, 1739–1743). When Clayton finally saw this work in print, the authorship had been attributed to Gronovius, who had often used his own phrase names, those of Linnaeus, or those from earlier works (mostly Banister's specimens as rephrased by Morison, Ray, Plukenet, or Petiver) for Clayton's specimens. Moreover, Gronovius had cited Clayton's phrase names only as synonyms. At the urging of Collinson and Catesby, Clayton obtained most of the available Linnaean works and began to study them, all the while sending more specimens to Catesby, now arranged in the Linnaean fashion. Once again Clayton was disappointed, for the second part of the flora, published in 1742, still carried Gronovius's authorship.

Linnaeus was interested in Clayton's fine specimens, for with them he was able to evaluate many of the names for American plants proposed previously by Ray, Plukenet, Petiver, Hermann, and others. To what extent he called on Dillenius to resolve the nomenclatural morass, or actually examined specimens at Oxford in 1736 to do so, is unknown. But resolve nomenclatural problems he did, and in *Flora Virginica* one finds signs of the same care that Linnaeus would give to the flora of the whole world two decades later.

Another of Catesby's English friends was the physician John Mitchell (1690–1768), who came to Virginia in 1735 and remained until 1746. Unfortunately for Mitchell, Spanish pirates plundered his England-bound ship, and most of his collection was lost. With the help of Collinson, Mitchell published several new genera, but Linnaeus never saw any of the surviving specimens. In New York, Cadwallader Colden (1688–1776) became a correspondent of Linnaeus via Collinson and sent him a manuscript on the plants of that colony. This was published in 1749 and 1751. Unfortunately, the same pirates took his specimens and none were recovered. John Bartram (1699–1777) of Philadelphia sent seeds and dried specimens to Sloane, Sherard, Collinson, Dillenius, and Gronovius, and through them to Linnaeus, promoting his botanical garden (the first in the United States, founded in 1731) and his desire for a royal appointment as the King's Botanist.

Pehr Kalm (1715–1779), who had studied with Linnaeus, arrived in New York in September 1748 and was soon collecting seeds and plants in New Jersey and Pennsylvania. The following year, he traveled through New York to southern Canada; in 1750 he went across western Pennsylvania to the Great Lakes. When Kalm left the United States in October of that year, he was heavily laden with all kinds of plants, not just those of interest to the gardener or physician.

The Prince of Botany

The stature of Linnaeus had grown significantly in the eyes of his fellow naturalists after he had left Holland for Sweden in 1738. His writings were eagerly awaited and carefully studied by naturalists. Specimens were sent to him, often via intermediaries such as Collinson or Gronovius, for identification and naming. The simplicity of his system attracted many, and anyone could readily determine if a given plant was known to the great man. His production was prodigious. Edition after edition of *Systema Naturae* and *Genera Plantarum...* (ed. 1, Leiden, 1737) was published. A summary of the world's flora at the species level, however, was more difficult to achieve.

Three problems appear to have troubled Linnaeus: the morass of names created by previous writers, the increasing number of new species being found, and the vast array of existing specimens and published literature. He had already developed a concise style of presentation, which he had used in *Hortus Cliffortianus...* (Amsterdam, 1737), and specimens and literature were being sent to him from all over the world, notably duplicates of Clayton's Virginia plants from Gronovius, and seeds and dried specimens from Collinson and Miller, and from Madrid, Paris, and Vienna. The most recent items were less troublesome. The older literature and collections, however, remained difficult. Linnaeus studied the herbarium of Joachim Burser (1583–1639) to understand the dispositions of names according to Bauhin's *Pinax,* consulted the herbarium and drawings of Hermann's Ceylon plants, and examined copies of the original drawings made by Father Charles Plumier (1646–1704) of plants observed in the West Indies.

Preventing Linnaeus from fully comprehending the past were the names associated with the specimens scattered among the vast holdings of Hans Sloane. Through his many friends, Linnaeus sought to obtain the older literature so that he might study it. Even so, as he struggled with writing a species plantarum, he encountered too many unanswered questions, and in 1748, with the manuscript only half finished, he gave up.

In June 1751 Kalm arrived at Uppsala with his American discoveries. He had kept a diary and wanted to publish therein the names of the plants he had found. He called on his mentor to identify the remainder of his plants because he was certain several were new. Earlier, in 1749 and 1751, Linnaeus had described several of Kalm's new genera, but the task of naming all of the species was a challenge.

To what degree Kalm's arrival prompted Linnaeus to attempt again a summary of the world's flora is not known, but the two events coincide so closely that it is likely that Kalm provided at least some impetus. This time, however, instead of trying to account for every species and all names, Linnaeus treated only those species he felt he had some degree of confidence to address, and he cited only a selected synonymy. Although he had not adopted the use of binomials in his first draft, the logic of using them became obvious in 1751:

each species would have a simple name that the uninformed could use, and by making each unique within a genus, indexing was simplified. Only 13 months after he had started the new draft, and in spite of an extended illness, he completed the task.

Some last-minute additions were inevitable. Pehr Osbeck (1723–1805), chaplain of a ship, returned from China with numerous boxes of plants just as the book was being readied for the printer, and garden material (mainly from Bartram via Collinson or Miller) continued to arrive. The novelties were inserted, and *Species Plantarum,* the long-sought summary of the world's flora, was published in two volumes at Stockholm in 1753. Linnaeus had accounted for some 8000 species from all parts of the known world. He revolutionized the science of taxonomy and ushered in a new era. A milestone had been set in the progress of knowledge, and no one knew it more than Linnaeus himself. He would thereafter be, as he christened himself, the Prince of Botany.

"A Child of Fortune"

The majority of botanists, although not all, readily adopted Linnaeus's innovative nomenclature and classification scheme, and modeled their work accordingly. The adoption of his system by the North American community was championed by Collinson and John Ellis (1710–1776) in London. From southeastern Canada to Georgia, plants suspected to be new were sent to Europe in the hope that they would prove to be novelties and that the great Linnaeus himself would name them. Some, such as Jane Colden of New York, Cadwallader Colden's daughter, labored in obscurity. Others, such as Alexander Garden (1730–1791), became famous, although he waited a long time before hearing that some of his new plants would be given names.

With Collinson's help, John Clayton prepared his own edition of a Virginia flora, but his hopes were ended when Gronovius published a second edition of *Flora Virginica* in 1762 (Leiden). Clayton's manuscript and associated drawings by Georg Ehret (1708–1770) were lost when Clayton's clerk's office was set ablaze by an escaped prisoner; the remainder of his correspondence and library were destroyed in a house fire around 1906.

The collections of John Bartram and his son William (1739–1823) fared better. They had been collected during John Bartram's travels from Lake Ontario to Florida. As they arrived in England, they were dispersed among Sloane and his friends. Collinson sent seeds and specimens to Linnaeus, but many of their new species remained undescribed. In 1768 Philip Miller named many of the Bartrams' curiosities, and Nicolaus Jacquin (1727–1817) named others from his gardens in Vienna.

Younger men were entering the field, especially in England, to carry on the botanical tradition started by Sloane and furthered by Collinson and Ellis. Foremost among them was Sir Joseph Banks (1743–1820). After his father's death, the young Banks settled in London where, as a "child of fortune," he became acquainted with Philip Miller and others associated with the Chelsea Physic Garden. In 1766 Banks sailed for Labrador and Newfoundland, where he made numerous collections that allowed him, as a newly elected member of the Royal Society, to become acquainted with others interested in the plant kingdom. This resulted in his voyage to Australia for botanical discoveries that were to earn him lasting fame.

Sailing with Banks in 1768 was Daniel Solander (1736–1782), a Linnaean student who had become the assistant librarian at the newly established British Museum in 1762. As Banks's personal librarian, Solander had access not only to the specimens the two had collected while traveling around the world from 1768 to 1771, but also to those available to him as Keeper of the Natural History Department at the museum. As a result, Solander finally gained access to the collections Sloane had acquired over his 93 years, and for the first time it was possible to study the original specimens used by Plukenet, Petiver, and others to describe species that had long puzzled Linnaeus. Because of his own great wealth and interest, Banks too obtained herbaria, including those of Gronovius and Jacquin.

As president of the Royal Society, a position he assumed in 1778 and retained until his death, Banks was able to direct the interests of the society into many corners of the world in search of botanical novelties. The world's largest collection of plants lay before Solander, and he set to work describing the thousands of new species he was finding. His early death prevented his publishing much, but through the work of his successors as Banksian librarians, Jonas Dryander (1748–1810) and Robert Brown (1773–1858), many of Solander's names were ultimately published. Unfortunately, the authorship of most of his names for American species must be credited to William Aiton (1731–1793), who, as head gardener for Princess Augusta of Saxe-Gotha at Kew, was the author of the three-volume *Hortus Kewensis...* (London, 1789), in which most of the names were formally established. Even the efforts of Francis Masson (1741–1805), one of Kew's foremost collectors, who worked briefly in Canada, were relegated to obscurity in a similar manner.

Views of Enlightenment

The Linnaean Society of London, established in 1788, had as its first president James Edward Smith (1759–1828), who, with Banks's encouragement, had pur-

Benjamin Smith Barton, 1766–1815. [From J. Ewan. 1988. Benjamin Smith Barton's influence on Trans-Allegheny natural history. Bartonia 54: 28–38. Reprinted with permission.]

Thomas Nuttall, 1786–1859. [Hunt Institute, Carnegie Mellon University, Pittsburgh, engraving by J. Thomson after W. Derby. Published by Henry Fisher Caxton. London. 1825.]

chased Linnaeus's library and herbarium in 1784. Smith championed the Linnaean system for the next half century, after it had outlived its usefulness. In England, Robert Brown (1773–1858) and John Lindley (1799–1865) led the opposition to it. In France, the changes in social values brought about by the Revolution of 1789 coincided with the acceptance of a natural system of classification. Antoine Laurent de Jussieu (1748–1836), nephew of Bernard de Jussieu and friend of Linnaeus, in his *Genera Plantarum...* (Paris, 1789), arranged the genera of the world into 100 families (ordines naturales) based on concepts developed by his uncle Bernard, in a continuation of the ideas proposed a generation before by Michel Adanson (1727–1806) in his *Familles des Plantes* (2 vols., Paris, 1763[–1764]).

Adanson's views were unconventional. As had Pierre Magnol (1638–1715), Sloane's professor, long before him, Adanson believed that plants could be arranged into natural families and genera. It was a classification free of a priori weighting and metaphysical themes, all entities circumscribed so as to reflect their affinities as determined by the sum total of their features, and a system derived only from an empirical search for similarities and discontinuities.

The differing views of classification had little impact in America, and new species from there continued to be described by Europeans. Essentially, no one in America was independent enough of European authority to risk expressing his own opinion. Linnaeus continued to revise his *Species Plantarum* and to add new American species to its pages. He occasionally recognized names proposed by others, but the many names of John Hill (1716–1775) and Miller, each of whom had described several North American species, were not taken into consideration. In Linnaeus's own view, binomials were trivial names, only for the uninitiated, and were never intended to be used by the knowledgeable naturalist. Therefore, when Gronovius published the second edition of *Flora Virginica* in 1762, he used polynomials, not binomials. Consequently, this work is now nomenclaturally irrelevant by virtue of the current code of botanical nomenclature.

To describe their new species, the Bartrams clearly depended on others, e.g., Aiton in *Hortus Kewensis,* Jean-Baptiste de Lamarck (1744–1829) in his 13-volume *Encyclopédie Méthodique* (Paris and Liège, 1783–1817) and Charles-Louis L'Héritier de Brutelle (1746–1800) in his *Stirpes Novae.* The first flora authored by an American, *Flora Caroliniana...* (London, 1788) by Thomas Walter (1740–1789), was published and heavily edited in England by John Fraser (1750–1811). The first wholly American contribution was *Arbustum Americanum...,* by Humphry Marshall (1722–1801), published in Philadelphia in 1785. Both floras were in the Linnaean style. Some early American taxonomic works appeared in the *Transactions of the American Philosophical Society.* Unfortunately most of the names proposed by H. Muhlenberg (1753–1815) are invalid, according to the modern code of nomenclature (W. Greuter et al. 1988). Likewise, the validity of the names mentioned by the younger Bartram in his 1791 *Travels...* (Philadelphia) is still in doubt.

The year 1789 marked the end of the domination of the Linnaean classification. Linnaeus's herbarium and library were in London, and binomial nomenclature was widely adopted. New editions of *Species Plantarum* were being written by others. Nevertheless, the Linnaean manner of presentation continued into the 1830s. Importantly for American naturalists, however, the War of 1812 with England broke the bond that demanded their unquestioned acceptance of the Linnaean method. This allowed North American naturalists to be influenced by the radical views of the French and the French naturalists. Also in 1789, Benjamin Smith Barton (1766–1815) began teaching natural history at the University of Pennsylvania, and an American flavor was added to taxonomy. Others soon followed, notably Samuel L. Mitchell (1764–1831) at Columbia University in 1792. The center of American botany was, nevertheless, firmly established in Philadelphia.

The Continent Is Opened

The establishment of the center of botany in Philadelphia was effected through the efforts of the Bartrams, their garden, and a host of others with scientific backgrounds who resided in that city after the Revolutionary War. Students interested in the native flora flocked to use Barton's large library and herbarium. Ultimately, Barton sponsored the fieldwork of Fredrick Pursh (1774–1820) and Thomas Nuttall (1786–1859) as part of his dream of writing a flora of North America. He also introduced William Baldwin (1779–1819), William Darlington (1782–1863), and Yale's Eli Ives (1779–1861) to taxonomy and aided them in their early careers.

While the American scene was developing, the French André Michaux (1746–1802) and his son, François André Michaux (1770–1855), began their explorations of temperate North America. In 1785, they traveled from Lake Mistassini to Florida in search of plants, with special emphasis on potential garden introductions. In 1803, a year after the elder Michaux had died, *Flora Boreali-Americana...* (2 vols., Paris and Strasbourg) was published with the aid of his Paris colleague Louis Claude Richard (1754–1821). At last, temperate North America had an account of its native plants.

Michaux's book used a modification of the Linnaean

and Jussieu styles as promoted by Augustin Pyramus de Candolle (1778–1841) in his revision of Lamarck's *Flore Françoise...* (ed. 3, 5 vols., Paris, 1805). Michaux illustrated some of his new genera and gave precise habitat information. He did not, however, include keys, the shorthand method for identification that Lamarck had developed in an earlier edition of the *Flore.* Michaux had talked with Barton and visited his garden. Consequently, unlike the works of Linnaeus that were still in common use, Michaux's was based on firsthand observation and a careful comparison of his specimens against those used by others who had named plants previously. This work prodded Barton to proceed with his flora, a small part of which had been published in London in 1787 as *Observations on Some Parts of Natural History...,* Part 1.

Criticisms of Michaux's work abounded. Barton was not satisfied, and he urged his students and colleagues into the field to collect the new species that he knew existed to the west. Pursh, then in London, published *Flora Americae Septentrionalis...*in 1814 (2 vols., London). Similar in format to Michaux's work, this one included the new species collected by Barton, by Pursh for Barton, and by Meriwether Lewis (1774–1809), President Jefferson's personal secretary, who was in the Far West as part of the Lewis and Clark expedition of 1804–1806. A few of the expedition's early specimens had reached the president, who sent them on to Barton for naming. All the specimens found in 1805 were lost, and only those collected on the hasty return trip of 1806 reached Philadelphia. Even so, these specimens were ignored, and had Pursh not purloined snippets of the plants, they might not have been consulted for years.

The Lewis and Clark specimens represented the first American contribution from the western part of North America. But Lewis was not the first to collect specimens from that area. Georg Steller (1709–1846) had managed, in six hours, to collect 141 specimens on a small island off the coast of Alaska in 1741. After his death, these specimens were sent to Linnaeus, who accounted for about 30 of them (all from Russia) in his 1753 work. Plants from coastal California were first collected in 1786 by a French gardener, Jean-Nicolas Collignon (1762–1788), and Lamarck described the first species from west of the Mississippi, *Abronia umbellata,* five years later.

The English, Spanish, and Russians all sent out early expeditions that touched land from San Diego to Alaska. In 1791, the Malaspina expedition visited Monterey in California, Nootka Sound in British Columbia, and Yakutat Bay in Alaska. Its naturalist, Thaddeo Haenke (1761–1817), made the first large collection of dried specimens to reach Europe from this part of the world. Many of his new species were described by Karl Presl (1794–1852) in the two-volume *Reliquiae Haenkeanae...* (Prague, 1825–1835). Archibald Menzies (1754–1842) was the first of many British naturalists to explore the Far West. He was with the Vancouver expedition when it surveyed the Pacific Coast from Chile to Alaska from 1792 to 1794. Many of Menzies's new species were named by James Edward Smith in articles written between 1802 and 1820 for Abraham Rees's *Cyclopaedia...* (39 vols., London).

Alaska was then a possession of the Russians, and when the von Krusenstern expedition arrived in Sitka in July 1805 during their round-the-world voyage, they found at anchor an American ship, the *Juno.* The Russians purchased the ship, and in March 1806, a small company set sail for the coast of California to obtain supplies. On board was Georg von Langsdorff (1774–1852), surgeon-naturalist, who botanized briefly near San Francisco in April. In his published remarks (*Bemerkungen auf einer Reise um die Welt,* London, 1813), however, Langsdorff did not account for a single California plant.

The first specimens from Greenland were collected by Hans Egede (1686–1758) and his son Poul (1708–1789), starting in 1728. Unknown to Linnaeus, these specimens went unreported. The first major collection was made by Morten Wormskjold (1783–1845), who led a naval expedition to Greenland in 1813. In the 1830s, Jens Vahl (1796–1854) made a large collection there.

The first botanical explorer of Newfoundland and St. Pierre and Miquelon was Bachelot de la Pylaie (1786–1856) who collected plants on trips in 1816 and 1818. His specimens are housed in Paris. The Moravian missionaries, compatriots of the Egedes of Greenland, were active along the coast of Labrador. They sent carefully prepared specimens across the ocean to Banks in the 1780s, and then to others in England and the United States in the nineteenth century. John Goldie (1793–1886) collected near the lower Great Lakes in 1819, and James McNab (1810–1878) collected several hundred specimens in southeastern Canada in 1834. Both men were mainly interested in plants of horticultural significance. Through the nineteenth century, many collectors sent specimens to German botanists, and until the 1890s, American recipients often obtained materials from German intermediaries.

The first Arctic expedition of botanical significance was led to Baffin Bay by John Ross in 1818, and it included the naturalist Edward Sabine (1788–1883). The three expeditions of Edward Parry (1790–1885), between 1819 and 1825, resulted in Robert Brown's *Chloris Melvilliana* (London, 1823), so named because many of the plants were collected on Melville Island.

Exploration of the interior of North America accelerated in the early nineteenth century. Thomas Nuttall was 22 years old in the spring of 1808 when he first met Barton in Philadelphia and began to make use of his extensive library. Nuttall proved to be a willing student and an even more willing explorer. He actively sought plants near Philadelphia, and in April 1810, when Barton offered him eight dollars a month plus expenses to collect in the Old Northwest, Nuttall eagerly accepted. Within days he was heading for the upper reaches of the Mississippi River. Later, Nuttall proceeded to St. Louis where he accepted an invitation to travel up the Missouri River the following March with a fur-trading expedition led by Wilson Price Hunt of the American Fur Company.

Also in St. Louis was a Scottish naturalist, John Bradbury (1768–1823). He too had accepted an offer to join Hunt, and for a while he and Nuttall traveled jointly but separately, up the Missouri, each collecting wherever possible. In June, Manuel Lisa and his Missouri Fur Company caught up with the Hunt party. Lisa was trying to reach the Arikara Indians first, and Bradbury sensed that if he joined Lisa, he could beat Nuttall to the botanical treasures. Both groups traveled their separate ways to the Mandan villages in what is now North Dakota before returning to St. Louis in the fall of 1811, loaded with a large number of new and exciting plants.

Shortly thereafter, Nuttall went to New Orleans, where he sailed for England, rather than Philadelphia as he had agreed with Barton. The War of 1812 was imminent, and Nuttall felt it was prudent to return to his native England. Meanwhile, Bradbury had remained in Louisiana to collect, although he had sent his Missouri specimens to London. Bradbury was trapped in America by the War of 1812, while Nuttall, his plants, and those of Bradbury sailed peacefully away.

The consequences were devastating for Bradbury. In England, Pursh gained access to Bradbury's collection and described most of the new species in 1813. Nuttall technically avoided being scooped by Pursh when some of his species were described in a catalog of new plants from "the river Missourie," issued by John Fraser earlier in 1813. Shortly after, both Pursh and Nuttall published new species in Curtis's *Botanical Magazine,* each description augmented by a colored plate.

Nuttall returned to the United States in 1815 and collected throughout much of the Southeast. He worked feverishly in Philadelphia to update the Michaux and Pursh floras in his *Genera of North American Plants...* (2 vols., Philadelphia, 1818).

Other works also appeared. From Charleston, John Linnaeus Shecut (1770–1836) published his *Flora Carolinaeensis...*in 1806. A decade later Stephen Elliott (1771–1830) followed with *A Sketch of the Botany of South-Carolina and Georgia* (2 vols., Charlestown, 1821–1824). In 1814 Jacob Bigelow (1787–1879) published *Florula Bostoniensis* (Boston). All of these works followed the Linnaean method. Amos Eaton (1776–1842) included a section on the natural system as envisioned by Jussieu in his *Manual of Botany...* (ed. 1, Albany, 1817), but he retained the plants in the familiar Linnaean sequence. Even Nuttall's work retained traces of the Linnaean era, although as early as 1808 he was using Jussieu's ordinal names. The first author of an American text to follow Jussieu's natural system was J. F. Correa da Serra (1750–1823), who published a *Reduction of All the Genera...in the Catalogues... of...Muhlenberg...*in 1815 (Philadelphia). C. S. Rafinesque (1783–1840) also used a natural system in his 1817 *Florula Ludoviciana...* (New York). Two years earlier, he had recognized more than 300 natural orders, many of which were adopted by others over the next thirty years without, as Rafinesque complained, giving him credit for their authorship.

As the second decade of the nineteenth century closed, the center of American systematics moved again, this time from Philadelphia northeastward. In New York, David Hosack (1769–1835) established the Elgin Botanic Garden, the first public garden in the United States, and published a catalog of its plants in 1806 (New York). A likable person, Hosack attracted several young people into botany, the most notable being Amos Eaton, then a student of law. While he was jailed for his dealings in a troublesome land sale, Eaton taught the young son of the prison's fiscal agent his first lessons in the Linnaean method; the youngster was John Torrey (1796–1873).

Only 10 Feet Less

Torrey had been educated and trained in the usual haphazard way of his generation. His training in medicine easily allowed him to assume teaching positions in chemistry, botany, and mineralogy. Torrey was an excellent teacher as well as a skilled botanist. Both he and Eaton were to share in the labors of the one person who came to dominate American taxonomy more than any other, Asa Gray (1810–1888). A half century later, two peaks in the Colorado Rockies were named for Torrey and Gray, with Torrey's Peak only 10 feet less in elevation than Gray's—a fitting tribute.

Eaton held fast to the Linnaean way; Torrey faltered, then adapted. Torrey held firmly to tradition through his 1826 *Compendium of the Flora of the Northern and Middle States* (New York) only to change that very year in an article on Rocky Mountain plants. When the first part of Torrey and Gray's *Flora of North America...* (2 vols., New York, London, and Paris, 1838–

John Torrey, 1796–1873. [Hunt Institute, Carnegie Mellon University, Pittsburgh.]

Asa Gray, 1810–1888. [Hunt Institute, Carnegie Mellon University, Pittsburgh.]

David Douglas, 1799–1834. [Hunt Institute, Carnegie Mellon University, Pittsburgh, lithograph published by R. Martin & Co.]

1843) appeared, the Linnaean era in America was ended, and Lindley's natural system was promoted by America's new voice of authority. Even Eaton gave in, in a fashion, urging Gray to take over his regional flora for the northeastern United States, which Gray eventually did. In many ways, the future of American botany was now firmly in the hands of Asa Gray.

When Gray went to Harvard University in 1842, following Nuttall, and Torrey to Columbia and Princeton, after years at West Point, the simple fact of distance lessened their joint efforts. Westward exploration and collecting intensified during this time, and the resulting specimens sent to them for naming overwhelmed them with responsibilities and an abundance of new species.

The Age of Exploration

The increase in plant collecting was stimulated in part by the preparation of a flora that addressed a significant portion of temperate North America, William Jackson Hooker's (1785–1865) *Flora Boreali-Americana...* (2 vols., London, Paris, and Strasbourg, [1829–] 1833–1840). This work was devoted primarily to the British-dominated portions of the continent and was based on extensive field studies by several explorer-naturalists, the foremost being David Douglas (1799–1834) of the Royal Horticultural Society of London.

Douglas, who was particularly interested in oaks, first came to North America in 1823, confining his botanical activities mainly to the middle Atlantic region and as far west as Michigan and southern Ontario. He no sooner had returned to London in January 1824 than he was urged to proceed to western North America. On 26 July 1824 Douglas left Gravesend for the mouth of the Columbia River. He collected extensively in the Oregon Country and crossed Canada to reach York Factory on Hudson Bay in 1827 for a brief trip back to England. In 1829 he returned to Fort Vancouver and made side visits to California and the Hawaiian Islands, where he died in July 1834.

Douglas's enthusiasm inspired others. Traveling with him to Northwest America in 1824 was a fellow Scot, Dr. John Scouler (1804–1871), a surgeon-naturalist. Scouler collected from Fort Vancouver on the Columbia River to Nootka Sound and made his specimens available to W. J. Hooker, his former professor at Glasgow. A Hudson's Bay Company surgeon, William Fraser Tolmie (1812–1886), collected mainly in Washington and perhaps British Columbia, climbing Mt.

Rainier in August 1833. Later, a fur trapper, John McLeod, collected along the Oregon Trail from Fort Vancouver to the 1837 rendezvous site on the Green River in Wyoming. He is "Tolmie's friend" to whom Hooker assigned so many unusual species in his flora.

In Canada, Dr. John Richardson (1787–1865) and Thomas Drummond (1780–1835) were associated with the second Franklin "Land Expedition to the Polar Sea" of 1825–1827, a part of the great search for the Northwest Passage involving the Parry and the Beechey expeditions. Both Richardson and Drummond made large collections from the Mackenzie River region south and eastward to the Rocky Mountains and Hudson Bay. The collections of Douglas, Richardson, and Drummond were significant for Hooker's flora.

Those made by Alexander Collie (1793–1835) and George Lay (d. 1845) on the *Blossom* expedition, commanded by Frederick Beechey, proved to be of lesser significance, although several new species were found in California, Mexico, and other regions of the world. W. J. Hooker and George Walker-Arnott (1799–1868) described the plants made on this 1825–1828 voyage over many years. Even so, most of the new species from North America for which they accounted in their *Botany of Captain Beechey's Voyage...* (10 pts., London, [1830–]1841) were based on specimens in the collections of Menzies and Douglas.

In Alaska, Carl Mertens (1796–1830), naturalist with the Russian Lutke expedition, visited Sitka in 1827 aboard the *Senjavin*. His plants were described by the German-born Russian botanist August Bongard (1786–1839) in 1833. Ilja G. Vosnesensky (1816–1871), who had collected at Fort Ross in California in 1842, collected plants during his visits to Alaska between 1840 and 1849, often in collaboration with others of the Russian-American Company.

In the frozen Arctic, John Rae (1813–1893) and David Lyall (1817–1895), among others, made many plant collections from previously unexplored islands. Meanwhile Richardson and others searched from 1847 to 1850 for some trace of the third Franklin expedition, which had ended in tragedy. Joseph Dalton Hooker (1817–1911) subsequently accounted for the Rae and Lyall specimens.

The success of the Lewis and Clark expedition was not repeated by the United States government for some 30 years. The Freeman expedition up the Red River of the South in 1806 was repulsed by the Spanish, as was the one led by Zebulon Pike into the Colorado Rockies the same year. In 1819, however, Nuttall ventured up the same Red River into what is now Oklahoma, and his findings were numerous. The same year, Dr. William Baldwin (1779–1819), who had studied with Barton, traveled up the Missouri River on the Yellowstone expedition, which included Stephen H. Long of the topographical engineers. Delays and the death of over a hundred men due to scurvy prompted Congress to halt the expedition's funding. Baldwin died at Franklin, Missouri, during his own attempt to reach St. Louis, with little to show for his efforts.

Edwin James (1797–1861) replaced Baldwin as surgeon-naturalist in 1820, joining a new Long expedition in St. Louis. With him was Thomas Say (1787–1834), a self-taught naturalist who became one of America's best-known zoologists and a distinguished entomologist. The route was westward to the Rocky Mountains of central Colorado; it was direct and rapid, and both men collected whenever they could. Once in the Rockies, James climbed Pikes Peak in June, where he found many alpine species. By September, the expedition was at Fort Smith in the Arkansas Territory on its return leg to St. Louis.

Baldwin and James's specimens were sent, eventually, to Torrey, who accounted for them in 1824 and 1827. Less delayed in their description were the plants collected by Captain David Douglass (1790–1849), then at West Point, who accompanied the Cass and Schoolcraft expedition on their search for the headwaters of the Mississippi River in 1820. Torrey published his summary of Douglass's plants in 1822.

Lewis David von Schweinitz (1780–1835), the noted mycologist and monographer of North American *Carex*, wrote a catalog of the small number of plants Say found along the Red River of the North in 1823 while on the Long and Keating expedition. Just before he died, von Schweinitz purchased Baldwin's herbarium from the latter's widow; this was eventually deposited at the Academy of Natural Sciences in Philadelphia, and many of its new species were described by Torrey or by Nuttall.

Westward Expansion

Torrey and Gray realized the importance of collectors exploring remote regions of North America, and if they were to account for its plants, skilled botanists had to be in the field whenever possible. The increasing number of government expeditions provided them with a potential source of novelties, and Torrey lobbied hard to ensure that naturalists loyal to him and to Gray were attached to as many expeditions as possible. Even when Congress seized on the notion of a grand, worldwide expedition, such as the British had conducted a half century earlier, Torrey saw to it that William Brackenridge (1810–1893) was on Wilkes's round-the-world voyage.

To be sure, private collectors, friends, and even remote acquaintances provided Torrey and Gray with

William Starling Sullivant, 1803–1873. [Hunt Institute, Carnegie Mellon University, Pittsburgh, copied from original photograph in de Candolle Collection, Conservatoire et Jardin Botaniques, Genève.]

exciting specimens, especially from the southeastern part of the United States. Nuttall's friends, Hardy Croom (1797–1837) and Harris Loomis (d. 1837), indirectly contributed plants, as did Rafinesque's replacement at Transylvania University, Charles Short (1794–1863), also one of Barton's students. Torrey and Gray's cadre of collectors in the 1830s was impressive, and specimens came to them from many who were to make substantial contributions in their own right: Lewis Beck (1798–1853), John Blodgett (1809–1853), Samuel Boykin (1786–1848), Samuel Buckley (1809–1884), Elias Leavenworth (1803–1887), John LeConte (1784–1860), Charles Pickering (1805–1878), Zina Pitcher (1797–1872), John Riddell (1807–1865), William Sullivant (1803–1873), and Edward Tuckerman (1817–1886).

In the still unexplored American West, other adventurers and naturalists were finding a wealth of novelties. Nathaniel Wyeth (1802–1856), a Boston fur trader, followed the Oregon Trail from St. Louis to Fort Vancouver in 1832. He returned the following year via the Clark River of northern Idaho to the Flathead Post in Montana, then south to Fort Bonneville in Wyoming. He submitted his large collection to Nuttall, then at Harvard, to name. This Nuttall did, but before the paper was published, Nuttall was headed west himself with Wyeth. Nuttall collected thousands of specimens during the next three years in the Northwest, California, and Hawaii. Several hundred species were described ultimately by Nuttall or by Torrey and Gray in *Flora of North America.*

During this time, Thomas Drummond left the Canadian north and collected from Florida to Texas from 1831 to 1835, his specimens going to W. J. Hooker in London. The specimens collected earlier in Texas by Jean Berlandier (1805–1851) had been sent to Alphonse de Candolle (1806–1893) in Geneva. Those of a later immigrant, the German Ferdinand Lindheimer (1801–1879), were studied by Gray, and the novelties were published with George Engelmann (1809–1884) of St. Louis between 1845 and 1850. Adolf Scheele (1808–1864) also published seven articles on Lindheimer's Texas plants, from 1848 to 1852, in the European journal *Linnaea.*

George Theodor Engelmann, 1809–1884. [Hunt Institute, Carnegie Mellon University, Pittsburgh, photograph by Brimmer & Kalb, St. Louis.]

The westward movement of Americans began in earnest in the 1840s, but the challenge of the great grassy plains (termed the "Great American Desert" by Pike) was one few wished to risk. Missouri Senator Thomas Hart Benton knew that to open the West, the United States must have a safe route to the Pacific Coast. Who better to find it than his son-in-law, John Charles Frémont (1813–1890)?

Frémont's expeditions are now schoolboy legends, complete with the exploits of Kit Carson and "Broken-hand" Fitzpatrick, the lost cannon, and the battle of San Pasqual. The name of Frémont is found in every western state, but few know that many of the plants on the peaks, near the rivers, or in the counties that bear his name were also found or named by Frémont. Yet he was but one of many explorers, naturalists, soldiers, and scientists for the U.S. Army's Corps of Topographical Engineers that Torrey and Gray urged to collect plants.

The war with Mexico saw Frémont in the West once again, conquering California and collecting plants. Even the military commander, Major William Emory (1811–1887), and his "Army of the West" collected plants as they crossed New Mexico and Arizona to reach the Pacific Coast and claim California in the name of the U.S. Army from Frémont, self-proclaimed military governor. After the war, the Mexican boundary survey, led by Emory, collected more western specimens with the assistance of Charles Parry (1823–1890), George Thurber (1821–1890), and Charles Wright (1811–1885).

The fulfillment of Benton's westward vision brought to the United States more territory than just that resulting from the war with Mexico. With the acquisition of the Oregon Country in 1846, the United States stretched across a continent. The activity associated with this expansion brought more new plants to the attention of the small American botanical community around Torrey and Gray.

While Torrey and Gray labored on their continental tome, manuals and regional floras were published. The eight editions of Eaton's *Manual* informed naturalists about the plants in the northern states. In 1841, John Darby (1804–1877) of Georgia published a companion *Manual* (Macon) for the southern states, followed by his *Botany of the Southern States* in 1855 (New York, Cincinnati, and Savannah). After Eaton's death in 1842, Gray began to work on his own regional flora, *A Man-*

ual of the Botany of the Northern United States. It was first published in 1848 in Boston, Cambridge, and London, and went through numerous editions.

Alvan Chapman (1809–1899) published the first edition of his *Flora of the Southern United States* in 1860 (New York), and it went through two more editions, the last appearing in 1899. Journals established in different parts of the United States allowed American naturalists, even amateurs, to publish their new discoveries. The results were overwhelming, and it was all Torrey and Gray could do to keep up with teaching and with writing their government reports. The long desired flora of North America was an impossibility as long as novelties arrived almost monthly from the West.

At Harvard, Gray's influence grew as that of Torrey's waned, and the center of North American botany moved from New York to Cambridge. Gray, the modernist, had readily adopted the arrangement of natural orders proposed by Lindley in 1830 (*An Introduction to the Natural System of Botany...*, London) and had developed keys to the families similar to those in Lindley's *Key to Structural, Physiological and Systematic Botany...*, published in 1835 (London). These were now standard in most American floras. Gray also included keys to the genera, taking the idea from the British naturalist Samuel Frederick Gray (1766–1828), who first used them in 1821 in his *Natural Arrangement of British Plants* (London). Asa Gray did not include keys to species in his works. Rather, species were arranged into groups under diagnostic headings, following the format designed by de Candolle for his *Prodromus...* (17 vols., Paris etc., 1823–1873).

As a professor, Gray realized that students needed a useful, concise textbook in botany. Many were available, but those in English were nearly always directed toward the British student. American textbooks were few, and Barton's *Elements of Botany...* (ed. 1, Philadelphia, 1803) was outdated. Once again, Lindley provided a model, and in 1836 Gray published his own *Elements of Botany* (New York). Its major competitor was *Class-book of Botany...*, by Alphonso Wood (1810–1881), which first appeared in 1845 (Boston and Claremont, N.H.). Almira Hart Lincoln (later Phelps; 1793–1884), better known as "Mrs. Lincoln," began publishing her *Familiar Lectures on Botany* in 1829 (Hartford, New York, Boston). The three works sold over two million copies!

The Patriarch

The trip to Europe that Gray took in the winter of 1839 was fundamental for many reasons besides allowing him to study critical type material. Although Gray paid his compliments to Robert Brown and met the elder Hooker and his son Joseph at the British Museum, it was George Bentham (1800–1884), who was closer to his age and temperament, with whom he felt a companionship. Bentham was rapidly becoming England's foremost botanist. He had recently completed a study of the North American eriogonums, a favorite group of Torrey's, and was interested in mints throughout the world.

Gray had been following established tradition, working on regional and national floras, reviewing only those species of a group as he came to them. Bentham, however, tended to study a group throughout its range, thereby gaining a greater understanding of it. The idea of examining a group throughout its range was reasonable, providing one had access to libraries and collections. This was not a realistic expectation for Gray who, at the time, was considering a position at the University of Michigan. Still, several wholly American genera could do with his attention, and most were in Asteraceae. Therefore, like Bentham, Gray began to write both monographic and floristic treatments, and he encouraged others to do the same.

When Gray returned to England in June 1850, he was fully established as America's leading plant taxonomist and perhaps its leading botanist, and Bentham's position was no less secure. With Bentham, Gray worked over the plants from the Wilkes expedition and continued his studies of type material. In Dublin, Gray visited William H. Harvey (1811–1866), the algologist who had accompanied Wilkes. When Joseph Hooker returned to London in the spring of 1851 with his collection from the Ross expedition to India, Gray was at hand to glimpse the wonders from another part of the world. Here, too, Gray met Richard Owen and Thomas H. Huxley, both of whom played important roles at the end of the decade in the debates on evolution.

Awaiting Gray at home in the fall of 1851 were specimens from more expedition botanists, friends, and colleagues. He published on the specimens collected by Wright and Thurber on the boundary survey; the publications from those of Wilkes and Emory were delayed; and no sooner had he started again on the flora than the railroad survey specimens began to accumulate.

When the theory of evolution began to have an impact in 1859, Gray was in the thick of it, and the time to devote to the flora was limited even more. Bentham was no less overwhelmed with his tasks. Bentham, however, persevered in his labors, and his tasks were completed while Gray's rarely were.

Other events conspired to slow the work both of Gray and the plant collectors. Aside from the fiscal problems at Harvard and the demands of students, the American Civil War was raging. Torrey was nearing 70, and Daniel C. Eaton (1834–1895) at Yale, the only other full-time

Léon Provancher, 1820–1892. [Archvies/ Collection Léon-Provancher, Université Laval, Sainte-Foy, Quebec.]

Robert Kennicott, 1835–1866. [Hunt Institute, Carnegie Mellon University, Pittsburgh, copied from original photograph at Academy of Natural Sciences, Philadelphia.]

botanical professor in the United States, was heavily committed to the study of ferns and their allies. Gray attempted to turn again to his much-neglected *Flora of North America* with the hope of producing another volume. It was not to be.

Decentralization of American botany had now begun. In Washington, D.C., the new Smithsonian Institution was considering the establishment of a botany section, and the newly instituted Department of Agriculture was supporting the idea of a National Herbarium. Businessman Henry Shaw (1800–1889) established a botanical garden in St. Louis, with the advice of George Engelmann, Asa Gray, and J. D. Hooker. On the West Coast, Albert Kellogg (1813–1877) and his friends had established the California Academy of Sciences in San Francisco in 1853, with the intent that their institution would be equal to any on the East Coast.

The Morrill Act of 1862, signed by President Lincoln during the war, said that the curricula of all land-grant colleges (i.e., state-supported colleges emphasizing agriculture and applied sciences) were to include botany, and in particular taxonomy. This resulted in numerous opportunities for young botanists to find work in some of the still relatively unexplored regions of the West. Therefore, as Gray considered retirement, the botanical community became fragmented, and the patriarch began to lose control.

Branching of Talent

Botanical exploration in western North America after the Civil War increased, albeit with limited governmental support. In Canada, John Jeffrey (1826–1854) arrived at York Factory on Hudson Bay in August 1850 and worked his way westward to British Columbia and eventually California, collecting for botanical gardens in Scotland. Eugène Bourgeau (1813–1877), with backing from the Royal Geographic Society, collected some 60,000 specimens from the Great Lakes to the Rocky Mountains over a two-year period (1857–1859) before going to Mexico, where he made even more extensive collections.

In 1858 George Lawson (1827–1895), the "father of Canadian botany," moved to Canada from Edinburgh and soon thereafter established the Botanical Society of Canada. In 1860 the first issue of the Society's *Annals* was published. Lawson also proved to be an inspiration for amateur botanists throughout the country, encouraging many to publish their observations. One such amateur, the Abbé Léon Provancher (1820–1892), published *Flore Canadienne...* (2 vols., Quebec, 1862[1863]), which treated mainly the plants of eastern Canada, although he accounted for the species from Ontario reported by W. J. Hooker. Provancher founded the journal *Le Naturaliste Canadien* in 1869.

To the north and in Alaska, Robert Kennicott (1835–1866) of the Chicago Academy of Sciences led a series of collecting forays from the Great Slave Lake region to Fort Yukon. On other trips he ventured to British Columbia, the Mackenzie River, and, in 1864, Sitka. Accompanying him on various trips were Joseph Rothrock (1839–1922), who later collected in the southern Rocky Mountains, and William H. Dall (1845–1927), who included a list of "useful plants" in a book he wrote in 1870 on Alaskan resources.

Geological surveys became the modern means for botanical explorations. In the United States, the survey of the fortieth parallel, led by Clarence King (1842–1900) from 1867 to 1874 across Nevada and Utah, resulted in the discovery of numerous new plants, most of which were found by the camp cook turned naturalist, Sereno Watson (1826–1892), in 1868 and 1869. A shy man with a checkered past, Watson took his collections to Cambridge where, with Gray's assistance, he worked up the results. By 1872 Watson was a permanent fixture, taking over much of Gray's correspondence with western botanists and serving, essentially, as curator of the herbarium.

The following year Gray formally retired from Harvard with the intent of devoting the remainder of his years to taxonomy. Unfortunately, Torrey had died shortly before Gray retired, and so had Louis Agassiz (1807–1873), Harvard's famed zoologist and fellow professor; Gray was now free to concentrate on the North American flora, but very much alone.

After the American Civil War, many ventured west and began to collect plants. Charles L. Anderson (1827–1910) collected in western Nevada from 1862 to 1867. He was amused that the clover by the privy proved to be a new species that Gray named for him. In 1871 Anderson wrote a catalog of Nevada flora for the state mineralogist's report. John Gill Lemmon (1832–1908) moved to California in 1866 and began his long botanical career there; most of his plants were named by Watson. The state-sponsored geological survey program in California had two fine collectors, William Brewer (1828–1910) and Henry N. Bolander (1831–1897). With Watson and Gray, Brewer published a two-volume flora of California (Cambridge, Mass., 1876–1880), the first modern flora for a western state.

The last military expeditions, as well as the expeditions made in the early years of the U.S. Geological Survey (established in 1879), provided Gray and Watson with a constant flow of new species from the Rocky Mountains. Joseph Rothrock, who later had a distinguished career in forestry, served as botanist/surgeon

Joseph Trimble Rothrock, 1839–1922. [Hunt Institute, Carnegie Mellon University, Pittsburgh.]

Sereno Watson, 1826–1892. [Gray Herbarium Archives, Harvard University.]

William Henry Brewer, 1828–1910. [Hunt Institute, Carnegie Mellon University, Pittsburgh, photograph by Warren of Cambridge, Mass.]

for the Whipple expedition from 1873 to 1875. The several expeditions of Ferdinand Hayden (1829–1887) and John Wesley Powell (1834–1902) employed many botanists. Engelmann generally published the new plants found by Hayden, and Watson published those found by Powell.

To the north, the Geological Survey of Canada began its boundary survey in 1873. George Dawson (1849–1901) served as geologist/botanist, although most of the plants were collected by Thomas J. W. Burgess (1849–1925). At this point, Canada's foremost collector, John Macoun (1831–1920), began his botanical career in earnest. From 1869 until his death, Macoun collected thousands of specimens throughout much of Canada. He summarized these records in his *Catalogue of Canadian Plants*, published between 1883 and 1902 (7 pts., Montreal and Ottawa).

For his part, Gray concentrated on writing his *Synoptical Flora of North America* (2 vols., New York etc., 1878–1897), the project of his retirement. The years between 1842, when the last part of *Flora of North America* appeared, and 1873, when Gray retired, had been expansive in terms of growth in the number of new species and knowledge of the continent's flora. Gray turned to others for help, asking several younger men to assist him while retaining the cooperation of his old friend Engelmann and his newly acquired assistant Watson. Added to Gray's group were John Merle Coulter (1851–1928), William Farlow (1844–1919), George Goodale (1839–1923), and Charles Sprague Sargent (1841–1927), all famous later. Likewise, many of Gray's students began to fill the botanist posts at the newly established land-grant colleges, thereby providing him with capable collectors in critical regions. The goals were set, the players in position, and the work commenced, but Gray had not anticipated the rise of western botanists whose work would only make his more difficult.

Systematic botany had been centered in Cambridge for 30 years, and its position had never been challenged. Few botanists attempted to work on the plants of North America without Gray's prior approval, and consequently he could influence what was done. To some extent, Gray dominated what was published by controlling the publications. By having one of the largest libraries and collections of American plants immediately at hand, Gray quickly settled identifications.

To many working in the American West, however, there was no joy in collecting a novelty if its publication was delayed because Gray or Watson already had specimens but had not yet published the name. Equally troublesome was the assumption that those in the East

Marcus Eugene Jones, 1852–1934. [Hunt Institute, Carnegie Mellon University, Pittsburgh, copied from original photograph owned by Mabel Jones Broaddus, his daughter.]

Nathaniel Lord Britton, 1859–1934. [Hunt Institute, Carnegie Mellon University, Pittsburgh, photograph by L. A. Charette.]

John Macoun, 1831–1920. [Hunt Institute, Carnegie Mellon University, Pittsburgh.]

could somehow prevent those in the West from publishing their findings. At the forefront of the western dissidents was Edward L. Greene (1843–1915), minister, college professor, and protagonist. Greene had a keen eye for botanical novelties and a good knowledge of the classical literature. He was a "splitter," the antithesis of Gray and his followers, and he was not opposed to taking up the earliest available, validly published name, particularly names that Gray had chosen purposely to ignore.

Another botanical irritant was Marcus Eugene Jones (1852–1934), mining engineer, botanist, and Latinist. Jones collected widely in the Intermountain West, finding many locally endemic plants among others.

Individually and in the problems they jointly caused, Greene and Jones initiated the end of Gray's domination. Their defiance took two forms: first, they worked with the Smithsonian and the California Academy of Sciences, both institutions trying to proclaim their own independence; and second, they established their own journals. They therefore had means to publish names, and they had institutions from which to work.

Gray's efforts were focused. He wrote his *Synoptical Flora* and let Watson and others deal with the troublemakers. As for the expanding influence of other institutions, this Gray encouraged. North America needed more botanists and botanical institutions, but he felt these should be dispersed among universities, with only the difficult problems coming to Cambridge for final review. The feelings of others to the contrary, Gray did not discourage others from publishing their results; he simply felt that editors should not needlessly let new synonymy be created. The influence Gray possessed was substantial, and his authority was final. Watson, Farlow, Sargent, and Coulter were not equal to Gray in this, and when Gray died in 1888, still working on his flora, no one among them was capable of dominating taxonomy.

"Bughole Botany"

When John Torrey died in 1873, his large library and rich collection were at Columbia University. In 1896, when Nathaniel Lord Britton (1859–1934) became the first director of the newly established New York Botanical Garden, he arranged to have transferred the university's herbarium of some 400,000 specimens, along with Torrey's library. Both arrived in 1899. New York was suddenly able to take a botanical lead.

In addition, with the backing of Addison Brown (1830–1913), Britton published *An Illustrated Flora of the Northern United States...* (ed. 1, 3 vols., New

Merritt Lyndon Fernald, 1873–1950. [Gray Herbarium Archives, Harvard University.]

York, 1896–1898) for the very region for which Gray's *Manual* had long been the primary authority. The sixth edition of the *Manual* (New York and Chicago, 1890), by Watson and Coulter, continued Gray's format of a single, tightly concise volume with keys to all taxa, brief descriptions, limited synonymy, habitat data, and distribution notes. Instead, Britton provided expanded keys, full descriptions, rather complete synonymy, and illustrations of each species. Furthermore, Britton produced his own one-volume *Manual...* in 1901 (New York). Unlike his competitors at Harvard, Britton accepted many of Greene's reestablished genera and used species names long since ignored by Gray. Britton's concept of species was similar to that of Gray, but wherever a split might be possible, Britton would make it, whereas Gray would not.

The publication battle was fierce. Britton issued new editions of his *Manual* in 1905 and 1907 (both New York); Benjamin L. Robinson (1864–1935) and Merritt Lyndon Fernald (1873–1950) answered with the seventh edition of Gray's *Manual* in 1908 (New York, Cincinnati, and Chicago). This was countered in 1913 by a second edition of Britton and Brown's *Illustrated Flora* (New York). Likewise, both sides began to marshal supporters at other institutions and began to stake claims to portions of the United States. In Washington, D.C., the Smithsonian Institution was doing likewise through its flora program, considering itself the sole authority on the flora of much of the West.

Under Gray's influence, Coulter had published his *Manual of the Botany...of the Rocky Mountain Region*...in 1885 (New York and Chicago), but by the time Aven Nelson (1859–1952) completed a revision in 1909 (*New Manual of Botany of the Central Rocky Mountains...*, New York, Cincinnati, and Chicago), the result was more a work of New York than of Cambridge. Smithsonian efforts on the West appeared as whole issues of the *Contributions from the United States National Herbarium.* Coulter published his flora of western Texas in 1891. Frederick Coville (1867–1937) published his on Death Valley in 1893. State floras followed for Washington (1906), written by Charles Piper

Aven Nelson, 1859–1952. [Hunt Institute, Carnegie Mellon University, Pittsburgh.]

Smith (1877–1955), and for New Mexico (1915), by Elmer O. Wooton (1865–1945) and Paul C. Standley (1884–1963). In 1925, Ivar Tidestrom (1864–1956) published his flora of Utah and Nevada. Most of these works consisted of little more than keys to species and genera, with annotations reporting distributions.

The group at New York was the most productive. John Kunkel Small (1869–1938) completed the first edition of his 1400-page *Flora of the Southeastern United States...* (New York, 1903), with full keys, good descriptions, habitat and distribution data, and an occasional illustration. Per Axel Rydberg (1860–1931) produced catalogs of the plants of Montana (New York, 1900) and Colorado (Fort Collins, 1906) in the abbreviated style of the Smithsonian publications. They were much in the manner of his first two floristic efforts, on the Sand Hills of Nebraska (1895) and on the Black Hills (1896), published in the *Contributions*. His *Flora of the Rocky Mountains and Adjacent Plains...* (New York, 1917) and his posthumous *Flora of the Prairies and Plains of Central North America...* (New York, 1932), however, were more akin to Small's works.

Small and Rydberg were both splitters, which appalled many, especially those of the Harvard tradition. Not only did Small and Rydberg recognize numerous species, they fragmented genera and families to a degree never before seen in American botany. Marcus E. Jones called Rydberg's effort "bughole botany" because, to him, Rydberg seemed to recognize species based on the number of bugholes in each leaf.

LeRoy Abrams (1874–1956) was in the New York camp after spending a year at the garden while finishing a flora for the Los Angeles area (ed. 1, Stanford, 1904). Under Britton's influence, Abrams began *An Illustrated Flora of the Pacific States* (Stanford). The first volume appeared in 1923, but the depression and war years delayed the second until 1944; the remaining two volumes were published in 1951 and 1960.

California still proved to be resistant to eastern domination. Certainly its isolation contributed, but so too did its long history of local botanical efforts dating to the 1850s. Following publication of the state flora by

John Kunkel Small, 1869–1938. [Hunt Institute, Carnegie Mellon University, Pittsburgh, photograph by George F. Weber.]

Per Axel Rydberg, 1860–1931 (left) with Alfred Rehder, 1863–1949. [Hunt Institute, Carnegie Mellon University, Pittsburgh, copied from original photograph at Field Museum of Natural History, Chicago.]

Willis Linn Jepson, 1867–1946. [Hunt Institute, Carnegie Mellon University, Pittsburgh.]

Brewer and Watson, a series of local floras began to appear. Volney Rattan (1840–1915) published *A Popular California Flora* in 1879 (ed. 1, San Francisco), and Greene, then at Berkeley, followed with *Flora Franciscana* (4 pts., San Francisco, 1891–1897). Willis Linn Jepson (1867–1946), the state's most acclaimed taxonomist, began his efforts in 1901 with *A Flora of Western Middle California...* (ed. 1, Berkeley), followed by his statewide *Manual...* (Berkeley, [1923–1925]). Jepson began the monographic style of flora writing in 1909 when he published the first volume of *A Flora of California...* (3 vols., San Francisco etc., 1909–1943).

Monographs and revisions began to appear more frequently, many the works of newly trained students. Several authors presented treatments on specific groups of plants, such as Sargent's *The Silva of North America...*, published between 1890 and 1902 (14 vols., Boston and New York), and various treatments of grasses beginning with George Vasey (1822–1893) and ending with Albert S. Hitchcock (1865–1935), with his *Manual of the Grasses of the United States* (ed. 1, Washington, 1935).

More state floras gradually became available as well. Tidestrom and Sister Mary Teresita Kittell (1892–?) treated Arizona and New Mexico in 1941 (Washington), followed by Thomas H. Kearney (1874–1956) and Robert H. Peebles (1900–1956), who wrote on the plants of Arizona (Washington, 1942; Berkeley and Los Angeles, 1951). Morton E. Peck (1871–1959) authored the first edition of his Oregon flora in 1941 (Portland). One for Idaho by Ray E. Davis (1895–1984) followed in 1952 (Dubuque). Events in the East were similar. In 1901 Charles Mohr (1824–1901) completed a flora of Alabama (*Contr. U.S. Natl. Herb.* 6), and Augustin Gattinger (1825–1903) published one for Tennessee (Nashville, 1901). In 1910 Forrest Shreve (1878–1950) wrote an account of the plants in Maryland (Baltimore).

Albert Spear Hitchcock, 1865–1935. [Hunt Institute, Carnegie Mellon University, Pittsburgh.]

On the Road

The period from 1870 to 1945 had many collectors, amateur and professional. They roamed widely in search of novelties, some publishing their own findings, some handing over their specimens for others to name. Local societies and clubs were formed, proceedings published, and activities sponsored. Institutional and university faculty were encouraged to be in the field, and students particularly so.

Botanical exploration of Alaska accelerated. Albert Kellogg of the California Academy of Sciences collected about 500 specimens in Alaska in 1867 while assigned to a survey party of the new U.S. Coast and Geodetic Survey. Other government officials began to collect plants as well. Mark W. Harrington (1848–1926) and Lucien M. Turner (1847–1909) sent specimens to Cambridge and to the U.S. National Herbarium in the 1870s. During the same decade, Edward W. Nelson (1855–1934), then a weather observer at St. Michael but eventually chief of the Bureau of Biological Survey, made a small collection. The foremost collectors of the period, however, were Frans Kjellman (1846–1907) and Ernst Almquist (1852–1946), both of Sweden, who made extensive collections along the coast of the Bering Sea.

James M. Macoun (1862–1920) and John Macoun (1831–1920), with the Geological Survey of Canada, explored much of western Canada, depositing their specimens at Ottawa. William C. McCalla (1872–1962), photographer, collector, and author of a wildflower text for high schools, also collected with J. M. Macoun.

Amateur botanists were also busy throughout Canada. David Watt (1830–1917), a businessman, amassed a large herbarium mainly through exchange. His dream, unfulfilled, was to write a flora of Canada. Braddish Billings Jr. (1819–1871), who worked for the St. Lawrence and Ottawa Railroad in the 1860s, published a local list of Ontario plants. In British Columbia, Joseph Kaye Henry (1866–1930), a professor of English, published a descriptive flora of the province in 1915 (Toronto). Robinson and Fernald from Harvard took an interest in the flora of eastern Canada in the 1890s,

Joseph Louis Conrad Kirouac (Frère Marie-Victorin), 1885–1944. [Hunt Institute, Carnegie Mellon University, Pittsburgh.]

and they botanized there well into the next century. Fernald himself remained active in Canada into the 1930s, discovering numerous new species, most of which he described in *Rhodora,* the journal of the New England Botanical Club.

The foremost collector in eastern Canada was Frère Marie-Victorin (Joseph-Louis Conrad Kirouac) (1885–1944). In 1920 he established the Institute Botanique at the Université de Montréal, and despite suffering a heart attack in 1923 while in the Shickshock Mountains on the Gaspé Peninsula, he was in the field for the next five years in some of the harshest environments. In 1931 he established the Jardin Botanique de Montréal. In 1935 he published *Flore Laurentienne...* (Montreal).

The northern part of the continent was also becoming accessible. Another Cambridge-based collector was Hugh Raup (b. 1901) of the Arnold Arboretum. With support from the National Museum of Canada, Raup collected well to the north of most, working mostly in the Northwest Territories and along the Alaska Highway. Malte Oskar Malte (1880–1933), originally an agrostologist with the Canadian Department of Agriculture, moved to the National Museum in 1920. He concentrated on the prairie flora as well as on the flora of boreal and arctic Canada. Morton P. Porsild (1872–1956) was then actively collecting in Greenland. Alf Erling Porsild (1901–1977), his son, joined the National Herbarium of Canada and was there from 1936 to 1966. He reported on collectors for *Flora of the Canadian Arctic* (1955, 1957). Nicholas Polunin (1909–1991) covered the eastern Canadian Arctic (*Bull. Natl. Mus. Canada* 92, 97, 104, 1940–1948).

The Abbé Ernest Lepage (1905–1981) began his botanical career with an interest in bryophytes and lichens. In 1943 he began to collect with Père Arthème-Antoine Dutilly (1896–1973), and together they botanized the arctic and boreal regions from Alaska to Labrador.

In Alaska, Frederick Funston (1865–1917), who had previously collected plants in Death Valley, made a large collection of plants from 1892 to 1894 that Coville described in 1895. At the same time, but continuing to 1902, Martin Gorman (1853–1926), who also col-

Morton Pedersen Porsild, 1872–1956. [Hunt Institute, Carnegie Mellon University, Pittsburgh, copied from original photograph at Herbarium, University of California, Berkeley.]

Ernest Lepage, 1905–1981. [Hunt Institute, Carnegie Mellon University, Pittsburgh.]

Edward Palmer, 1831–1911. [Hunt Institute, Carnegie Mellon University, Pittsburgh.]

lected in Mexico, made several trips to Alaska. Most of his specimens went to the U.S. National Herbarium. Thomas Howell (1842–1912), who is better known for *A Flora of Northwest America...* (Portland, 1897–1903), was with Gorman in 1895. The Alaskan collections of Walter H. Evans (1863–1941), Arthur L. Bolton (1876–?), Frank C. Schrader (1860–1940), and botanical artist Frederick Walpole (1861–1904) also eventually reached Washington, D.C.

The 1899 Harriman expedition to Alaska had Coville, William Trelease (1857–1945), Brewer, and Thomas Kearney collecting plants (later described in vol. 5 of the 1904 report on the expedition). The most active of the Alaskan collectors, however, was Jacob P. Anderson (1874–1953), who collected extensively throughout the state from 1914 to 1940 while employed by the federal government. Also active was Eric Hultén (1894–1981) of the University of Lund, who amassed several thousand specimens in preparation for his many publications on the Alaskan flora.

Across the conterminous United States, numerous collectors were in search of novelties. Fernald was active in the Northeast until the late 1940s, making a series of expeditions with friends, all duly reported in *Rhodora*. Small continued to collect in the Southeast, especially in Florida after 1903, discovering several new species that were incorporated into a 1933 revision of his flora (*Manual...*, New York).

Many came to the Rocky Mountains, some even collecting an occasional plant on the Great Plains portion of their trip. By the 1870s, Charles Parry and Edward Palmer (1831–1911) were well established in the West, collecting throughout the Rockies and the American Southwest for the Department of Agriculture and Smithsonian Institution, respectively.

Theodore D. A. Cockerell (1866–1948) collected plants on all of his travels. In 1889 he met Alice Eastwood (1859–1953), a Denver school teacher, collecting in the Rockies. Three years later she was an assistant curator at the California Academy of Sciences and returned to Colorado, searching the western Mancos clay hills for undiscovered wonders. Over the next 60 years, Miss Eastwood, in her long dresses and broad-brim hats, looked for novelties to collect. In the 1930s and 1940s her traveling companion was her colleague, John Thomas Howell (b. 1903), who has continued the study of California plants.

Across the Intermountain West came collectors like Carl Purpus (1851–1941), who also collected plants in Canada and Mexico, and the team of Dwight Ripley (1908–1973) and Rupert C. Barneby (b. 1911), who discovered many narrowly endemic species. H. Theo-

Jacob Peter Anderson, 1874–1953. [Hunt Institute, Carnegie Mellon University, Pittsburgh, photograph by Maxine Williams.]

Alice Eastwood, 1859–1953. [Hunt Institute, Carnegie Mellon University, Pittsburgh.]

John Thomas Howell, 1903–. [Hunt Institute, Carnegie Mellon University, Pittsburgh, photograph by California Academy of Sciences.]

dor Holm (1854–1932) collected in Greenland, Canada, and the Rocky Mountains, ending his days concentrating on grasses and sedges at the Catholic University of America in Washington. Isaac Martindale (1842–1893) amassed a huge herbarium through exchange and purchase, as well as through his own efforts. The collection is now at the U.S. National Arboretum.

Many students went on to distinguished careers. Aven Nelson encouraged many students during his long tenure at the University of Wyoming (1887–1952): Elias Nelson (1876–1949), Leslie Goodding (1880-1967), James Macbride (1892–1976), Edwin Payson (1893–1927), Louis Butler Payson (1895–1969), George Goodman (b. 1904), Louis O. Williams (1908–1990), and Reed C. Rollins (b. 1911). Jepson, at the University of California, fostered the careers of Herbert Mason (b. 1896), Ivan Johnston (1898–1960), David Keck (b. 1903), Conrad V. Morton (1905–1972), and Lincoln Constance (b. 1909). Many students took their degrees in St. Louis, working through a graduate program at the Missouri Botanical Garden. The dissertations of Carl C. Epling (1894–1968), Robert E. Woodson (1904–1963), Mildred E. Mathias (b. 1906), Julian A. Steyermark (1909–1988), C. Leo Hitchcock (1902–1986), and many others were published in the *Annals of the Missouri Botanical Garden*.

By the turn of the century, Cornell University was attracting many students, who on finishing went far and wide to educate the next generation. One was Bassett Maguire (1904–1991). He went to Utah State University in 1931 and remained until 1943, collecting from Montana to Arizona, taking a year out in 1938 to attend Cornell and complete his doctoral degree. Although he continued to collect in the West after he joined the staff of the New York Botanical Garden, his enthusiasm for the tropics was greater, and so his skill and drive were directed there.

The combination of world wars and economic depression had slowed systematic botany. Publications were less frequent, and collecting was limited. Also fewer in number were the small, privately published journals that were largely for the exclusive use of their editor, such as those promoted by Greene, Jones, and Amos A. Heller (1867–1944), and those restricted to friends, such as those published by Charles Orcutt (1864–1929) and the team of Mary Katharine (Curran) Brandegee (1844–1920) and her husband, Townshend Brandegee (1843–1925). Journals associated with institutions and societies, however, became much more numerous.

Mary Katherine Layne Curran Brandegee, 1844–1920. [Hunt Institute, Carnegie Mellon University, Pittsburgh.]

The growth of university-based systematics increased significantly, even during the difficulties of the Great Depression. When World War II ended, universities welcomed a generation of mature students, and for the next three decades systematic botany in the United States and Canada was dominated by individuals trained during or shortly after the war.

Where Have All the Flowers Gone?

With the end of the war, taxonomy as understood by Torrey and Gray was gone. Searching for novelties or trying to find unreported species was no longer the primary focus of the field. Taxonomy broadened into systematics and was rapidly becoming a laboratory science. Walking across a meadow enjoying the flowers was often replaced by white coats, laboratory walls, and the thrill of a good chromosome squash or a well-resolved chromatograph.

This was to be expected. Since the 1930s, more and more taxonomists had become interested in experimental taxonomy. The long-debated questions of relationships could be addressed and reasonable conclusions could be drawn using the new technology, whereas earlier any solution was open to doubt. Transplantation studies, championed in the New World by Clausen, Keck, and Heisey, proved informative, and new insights were possible concerning the relationship of a plant to its habitat.

Over the past half century, the working lifetime of many still very active systematists, new methods have evolved, and problems once thought impossible to address are being routinely resolved. Even such mundane things as the rules of nomenclature have changed. Significant changes may still come, as today's cladists and molecular systematists challenge the fundamental way all systematists think, much as Darwin did in 1859.

Systematic centers are now more dispersed, although Cambridge, New York City, Ithaca, Washington, St. Louis, and Berkeley are still major players, and now many other institutions, from Austin to Ann Arbor, and Chapel Hill/Durham to Seattle have joined them, each offering its own particular brand of systematics.

Fieldwork is still needed, but now few remote areas remain in North America north of Mexico where one might find large numbers of undescribed vascular plants. New genera, such as *Apacheria, Shoshonea,* or *Dedeckera,* are increasingly less likely to be found, although new genera carved from older groups will continue to be proposed.

Many new floras and revisions of older ones have been completed since 1945. The style of manuals has improved over that of Britton's in 1901. Jepson's use of detailed descriptions is now the norm, as are full distribution and ecological statements, data on flowering times, and notes on uses. Distribution maps are more common, and illustrations are becoming standard. The style of regional floras has changed too, most being illustrated and monographic.

And funding has changed. Research support for floristics was minimal before World War II, but in the early 1950s federal funding in both the United States and Canada became available for study of the plants of large portions of North America. Floristic projects outside of North America now share the available funds with North American projects, which still play a vital role.

Even the reasons for doing taxonomy have changed. While the identification, naming, and classification of plants are still fundamental needs for all botanical science, they have, in recent years, become important to governments also. Concern for endangered plant species, regulation of trade and commerce, and the ecological well-being of the environment have awakened a broad array of citizens who truly are concerned about where all the flowers have gone.

The dream of Barton, Torrey, and Gray to have a flora of North America has never left the botanical community. Over the past 50 years various attempts have been made to accomplish the task. At least now we are at a stage familiar to Torrey and Gray—one volume is completed. Certainly when Torrey talked to Hooker about the task in 1833, the time frame was realistic relative to what was known. He and Gray did not realize they might be overwhelmed with new species. We know that we may not find many new species; rather, we now recognize the need to understand more fully what species we have.

No modern flora can be written without acknowledging the legacy of those who went before us. We must try to make sense of the diversity that exists; that is our duty as systematists. Those devoted to the task must be willing to set aside the multitude of competing requirements and spend their lives at it; no doubt, Asa Gray would understand.

Review

A detailed history of botanical explorations and the development of floristics in North America north of Mexico has yet to be written. The literature is very scattered, and as with most history, much of it is hard to find. Nonetheless, some portions of the subject have been reviewed (E. B. Davis 1987), and Reveal (1990, 1992) has broadly covered many of the events.

The colonial era is summarized in R. P. Stearns (1970), and this is the basic source for the period. Major bibliographies exist on many of the Old World naturalists who described American plants, and these may be found easily in F. A. Stafleu and R. S. Cowan (1976–1988). As for those who actually collected in North America, works are available on John Banister (J. Ewan and N. Ewan 1970), Mark Catesby (G. F. Frick and R. P. Stearns 1961), Alexander Garden (E. Berkeley and D. S. Berkeley 1969), and John Clayton (E. Berkeley and D. S. Berkeley 1963). John and William Bartram have been the subject of numerous books, but the one by E. Berkeley and D. S. Berkeley (1982) is most readable. An essential reference on French botanists in North America before 1850 is J.-F. Leroy (1957). A good bibliographic and biographic source on Marie-Victorin is D. Barabé (1985). An augmented autobiography of Macoun is now available (J. Macoun 1922). J. S. Pringle (forthcoming) will provide access to Canadian floristic history.

Linnaeus's contribution is complex, and works are available on nearly every aspect of his life and times. F. A. Stafleu (1971) wrote one that is perhaps the most useful to systematists trying to understand eighteenth-century taxonomy, but that by W. Blunt and W. T. Stearn (1971) is most informative on Linnaeus.

B. Henrey's (1975) three-volume work on British botanical and horticultural literature before 1800 is indispensable and wonderful reading. More than in any other single source, one may find here the magic of discovery.

An excellent recent work by H. B. Carter (1988) on Joseph Banks covers not only the man but his times as well. This is a nice companion to D. J. Mabberley's study (1985) of Robert Brown.

A biography has finally appeared on André and François André Michaux (H. Savage Jr. and E. J. Savage 1986), but none on Pursh has yet been published. The standard references for Nuttall (J. E. Graustein 1967), Torrey (A. D. Rodgers III 1942), and Gray (A. H. Dupree 1959) remain. Likewise, the brief bibliographic sketches given by H. B. Humphrey (1961) for several dozen of North America's botanists provide a useful introduction. The older book on Philadelphia botanists by J. W. Harshberger (1899) contains odd bits about personalities and characteristics that more recent writers tend to ignore or, out of respect, not mention.

The classical work on western explorations in the United States remains S. D. McKelvey's opus (1955). This work is basic for an understanding of the early collectors and botanists. For an overview of western American history, the texts by R. A. Billington (1982) and L. R. Hafen et al. (1970) are the best. A more detailed study of westward expansion, and the role played by scientists, may be found in W. H. Goetzmann (1966).

Publications on the history of botanical explorations of our northern regions are extremely scattered. E. Hultén's 1940 article on Alaska is still the best, but for Canada, the forthcoming articles by J. S. Pringle will improve the situation immeasurably.

L. W. Lenz (1986) has published an excellent work on Marcus E. Jones, and R. L. Williams (1984) examined the life of Aven Nelson. Both provide a broad overview of taxonomy at the end of the last century and the first few decades of the present one. With the nice biography on Rydberg by A. Tiehm and F. A. Stafleu (1990), the chronicling of Greene remains to be tackled.

Important information on botanical history is available. First, most floras have, in their introductions, a section on the individuals who have collected in the area, and these are often informative. Second, botanical journals are a rich and never-ending source of articles and tidbits on the past, and even revisions and monographs can be well worth perusing because historical items are frequently scattered throughout them.

Two recent systematics symposia at the Missouri Botanical Garden have concentrated on historical subjects. In 1984 George Engelmann served as the focal point; in 1989, the period from 1889 through 1989, only briefly mentioned in this volume, was examined in some detail. Papers delivered at those symposia were published in the *Annals of the Missouri Botanical Garden* in 1986 and in 1991.

Finally, there are the contributions of a few individuals who, as a result of their life's work, have made understanding the past that much easier. First and foremost is Joseph Ewan (K. Crotz 1989). Anyone wishing to know something about American botanical history should do a computer search on his name. The results are staggering. Likewise, search on the name William T. Stearn. His forte has been the Linnaean era. And then there is Ronald L. Stuckey. His name is less commonly seen, but his subject—botanists in the East during the Torrey and Gray years—is more difficult.

Most who pick up this and the subsequent volumes will find the history of systematics and floristics on every page, for each plant name has a story to tell. Those who look into that story will find wonderful rewards and an even greater appreciation of systematics.

8. Weeds

Ronald L. Stuckey
Theodore M. Barkley

Concepts

The flora of North America includes a large number of conspicuous plants that are called "weeds." The concept of weed is not precisely defined, for it has both a sociological and a biological component. From the sociological perspective, a weed is simply a plant that is growing where someone wishes it were not, and therefore, a weed may be regarded casually as a "plant-out-of-place." By that definition, a rose growing in a wheat field would be a weed; a rose in a garden would not. Some plants, however, have the genetic endowment to inhabit and thrive in places of continual disturbance, most especially in areas that are repeatedly affected by the activities of humankind. These plants are biologically "weedy," and they are sometimes termed colonizing or invasive plants. These biological weeds are the focus of this essay.

Weeds have a measurable effect on the affairs of society, and therefore they have attracted much attention. Several compendia on the subject published since the mid-1960s form the basis of this essay: H. G. Baker and G. L. Stebbins 1965; J. R. Harlan 1975; J. G. Hawkes 1983; W. Holzner and M. Numata 1982; L. J. King 1966; H. J. Lorenzi and L. S. Jeffery 1987; H. A. Mooney and J. A. Drake 1986; S. R. Radosevich and J. S. Holt 1984; and H. A. Roberts 1982.

Weeds occur in all growth forms and in many lifestyles. The majority of weeds are flowering plants, and a high proportion of them share some or all of the following characteristics: short life cycle, rapid growth rate, high level of energy allocated to reproduction, efficient dispersal mechanisms, high population growth rate, wide distribution, seeds with long life spans, and flexible use of environmental resources (H. G. Baker and G. L. Stebbins 1965; F. A. Bazzaz 1986). A catalog of the attributes of weediness was outlined by H. G. Baker (1965, 1974) for "the ideal weed" (table 8.1). He noted that a plant with but few of these attributes is less likely to be successful as a weed than is a plant with all or most of them; therefore, the variation ranges from casual, local weeds to aggressive, widespread weeds.

The scientific study of weeds is driven largely by economic considerations, but some interest in these plants also exists because of their biological nature. For example, why do some plants respond as colonizers, whereas others do not have that intrusive ability?

Control

Weeds have been traditionally controlled by pulling, hoeing, burning, smothering, or otherwise damaging them, or by poisoning the soil. Studies in plant physiology, however, have led weed scientists to a good grasp of both the chemical descriptions and the modes of ac-

Contribution no. 92-126-B, Kans. Agr. Expt. Sta., Manhattan—*T. M. B.*

TABLE 8.1. Ideal Weed Characteristics

1. Germination requirements fulfilled in many environments
2. Discontinuous germination (internally controlled) and great longevity of seeds
3. Rapid growth through vegetative phase to flowering
4. Continuous seed production for as long as growing conditions permit
5. Self-compatible, but not completely autogamous or apomictic
6. When cross-pollinated, unspecialized visitors or windborne
7. Very high seed output in favorable environmental circumstances
8. Produces some seed in wide range of environmental conditions; tolerant and plastic
9. Has adaptations for short- and long-range dispersal
10. If a perennial, has vigorous vegetative reproduction or regeneration from fragments
11. If a perennial, has brittleness, so not easily drawn from the ground
12. Has ability to compete interspecifically by special means (rosette, choking growth, allelochemics)

After H. G. Baker 1965, 1974.

tion of growth-regulating substances, which made possible the development of modern chemical weed control, i.e., the use of selective herbicides. These herbicides are biologically active agents that are introduced into the environment for the laudable purpose of controlling weeds, with minimal evident effects on the crop plants.

Unfortunately, herbicides may have unexpected and unwanted consequences, and they can contribute to unexpected environmental problems. Safe and effective use of herbicides requires both technological skill and planning, and questions must always be asked about the residual effects of applied herbicides and their degradation products in the ecosystem. The sale and use of herbicides are regulated by the appropriate branches of federal and regional governments in both Canada and the United States to help ensure the proper use of these chemicals.

Many agricultural-chemical firms manufacture and distribute herbicides, and most have research programs aimed at increasing the usefulness of their products by developing herbicides with narrow ranges of target plants, and with little or no measurable effect on other plants and the environment. The ideal herbicide would provide complete and rapid control of the target weed, while not affecting the crop plants or the environment, and it should be available at reasonable cost. Such a paragon does not exist, of course. As research progresses, however, the herbicides entering the market are more restricted in their spectra of target plants and ever more demanding of sophisticated application procedures. These factors, plus the costs of new herbicides, make it essential to have accurate identification of weeds.

Plant identification services are provided free or at nominal cost by one or more units of the governmental agricultural agencies in both the United States and Canada, chiefly through the state/provincial agricultural extension services or their regional equivalents. In addition, many regions have "user-friendly" weed manuals that are tailored to the needs of agriculturalists (e.g., T. M. Barkley 1983; Illinois Agricultural Experiment Station 1981; C. Frankton and G. A. Mulligan 1970).

The Weed Science Society of America (WSSA) is the focal group for the study of weeds in North America. It publishes two scientific journals, *Weed Science* and *Weed Technology,* and also the "Composite list of weeds" (1984), which lists 1934 species of currently or potentially important weeds in Canada and the United States. A supplement (1988) adds 58 species. The list provides accepted Latin names for each entity, plus a recommended colloquial name. A five-letter code name (the "Bayer code") is given for each species and is used as a bibliographic shorthand in the society's publications, e.g., *Cardaria draba,* "hoary cress," is given the code CADDR.

Origin

A topic of endless fascination is the origin of weeds. Reason suggests that weeds may be chiefly by-products of the disturbances of the earth at the hands of humankind. Areas of continual natural disturbance, however, have occurred throughout geohistorical time. Common causes of both local and regional disturbances include fire, flood, volcanism, and wave and ice action, among others. Some of these events are seasonal and predictable, and some are not. The activities of animals, particularly the abundant large grazers, have maintained areas of continual disturbance that are also comparatively rich in available nitrogen. It is in such continually disturbed sites that colonizing plant species are thought to have evolved.

Weeds that thrive in areas of continual disturbance may be treated as two intergrading components, ruderals and agrestals. Ruderals are those plants that occur in areas of irregular or inadvertent disturbance and are not intentionally manipulated for agriculture. Ruderals such as pepper grass, *Lepidium virginianum,* are associated with the early stages in natural succession. Agrestals are plants adapted to agricultural practices and are often associated with a particular crop. For example, goatgrass, *Aegilops cylindrica,* often grows with wheat, *Triticum aestivum;* gold-of-pleasure or false-flax, *Camelina sativa,* grows with flax, *Linum usitatissimum.* Agrestals are the weeds of evident economic importance in that they lower agricultural productivity by competing with crop plants for available resources and for standing room, and they obstruct tillage and harvest. As a consequence, agrestals are the subjects of most of the efforts in scientific weed management.

Agriculture in North America functions on models that were imported from Europe, which, in turn, developed from ancient practices. Even crops of New World ancestry (among them, corn and potatoes) have been adapted to European farming techniques in temperate North America. This style of farming involves regular tilling of the soil with the crops grown in monoculture, and it provides the proverbial plenitude of pleasant places for a plethora of pliant plants.

For over a thousand years, beginning with travel between Scandinavia and Greenland, regular traffic has flowed between the Old World and the New World. Therefore, the plants that are adapted to the activities of humankind have had ample opportunity and time to be transported from place to place. North America now has about 2000 species of plants that are variously called "weeds." Most of them are ruderals, but some, perhaps 200, are agrestals associated with crops. The majority of these weeds are clearly the same species that are associated with human activities in the Old World.

Furthermore, the weeds of the various climatic regions of North America are associated with the weeds of similar climatic regions of the Old World. For example, most weeds in eastern North America are those of western and central Europe; the weeds of California are those of the Mediterranean region; the weeds of the Intermountain region are those of southeastern Europe and southern Asia. And, as one would expect, the agricultural crops and practices of these North American regions parallel those of the climatically similar Old World regions.

The implication seems clear and is held as conventional wisdom, namely that most of our North American weeds originated in close proximity to human activities, particularly to those of early agriculturalists. As a sedentary agricultural lifestyle developed, and plants and animals were domesticated, the continual disturbance created by humans provided a reliable niche in which colonizing species evolved and found a permanent home. It is difficult to prove this simple explanation of the origin of weeds, but it fits the information at hand.

Invasions

The most troublesome and aggressive weeds are those foreign or alien species that have invaded the North American continent from regions elsewhere in the world. By comparison, fewer and less aggressive weeds are native species. Analysis of the geographical components of a large number of weeds usually shows over 60% to be foreign species. The distinction between foreign species and native species is not always clear, and it is not easy to measure the impact of those foreign or alien plants on the native vegetation. Several factors contribute to this lack of precision.

Botanists assume that species have a "place of origin," where at some time the species are differentiated from the ancestral entities. As time passes, a newly formed species migrates into new areas and/or expands its range, through the routine mechanisms of seed dispersal, seedling establishment, and other factors. Undoubtedly, some botanical traffic has occurred between North America and other continents since antiquity, but clearly colonization following Columbus's voyages to America initiated a significant number of invasions. Some of the historical aspects of plant migration at the hands of humankind are reviewed by V. Muhlenbach (1979).

Foreign or alien species are usually regarded as those that have been brought to North America by human activities in post-Columbian times, while native species either originated in North America or had arrived by various means in pre-Columbian times. Although botanists frequently use the term "introduced" for these foreign or alien species, in this chapter the term has a more restricted meaning and refers to those species deliberately brought by people into a new region, where the plants grow without cultivation (M. L. Fernald 1950, p. viii; G. H. M. Lawrence 1951, p. 279). How many species have been transported from their putative places of nativity to North America in post-Columbian times is, of course, unknown. The historical documentation for these plant movements is often not well known or not yet researched, and many times what is known is based on circumstance and inference. A species is regarded as foreign or introduced if the following conditions apply:

1. The species in question has been known and specimens have been deposited in herbaria repeatedly from areas elsewhere in the world, where it is presumed to be native.

2. The species was not known in North America to the earliest naturalists and botanists who maintained

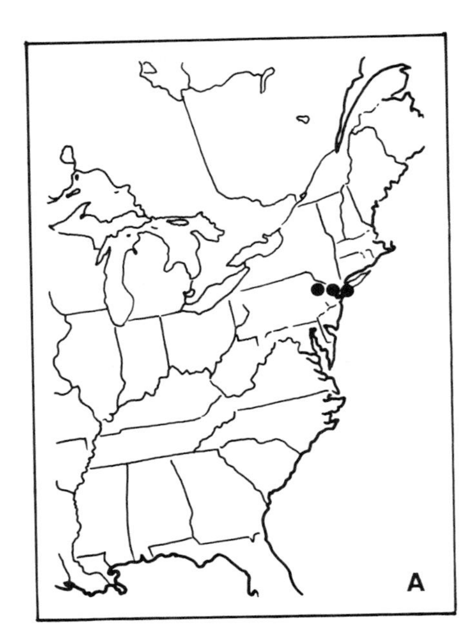

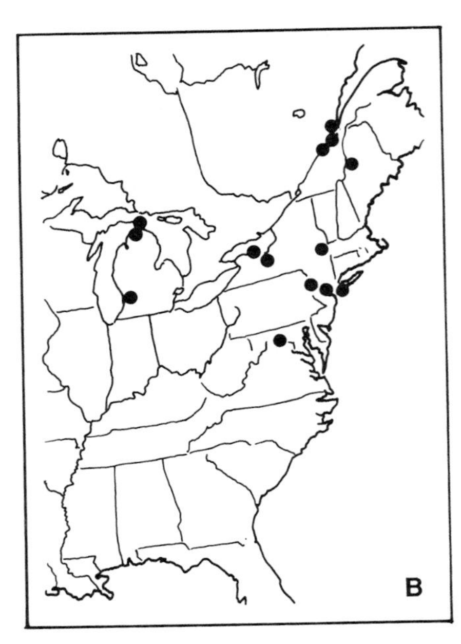

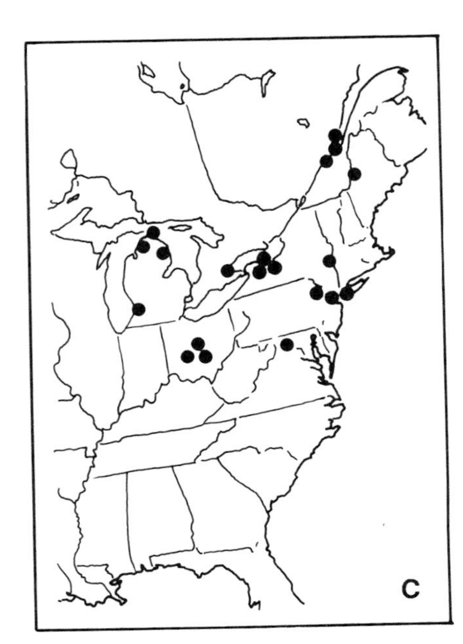

FIGURE 8.1. The distribution of *Veronica beccabunga* in eastern North America during three consecutive time intervals. A, distribution before 1900; B, distribution from 1900 to 1940; C, distribution from 1940 to 1985. Data derived from herbarium specimens. [Modified from D. H. Les and R. L. Stuckey 1985, with permission from *Rhodora.*]

herbaria, and/or it was known only as a recent immigrant.

3. The species has become known and was seen or obtained from its new location only after a given date (usually in an approximate decade), when it presumably arrived in the new area.

Herbarium records are the reliable sources of data on plant invasions and establishment of these foreign species or weeds. These records are by no means complete, for a species may be present in an area for a while before naturalists, botanists, or agriculturalists notice it and collect specimens. The time-lag, however, between the invasion of a species and its collection from the field is not likely to be great. Detective work in the herbarium can provide information as to when a plant had invaded or was introduced, how rapidly it spread into new regions, and in some cases, when it began either to migrate and expand or to decline in abundance or range, or both.

The invasion and spread of several foreign species, for example, have been documented by herbarium sources in papers by R. L. Stuckey and collaborators. Examples are the invasion and establishment of the alien species, *Epilobium hirsutum,* great-hairy-willow-herb; *Lycopus europaeus,* European water horehound; *Lythrum salicaria,* purple loosestrife; *Potamogeton crispus,* curly pondweed; *Rorippa sylvestris,* creeping yellow cress; and *Veronica beccabunga,* European brooklime. These European species all entered seaport cities along the east coast of North America and gradually moved from east to west across the continent. *Lythrum salicara* is the most prominent aggressive wetland weed known in the northern United States and sometimes Canada (R. L. Stuckey 1980). This pattern of invasion is also illustrated by *Veronica beccabunga* (fig. 8.1; D. H. Les and R. L. Stuckey 1985; see for references to other studies).

Various estimates and counts are available for the numbers of foreign species that have invaded North America and that now persist outside of intentional cultivation. Many, but not all, of these alien species persist as ruderal or agrestal weeds.

In Canada and the northeastern United States, the percentage of foreign species ranges from 20% to 30% (table 8.2). In the United States, those states in the northeast that have been occupied the longest by European colonists have percentages near or over 30%, the highest percentage known being 36% from the state of New York. A percentage range of between 20% and 30% occurs in those midwestern states most exten-

TABLE 8.2. Percentage Comparison of Numbers of Foreign Species to Total Numbers of Species since 1950 for Canada, Northeastern United States, and Various Other States of the United States

Geographic Unit	*Total Number Species*	*Number Foreign Species*	*Percentage*	*Reference*
Regions				
Canada	4153	884	27	H. J. Scoggan 1978–1979
Canada	3990	930	23	B. Boivin 1968
Northeastern United States and Southeastern Canada	5523	1098	20	M. L. Fernald 1950
New England	2882	877	30.4	F. C. Seymour 1969
States and Provinces				
New York	3022	1082	36	R. S. Mitchell 1986
Maine	2137	699	32.6	R. C. Bean et al. 1966
Illinois	3100	871	28.7	R. D. Henry, A. R. Scott 1980
Vermont	1927	533	27.6	F. C. Seymour 1969
Ohio	2518	599	23.8	C. Weishaupt 1971
Missouri	2438	557	23	J. Steyermark 1963
Iowa	1943	434	22.3	L. Eilers, pers. comm.
Minnesota	1618	392	19.5	G. B. Ownbey, T. Morley 1991
California	6400	1160	18	M. Rejmanek et al. 1991
Kansas	1872	326	17.4	F. C. Gates 1940
Colorado	3091	492	16	A. Weber, pers. comm.
Manitoba	1224	193	16	R. D. Henry, A. R. Scott 1980
North Dakota	1143	171	15	O. A. Stevens 1950
Washington, se, and adjacent Idaho	—	—	14	R. D. Henry, A. R. Scott 1980
Arizona	3438	231	7	R. D. Henry, A. R. Scott 1980

sively involved in agriculture, with outlying, more northern and western states having values below 20%. These noted variations reflect the history of migration and settlement of the European people as they moved westward across the North American continent, and of the agricultural, industrial, and recreational practices that have been developed since then.

A reference for foreign species of continental scope is the *National List of Scientific Plant Names* (NLSPN), which was prepared by the Soil Conservation Service of the U.S. Department of Agriculture (1982). This publication is a compilation from many sources of those taxa that occur in North America north of Mexico (including Greenland), with the nomenclature brought into accord with current usage. The NLSPN serves the useful purpose of offering Latin names and relevant synonymies for governmental publications. The compilers of the NLSPN have noted each taxon as native or foreign to the flora of North America, and the sources of their information are cited.

Summary

Foreign species are among the most conspicuous and abundant species in places that are heavily affected by the workings of humankind. In the farming regions of the American heartland, the crops are nearly all derived from elsewhere, an exception being the native sunflowers, which are grown for their seeds. The agrestal weeds often come from the places of development of the crops, and the roadside and waste-ground ruderal weeds are mostly of Old World origin. For example, in certain regions the various species of *Bromus* and other grasses may form the bulk of ruderal biomass. Moreover, most of the ornamental species in parks and gardens are alien, e.g., lawn grasses, rose bushes, lilacs. Therefore, with as many as one-fifth to one-third of the species in the flora foreign, they dominate the visual impact of the flora in much of middle North America. In other words, the showiest and most conspicuous plants in the flora are frequently those whose evolutionary development occurred elsewhere in the world, but whose arrival in North America was instigated by the activities of humankind and whose presence in the flora is maintained by continuing human activities.

Completion of the *Flora of North America North of Mexico* will provide a database that includes, among other things, information on each taxon's distribution, habitat preference, and provenance. For all taxa, and particularly for weeds, this database will change as more information is accumulated. The ranges for many weeds will expand (or contract). Through continuing analysis of herbarium materials and field studies, we will gain new insights on the provenance of weeds and their role in the natural system.

9. Ethnobotany and Economic Botany

Charles B. Heiser Jr.

When Europeans first arrived, in both eastern and southwestern North America north of Mexico, they found people who were practicing agriculture, much of it with crops from Mexico. Consequently, the use of native wild plants received scant attention. This changed, however, when the Europeans penetrated the areas inhabited by hunters and gatherers. According to R. I. Ford (1986), "the traditional use of plants and animals by American Indians is better documented than for the early peoples of any other continental area of the world." Ford has brought together a number of the significant papers dealing with the use of plants and animals by the native people. Furthermore, archaeological investigations, particularly in the last half century, have also contributed greatly to our understanding of the plants used by the native North Americans.

The immigrants to North America from Europe brought the Old World crops to North America, and those plants soon came to be the dominant cultivated crops in northern North America. Many weeds, a few of which were found to serve useful purposes, were also introduced unintentionally.

Wild Food Plants

The American Indians utilized a large number of plants for food or beverages. E. Yanovsky (1936) listed 1112 species as so used. Many of these plants were of minor use, and a few were not natives but introduced species. The vast majority of the plants in the list are angiosperms, of course, but algae, fungi, lichens, seedless vascular plants, and gymnosperms are also represented.

One of the most widely distributed sources of food was provided by oaks (*Quercus* spp.). Acorns of over 25 species were used, with those of various western species providing the basic food for some California Indians. Various ways were devised to remove the bitter principles to make the acorns palatable. Nuts of other genera, such as chestnut (*Castanea* spp.), hickory (*Carya* spp.), and walnuts (*Juglans* spp.), were also widely used.

Camas *(Camasia quamash)* was widely used in western North America. The bulbs were cooked in various ways and then eaten or ground and made into a flour that was used for making a "bread." In the central and eastern area, the tubers of the potato bean *(Apios americana)* were cooked for food.

Fruits or "berries" belonging to various families were gathered throughout most of North America. Rosaceae provided the greatest number of contributions. Grains of a number of grasses were collected for food. One of the grasses, wild rice *(Zizania palustris)*, was a staple in the Great Lakes region. The tapping of maple (*Acer saccharum* and other species) for sugar was practiced in the central and eastern areas as it still is today. Less widely known is that the pounded bark of these species was also used for food. Among the gymnosperms, the edible seeds of the pinyon *(Pinus edulis)* and of several other western pines stand out as important food sources.

Many of these wild plants were used by the Old World immigrants in the early historical period, and even now

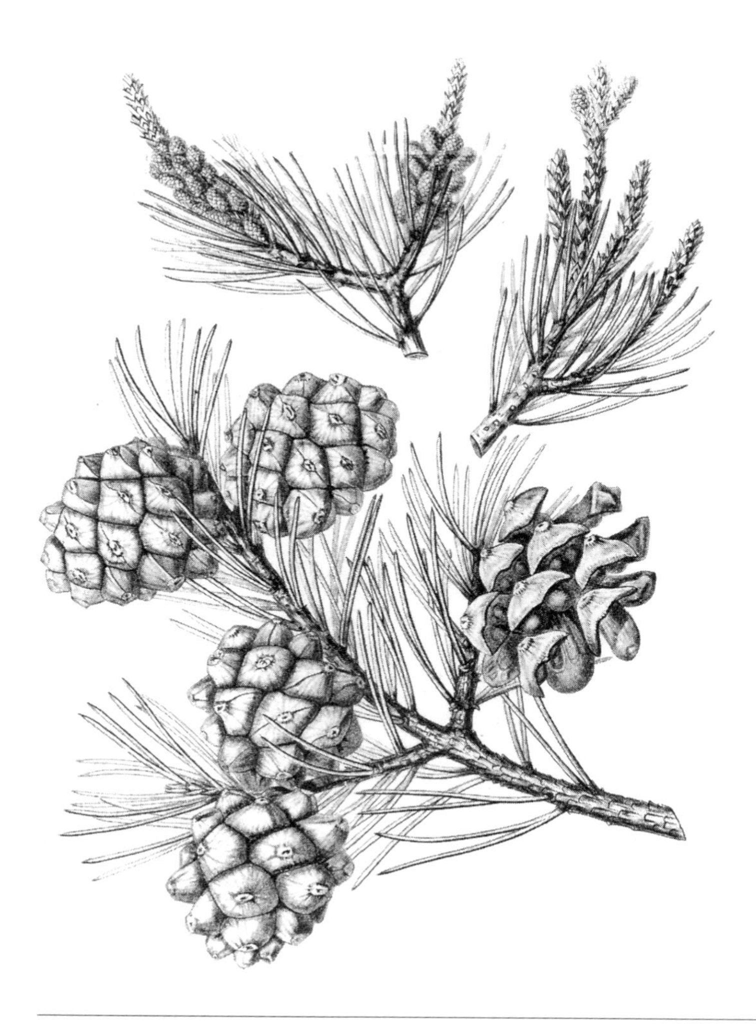

Pinus edulis. [From C. S. Sargent. 1897. The Silva of North America. Boston and New York. (Facsim. ed. 1947, Magnolia, Mass.) Vol. 11, plate 552, by C. E. Faxon.]

some people collect wild foods, but more as a hobby than as a necessity. Many books, often regional in their treatment, such as M. L. Fernald et al. (1958), list edible plants and give recipes for their use. A number of the plants gathered from the wild at present are introduced weeds or escapes from cultivation, including favorites such as asparagus *(Asparagus officinale),* burdock *(Arctium lappa),* chicory *(Cichorium intybus),* wintercress *(Barbarea vulgaris),* daylily *(Hemerocallis fulva),* and watercress *(Rorippa nasturtium-aquaticum).*

Domesticated Food Plants

The archaeological record indicates that food production was being practiced in northern North America over 3000 years ago. The principal food crops in eastern North America at the time of contact with Europeans were corn, also called maize *(Zea mays),* squash *(Cucurbita pepo),* and beans *(Phaseolus vulgaris).* All of these plants are known much earlier in archaeological records from Mexico, and for some time it was naturally assumed that food production came to northern North America, along with these plants, from Mexico. The possibility that there had been an independent domestication of plants in northern North America was mentioned as early as 1924 (B. D. Smith 1987), and for some time evidence has indicated that the sunflower *(Helianthus annuus)* and sumpweed *(Iva annua)* were domesticated in eastern North America before the arrival of corn and beans from Mexico.

More recently B. D. Smith (1985) has shown that a domesticated chenopod *(Chenopodium berlandieri* or *C. bushianum)* was also present in eastern North America. Other plants, such as May grass *(Phalaris caroliniana),* were being cultivated, but there is no evidence that they were domesticated (C. W. Cowan 1985). When the Europeans arrived, the sunflower was seen in cultivation, but both *Iva* and *Chenopodium* had apparently already disappeared as domesticated plants. Two rather cryptic references in the literature, however, might

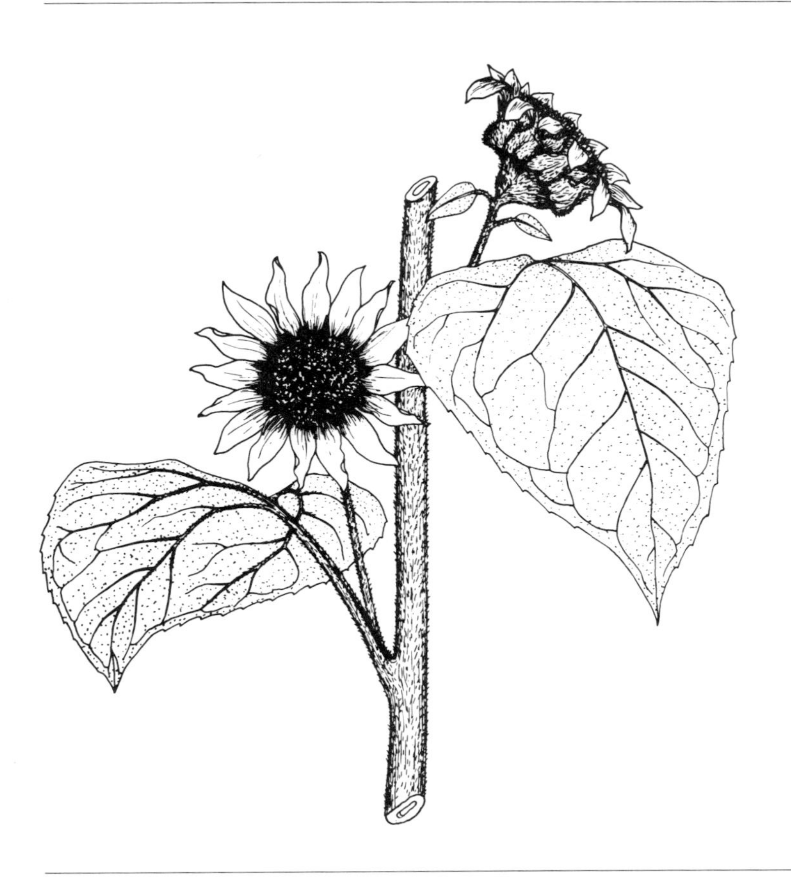

Helianthus annuus. [From J. A. Steyermark. 1963. Flora of Missouri. Ames. Plate 369, fig. 5. Reprinted with permission from Iowa State University Press.]

indicate that *Chenopodium* was still cultivated at the time Europeans settled the continent (D. L. Asch and N. E. Asch 1977).

The hypothesis that plants were domesticated in eastern North America prior to the introduction of Mexican crops seemingly became untenable when both squash *(Cucurbita pepo)* and bottle gourd *(Lagenaria siceraria)* were reported at a site in Missouri dated at over 4000 years old (M. Kay et al. 1980), because it was thought that both of these plants came from Mexico where they were cultivated at much earlier dates. It now seems likely, however, that *C. pepo* was a natural element in northern North America and that it is still represented there by the plant usually known as *C. texana*. Therefore it is possible that *C. pepo* was domesticated independently in Mexico and farther north in North America (D. S. Decker 1988; C. B. Heiser 1989). The bottle gourd cannot be considered an indigenous plant of North America; its entry could have been as a weed, with or without human aid. The possibility even exists that it came to Florida by ocean currents from South America (C. B. Heiser 1990).

In addition to the plants mentioned previously, the Jerusalem artichoke *(Helianthus tuberosus)* was also cultivated in eastern North America. Because no archaeological material is yet known, it is impossible to assign any dates to its domestication.

In the Southwest about 3000 years ago, agriculture began with what has been designated as the Upper Sonoran agricultural complex (R. I. Ford 1985b). Maize (corn), squash *(Cucurbita pepo)*, beans *(Phaseolus vulgaris)*, and the bottle gourd, all of which apparently came from Mexico, were represented. Some 1500 years later the Lower Sonoran agricultural complex developed. In addition to the plants from the earlier agricultural complex, it included other squashes *(Cucurbita mixta* and *C. moschata)*, tepary bean *(Phaseolus acutifolius* var. *latifolius)*, lima bean *(P. lunatus)*, and jack bean *(Canavalia ensiformis)*, and a panic grass *(Panicum sonorum)*. These plants all could have come from Mexico, but possibly some of them, such as the tepary bean and the panic grass, were also domesticated in an area that is now within the boundaries of the United States (R. I. Ford 1985b; G. P. Nabhan 1985).

Following the settlement by Europeans, the agriculture of North America soon underwent a drastic change as crops brought from the Old World spread. The present dependence of North American agriculture on crops of foreign origin is indicated in table 9.1. Of the food crops listed in table 9.1, only the sunflower is indige-

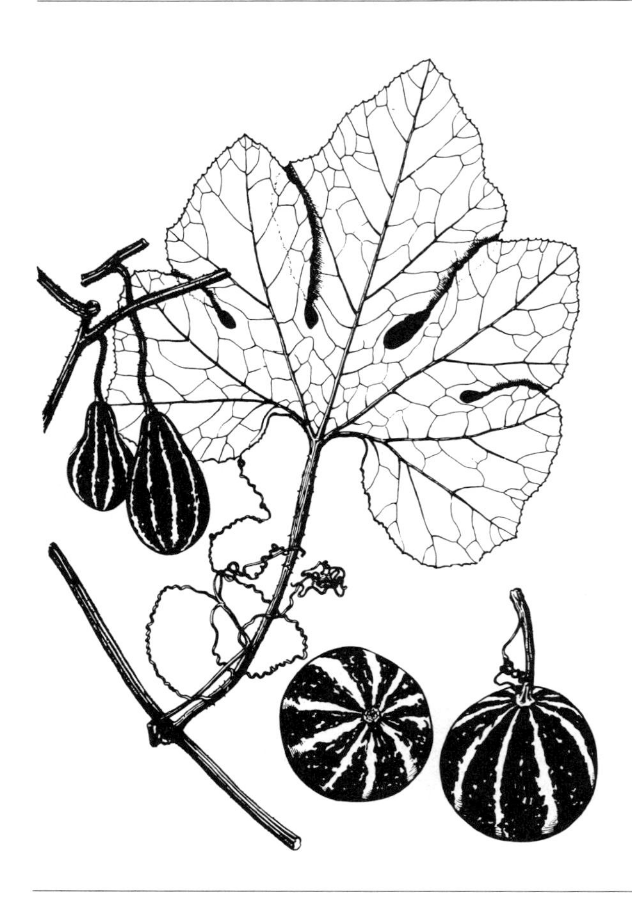

Cucurbita texana (C. pepo subsp. *ovifera* var. *texana),* the likely progenitor of the squashes and pumpkins of eastern North America. [From L. H. Bailey (1943). Reprinted with permission from the Bailey Hortorium, Cornell University.]

nous to North America north of Mexico, and only two of the others, maize and beans, were being grown in northern North America at the time of the discovery. Seven other crops of those listed in table 9.1—potato, sweet potato, tomato, peanut, avocado, tobacco, and cotton—are also native to Latin America; all of the others are Old World in origin.

Today the United States is the leading country of the world in the production of maize (corn), oats, sorghum, tomatoes, soybeans, and peaches, and second for six other crops. Canada is first in the production of linseed and second for rapeseed (table 9.1).

Narcotics, Hallucinogens, Stimulants, and Alcoholic Beverages

Tobacco, usually smoked, was widely used in northern North America. The domesticated *Nicotiana rustica* was cultivated in the East in prehistoric times and in the Southwest in the early historical period. In other regions tobacco from several native species was gathered from the wild, or occasionally was cultivated by the Native Americans. All of these tobaccos were replaced by *Nicotiana tabacum,* which was introduced from Latin America by John Rolfe in 1610 or 1611 and soon became the basis of the tobacco industry in Virginia and, later, in the other colonies (C. B. Heiser 1969).

Although many plants were used as hallucinogens in Mexico, only a very few were employed in northern North America. Species of *Datura* were employed prehistorically in ceremonies in the Southwest. *Datura inoxia,* used in Arizona, New Mexico, and California, was apparently the major species. Seeds of the mescal bean, *Sophora secundiflora,* a species found naturally in New Mexico and Texas, were also used prehistorically as a hallucinogen; a cult involving their use developed in the Plains area in modern times. Peyote *(Lophophora williamsii),* which has its northernmost limit in the lower Rio Grande valley of Texas, was used by Indians in Texas as a hallucinogen in 1760. Its use among Native Americans began to spread about 1880 and reached Canada in this century. Peyote has been legalized for use in the Native American Church. More details on these plants may be found in R. E. Schultes and A. Hofmann (1980). Today, the introduced marijuana *(Cannabis sativa)* is the most widely grown hal-

TABLE 9.1. Production of Important Crops Grown in Canada and the United States and Total Production for the World*

Crop	Region	Production
Grains		
Maize (corn)		
	Canada	6,400
	U.S.	191,197
	World	470,318
Wheat		
	Canada	24,383
	U.S.	55,407
	World	538,056
Sorghum		
	Canada	—
	U.S.	15,694
	World	57,976
Barley		
	Canada	11,672
	U.S.	8,784
	World	168,964
Rice		
	Canada	—
	U.S.	7,007
	World	506,291
Oats		
	Canada	3,549
	U.S.	5,425
	World	42,197
Rye		
	Canada	835
	U.S.	343
	World	34,893
Legumes		
Soybeans		
	Canada	1,219
	U.S.	52,440
	World	107,350
Peanuts		
	Canada	—
	U.S.	1,828
	World	22,594
Beans		
	Canada	77
	U.S.	1,104
	World	15,872
Peas		
	Canada	274
	U.S.	177
	World	16,447
Lentils		
	Canada	105
	U.S.	54
	World	2,242
Oil Crops		
Cottonseed		
	Canada	—
	U.S.	8,986
	World	49,085
Sunflowers		
	Canada	69
	U.S.	813
	World	21,867
Rapeseed		
	Canada	3,058
	U.S.	56
	World	22,302
Linseed		
	Canada	531
	U.S.	34
	World	2,121
Earth Vegetables		
Sugar beets		
	Canada	805
	U.S.	23,547
	World	305,882
Potatoes		
	Canada	2,754
	U.S.	16,659
	World	276,740
Onions		
	Canada	131
	U.S.	2,168
	World	26,319
Sweet potatoes		
	Canada	—
	U.S.	542
	World	133,234
Fruits		
Oranges		
	Canada	—
	U.S.	8,149
	World	50,630
Grapes		
	Canada	63
	U.S.	5,334
	World	59,158
Apples		
	Canada	495
	U.S.	4,367
	World	40,226
Plums		
	Canada	6
	U.S.	786
	World	6,518
Peaches		
	Canada	47
	U.S.	1,205
	World	8,586
Pears		
	Canada	25
	U.S.	764
	World	9,675
Avocados		
	Canada	—
	U.S.	175
	World	1,549
Miscellaneous		
Tomatoes		
	Canada	548
	U.S.	10,255
	World	69,328
Tobacco		
	Canada	74
	U.S.	670
	World	7,293
Hops		
	Canada	450
	U.S.	25,000
	World	112,038

Excerpted from Food and Agricultural Organization of the United Nations. 1990. FAO Yearbook [for 1989]. Rome.

* All figures are in 1000s of metric tons except those for hops, which are in metric tons.

Sophora secundiflora. [From C. S. Sargent. 1890. The Silva of North America. Boston and New York. (Facsim. ed. 1947, Magnolia, Mass.) Vol. 3, plate 121, by C. E. Faxon.]

Diospyros virginiana. [From C. S. Sargent. 1890. The Silva of North America. Boston and New York. (Facsim. ed. 1947, Magnolia, Mass.) Vol. 6, plate 253, by C. E. Faxon.]

***Rhamnus purshiana*. Pencil drawing by Frederick Walpole (1861–1904). [Hunt Institute, Carnegie Mellon University, Pittsburgh, indefinite loan from Smithsonian Institution.]**

lucinogen in North America. Because it is illegal, exact production figures are not available.

Leaves of yaupon, *Ilex vomitoria,* were the principal ingredient of a drink used by the Indians of the Southeast to induce vomiting as part of a purification ritual (H. E. Driver 1961). A tea was also made from this and other species of *Ilex* (E. Yanovsky 1936). *Ilex* is known to contain caffeine.

The Indians of the Americas never distilled alcohol from plants; however, naturally fermented beverages were utilized in some places. Fruits of various cacti were sometimes fermented by Indians in Arizona to make a wine (H. E. Driver 1961). Unlike the Indians in many other parts of the Americas, those of northern North America did not adopt maize (corn) as a source for an alcoholic beverage before the conquest. This was remedied sometime later by people of European descent who used maize to prepare a whiskey that became known as bourbon (after Bourbon County, Kentucky), and this is the only distinctive important contribution of northern North America to the world's alcoholic beverages. A beer from persimmon *(Diospyros virginiana)* was reportedly made by Indians in the East (U. P. Hedrick 1950). The early colonists tried to make wine from the native grapes in eastern North America without much success. The Old World grape *(Vitis vinifera)* was successfully introduced into California by the Spanish and became the basis for the development of the wine industry in the United States.

Medicines

The native North Americans used more plants for medicines than they did for food. D. E. Moerman (1982) listed 2147 species as employed medicinally. Some of the same plants were used, of course, for both food and medicine. Many medicinal plants of Native Americans were later adopted by the European colonists (V. J. Vogel 1970), who also made medicinal use of introduced species that they had brought with them. C. F. Millspaugh (1892) described 180 plants as so used. T. Arnason et al. (1981) have provided a recent treatment of the plants used for medicine in eastern Canada. C. Prescott-Allen and R. Prescott-Allen (1986) reported 67 active ingredients from wild plants as occurring in 10 or more drug products sold in Canada; about one-fourth of the plants are native only to North America.

The principal wild plant of northern North America in terms of number of drug products produced from it is cascara sagrada, *Rhamnus purshianus,* native to the northwestern United States and British Columbia. In 1977 the annual retail value of the bark of cascara sagrada in the United States was 75 million dollars; it is generally thought to be the world's most widely used cathartic.

A second medicinal plant that deserves particular attention is ginseng, *Panax quinquefolius,* of eastern North America. Although of limited importance to Native Americans, in 1980 some 575,000 pounds of roots with a value of 39 million dollars were exported. About 85% went to Hong Kong. Approximately 70% of the exported roots were derived from cultivated sources. Destruction of its habitats and overcollecting have contributed to a decline in the supply of the wild plants, whose roots are considered more desirable than those of cultivated plants.

Wood

Forests, much of them coniferous, once covered much of North America. Trees (as sources of timber and, especially, as sources of fuel) were, in fact, northern North America's most important raw material until the twentieth century. The United States is still the world's leader in timber production by volume, and Canada is sixth; by value Canada is the world's top timber producer, and the United States is second (C. Prescott-Allen and R. Prescott-Allen 1986).

Wood is consumed in four main ways: as fuel, as pulp for the manufacture of paper, for construction (including furniture), and for by-products, such as rosin, turpentine, and terpenes. The use of wood for fuel has declined greatly over the years and now only about 10% of it is burned, whereas its use for pulp has greatly increased, over 30% now being so used. Nearly 60% is used for construction.

Silviculture has slowly taken hold in North America. About 800,000 hectares have been planted to trees each year in the past three decades in the United States. Most of the planting is of soft woods, principally pines (*Pinus* spp.). The principal hardwoods planted on a commercial scale are cottonwood (*Populus* spp.) and black walnut *(Juglans nigra).* Much of the farmland in the southeastern United States is now devoted to production of timber and/or wood for pulp. Most of the information in this section is derived from C. Prescott-Allen and R. Prescott-Allen (1986).

Other Early Uses of Plants

Native Americans used many wild plants both for dyes and for fibers. Species of *Apocynum, Urtica,* and *Linum* were most widely used for their fibers. Cotton *(Gossypium hirsutum)* was the only domesticated fiber plant in prehistoric times. It came to northern North America from Mexico. In precontact times, its cultivation within the flora was limited to a small area in the Southwest. In historical times, of course, cotton became widely cultivated in the southern United States. The devil's-claw *(Proboscidea parviflora),* used for a fiber in basketry, was domesticated in the Southwest. Its domestication, however, is apparently rather recent (P. K. Bretting and G. P. Nabhan 1986).

Ornamentals

Plants from northern North America are now used for their ornamental value in landscape plantings or in gardens both in North America and other parts of the world. Many woody species, often little or not at all changed from the wild types, are planted, and some, such as *Magnolia grandiflora,* have particularly extensive distributions as cultivated plants. A number of native and horticulturally unchanged herbaceous angiosperms and ferns are used in wildflower gardens, and cultivars of many of the native species are offered in the trade for use in more formal gardens. Some species, such as those of goldenrod (*Solidago* spp.) and sunflower (*Helianthus* spp.), are more appreciated in Europe than in their homeland. Descriptions of many of the native North American plants grown as ornamentals may be found in *Hortus Third* (L. H. Bailey et al. 1976).

New and Incipient Domesticates

Certain plants consumed by Native Americans for food have become domesticated only in the past century or so; these include the pecan *(Carya illinoinensis),* the blueberry *(Vaccinium corymbosum),* the cranberry *(V. macrocarpon),* and most recently, wild rice *(Zizania palustris).* According to U. P. Hedrick (1950), 11 kinds of plums (*Prunus* spp.), 15 kinds of grapes (*Vitis* spp.), 6 species of blackberries (*Rubus* spp.), and 4 species of raspberries (*Rubus* spp.) have also become commercially valuable cultivars derived from native species. Research is proceeding with several other native plants, particularly those for cultivation in semiarid regions, and particularly for products that might have industrial uses (e.g., jojoba [*Simmondsia chinensis*] for oil, buffalo gourd [*Cucurbita foetidissima*] for starch and oil, guayule [*Parthenium argentatum*] for rubber, and gumweed [*Grindelia camporum*] for resin). Some of these have already proved to be of value.

10. Plant Conservation

George Yatskievych
Richard W. Spellenberg

Settlers from Europe who began the modern colonization of North America in the sixteenth century brought rapid changes to the landscape of the continent. The natural biological systems of the lands from which these colonists originated had already sustained massive alterations. By that time, many tracts of European forests had been lumbered for fuelwood or cleared for farming, a trend dating back to the Greek and Roman periods and earlier (J. Perlin 1989; A. S. Mather 1990). Food chains had been altered by overhunting of game species and extirpation of top predators. Large areas had suffered from overgrazing by domestic animals, and erosion was a widespread problem, especially in southern Europe. Various types of pollution were also affecting the environment around settled areas. As colonists from the Old World arrived in North America and moved westward, they naturally implemented the land ethics of their forefathers, resulting in the gradual degradation of many natural habitats and their constituent plants and animals.

It was not until the latter half of the nineteenth century that the American conservation movement gained momentum. The nature writings of Henry David Thoreau in the mid-1800s did much to sensitize readers to the wonders of the outdoors. His writings influenced the later works of John Muir and others, which had a more direct effect in stimulating conservation in the United States (S. Fox 1981). The first state plant protection law was passed in Connecticut, in 1869, to protect dwindling populations of *Lygodium palmatum* (Hartford fern, Schizaeaceae) from overcollection for horticultural purposes (W. D. Countryman 1977). Although no longer in effect, this law set a precedent for future rare plant legislation in North America. At about the same time, in 1872, the world's first national park was established at Yellowstone in the Rocky Mountains, now a part of an extensive system of parks throughout North America (fig. 10.1). Banff National Park, the first in the Canadian system, was established in 1885 (M. Finkelstein 1990).

Today, conservation is a familiar topic. North Americans are inundated with daily reminders to recycle waste materials and conserve resources. The news media are filled with reports on deforestation, pollution, global warming, endangered species, and other environmental issues. Review literature is beginning to summarize the issues (e.g., G. Lean et al. 1990). Conservation organizations are ever more active in soliciting membership and involvement from the public and in raising funds to address specific issues in nature. Much of the public's attention has been focused on critical concerns in the tropics, but the loss of biodiversity in the temperate zone is every bit as important and timely. In the continental United States alone, an estimated 90 species of vascular plants have become extinct during the past 200 years, and many additional species are in imminent danger of extinction (E. S. Ayensu and R. A. DeFilipps 1978).

Plant conservation in North America is best treated as a series of interrelated topics. These include a gen-

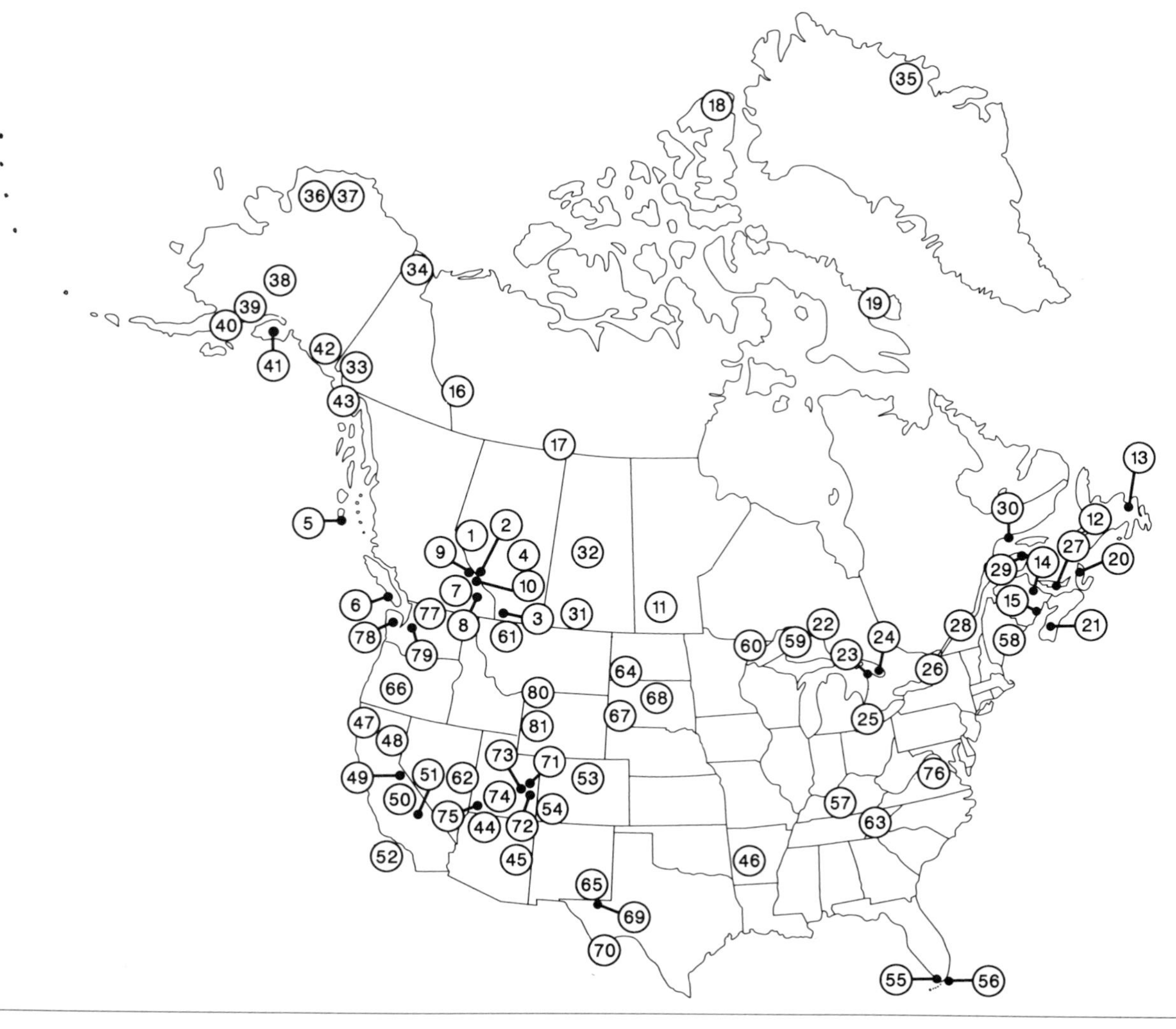

Canadian National Parks (NP) and National Park Reserves (NPR):

ALBERTA
1. Jasper NP
2. Banff NP
3. Waterton Lakes NP
4. Elk Island NP
17. Wood Buffalo NP

BRITISH COLUMBIA
5. South Moresby Island/ Gwaii Haanas NPR (pending)
6. Pacific Rim NPR
7. Mount Revelstoke NP
8. Glacier NP
9. Kootenay NP
10. Yoho NP

MANITOBA
11. Riding Mountain NP

NEWFOUNDLAND
12. Gros Morne NP (pending)
13. Terra Nova NP

NEW BRUNSWICK
14. Kouchibouguac NP
15. Fundy NP

NORTHWEST TERRITORIES
16. Nahanne NPR
17. Wood Buffalo NP
18. Ellesmere Island NPR (pending)
19. Auyuittuq NP

NOVA SCOTIA
20. Cape Breton Highlands NPR
21. Kejimkujik NP

ONTARIO
22. Pukaskwa NP (pending)
23. Bruce Peninsula NP (pending)
24. Georgian Bay Islands NP
25. Point Pelee NP
26. St. Lawrence Islands NP

PRINCE EDWARD ISLAND
27. Prince Edward Island NP

QUEBEC
28. La Mauricie NP
29. Forillon NP
30. Mingan Archipelago NPR (pending)

SASKATCHEWAN
31. Grasslands NP (pending)
32. Prince Albert NP

YUKON
33. Kluane NPR
34. Northern Yukon NP

GREENLAND NATIONAL PARKS
35. North East Greenland NP

UNITED STATES NATIONAL PARKS

ALASKA
36. Kobuk Valley NP
37. Gates of the Arctic NP
38. Denali NP
39. Lake Clark NP
40. Katmai NP 41. Kenai Fjords NP
42. Wrangell-St. Elias NP
43. Glacier Bay NP

ARIZONA
44. Grand Canyon NP
45. Petrified Forest NP

ARKANSAS
46. Hot Springs NP

CALIFORNIA
47. Redwoods NP
48. Lassen Volcanic NP
49. Yosemite NP
50. Kings Canyon NP
51. Sequoia NP
52. Channel Islands NP

COLORADO
53. Rocky Mountain NP
54. Mesa Verde NP

FLORIDA
55. Everglades NP
56. Biscayne NP

KENTUCKY
57. Mammoth Cave NP

MAINE
58. Arcadia NP

MICHIGAN
59. Isle Royale NP

MINNESOTA
60. Voyageurs NP

MONTANA
61. Glacier NP

NEVADA
62. Great Basin NP

NORTH CAROLINA
63. Great Smoky Mountains NP

NORTH DAKOTA
64. Theodore Roosevelt NP

NEW MEXICO
65. Carlsbad Caverns NP

OREGON
66. Crater Lake NP

SOUTH DAKOTA
67. Wind Cave NP
68. Badlands NP

TENNESSEE
63. Great Smoky Mountains NP

TEXAS
69. Guadalupe Mountains NP
70. Big Bend NP

UTAH
71. Arches NP
72. Canyonlands NP
73. Capitol Reef NP
74. Bryce Canyon NP
75. Zion NP

VIRGINIA
76. Shenandoah NP

WASHINGTON
77. North Cascades NP
78. Olympic NP
79. Mt. Rainier NP

WYOMING
80. Yellowstone NP
81. Grand Teton NP

FIGURE 10.1. Listing and approximate locations of national parks in North America.

eral rationale for preserving biodiversity, a discussion of problems in preserving individual species and their habitats, some definitions for terms describing species rarity, a summary of private and governmental contributions to plant conservation, and notes on present and future challenges. Some aspects of these topics are considered below.

The Importance of Conserving Biodiversity

Even with increasing attention given to environmental problems, the public in general is mostly unaware of the magnitude of the impact by humans on the environment, of the negative aspects of this often irreparable damage, and of its effects on the quality of life. A vivid general review of the multifaceted nature of the human assault on the natural world is presented in G. Lean et al. (1990). The value of biodiversity in economic terms remains questionable for most people when it is considered, for example, that the extinction of at least 90 plant species since 1800 in the United States seems to have had no negative effect (J. M. Moran et al. 1986).

A number of excellent reasons exist for preserving the organisms that share the planet with humankind. Most people actually involved with conservation of natural resources will cite moral and aesthetic criteria as principal reasons for preserving biodiversity (B. G. Norton 1986b). Other important arguments can be made to justify this need. The four main lines of reasoning in regard to preservation of biodiversity include: ethical and moral obligations, aesthetic and recreational potential, ecological impact, and economic impact (J. T. Miller 1985; Smithsonian Institution 1975). The last two topics, in particular, are inextricably related to one another.

Ethical and Moral Obligations

This argument is difficult to defend because of the widespread perception of a need to "develop" the natural world constantly for the immediate gain of humankind. It states that the extinction of any species diminishes the environment in both real and intangible ways and, from a moral standpoint, diminishes the human race, which is responsible for the welfare of every species. The concept that species have some sort of intrinsic value, and not only instrumental value, as discussed by J. B. Callicot (1986), has been rejected by many individuals, who retain a deep-seated belief that the natural world was created only to serve humankind. In contrast, others hold the view that humankind, as a species of thought and conscience, has a moral obligation for rational stewardship of nature. As Callicot discusses, doubts about the ethical-moral position very likely derive from the Judeo-Christian tradition, which gives humanity dominion over Creation and simultaneously believes that Adam was to "dress the garden and keep it."

Aesthetic and Recreational Potential

This line of reasoning is well understood by those who treasure time spent in the wild, but less so by individuals concerned more with economic interests. Diversity in nature and the presence of unspoiled places are crucial to humankind's general view of its place in the world and more specifically to the well-being of individuals. For many individuals, wild places provide a needed psychological refugium from the pressures of the civilized world.

Ecological Impact

This rationale is particularly associated with the economic argument and also not fully separable from the aesthetic-recreational rationale for the preservation of diversity. It is the least well understood by most nonbiologists. For this line of reasoning to be effective, the value of maintaining complexity in ecosystems must be defined as an economic and political issue. Local extinctions may be viewed as indicators of the declining health of natural communities and the reduced potential of these communities to provide free services to humankind.

Societies are generally unaware that most of the energy used by humans flows through giant, complex, little understood systems of natural communities, not through technology and agriculture (G. M. Woodwell 1977). Humans, as one species among millions globally, appropriate a disproportionately large amount (up to 40%) of the net primary productivity of terrestrial ecosystems (P. M. Vitousek et al. 1986). In these ecosystems, the loss of a single organism may affect the interactive controls on population levels of other organisms in the system, especially if the extirpated organism occupies a high place in the food chain or is a dominant member of the plant community (G. J. Vermeij 1986). At the local level, the simplification of a community may mean a reduction in productivity, a portion of which might be harvested directly by humankind, as in native grasslands or forests. Healthy, functioning ecosystems provide countless free services (P. R. Ehrlich 1980), which include climate regulation, waste disposal, and regulation of air and water quality.

Many scientists hold the view that extreme simplification of inherently stable, complex systems results in growing ecological instability and a reduction in the ability of these systems to withstand extreme environmental conditions, whether physical or biological in nature. This has been called the "diversity-stability hypothesis" (B. G. Norton 1987). Not all environments contain equal biotic complexity, and some evidence exists that this line of reasoning is not always applicable (B. G. Norton 1987). There are strong correlations, however, among complexity of an ecosystem, species diversity in that system, and the number of taxa in the system that are likely to become imperiled from any environmental perturbations. Moreover, evidence indicates that once extirpated from a particular site, species tend to be replaced by different taxa, and the ecosystem tends to become permanently altered (G. M. Woodwell 1977; B. G. Norton 1987). Thus, whether or not one accepts the diversity-stability hypothesis, the simplification of complex environments is clearly a detrimental phenomenon.

In a concept titled the "environmental trigger," G. Lean et al. (1990, p. 109) point out that environmental degradation, such as deforestation or desertification, makes some environmental disasters worse, and may trigger others. Unquestionably, in direct economic terms a biologically simpler world will become ever more expensive to maintain. Humanity must make the choice in the next decade either to continue as usual or to work effectively to maintain complex ecosystems and biological diversity; that is, to face squarely the environmental question. This choice will result in either a continuingly rich or an increasingly impoverished existence (T. E. Lovejoy 1980, 1986; G. Lean et al. 1990). The loss of a single species may seem insignificant, but when threatened species are taken as a whole, they indicate a severely stressed environment (T. E. Lovejoy 1979) and portend ecological problems of increasing severity.

Economic Impact

This rationale has been most usefully applied when there has been resistance to using public revenues for the protection of plants and animals that seemingly have no direct material value. Wealthy, highly developed countries can more easily afford protection of biodiversity than can developing countries, where much of the populace lives at subsistence level. These less developed countries must first provide opportunities for their citizens that do not require immediate exploitation of easily utilized resources.

Central to the economic argument, according to M. L. Oldfield (1984), is that "genetic resources from wild species and primitive forms of domesticated plants and animals provide the biotic raw materials that underpin every major type of economic endeavor at its most fundamental level." Once lost, a species can no longer benefit humans. In addition to presently utilized resources, extensive biodiversity provides humankind with the opportunity to discover new foods, fibers, drugs, plants for horticultural use and land reclamation, and so on.

Wild relatives of crop plants are important for the improvement of domesticates through research programs by plant breeders. The maintenance of diverse gene pools in other wild plants is taking on new importance in pest management and for the development of new agricultural crops, as the ability to move genes between species becomes feasible by modern methods of genetic engineering. The study of plant species provides information of direct economic value or of indirect use in the solution of numerous biological problems. The preservation of species as data sources for future research may be one of the most important contributions that can be made by biologists today (T. E. Lovejoy 1979). It now borders on cliché, but to lose species before they are studied for their biochemistry, physiology, evolutionary relationships, and ecology is tantamount to tearing pages from a book before it is read.

A variety of North American plants provide examples where native species individually are, or may be at a later time, commercially exploitable. Uses of these plants range from horticultural to medicinal and include such categories as plantings for erosion control (several families) and waste water filtration (several families), alternative crops for human foods (e.g., *Zizania* species, wild-rice, Poaceae) or animal fodder (e.g., *Cucurbita foetidissima,* buffalo gourd, Cucurbitaceae), and plants cultivated for unusual chemical constituents, such as the liquid wax found in seeds of jojoba (*Simmondsia chinensis,* Simmondsiaceae).

The number of North American species used in outdoor plantings has grown in recent years, as garden clubs, native plant societies, and nurseries have sensitized the public to the aesthetic and economic advantages of well-adapted, low maintenance alternatives to traditional garden plants. Successful use of native plants has increased public awareness, reducing the perception of the "weediness" of native species in the garden setting and reinforcing their desirability. Widespread acceptance of many species by the nursery industry as a whole was at first fairly slow but is now increasing. The establishment of native wildflower nurseries in many states (National Wildflower Research Center 1992) has helped to create a viable market, thereby making the widespread cultivation of several species by larger business concerns economically attractive.

Some plant types, such as drought-resistant species, plants native to prairies, and species that attract wildlife, have become particularly popular in various re-

Taxus brevifolia. Pen-and-ink drawing by Frederick Walpole (1861–1904). [Hunt Institute, Carnegie Mellon University, Pittsburgh, indefinite loan from Smithsonian Institution.]

gions. In the Southwest, for example, Xeriscaping™ (the use of drought-tolerant, native plants for landscaping) has been effective, for it requires little maintenance and depends on natural rainfall, thereby conserving the limited water supplies (C. Ellefson et al. 1992). This has not been without its problems, however, as some native plant populations, especially those of cacti and yuccas, have suffered severe declines following removal of plants by unscrupulous collectors for transplanting to yards. Presently, however, laws and conservation education are reducing horticultural predation on native plant populations, and an increasing number of commercial plant growers are investing the time and resources necessary to grow such plants from seeds or other propagules. In a few situations, regional endemics such as *Washingtonia filifera* (desert fan palm, Arecaceae) are actually more common in cultivation than in the wild.

A large number of plants also have potentially useful medicinal properties. Although plants have provided chemical compounds for use in numerous medicines, the majority of species have never been tested for potential medicinal, antifungal, or antibiotic properties. Even with several private and federal projects currently underway to test samples of large numbers of species, a long time will elapse before a significant percentage of the North American flora has been screened even cursorily by biochemists.

Success stories regarding the discovery of economically valuable, medicinal compounds among North American plants are not difficult to find. *Taxus brevifolia* (Pacific yew, Taxaceae), a small understory tree endemic to old-growth forests in the Pacific Northwest, provides an example of a species threatened by several factors, including its value for unusual chemicals. Although trees are too small for commercial lumber, they are damaged directly by logging operations that remove other commercially valuable conifers around them. The remaining yews become less vigorous after the shade-providing canopy of old growth is gone. In addition, yew wood is dense and red, making it desirable for fine wood crafts, and these slow-growing trees are often removed for this purpose. Recently, the National Cancer Institute's plant screening program discovered that an extract from the bark of this species yields the terpenoid taxol, which has been shown to be effective in the treatment of some kinds of cancer (D. Hinkley 1990).

The use of *Taxus brevifolia* in medicine, however, is not without problems. Biochemists have been unable to synthesize taxol successfully, so availability of the compound still depends on the extraction of plant materials from Pacific yew and some other *Taxus* species known to possess lower concentrations of taxol in their tissues. The harvest of bark from the plants, resulting in death for the plants and a relatively low yield of taxol, coupled with the need for large quantities of the compound for clinical trials, has alarmed some conservationists. During 1987–1990, as many as 62,000 mature specimens of Pacific yew were harvested for the National Cancer Institute (D. Hinkley 1990). This process yielded taxol in quantities sufficient only for short-term clinical tests. The remaining natural populations of this slow-growing species clearly could never sustain the level of collection necessary if taxol gains federal approval as a cancer treatment, and the future of this compound in medicine may ultimately depend on large-scale cultivation efforts.

Preserving Species and Habitats

Extinction is theoretically the ultimate fate for all species and, in that respect, it is a natural part of ecological and evolutionary processes. The world is now in peril, however, of entering into an extinction event of great magnitude (P. R. Ehrlich 1986; D. Simberloff 1986; N. Myers 1987; P. H. Raven 1987; E. O. Wilson 1988b). Extinction rates are estimated today to be 40 to 400 times greater than those that occurred in the past. By the year 2000, at least two-thirds of the tropical forests may be gone, and up to one million of the species that presently inhabit the earth may have perished (T. E. Lovejoy 1980, 1986), most of them totally unknown to science.

P. H. Raven (1987) has estimated that plant extinctions may be occurring presently at the rate of one or two species per day, and that during the next two to three decades global extinction rates for plants may average about 2000 species annually. Massive extinction events mark the ends of the Ordovician, Permian, Triassic, and Cretaceous periods, but at these geologically abrupt transitions, groups that underwent extinction were replaced by other newly evolving organisms. Now, with human-induced degradation of the environment, the same magnitude of change that accompanied these earlier extinction events is occurring again, but in a matter of a few centuries, rather than over many thousands of years. New groups of species are not able to evolve quickly enough to replace the old, and an unprecedented loss of biodiversity is permanently altering the course of evolution.

Although levels of plant extinction are of lesser magnitude in temperate regions than in tropical forests, which contain 50% (N. Myers 1980; E. O. Wilson 1988b) to 90% (T. L. Erwin 1983, 1988) of all species of organisms, the threats to species of North American flora are still a topic of great concern. As is illustrated by appendix 10.1, many different kinds of plants are affected by human impact on the environment. No particular class of life form seems to merit exemption. Understandably, regional endemics are most sensitive to environmental impacts, particularly when they occur in and near areas of urban development, as along the eastern seaboard and the West Coast.

Estimates by T. S. Elias (1977) and G. W. Argus (1977), among others, indicate that approximately 10% of species in the floras of the United States and Canada are endangered and rare. Elias further suggested that extinction rates in the United States increased during the period of 1800–1950. As noted earlier, about 90 plant species became extinct in North America during that time. A more recent survey suggests that as many as 475 additional continental United States taxa may become extinct by the year 1998 (L. Roberts 1988; Center for Plant Conservation, pers. comm.). As compared with the 1800–1950 period, that would be five times as many extinctions in one-third the length of time.

There are many reasons why species become imperiled in nature, aside from natural causes. All too often, complete and irreversible obliteration of habitat results from activities with direct economic purpose, such as inundation because of large dams, loss of habitat by urban expansion, and so forth. Reductions in diversity and in the health of habitats also commonly occur with more diffuse economic activities, such as ranching and logging, which often create a disturbance that allows other nonnative or "weedy" species to become established and spread. Highly focused (genus- or species-specific) activities imperil the existence of many species, as with excessive commercial and personal collecting for horticultural purposes, most notably among cacti and other succulents, orchids, and carnivorous plants (N. T. Marshall 1993). Some nursery catalogs continue to include offers of field-collected plants, although some of these species are protected by state law where they are collected. Ironically, many customers report high mortality rates among field-collected plants in their gardens and homes.

A related problem involves the overcollection of various native plants for the herbal trade. A number of unrelated species, such as *Hydrastis canadensis* (goldenseal, Ranunculaceae) and *Echinacea* species (coneflowers, Asteraceae), are collected in large quantities, usu-

Echinacea pallida. [From J. A. Steyermark. 1963. Flora of Missouri. Ames. Plate 369, fig. 1. Reprinted with permission from Iowa State University Press.]

ally by local "diggers," who sell their plants to regional dealers for processing and subsequent resale domestically or abroad. Although few of the most desirable taxa are uncommon enough to warrant federal listing (see below), the insatiable demand for herbal remedies, both in the Flora of North America region and around the world, has resulted in regional declines in population size and number for several species.

The best-studied example of this phenomenon is *Panax quinquefolius* (American ginseng, Araliaceae), which is presently a candidate for federal protection in the United States as an endangered or threatened species, and which has been treated as threatened in Canada in a Committee on the Status of Endangered Wildlife in Canada (COSEWIC) status report (D. J. White 1987). International trade in this species is regulated by virtue of its inclusion in appendix II of the Convention on International Trade in Endangered Species of Wild Fauna and Flora (CITES; see below). The genus *Panax* has a long history of use in herbal remedies, particularly in Asia. By the early eighteenth century, *P. quinquefolius* was already being harvested for export to China by colonists and American Indians in both southern Canada and the northeastern United States (A. W. Carlson 1986). The high demand for wild-collected plants has not diminished since that time.

Efforts at commercial cultivation of the plant date to the mid-nineteenth century, but so far they have not kept pace with demand. According to Carlson, over 340,900 kg (750,000 lb) of wild ginseng were exported from the United States in 1822, which amounts to an estimated 18–22 million plants harvested from natural populations during that year alone. The total number of ginseng specimens removed from natural populations during the past 250 years staggers the imagination, and regional decline or extirpation has been reported in many parts of the range of the species (W. K. Smith 1988). Current laws mandate monitoring of harvest and sales by the individual states, but no comprehensive and detailed studies have been undertaken to correlate relative degrees of impact on the species as a whole with local levels of exploitation. Available biological evidence suggests that further exploitation of wild ginseng should be halted in at least some portions of its range, but economic pressures have delayed actions to conserve this species properly.

Ultimately, however, the fate of most species in nature is tied to the health of the habitats in which they

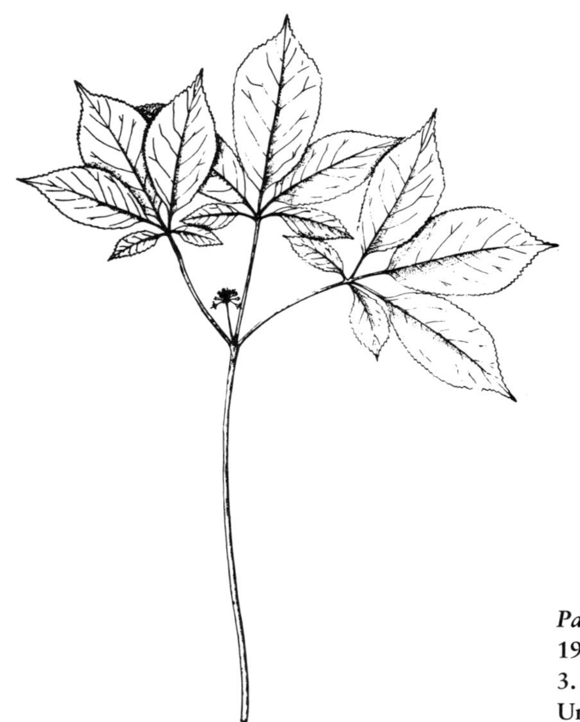

Panax quinquefolius. [From J. A. Steyermark. 1963. Flora of Missouri. Ames. Plate 264, fig. 3. Reprinted with permission from Iowa State University Press.]

live. Most conservationists feel strongly that preservation of endangered species can only be brought about through effective conservation of ecosystems (G. T. Prance 1977; B. G. Norton 1986b, 1987). G. M. Woodwell (1977) noted that many earlier studies indicated a predictable pattern of shifting species arrays following disturbance in ecosystems. Disturbance-sensitive species, usually habitat specialists with high fidelity to a particular undisturbed community type, are extirpated or reduced in abundance, followed by rapid invasion of "weedy" taxa. Once established, the newcomers may be difficult or impossible to displace, and the original habitat often is, at best, difficult to reconstruct. Thus, preservation of natural ecosystems is essential to the conservation of overall biodiversity, as well as to the long-term survival of many individual species. Preservation of entire sites can be more efficient and cost-effective than attempting to deal with individual species on a one-by-one basis.

Habitats that are not unique or especially scenic, however, may be very difficult to preserve. In this respect, protection of individual species of plants has been valuable, for to preserve a species requires habitat preservation. Often several other species that may be in decline, but that are not yet threatened with extinction, are also given protection in this way. Thus, many of the governmental and private organizations engaged in conservation issues have focused on this approach, preserving habitat for a suite of species, for both economic and biological reasons (see below).

In the long term, however, the maintenance of species diversity must be achieved by the preservation of sufficiently diverse habitats of adequate size to ensure the persistence of complex, interacting communities. Toward this end, botanical gardens have developed an increasingly important role in educating the public on the importance of habitat preservation, and in studying the causes that contribute to rareness, endemism, and population decline. Among these important contributions is the feasibility of preserving plant species in gardens and seed banks in the face of unavoidable habitat loss. Such programs of preservation, however, should be viewed as short-term measures. When species are brought into gardens or seed banks for preservation, attempts must be made to retain samples containing sufficient genetic diversity to permit the eventual reestablishment of vigorous populations with long-term viability, when suitable habitat is again available in the wild (D. A. Falk and K. E. Holsinger 1991). These programs form a crucially important bridge between habitat destruction, potential extinction, and eventual species recovery.

In most of North America, a large proportion of the environment has been compromised by urbanization, extraction of commercial timber species from forests, plowing of prairies, grazing of rangelands, draining of

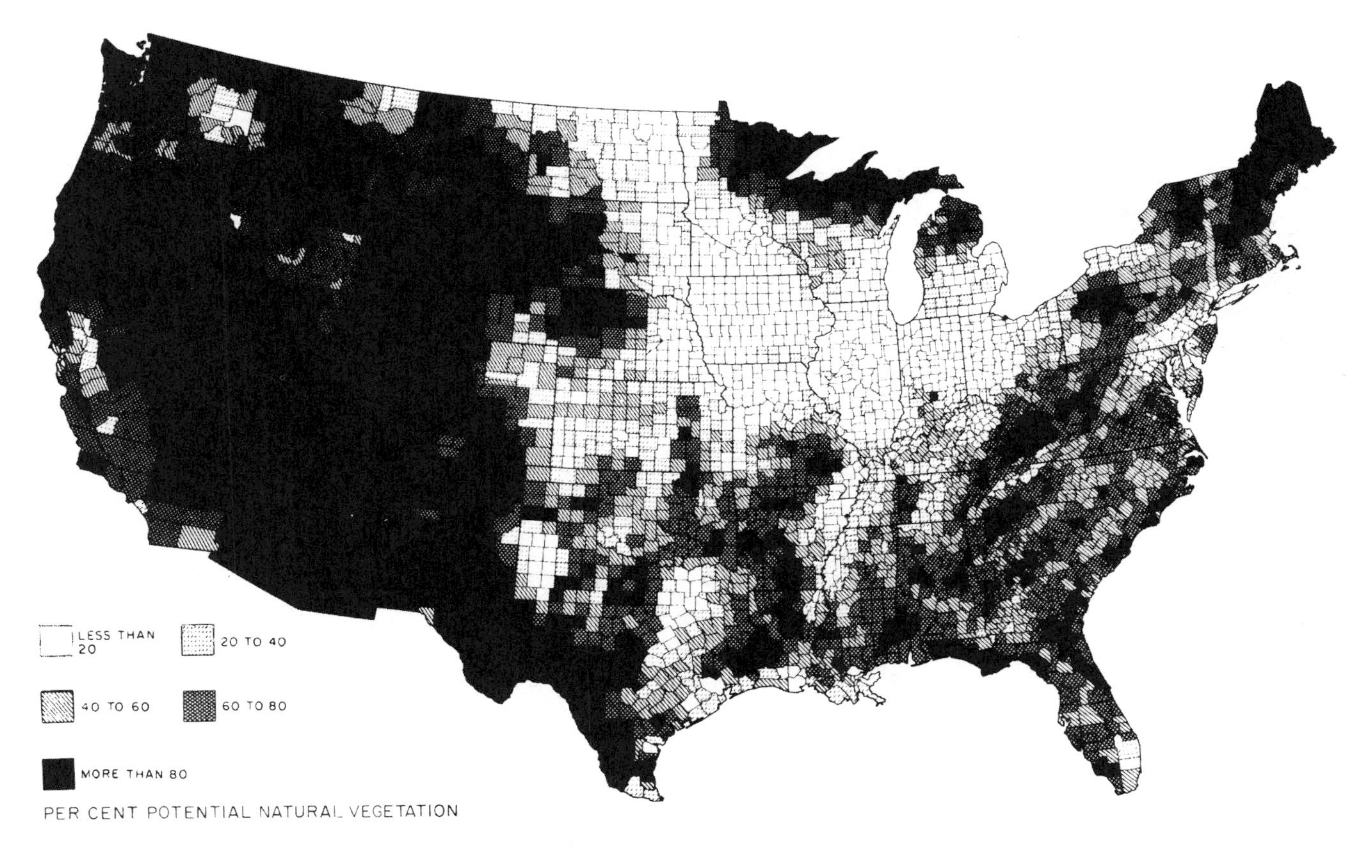

FIGURE 10.2. Percentage of area potentially covered by natural vegetation for counties in the conterminous United States. [Adapted from J. M. Klopatek et al. 1979, with permission from the Foundation for Environmental Conservation.]

wetlands, pollution, and related causes. J. M. Klopatek et al. (1979) examined the natural vegetation potentially remaining in the conterminous United States at the level of individual counties (fig. 10.2). Their analysis stressed how little natural vegetation remains in much of the eastern half of the country, particularly in the central states.

In states such as Iowa, Illinois, and Indiana, over half of the counties surveyed had less than 20% remaining potential natural vegetation. The situation is somewhat better in Canada (except in the prairie region), which has fewer people and a greater proportion of relatively intact vegetation, but areas of the country adjacent to its border with the United States have sustained large amounts of habitat destruction, particularly around the Great Lakes (G. W. Argus and K. M. Pryer 1990). Much of Greenland lacks vegetation because of permanent ice, but in those areas supporting plants, significant portions of the original vegetation remain intact, and human population density is generally low.

Very little is known about the life history, phenology, or autoecology of most plants, and our perceptions of species interrelationships in an ecosystem often result from hindsight after some taxa are diminished in abundance. The effects of various levels of disturbance of different kinds, including such seemingly innocuous aspects as the introduction of foreign pollinators, are usually not predictable based on currently available understanding. For example, the decline of the American bison in the nineteenth century and the general depletion of large game mammals in the eastern half of the United States by the early twentieth century are well-known facts, but the correlation between mammalian depletions and the concurrent decline of a native plant species, *Trifolium stoloniferum* (running buffalo clover, Fabaceae), has only recently been appreciated (R. E. Brooks 1983). This clover, which once grew from Kansas to West Virginia, apparently was dependent on large mammals both for seed dispersal and for the creation of lightly disturbed microhabitats necessary for its establishment and survival (J. J. N. Campbell et al. 1988). Although populations of such herbivores as deer have

Trifolium stoloniferum. **Pen-and-ink drawing by John Myers (1955–). [Missouri Botanical Garden.]**

recovered dramatically, *Trifolium stoloniferum* has only recently been rediscovered at a handful of widely scattered populations in its former range.

Invasive foreign taxa can also pose serious threats to the integrity of some habitats and their constituent species. Introduced species are a large component of the floristic diversity in most of North America. For instance, T. S. Elias (1977) estimated that 22% of the species found in the northeastern United States are exotics. Examples of the often rapid spread of "weedy" invaders, such as *Salsola* species (Russian-thistle, Chenopodiaceae), *Alliaria petiolata* (garlic mustard, Brassicaceae), and *Tamarix* (saltcedar, Tamaricaceae), are common. Once established in a community, exotic taxa can permanently alter it. They also may exert a selective influence on nearby communities as their seeds repeatedly and persistently compete for establishment there.

One of the best studied and most detrimental cases of habitat alteration by an exotic is that of *Lythrum salicaria* (purple loosestrife, Lythraceae). This Eurasian wetland species was introduced in the eastern United States in the early nineteenth century, probably as an ornamental and also by way of seeds from ship ballast (D. Q. Thompson et al. 1987). It rapidly spread westward in the United States and Canada, reaching the West Coast by the 1930s. Particularly since 1940, the species has become abundant in North American wetlands (D. Q. Thompson et al. 1987), forming dense stands to the exclusion of native species, including such widespread dominant and tenacious plants as *Typha* (cattail, Typhaceae).

This destructive species can invade relatively undisturbed sites, displacing all other plants but providing little in the way of resources for wildlife that live in, or migrate through, these wetlands. It undoubtedly also has contributed to the decline of many less common aquatic plant species, although no quantitative study has been published on this topic. In recent years, *Lythrum salicaria* has been declared a noxious weed in several states, with laws banning its importation and cultivation, and providing measures for its control.

Comments on Terminology

The vocabulary that relates to plant species that are in decline or faced with the prospect of extinction usually includes such terms as "threatened," "vulnerable," and

Lythrum salicaria. [From H. Baillon. 1877. Histoire des Plantes. Paris and London. Vol. 6, fig. 386.]

"endangered." These categories and others relating to imperiled species have been used in different contexts by various governmental agencies, conservation organizations, and nature writers. The official International Union for Conservation of Nature and Natural Resources (IUCN) Red Data Categories (G. Ll. Lucas and H. Synge 1978; S. D. Davis et al. 1986) are the most widely used terms worldwide. In this classification, "endangered" taxa are those in danger of extinction and whose survival is unlikely if the causal factors continue operating, and "vulnerable" taxa are those believed likely to move into the "endangered" category in the near future if the causal factors continue operating. Taxa at risk and that have small world populations, but which are not yet sufficiently imperiled to be classified as "endangered" or "vulnerable," are placed into the category of "rare." Other categories accommodate taxa that are extinct, that were formerly imperiled but are now out of danger, or that are insufficiently known but suspected of being at risk.

In the United States, a somewhat different system is used to provide legal designations for officially listed taxa under the Federal Endangered Species Act (J. A. Bartel 1987). In this classification, an "endangered" taxon is in danger of extinction throughout all or a significant portion of its distributional range. Taxa are considered "threatened" if they are considered likely to become "endangered" within the foreseeable future. According to Bartel, because these terms denote a legal status for plants that have completed the listing process, they cannot be applied in the same sense to candidates for listing, or to those candidates for which a proposed regulation has been published.

Apart from these two widespread and official systems of rare plant classification, descriptive terms such as "endangered" and "threatened" have been used qualitatively and relatively interchangeably in various contexts (J. L. Reveal 1981; J. A. Bartel 1987; H. Koopowitz and H. Kaye 1990), so that their meanings have converged on a common definition in the minds of many members of the public. In addition, definitions have occasionally been modified for specific reasons, such as by administrative agencies concerned with the protection of species that barely enter the geographic areas under their jurisdiction. Classifications of conservation status may or may not imply legal status, depending on the situation at hand. In addition, classification as "endangered" for any one species may change in a few decades if recovery efforts are effective.

Other terms, such as "critically imperiled," "imperiled because of rarity," "rare," and "uncommon," are also used. The Nature Conservancy has established a numerical system of global and state-level element rankings with category definitions that are based on taxon range, populational numbers and sizes, and vulnerability, and that include these terms (G. W. Argus and K. M. Pryer 1990). "Sensitive" is commonly used by land managers in the western United States for plants that are being monitored but that may or may not be protected by state or federal law. "Rare" and "uncommon," used in a number of situations, are also variably defined. All of these terms have many different meanings and are not consistently applied in North America with regard to legal status or conservation needs. Therefore, the authors of taxonomic treatments in the *Flora of North America North of Mexico* have described individual situations for taxa that are of concern to conservationists, rather than attempting to use any "standard" terms without explanation.

Contributions to Plant Conservation Efforts

An analysis of the full spectrum of efforts to preserve North American flora at local, state/provincial, national, and continental levels is beyond the scope of the present chapter. Some aspects of this broad and complex topic were reviewed by various contributors in H. Synge (1981) and were summarized briefly by S. D. Davis et al. (1986). As noted above, the efforts of some private and governmental groups are actually aimed at preservation of habitats, rather than individual species, but the boundary between the two approaches is often indistinct.

International Contributions

As developed nations with well-established, diverse conservation programs, the North American countries are in a better position to assist the activities of international organizations in other less-developed countries than to require such assistance themselves. Nevertheless, several organizations with an international focus are engaged in plant conservation projects dealing with North American plants.

The activities of the Nature Conservancy and of the United Nations Man and the Biosphere Program are described below. The International Union for the Conservation of Nature and Natural Resources (IUCN), a private international group, has published several summaries on various aspects of world conservation that have included data on North America, such as its "red data" book on selected plant species (G. Ll. Lucas and H. Synge 1978) and its summary of national parks and protected areas (I.U.C.N. 1985). This organization, together with the World Wildlife Fund and several other organizations, has also formalized a strategy for plant conservation at botanical gardens (V. H. Heywood 1989). The Species Survival Commission of IUCN maintains a North American Plant Specialist Group to address regional issues involving plant conservation. The World Wildlife Fund, which originated as an offshoot of the IUCN, has become increasingly active in plant-related conservation efforts, including the publication of a summary of available information on all federally listed taxa of the United States (D. W. Lowe et al. 1990).

The countries of the Flora of North America region are all party to CITES. This trade agreement, which controls the international trade of rare plants and animals, was originally enforced in 1975 by 10 countries and now has nearly 100 signatory nations. Three appendices to CITES separate taxa into groups (categories I, II, and III), based on degree of rarity and threat from international trade. Often whole genera or families are listed, rather than individual species. Plants native to North America that are listed by CITES include some or all members of the Agavaceae, Araliaceae, Cactaceae, Crassulaceae, Ericaceae, Orchidaceae, Portulaceae, Sarraceniaceae, and Zamiaceae, which are among those families in the region most prone to overcollection for horticultural or medicinal purposes and for international trade (D. S. Favre 1989).

Efforts in Greenland

Within the Flora of North America region, the protection provided to plants and habitats varies. Greenland, officially Kalâtdlit-Nunât, has an arctic flora of about 500 species, of which 15 are endemic. It has no national legislation specifically designed to protect plants (S. D. Davis et al. 1986). The country, however, contains the largest national park in the world (S. D. Davis et al. 1986; fig. 10.1), the North East Greenland National Park, which covers more than 700,000 km^2, and by its size alone protects significant portions of the botanical diversity of the country. The Melville Bay Nature Reserve (10,500 km^2) also preserves a large area of native landscape. The distribution and ecology of the relatively few endemic taxa have been well studied, but further research is necessary to assess the conservation status of most of these species.

FIGURE 10.3. Species of conservation concern by region in the Flora North America region. *Canada:* Number of rare endemic taxa in each of the provinces and territories; some species occur in more than one region (from G. W. Argus and K. M. Pryer 1990). *United States:* Numbers of federally listed endangered and threatened (plotted on map as E:T) plant species that occur in each Fish and Wildlife Service region as of the end of 1990. Some species occur in more than one region (Fish and Wildlife Service, pers. comm.)

Efforts in Canada

In Canada, with approximately 3500 native species of vascular plants, the situation is more complex. Lists of rare and endangered plants are available for all the territories and provinces through the Rare and Endangered Plants Project at the Canadian Museum of Nature, in Ottawa (G. W. Argus 1977). A recent, more comprehensive volume (G. W. Argus and K. M. Pryer 1990) summarizes the status of 1010 rare species on a countrywide basis. The assessment of the "endangered" or "threatened" status and the compilation of detailed status reports on approximately 440 high-priority taxa in the Canadian flora is also underway (for an example, see D. J. White 1987). These reports are being prepared under the auspices of the Subcommittee on Plants of COSEWIC, a consortium of provincial and territorial governmental agencies, the National Museum of Natural Sciences, the Fisheries and Oceans Department, the World Wildlife Fund Canada, the Canadian Nature Federation, and the Canadian Wildlife Federation that was formed in 1977 (F. R. Cook and D. Muir 1984).

No national legislation has been enacted to protect endangered plant species, for such protection is the jurisdiction of the provinces (A. L. Maurer 1985). All provinces have ecological reserve acts, and four have endangered species acts. Some sentiment exists, however, that the numerous regional laws and levels of administration are insufficient to preserve the nation's imperiled flora (G. W. Argus 1977). As explained by G. W. Argus and K. M. Pryer (1990), the greatest concentrations of rare plants in Canada occur in the topographically diverse mountain provinces of the west, and in the Great Lakes region of Ontario. Many of the plants of conservation concern to the country are confined to floristic provinces whose northern extremes barely enter Canada, and the main geographic distributions of such species are further south, in the United States. In Canada, the distribution of rare endemic plant species is shown on a province by province basis in figure 10.3, which implies the importance of the western provinces with regard to plant diversity.

In addition to a large number of local, provincial, and territorial reserves and parks, Canada contains large areas of federal reserves. Thirty-four national parks and national park reserves (fig. 10.1), some of them adjoining similar parks in the United States, protect nearly 2% of the 10 million km^2 total area of Canada (M. Finkelstein 1990). A comprehensive development plan for national parks in each of the 39 Canadian National Park Natural Regions has been completed (M. Finkelstein 1990) and projects an expansion of the national park system to protect the diversity of environments

Pedicularis furbishiae. [Reprinted with permission of Macmillan Publishing Company from *Where Have All the Wildflowers Gone?* by R. H. Mohlenbrock. Copyright 1983 by R. H. Mohlenbrock. New York and London.]

found in the country. Potentially, about 2.8% of Canadian lands will be protected in the national park system, if the plan is completed.

Efforts in the United States

Of the countries in the Flora of North America area, the United States has the best developed national legislation for plant protection. The federal Endangered Species Act was passed in 1973, and it has since been renewed and amended regularly. It is administered by the United States Fish and Wildlife Service. The Smithsonian Institution (1975) was charged with preparing an initial list of sensitive species, which was revised and made available to a broader audience (E. S. Ayensu and R. A. DeFilipps 1978). These initial lists contained 2099 and 2140 imperiled taxa, respectively, from the continental United States (3200 total taxa listed when Hawaii and Puerto Rico are included). These plants subsequently became candidates for potential listing as endangered or threatened under the Endangered Species Act (fig. 10.3). The list of federally protected and candidate species is periodically reviewed and published in the *Federal Register* (Fish and Wildlife Service 1990, 1992).

The federal Endangered Species Act is a powerful law in the sense that species in critical endangerment of extinction are evaluated for inclusion on a scientific basis, without regard for extrinsic factors. The act provides legal protection for listed taxa on federal lands, in cases of other land use involving federal funding, and particularly for horticulturally desirable taxa in interstate commerce. Other provisions of the act include the definition of "critical habitat" for taxa during the listing process, thereby guiding agencies in the selection of sites needed to protect each species, and also a mandate for formulation of a "recovery plan" for endangered and threatened taxa, aiming to restore eventually the numbers of individuals and populations of each listed plant so they are no longer in danger of extinction.

A well-known example of the power of the Endangered Species Act involves *Pedicularis furbishiae* (Furbish's lousewort, Scrophulariaceae), which was initially thought to be extinct but was rediscovered in 1976 growing near the type locality in Maine. (The plant is also known to grow in adjacent New Brunswick, Canada). It was listed as federally endangered in 1978. The discovery of several populations of this species along the St. John River in an area that was to be flooded as part of a federally funded dam project was instrumental in the denial of a federal permit for that project (D. W. Lowe et al. 1990).

The evaluation and listing process for individual species is lengthy and complicated, requiring extensive consideration and documentation before a species becomes officially protected. Where the need arises, a means exists to provide "emergency listing" for plants in danger of immediate extinction. Most often, however, the protection process moves slowly. Of the more than 2000 taxa initially accepted as candidates in 1975, only 58 had actually been listed as endangered or threatened by 1981 (J. F. Fay 1981); the most recent summary of officially listed taxa (Fish and Wildlife Service 1992) has only 169 taxa from the continental United States. Additional species have also been nominated and added to the original list of candidates (Fish and Wildlife Service 1990), and some of the original candidates have been deemed unworthy of listing after careful study.

Presently, after consideration at the regional level (fig. 10.2), a species may be entered onto the list of candidate plants for federal protection as "category 2." Here taxa may remain for variable periods of time as more information is gathered. If found not to be in need of listing because of persuasive evidence of extinction, taxonomic reevaluation, or reassessment of abundance, a taxon is moved to category 3 and is dropped from active consideration. If evidence accumulates for potential imminent extinction of the taxon, it is moved to category 1, preparatory to final listing as "threatened" or "endangered."

Prior to final listing, a proposal for each taxon must be published in the *Federal Register,* and governmental agencies, scientists, and the public are given a period of time for comments prior to a final ruling. Taxa for which a proposal package has been published receive a limited form of protection, which is focused on notification of potential threats on federal lands or federally funded projects and nonmandatory recommendations concerning protection (J. A. Bartel 1987). Proposed taxa and those in categories 1 and 2, however, often receive considerable protection, for state and federal land managers often incorporate consideration of these plants into local management programs.

Not all states in the United States have laws protecting sensitive plant species. Those that do, however, contribute to a bewildering variety of laws, ranging from legislation designed to protect particular species or families to more general laws against picking or injuring native plants on public property (P. Olwell et al. 1992). In some states, federally listed species are also given protection on state lands. Although written with good intention, many state plant protection laws in effect are generally considered inadequate or unenforceable (W. D. Countryman 1977; J. W. Hardin 1977) and are undergoing revision.

The United States has large areas of public lands, on which natural plant communities receive varying levels of protection. Each state has many state and local parks and preserves, the latter often coordinated through some form of natural areas system. About one-third of the country's 9,363,000 km^2 is owned by the federal government (S. D. Davis et al. 1986). R. E. Jenkins (1975) estimated that approximately two-thirds of the federally endangered plants occur on federal lands. Areas receiving the greatest degree of protection include the national parks (fig. 10.1), monuments, lakeshores, seashores, scenic rivers, and wildlife refuges, which total nearly 650,000 km^2 (I.U.C.N. 1985).

Most other federal lands, such as those in national forests or those controlled by the Bureau of Land Management, are managed under a policy of "multiple use management." Permission for grazing, lumbering, mining, and other practices that by their nature result in environmental modification are evaluated on a per parcel basis. It is on these lands that many plant species gain protection. An increasing number of "research natural areas" and "wilderness areas" are being designated on these and other federal lands, to set aside the least disturbed parcels for stricter preservation of their natural landscapes (E. S. Ayensu 1981). With regard to plants, most administrative units of land-managing federal agencies now also have staff botanists whose responsibility is monitoring sensitive species and preparing management plants for their conservation.

Natural Heritage Programs

In the United States and Canada, all states and provinces have agencies designed to deal with management of natural resources. Conservation programs for nongame species or species of little commercial importance often comprise minor elements in agencies whose principal missions concern management of timber resources or fish and game animals. An increasingly important contribution of these state agencies to plant conservation is the assessment and monitoring of regionally and globally imperiled taxa through the efforts of a network of Natural Heritage Programs (L. E. Morse 1981).

These programs were first developed in 1974 in the United States, in 23 states, by the Nature Conservancy (see below), and now exist in every state. They are also being implemented in Canada, where they are called Conservation Data Centres. The first such centre was established in Quebec, in 1988, and two others, in British Columbia and Ontario, followed in 1990 (P. Hoose and S. Crispin 1990). Saskatchewan added a centre in 1992. We hope similar programs will be implemented in the remaining provinces and territories in the near future.

Most of the Natural Heritage Programs and Conservation Data Centres were initiated in cooperation with the relevant state or provincial department in charge of fish and game, conservation, or natural resources, and then turned over to these governmental agencies for administration after about two years of operation. Their mission involves completion of relatively detailed inventories of various natural features, including rare plants, with specific site information on each element stored in a database and on topographic maps.

This information is then made available to land managers, scientists, and project planners, for use in environmental impact assessments and other conservation-related activities. The data thus generated have facilitated compilation of more accurate lists of regionally and globally endangered species occurring in each state. Many of these lists have been published (E. S. Ayensu 1981; P. Olwell et al. 1992). They have also helped to identify the regions most critically in need of further exploration. Even in well-developed countries such as Canada and the United States, knowledge of the biota is very uneven, for there are areas that are relatively inaccessible and biologically poorly known. Considerable botanical survey work remains to be completed in such areas as northern British Columbia, much of the eastern portion of the Northwest Territories, parts of Labrador and Newfoundland, and remote areas of the southwestern United States.

Contributions by Private Organizations

In addition to local, state, and national governmental efforts for plant conservation, numerous private organizations with similar goals are active in each country. The activities of such organizations range from stimulating public awareness and education to legislative lobbying and even the preparation and filing of lawsuits. Some also purchase properties for preservation or collect germplasm of imperiled species for storage or cultivation.

The largest and best known of such organizations is the Nature Conservancy, which was started in 1950 by a group of concerned scientists and environmentalists (L. E. Morse 1981). The group began in the United States, where there are now state chapters throughout the country. It is increasingly active internationally, particularly in Canada, Central America, and South America. The Nature Conservancy practices habitat and endangered species conservation primarily through the purchase of outstanding parcels of land, which are then either retained or sold or donated to various public agencies charged with environmental protection. Unfettered by extensive governmental procedures and conflicting public pressures, the Conservancy's ability to respond quickly to opportunities has resulted in the preservation of many parcels of land that might otherwise have been altered permanently by human activity.

The Nature Conservancy has also been active in several other ways. The organization has been a leading force in research on various types of active versus passive management for preservation of natural landscapes and endangered elements on its lands (L. E. Morse 1981). It has established land registry programs in most states, in which private landowners receive assistance and recognition for important habitats on lands that they own and agree to manage for preservation of natural features. The state Natural Heritage Programs initiated by the Conservancy have been instrumental in the accumulation of detailed site information on special elements occurring in each state. Because each state program uses consistent nomenclature, identical databases, and the same definitions for ranking global and regional endangerment (as reviewed in G. W. Argus and K. M. Pryer 1990), the Conservancy has been able to pool regional data into a national database, the most complete and informative of its kind. This national information system, known as the Heritage Data Center Network, has become an extremely important tool for summarizing needs and priorities in both species and habitat conservation.

Several groups have focused on the study of endangered species through the preservation of germplasm in storage and cultivation, including the Center for Plant Conservation, the Canadian Plant Conservation Programme, the National Wildflower Research Center, and various individual botanical gardens and arboreta. The best known of these is the Center for Plant Conservation (CPC), which was organized in 1985, and coordinates the efforts of more than 25 regional member institutions to maintain in cultivation critically imperiled taxa for conservation purposes (L. R. McMahan 1991).

Germplasm from more than 300 taxa is stored in a seed bank (in coordination with the United States Department of Agriculture's National Seed Storage Laboratory), and cultivated samples at each botanical garden and arboretum are grown for research and educational purposes. One objective of the CPC's program is the reintroduction of subject taxa into the wild, and several such projects are already underway (L. R. McMahan 1991). The group has promoted research on various sensitive taxa as well. The CPC also has developed a comprehensive database on rare plants to complement its other work.

Present and Future Challenges

Certainly North America is not unique in regard to the degradation of the natural environment and the loss of biological diversity. Human impact on the environment is relentless throughout the world, and loss of natural habitat in the tropics may ultimately have a greater effect on stability of global ecosystems than will modifications of the environment in the temperate regions. The developed nations of North America, however, bear a responsibility to take lead positions in responsible conservation within their countries and worldwide.

In spite of the programs outlined above, much remains to be done on the short term to preserve biotic diversity in North America's shrinking natural landscape. Many of the approaches that have been identified as priorities have already been given some attention, but they require further expansion and completion. One extremely important project that has long been cited as necessary to plant conservation efforts is the completion of a single floristic treatment that summarizes and correlates baseline data on taxonomy, biogeography, and ecology of plants for a major portion of the continent in a way that no regional account can. The completion of a *Flora of North America North of Mexico* will result in a comprehensive account of the plants of the region that will help public land managers, private citizens, biologists in academic and research institutions, and legislators to arrive at sound decisions regarding matters of important environmental concern (N. R. Morin et al. 1989).

In addition to this great synthesis, considerable primary research is still necessary on individual plant species of concern. For many endangered taxa, detailed studies of population genetics, taxonomy, ecology, biogeography, and life history factors are still necessary to supplement existing observations that these taxa are uncommon in nature (E. S. Ayensu 1981). Such information is necessary to assess why a species is restricted in range, what the causes of endangerment are, how much genetic variation is left in the remaining populations, and what the success of certain recovery efforts might be. Research programs at various universities and colleges in North America, as well as the efforts of botanical gardens and arboreta, already are making important contributions to our knowledge of some critical species, often in coordination with governmental agencies (L. R. McMahan 1991).

In Canada, the World Wildlife Fund Canada, and the Canadian Fish and Wildlife Service have supported the development of the COSEWIC status reports, which were initiated in 1977. Some of these funds have supported needed basic research. In the United States several federal agencies, among them the United States Forest Service, the Bureau of Land Management, and particularly the United States Fish and Wildlife Service, have allotted public funds for status reports and the development of information regarding plant species on the federal candidate list. The Endangered Species Act also mandates research on various means to effect recovery of federally endangered and threatened taxa to the point where they are no longer at risk. The Center for Plant Conservation has been instrumental in the promotion of research on various critically endangered species under cultivation at botanical gardens and arboreta.

The question of species recovery also requires much further research. The assessment and monitoring of endangered species in nature does not ensure the perpetuation of these taxa. In many cases, cessation of habitat degradation may be sufficient to ensure continued existence of a declining species. In other situations, active management of a habitat, such as controlled burning, may be necessary to effect increases in population size or to alter existing habitats to render them more favorable to colonization by a particular species. For some species, measures to reintroduce populations into nature from cultivated sources may be necessary to ensure that adequate numbers of individuals are present for successful reproduction or simply to prevent extinction in the wild.

The process of plant reintroduction first requires collection and maintenance of samples from existing populations, for which botanical gardens and arboreta are best suited. The concept of plant reintroductions is not without controversy. Questions arise about the effectiveness of introducing germplasm from one portion of the range of a species, where genetic differentiation in response to ecological factors may have occurred, into a geographically distant area, where selection factors may differ sufficiently to thwart long-term success. The potential effects of such foreign germplasm on the fitness of remaining, nearby, natural populations of the same taxon also remain controversial.

No matter how much information about endangered species is generated by scientists, the fate of preservation efforts ultimately depends on the ability of conservation organizations and agencies to convince private industry, governments, and the general public of the critical need to set aside sufficiently large and numerous areas of suitable habitat and to support philosophically and financially the work required to achieve conservation goals. Although initially costly, in the long run humanity will benefit. Education programs designed to inform the public and sensitize people to critical problems in the environment on which they depend require continual expansion and innovative revision to remain effective. Botanical gardens and arboreta have unique opportunities to focus public attention on plant conservation by bringing visitors and members into personal contact with the individual species, in helping them to understand the problems through the use of interpretive and educational materials, and even by getting the public involved in volunteer programs aimed at plant and habitat conservation.

An informed public has political power. Through the democratic process, voters are now beginning to insist that land owners and managers alter land-use practices toward more environmentally sustainable harvests of natural resources, toward minimization of impacts of erosion and nutrient loss in soils, and toward reductions in other destabilizing forces that accompany extensive development without ecological consideration.

The role of biosphere reserves is especially worthy of note in this regard. These reserves are part of the Man and the Biosphere Program, begun in 1971 by the United Nations Educational, Scientific, and Cultural Organization (UNESCO). Unlike other nature preserves, the biosphere reserves are generally under multiple ownership. They have core areas (usually portions of some existing national park or wilderness lands) managed strictly for preservation of natural ecosystems, a surrounding buffer zone whose management objectives may include research, recreation, tourism, etc., and a transition zone in which the land management is intended to facilitate harmonious integration of the reserve into the existing land-use practices of the surrounding region (J. R. Vernhes 1989).

The often complex tapestry of adjacent areas ascribed to different zones theoretically should make biosphere reserves useful for research on both pristine and disturbed ecosystems, as well as for public education on land management. Obviously, the selection of areas whose natural landscapes and land-use histories are adequate for this sort of program is a key factor. At present, in the Flora of North America region, four Canadian areas and about 40 areas in the continental United States, mostly in conjunction with existing national parks, have been designated as biosphere reserves. Several potential reserves are presently under evaluation.

Public perceptions also need to be balanced between short-term goals for preservation of presently endangered species and the need for a long-term perspective, to allow treatment of both the individual symptoms of each species and the more devastating general loss of natural habitats and biodiversity. Although the various programs outlined above present evidence of widespread interest and actions for plant conservation in North America, a feeling among many conservationists suggests that much more needs to be done, quickly, to address biodiversity issues in the region properly.

Often, the sense of impending doom expressed by some environmentalists stems from dissatisfaction with seeming priorities of society, particularly the continuous destruction of natural resources through exploitation for commercial purposes. For industrially developed countries, the demands on big business to remain competitive in national and international markets often result in the elevation of economic priorities in direct conflict with needs required to maintain environmental health. In underdeveloped countries the situation is often much worse, for the survival of a large percentage of the inhabitants often depends on the progressive destruction of pristine and complex ecosystems. Ultimately, North American countries and those elsewhere in the world are shackled by a series of constraints involving growing populations, fluctuating economies, and the complexities of politics. Sometimes, short-term goals for various kinds of environmental conservation are deemphasized in the fight for economic and personal survival. This short-sighted approach to problem solving cannot, however, continue indefinitely without virtually the complete loss of biological diversity.

APPENDIX 10.1. Example of the Diversity of Plants in Danger of Extinction

In this list plants are arranged by general life form and then by family. For this example, only those plants federally listed as "endangered" in the continental United States as of 1989 are included (extracted from and following the nomenclature of D. W. Lowe et al. 1990). Many additional taxa have been listed as endangered since 1989, or are listed as threatened, or are currently proposed for listing.

Nonseed Plants

Quillworts
- Isoëtaceae
  - *Isoëtes melanospora*
  - *Isoëtes tegetiformans*

Ferns
- Dryopteridaceae
  - *Polystichum aleuticum*

Seed Plants—Gymnosperms

Trees
- Cupressaceae
  - *Cupressus abramsiana*
- Taxaceae
  - *Torreya taxifolia*

Seed Plants—Angiosperms

Trees—all dicots
- Betulaceae
  - *Betula uber*

Shrubs—all dicots
- Annonaceae
  - *Asimina tetramera*
  - *Deeringothamnus pulchellus*
  - *Deeringothamnus rugelii*
- Anacardiaceae
  - *Rhus michauxii*
- Berberidaceae
  - *Mahonia sonnei*
- Ericaceae
  - *Arctostaphylos pungens* var. *ravenii*
  - *Rhododendron chapmanii*
- Fabaceae
  - *Amorpha crenulata*
  - *Lotus dendroides* var. *traskiae*
- Frankeniaceae
  - *Frankenia johnstonii*
- Lamiaceae
  - *Dicerandra christmanii*
  - *Dicerandra cornutissima*
  - *Dicerandra frutescens*
  - *Dicerandra immaculata*
- Lauraceae
  - *Lindera melissifolium*
- Malvaceae
  - *Malacothamnus clementinus*
- Oleaceae
  - *Chionanthus pygmaeus*
- Rhamnaceae
  - *Ziziphus celata*
- Rosaceae
  - *Cowania subintegra*
  - *Prunus geniculata*
- Styracaceae
  - *Styrax texana*

Succulents
- Monocots
  - Agavaceae
    - *Agave arizonica*
- Dicots
  - Cactaceae
    - *Ancistrocactus tobuschii*
    - *Cereus eriophorus* var. *fragrans*
    - *Cereus robinii*
    - *Coryphantha minima*
    - *Coryphantha sneedii* var. *sneedii*
    - *Echinocactus horizonthalonius* var. *nicholii*
    - *Echinocereus chisoensis* var. *chisoensis*
    - *Echinocereus engelmannii* var. *purpureus*
    - *Echinocereus fendleri* var. *kuenzleri*
    - *Echinocereus lloydii*
    - *Echinocereus reichenbachii* var. *albertii*
    - *Echinocereus triglochidiatus* var. *arizonicus*
    - *Echinocereus triglochidiatus* var. *inermis*
    - *Echinocereus viridiflorus* var. *davisii*
    - *Pediocactus bradyi*
    - *Pediocactus despainii*
    - *Pediocactus knowltonii*
    - *Pediocactus peeblesianus* var. *peeblesianus*
    - *Pediocactus sileri*
  - Crassulaceae
    - *Dudleya traskiae*

Leafy, herbaceous plants (excluding carnivorous plants and grasses, listed below)
- Monocots
  - Liliaceae
    - *Erythronium propullans*
    - *Harperocallis flava*
    - *Trillium persistens*
    - *Trillium reliquum*
  - Orchidaceae
    - *Isotria medeoloides*
    - *Spiranthes parksii*
- Dicots
  - Acanthaceae
    - *Justicia cooleyi*
  - Alismataceae
    - *Sagittaria fasciculata*
  - Apocynaceae
    - *Amsonia kearneyana*
  - Apiaceae
    - *Eryngium constancei*
    - *Eryngium cuneifolium*
    - *Lomatium bradshawii*
    - *Oxypolis canbyi*
    - *Ptilimnium nodosum*
  - Asteraceae
    - *Chrysopsis floridana*
    - *Echinacea tennesseensis*
    - *Erigeron maguirei* var. *maguirei*
    - *Hymenoxys texana*
    - *Liatris ohlingerae*
    - *Pityopsis ruthii*
    - *Solidago shortii*
    - *Stephanomeria malheurensis*
    - *Thymophylla tephroleuca*
  - Boraginaceae
    - *Amsinckia grandiflora*
  - Brassicaceae
    - *Arabis macdonaldiana*

APPENDIX 10.1 (Continued).

Arabis serotina
Cardamine micranthera
Erysimum capitatum var. *angustatum*
Glaucocarpum suffrutescens
Lesquerella filiformis
Lesquerella pallida
Thelypodium stenopetalum
Warea amplexifolia
Warea carteri

Campanulaceae
Campanula robinsiae

Caryophyllaceae
Arenaria cumberlandensis

Chenopodiaceae
Nitrophila mohavensis

Cucurbitaceae
Tumamoca macdougalii

Euphorbiaceae
Euphorbia deltoidea

Fabaceae
Astragalus humillimus
Astragalus osterhoutii
Astragalus robbinsii var. *jesupii*
Baptisia arachnifera
Galactia smallii
Hoffmannseggia tenella
Lupinus aridorum
Trifolium stoloniferum

Hydrophyllaceae
Phacelia argillacea
Phacelia formulosa

Hypericaceae
Hypericum cumulicola

Lamiaceae
Acanthomintha obovata subsp. *duttonii*
Hedeoma todsenii
Pogogyne abramsii

Malvaceae
Callirhoë scabriuscula
Iliamna corei
Sidalcea pedata

Nyctaginaceae
Abronia macrocarpa
Mirabilis macfarlanei

Onagraceae
Oenothera avita subsp. *eurekensis*
Oenothera deltoidea subsp. *howellii*

Papaveraceae
Arctomecon humilus
Argemone pleiacantha subsp. *pinnatisecta*

Polemoniaceae
Eriastrum densifolium subsp. *sanctorum*

Polygalaceae
Polygala smallii

Polygonaceae
Centrostegia leptoceras
Eriogonum ovalifolium var. *williamsiae*
Eriogonum pelinophilum
Polygonella basiramia

Primulaceae
Lysimachia asperulaefolia

Ranunculaceae
Clematis socialis
Delphinium kinkiense
Ranunculus acriformis var. *aestivalis*
Thalictrum cooleyi

Rosaceae
Potentilla robbinsiana

Scrophulariaceae
Agalinis acuta
Castilleja grisea
Cordylanthus maritimus subsp. *maritimus*
Cordylanthus palmatus
Pedicularis furbishiae
Penstemon haydenii
Penstemon penlandii
Scutellaria montana

Carnivorous plants—all dicots

Sarraceniaceae
Sarracenia oreophila
Sarracenia rubra subsp. *alabamensis*
Sarracenia rubra subsp. *jonesii*

Grasses—all monocots

Poaceae
Swallenia alexandrae
Tuctoria mucronata
Zizania texana

PART V CLASSIFICATION AND CLASSIFICATION SYSTEMS

11. Concepts of Species and Genera

G. Ledyard Stebbins

A Brief Recent History of Species Concepts and Methods of Analysis

Before 1838, the date of publication of the first widely used regional flora by a North American botanist, species and species concepts were shaped by contemporary European botanists. As specimens poured into eastern herbaria from explorers returning from the West, however, American botanists cataloged and described them. Engelmann, Gray, and Torrey followed standards set by Europeans, producing monographs and floras similar to previous publications. Others, like E. L. Greene, Aven Nelson, P. A. Rydberg, and M. L. Fernald, as well as M. O. Malte and A. E. Porsild in Canada, adopted narrower concepts of species, describing as a new species any specimen that appeared different from others with respect to one or more characteristics of external morphology.

Consequently, regional floras appeared that differed strikingly in the number of species recognized and the morphological differences used to define them. Floras or partial floras that followed the more classical concepts included B. L. Robinson and M. L. Fernald (1908), N. L. Britton (1901), W. L. Jepson (1925), and L. Abrams and R. S. Ferris (1923–1960). Those that followed the narrower concept of species included P. A. Rydberg (1917) and various monographs, written mostly by Rydberg, that formed the basis of the series North American Flora.

In 1930 the Plenary Symposium on the Nature of Species held at the 5th International Congress of Botany, Cambridge, England, addressed the problem of species. Despite a lack of consensus, the dominant belief emerged that a species is any group of individuals (or of dried specimens) that an experienced taxonomist decides to call a species. Among some taxonomists, this opinion still persists (A. Cronquist 1978).

To resolve the dilemma and establish better connections between taxonomy and biology, two new disciplines came into being: experimental taxonomy and biosystematics. Both were favored by geneticists and cytologists who were increasingly aware of the potential impact of cytogenetic research on taxonomic classification.

Experimental taxonomists sought to determine to what extent visible morphological differences are based on environmental modification, and to what extent these differences reflect genetic differences. In Sweden, G. Turesson (1922) conducted several experiments in which adult plants derived from a single seed (a genet) were divided into several separate divisions (ramets) and raised side by side under the same controlled conditions. He concluded that every widespread species consists of a number of genetically different races, or ecotypes, each of which is adapted to a particular environment. In this ecotype concept, Turesson emphasized the discontinuities between these ecotypes, but other experimental botanists recognized that in most natural species, genetically different individuals can be arranged into a continuous series, within which discontinuities in the environment are reflected to a greater extent than those based on genetic differences between adjacent individuals.

J. Clausen, D. D. Keck, and W. M. Hiesey (1940), following a design set up by H. M. Hall, considered the method valid only if the experiments were frequently repeated in different environments. Consequently, they conducted tests at three field stations (Stanford, 50 m above sea level; Mather, 1200 m; and Timberline, 3000 m). Their results have been widely quoted in the evolutionary literature. For taxonomic classification, the following conclusions are significant: (1) The most modifying effect of the environment is on general vigor and survival, followed by time of flowering and length of winter dormancy. (2) The environment also greatly affects characteristics of general form, such as stem length, leaf size, overall leaf shape, and overall architecture of the inflorescence, including number of flowers. (3) Least affected are more detailed characters such as dentition of leaves, nature of pubescence (glandular versus nonglandular), shape of sepals, size, shape, and color of petals, and size and shape of fruits.

The second series of experiments concerned morphological discontinuities between populations that form the primary basis for recognizing different species. Planned programs of artificial hybridization were pioneered by J. Clausen (1926) in *Viola,* E. Baur (1932) in *Antirrhinum,* R. E. Clausen and T. H. Goodspeed (1925; T. H. Goodspeed and R. E. Clausen 1928) in *Nicotiana,* as well as J. Clausen et al. (J. Clausen 1951) in the Compositae subtribe Madiinae. The possible factors considered were ease versus difficulty of producing hybrids, pollen and seed fertility of F_1 individuals, chromosomal behavior, and in some instances, vigor and fertility of subsequent generations. Their interpretations were aided by cytogenetic analyses of hybrids between chromosomally different individuals belonging to the same species (performed by B. McClintock [1932], as well as others [H. B. Creighton and B. McClintock 1931]). As a result, the principal chromosomal differences that cause hybrid sterility were identified as differences in chromosome number and in major patterns of structure, e.g., translocations and inversions.

Biosystematics (W. H. Camp and C. L. Gilly 1943, although the term was often used informally during the 1930s) encompasses all uses of biological information, including observations of gross morphology made on dried specimens, that aid the understanding of species and their relationships. It therefore includes experimental taxonomy and interdisciplinary observations, and it requires a careful synthesis of knowledge.

Important biosystematic methods include observations designed to aid in understanding species as systems of populations rather than as groups of individuals. Representative samples ("mass collection") of individuals from a single population are collected and used to estimate the amount and limits of variability with respect to various morphological characters that exist in a single population. For complex characters such as the color patterns of petals, carefully designed diagrams that lead to easily measured parameters are often helpful. An outstanding example of this method is the investigation of the blue flag *Iris* species of eastern North America (E. Anderson 1936).

Also important are observations and experiments on pollination dynamics, particularly determining the frequency of self- versus cross-pollination and the presence versus absence of asexual or apomictic seed production. An early effort in this direction was the investigation by J. Crosby (1949) on the distribution of heterostyly versus homostyly in English populations of *Primula.* More extensive observations by H. G. Baker (1948, 1953, 1955) on these characteristics in Plumbaginaceae and other families led to the generalization that obligate outcrossing is most common in populations of a species that occur near its center of origin, while self-fertilization and apomixis are more common in peripheral populations, particularly those that have migrated over long distances.

Perhaps the most commonly gathered biosystematic data are chromosome numbers and, in species having relatively large chromosomes, gross structural features such as size and length of arms. During the 1920s and 1930s, G. A. Levitzky (1931) and others investigated these characteristics in many taxa: *Crepis* (E. B. Babcock 1947), *Nicotiana* (R. E. Clausen 1932; T. H. Goodspeed 1934), Poaceae (N. P. Avdulov 1931), and Ranunculaceae (W. C. Gregory 1941).

The importance of chromosome numbers and morphology goes beyond that of individual characters of gross morphology. For the purpose of placing names on dried specimens, chromosomes can be regarded as "just another character." For biosystematics and evolutionary botany, however, this point of view is too superficial. Although by themselves they rarely provide definitive answers, chromosome numbers can be combined with other characteristics to reach syntheses that are far more significant than those based solely on gross morphology.

Other biosystematic characteristics of particular importance in certain families are relationships to various pollinators, as recorded for the Polemoniaceae (V. Grant and K. A. Grant 1965) and for the Orchidaceae (L. van der Pijl and C. Dodson 1966). In the Poaceae, both N. P. Avdulov (1931) and J. R. Reeder (1957) showed that the biology of seed germination and seedling structure are of primary importance. The biosystematics of

J. C. Clausen, W. M. Hiesey, and D. D. Keck. [Courtesy of Carnegie Institution of Washington.]

Asclepiadaceae are intimately bound up with their relationships to the insects and their larvae that feed on these plants (P. R. Ehrlich and P. H. Raven 1964). The research of D. B. O. Savile (1979) has shown that the nature of fungal parasites can provide valuable biosystematic information about affinities between various groups of angiosperms.

Biochemical kinds of biosystematic information have risen most dramatically in importance during recent years. Much of this information, particularly that concerned with immunological relationships and with gene sequences in chloroplast DNA, primarily concerns relationships above the species level. Significant information with respect to variation within and relationships among species is provided by secondary organic compounds, particularly phenolics and alkaloids, by electrophoretic properties of proteins, particularly enzymes, and to a lesser extent, by some kinds of nuclear DNA.

The importance of phenolics for population studies first became evident from the research on species and hybrids of *Baptisia* (R. E. Alston and B. L. Turner 1963, 1963b; for a review see J. B. Harborne and B. L. Turner 1984). Variation detected by means of extracting proteins and comparing their mobility on an electrophoretic grid was first recognized as important in humans and *Drosophila* during the 1960s, and shortly thereafter the method was applied to populations of an introduced grass species, *Avena barbata* (A. Kahler and R. W. Allard 1970). Since then, the technique has been used with outstanding success (L. D. Gottlieb 1986; D. E. Soltis and P. S. Soltis 1989; D. J. Crawford 1990).

After a few simple hybridizations, progeny analyses of a strictly Mendelian nature were made. Electrophoresis is an accurate method of determining genetic diversity in populations with respect to those proteins that can be extracted and subjected to the technique. These proteins contribute little or nothing to observable morphological differences, but in one way this helps rather than hinders an understanding of the evolution of phenotypic diversity. Although many electrophoretic differences do have some adaptive significance, it is generally lower than that of many other differences, so that correlation of genetic distance between populations with the time elapsed since their descent from a common ancestor can be expected to be relatively high. At the intra- and interspecific levels, protein electrophoresis is currently the most accurate method for estimating rates of evolution.

The use of enzymes that cut the DNA molecule at

particular positions in the nucleotide sequence has rendered possible comparisons at the level of the gene structure itself. These methods are relatively laborious, however, and most of the results obtained to date on plant materials are more relevant to differences at the level of genera and families than of populations and species. At this lower level, mitochrondrial DNA has been very useful in animals, but for various reasons, less so in plants. Nevertheless, investigations of the repeated sequences that code for ribosomes have been useful for distinguishing between cultivated varieties of barley (*Hordeum vulgare,* R. W. Allard et al. 1990) and are likely to become increasingly important in the future.

Chloroplast DNA has, to the present, yielded more interesting results about plant species relationships than has mitochondrial DNA, ribosomal DNA, or the coding DNA of various genes. At the level of higher categories, chloroplast DNA has been a valuable aid to studies of phylogeny in both the Fabales and Asterales due to the presence of inversions, while differences in nucleotide sequence have yielded quite unexpected affinities between the genus *Clarkia* and the morphologically very different monotypic "genus" *Heterogaura,* which, if DNA similarity is regarded as the ultimate test, must now be subsumed within the genus *Clarkia* (K. Sytsma and L. D. Gottlieb 1986).

Important results emerging from the application of protein electrophoresis have emphasized the widespread occurrence of mosaic evolution: radically different rates of evolutionary change of different characters within the same evolutionary line during the same time period (G. L. Stebbins 1983, 1983b). G. De Beer (1954) showed that the most famous intermediate between two classes of animals, *Archeopteryx,* is not intermediate between reptiles and birds with respect to all of its individual characteristics but is rather a mosaic of different characteristics, some of which are typically reptilian and others typically avian. The evolution of birds, as well as of other vertebrate classes, has not been a steady, even progression with respect to all characteristics, but rather a mixture of fast and slow rates, depending on the trait involved. Although the fossil record of intermediate forms between classes of vascular plants is too deficient in diagnostic characteristics to provide a firm basis for determining whether or not mosaic evolution has dominated macroevolutionary changes in plants, the record can be interpreted in this way.

Evolutionists have naturally asked if the evidence from genetic diversity within and differences among populations, as determined by their electrophoretic properties, coincides with or differs from that derived from gross morphology with respect to the rates of evolution that are suggested. Four different comparisons provide an equivocal answer. The first two are in the genus *Clarkia* (Onagraceae), which consists of many annual species, the majority of which are endemic to California. Upon completion of the first cytotaxonomic investigation of this genus, H. Lewis (1962) proposed that it includes several examples of the sudden origin of new species, which he termed catastrophic speciation, without intervention of polyploidy, which was then widely recognized as a source of this phenomenon.

Two of these examples, *Clarkia lingulata* (H. Lewis and M. H. Roberts 1956) and *C. franciscana,* were reexamined by L. D. Gottlieb (1974, 1974b), using electrophoretic techniques. He showed that small differences between *C. lingulata* and its close relative *C. biloba* support Lewis's hypothesis (fig. 11.1), but that enzyme differences between *C. franciscana* and its two nearest relatives, *C. rubicunda* and *C. amoena,* which Lewis believed to be older and ancestral, are in fact great enough to justify the assumption that all three species diverged from a common ancestor at about the same time (figs. 11.2, 11.3).

S. I. Warwick and L. D. Gottlieb (1985) investigated enzyme differences among six species of *Layia* (Asteraceae), another California genus that had previously been carefully investigated from the cytogenetic viewpoint (J. Clausen 1951). In this group they found that evidence from enzyme genetics with respect to the age and relative affinities of the species supported completely that obtained from cytogenetics.

Another significant investigation was a comparison by K. Helenurm and F. R. Ganders (1985; Ganders 1989) of differences among species of *Bidens* (Asteraceae) found in Hawaii as compared to those found on the North American mainland. The Hawaiian species differ strikingly from each other with respect to gross morphology, particularly vegetative characteristics, but enzyme differences between recognized species are no greater than between different populations assigned to the same species. On the other hand, mainland species show the expected pattern: differences between species are considerably greater than those between populations.

These results are to be expected on the assumption that enzyme differences chiefly reflect (1) the length of time during which two populations or species have evolved independently since their divergence from a common ancestor, and (2) fluctuations in population size, since small size of populations favors accumulation of neutral differences via chance events. Meanwhile many characteristics of gross morphology are adaptive and are therefore subject to selection pressures that act independently of size; these pressures may produce changes rapidly or slowly, depending on whether they are strong or weak.

FIGURE 11.1. Inflorescence, petals, and chromosomes of (A) *Clarkia biloba* ($n=8$) and (B) *C. lingulata* ($n=9$), the latter restricted to a small area near the southern limit of the distribution of *C. biloba*. Although the two species have different chromosome numbers and form sterile F_1 hybrids (C), patterns formed by electrophoresis of allozymes in individuals of *C. biloba* resemble those of *C. lingulata* as closely as do individuals belonging to different populations of *C. lingulata*. The latter evidence confirms the hypothesis that *C. lingulata* was recently derived from *C. biloba* via a "genetic revolution." [From H. and M. E. Lewis 1955, and H. Lewis and M. H. Roberts 1956. Reprinted with permission.]

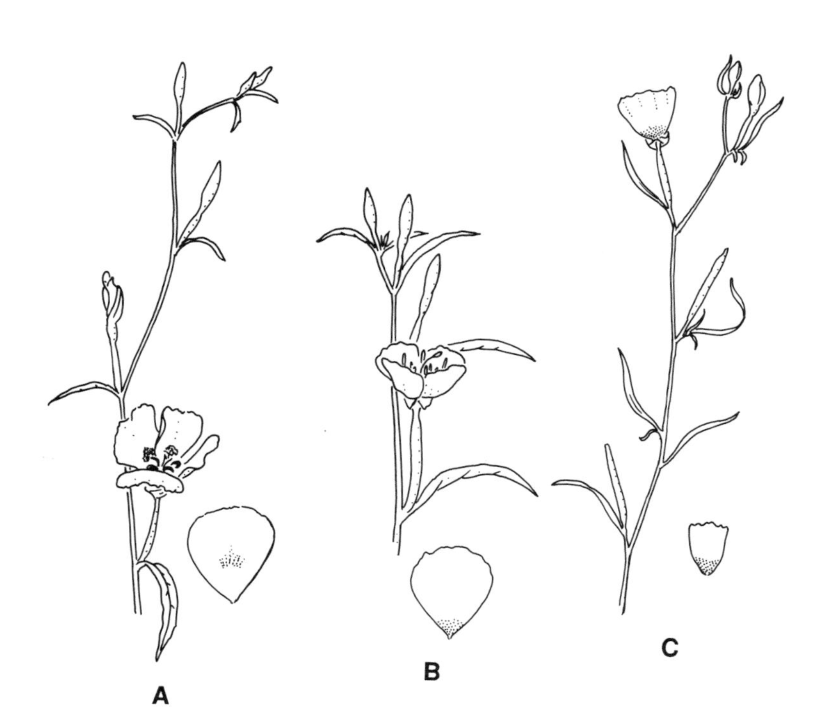

FIGURE 11.2. Inflorescences and petals of (A) *Clarkia amoena*, (B) *C. rubicunda*, and (C) *C. franciscana*. Although all three species have the same chromosome number ($n=7$), hybrids between them are completely sterile. [From H. and M. E. Lewis 1955, and H. Lewis and P. H. Raven 1958. Reprinted with permission from the New York Botanical Garden.]

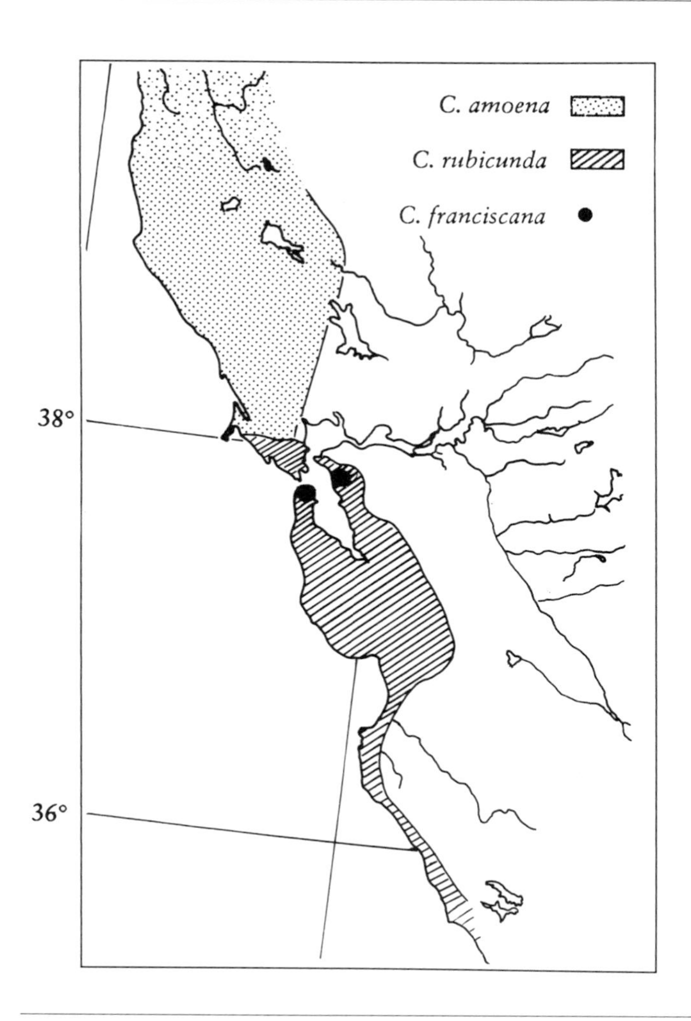

FIGURE 11.3. Distribution of three species of *Clarkia* in west central California. The highly restricted species *C. franciscana*, found only in two populations on opposite sides of San Francisco Bay, differs from the other two species by virtue of a high degree of inbreeding and homozygosity and restriction to serpentine soils. Their allozyme patterns include 8 alleles at 6 loci not found in *C. rubicunda*, a species sympatric with *C. franciscana*.This supports the hypothesis that *C. franciscana* is as old as the two other species and derived from neither of them. [Based on data from H. Lewis and P. H. Raven 1958, with permission from the New York Botanical Garden, and L. D. Gottlieb and S. W. Edwards 1992.]

The Problem of Defining and Delimiting Species

The usefulness of a continental flora may depend to some extent on the degree to which contributors are able to adopt similar standards for delimiting species. Ideally, this essay would be most useful if it could prescribe specific standards that all contributors would be expected to follow. Practically, this is impossible for several reasons, the chief one of which is the diversity of biological factors that are responsible for the diverse patterns of intra- and interspecific morphology that exist in the various groups that are treated. On the other hand, subjective opinions differ so much from one botanist to another as to produce anarchy if every contributor were left to his or her own devices. The objective of the present section is not to arrive at a hard-and-fast definition of species, but to review criteria that have been used by various systematists, pointing out their advantages and disadvantages.

Like all floras, the present one is based on the assumption that all populations of the plants that it treats must be accommodated into one of several thousand basic units designated as species. Moreover, with the exception of a few rare, localized entities, each species is a system of populations that differ from each other at least quantitatively with respect to some of their visible and measurable characteristics. Although some evolutionary botanists might wish to recognize as distinct species populations known as "sibling species," which are morphologically alike but either cannot be crossed or produce sterile hybrids if crosses are obtained, recognition of such "species" in a large flora is impractical. Botanists who are treating groups in which siblings have been experimentally demonstrated, as in the genus *Holocarpha* (J. Clausen 1951), might do well to state that they exist, but not otherwise include them. The current practice of delimiting species chiefly on the basis of morphological discontinuities between several visible traits is justified. Difficulties arise, however, with respect to the extent and nature of these discontinuities.

Many biologists, particularly zoologists, believe that species are nonarbitrary objective entities that exist in nature whether humans are there to describe and delimit them or not. If accepted, this concept assumes that if enough has been learned about patterns of population diversity in nature, the limits of species will be obvious to everyone.

The biological species concept is based on the belief that the great majority of populations consist of groups among which free exchange of genes via hybridization in nature is possible, but between different "species groups" exchange of genes is prevented by the action of one or usually several isolating mechanisms. This appears in nearly all textbooks of evolution (G. L. Stebbins 1950), as well as plant speciation (V. Grant 1971, 1981). The minority of exceptions to this rule are usually regarded as species that will evolve into distinct species in the near future, based on the evolutionary time scale.

For many groups of animals, this belief may be completely justified because of the much greater effectiveness in animals than in plants of two kinds of isolating barriers. One, highly specific courtship patterns, prevents sexual union even between entities that, if cross-fertilized artificially or allowed to mate under unnatural conditions, can produce viable and fertile hybrid offspring. A second is the far more complex patterns of development in animals than in plants. Therefore in hybrids, disharmonious gene action in development is far more common and is likely to produce complete inviability or sterility in at least one sex, that in which the sex-determining mechanism is heterochromosomal (X–Y in males, or W–Z in females).

Another difference in higher animals is the almost universal presence of separate sexes, and the condition that sexual reproduction must always be biparental, while in higher plants, which are usually hermaphroditic (with perfect flowers or monoecy), uniparental reproduction via self-fertilization is often possible. In some evolutionary lines, particularly weedy annuals, it predominates. A cursory review of North American genera suggests that in about 25% of them, most or all species are distinct as judged by the presence of actual or potential gene flow among different populations, balanced by the absence or rarity of gene flow between populations of different species because of hybrid inviability or sterility. In another 25%, the common presence of self-fertilization and/or partial fertility of F_1 hybrids, and apomixis in a few examples (e.g., *Crataegus*, *Rubus* subg. *Eubatus*), often renders clear definitions of species difficult. In about 50% of the genera, two or three of these different conditions exist in different species groups of the same genus.

Data are not available that would allow us to determine whether or not these deviations from expectation, according to the biological species concept, are correlated with different evolutionary ages of the groups. The conditions mentioned exist among recently evolved groups. The "normal" pattern found in *Layia* (Asteraceae) probably reached its present condition during the past 10,000 years, with the desiccation of climate that began after the Pleistocene pluvial (glacial) epoch. The same is probably true of many groups of self-fertilizing annuals in California and the Southwest. Most examples of extensive hybrid swarms are also recent; many have occurred since the onset of disturbances caused by human agriculture.

On the other hand, fossil evidence indicates that in woody species the parental taxa that entered into the hybrid swarms have been distinct from each other since the middle or late Tertiary period, 15–25 million years before the present (M.Y.B.P.) (D. I. Axelrod 1958). No convincing evidence exists to confirm that the majority of species that are poorly defined on the basis of observations of natural populations are more recent than are the better defined species. In plants, the condition of arrested speciation or stasis can, in some groups, persist for millions of years.

From these observations, two questions arise. First, what, if any, are the factors that arrest speciation versus those that permit the completion of the process via reproductive isolation? Second, how can these situations be incorporated into species concepts and taxonomic treatments that will both be useful for identification and produce evolutionary information?

Growth Habit, Population Structure, and Reproductive Isolation

Ever since the formulation of the synthetic theory, plant evolutionists have known that in some genera, species recognized by most taxonomists often hybridize to form interspecific hybrid swarms, while in others barriers of reproductive isolation exist between populations usually assigned to the same species (G. L. Stebbins 1950; J. Clausen 1951; V. Grant 1971, 1981). Most genera in which conventional species boundaries may be blurred (*Quercus, Arctostaphylos, Ceanothus*) are woody, but some are herbaceous (*Iris, Apocynum, Elymus*). The latter, however, consist of individuals that are relatively long-lived, so that the partial sterility of F_1 hybrids is not as serious a detriment to reproductive fitness as it is in short-lived perennials or annuals.

On the other hand, clusters of sibling species that are not usually recognized by taxonomists have been found in annuals, both self- and cross-fertilizing. These correlations suggest by themselves that biological properties other than the fertility versus sterility of hybrids should be recognized when species concepts are being formulated and species are being delimited.

Particular attention must be given to examples of species pairs that are sympatric over large areas and differ with respect to absence or presence of reproductive isolation in different parts of their areas of contact. Situations of this kind are particularly common in *Quercus* (J. M. Tucker 1952; T. M. Cooperrider 1957; W. B. Brophy and D. R. Parnell 1974; R. J. Jensen and W. H. Eshbaugh 1976; V. Grant 1981). *Quercus* is particularly important because its Tertiary fossil record is very good and indicates that the species involved have been distinct from each other for millions of years, except for local hybridization (D. I. Axelrod 1958). Where evidence of this kind is available, longtime persistence of distinctive morphological and ecological characteristics should be considered as part of the biological properties of species.

A complete analysis of reasons why these differences exist is impractical. Nevertheless, one point appears to be clear. The relationship of growth habit to patterns of speciation is not direct, but acts via the different kinds of population structure, both spatial and temporal, that result from different growth habits. Annual species, particularly in regions where climatic stress varies from one season to another, are most successful in the long run if they always produce the maximum amount of seed that the environment permits. In poor seasons the individual plants are few and small; in good seasons, numerous and large. Great fluctuations in population size are the rule, not the exception. On the other hand, woody plants and long-lived perennials respond to stress conditions by producing little growth and often no seeds at all.

A similar situation holds for migration and the colonization of new habitats. If the colonists are annuals and self-fertilizers or apomicts, the first individual colonist will usually produce a large crop of seeds, and its descendants will capture the area via a rapid increase in population size. In contrast, if the new colonist is a tree, shrub, or rhizomatous herb, it may survive by vegetative growth and propagation for many years, and its descendants may capture the environment by a relatively slow increase in population size, accompanied by high, aggressive competitiveness of individual plants.

The final connection between growth habit and degree of reproductive isolation can be established by the properties of populations first proposed by S. Wright (1931) and reaffirmed by many other population geneticists. Repeated reductions of populations to small size promote the establishment of neutral differences between them both in genes and in chromosome structure. As many experiments of interspecific hybridization have shown, the latter are the commonest source of reproductive isolation between plant species. In this connection, the same species groups that contain few and weak reproductive isolation mechanisms also are usually homoploid and relatively homogeneous in karyotype. The same ecological characteristics that are associated with strong reproductive isolation barriers between species having the same chromosome number are also associated with polyploidy.

Toward a Broader Biological Species Concept

Before considering the bearing of the above information on species concepts, two specific examples need to be reviewed. They represent situations that are not uncommon and that constitute serious barriers to the acceptance of the biological species concept as usually formulated.

The first of these is the complex of *Vicia sativa*. It consists of weedy, self-fertilizing annuals, native to southern Europe, but extensively introduced throughout much of North America. Most American manuals recognize two introduced species, *V. sativa* and *V. angustifolia*, and one or more varieties. In *Flora Europaea* (P. W. Ball 1968), both of these, plus several other entities often regarded as species, are placed as subspecies in the single species *V. sativa*.

A thorough review by D. Zohary and U. Plitmann (1979) of the cytogenetic literature on this group states that several entities within it are separated from each other by barriers that are usually regarded as delimiting species. Three different somatic chromosome numbers exist, $2n = 10$, 12, and 14. Furthermore, entities that have the same number may differ greatly with respect to gross chromosome morphology. Not unexpectedly, therefore, F_1 hybrids between such entities have irregular meiosis and low fertility of both pollen and seeds. Nevertheless, the plants are so vigorous that they produce some seeds via self-fertilization that give rise to progeny of F_2 and subsequent generations. These are derived from individuals that quickly recover full fertility while retaining intermediate phenotypes with respect to the characteristics by which the original parents differ.

This result is possible because the genes responsible for chromosome differences segregate independently of those that code for different morphological characters, a condition that has been well known ever since progeny derived from partially fertile interspecific hybrids were analyzed (G. L. Stebbins 1950). The persistence of such intermediates is the basis for treatments by P. W. Ball (1968) as well as by D. Zohary and U. Plitmann (1979), in reducing to subspecies rank *Vicia sativa* and *V. angustifolia.* Zohary and Plitmann pointed out further that some of the subspecies that they recognized include nonweedy natural populations that differ from each other more widely than do the different subspecies that are more common and weedy. Apparently, adaptation to habitats created by humans, by favoring hybridization plus establishment of fertile derivatives, has promoted evolutionary convergence rather than divergence.

The second example is from the genus *Quercus* (fig. 11.4). *Quercus douglasii* is the commonest species of tree growing on the foothills that surround the Central Valley of California. In various parts of its range, it is sympatric with at least four other species of *Quercus,* and hybrids with all of them have been recorded with varying frequency. Its leaves are deciduous, moderate in size, shallowly lobed, and bear trichomes.

For 250 km along the western margin of its range, extending from northwest to southeast, *Quercus douglasii* is sympatric with *Q. dumosa* (including *Q. turbinella,* which hybridizes freely with *Q. dumosa* and also with *Q. berberidifolia* and others), shrubs having evergreen leaves with spiny margins and stellate trichomes having 2–9 branches, as compared to 3–5 in *Q. douglasii.** This species pair was studied by J. M. Tucker (1952b). He found that in the northern part of the range of sympatry, the two species form occasional hybrids, without producing hybrid swarms. Therefore, the two sets of populations are completely discontinuous and easily separated from each other both morphologically and ecologically (*Q. douglasii* grows on open grassy hillsides, and *Q. dumosa* among other shrubs on shady canyon sides). In the central and southern parts of their ranges, however, hybridization is more common, and the distinctness of the two species depends on the degree of overlap.

No evidence suggests that these species are becoming merged into a single population system, in spite of the extensive hybrid swarms that occur in certain areas. Fossil leaves that closely match each species are known from formations that date back to the early Pliocene and late Miocene epochs, 15–9 M.Y.B.P. Moreover, the other woody species associated with these fossils are the same as those that are associated with each species in its typical modern habitat, indicating that the adaptive properties of the late Tertiary ancestors of the modern species were the same as those of their present descendants.

These two examples show that over evolutionary time, given certain habitat changes, well-developed barriers of reproductive isolation may not promote the divergence of populations. On the other hand, poorly developed genetic barriers may persist for long periods of time, even when populations are spatially sympatric, while the populations involved still remain, to a large degree, distinct from each other.

Each of the two examples described above has counterparts in several genera characteristic of the North American flora. Perhaps the largest species group that resembles *Vicia* is *Panicum* subg. *Dichanthelium* (R. Spellenberg 1975). *Ceanothus,* a woody genus having large numbers of species in California, is similar to *Quercus* (H. McMinn 1944; M. A. Nobs 1963). A striking example is *Mimulus* in a broad sense. The annual species belonging to the complexes of *M. guttatus* and *M. glabratus* (figs. 11.5, 11.6) show a pattern similar to *Vicia,* but the shrubby species that constitute subgenus *Diplacus* resemble *Quercus* (V. Grant 1971; M. R. Macnair and P. Christie 1983). This difference is particularly important with relation to the explanation given above, that stronger barriers of reproductive isolation in short-lived plants are due to their occupation of pioneer habitats and a greater influence of chance events, while shrubs, often predominating in climax situations, are more strongly and directly affected by natural selection. Although all species of *Mimulus* are cross-pollinated by insects, annuals of the *M. guttatus* complex are often self-pollinated and can rapidly form new populations from a single founder individual, while the shrubs belonging to subgenus *Diplacus* are rarely selfed, and each flowering shrub is accompanied by several neighbors.

Botanists do not need to revert to a purely subjective species concept, nor do they need a concept that is based entirely on the presence versus absence of reproductive isolation. Therefore, this author agrees with opinions expressed recently by D. Levin (1979), L. D. Gottlieb (1986), B. D. Mishler (1985), B. D. Mishler and M. J. Donoghue (1982), P. H. Raven (1986), and others. The divergence between zoologists and botanists is justified by differences between animals and plants and the populations that they form. In animals, the difference between presence versus absence of gene exchange be-

*Stebbins is treating *Quercus dumosa* in a relatively broad sense, including oaks usually treated in four different taxa (species and subspecies) by other authors (J. M. Tucker 1952; E. L. Little Jr. 1979; K. C. Nixon and K. P. Steele 1981). The hybridization described in the text is between *Q. douglasii* and *Q. turbinella* subsp. *californica.* A full taxonomic treatment of this group will be presented in vol. 3.—Ed.

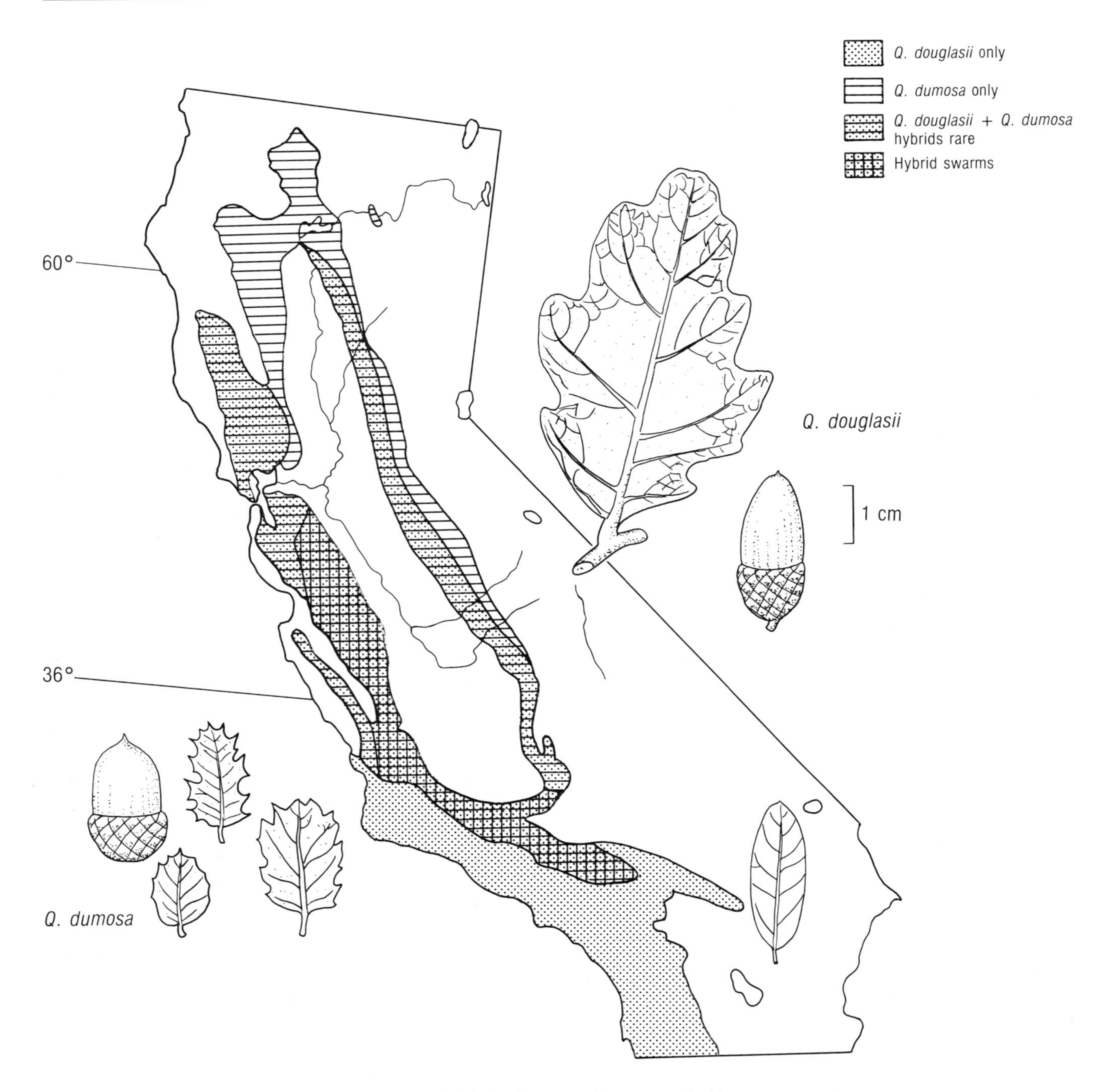

FIGURE 11.4. Localized differences in successful hybridizataion between a deciduous tree speices, *Quercus douglasii,* and an evergreen shrubby species, *Q. dumosa* sensu lato (including *Q. turbinella*), in northern and central California. In northwestern interior California, the two species are sympatric, retain their identity, and occasionally hybridize but do not form hybrid swarms. In the central, western, and southwestern parts of the San Joaquin Valley, hybrid swarms are extensive, and in some populations hybrid genotypes far outnumber those characteristic of the parental species. [Based on data cited in H. McMinn 1935 and 1951, supplemented with data from L. D. Benson 1962, with permission from the University of California.]

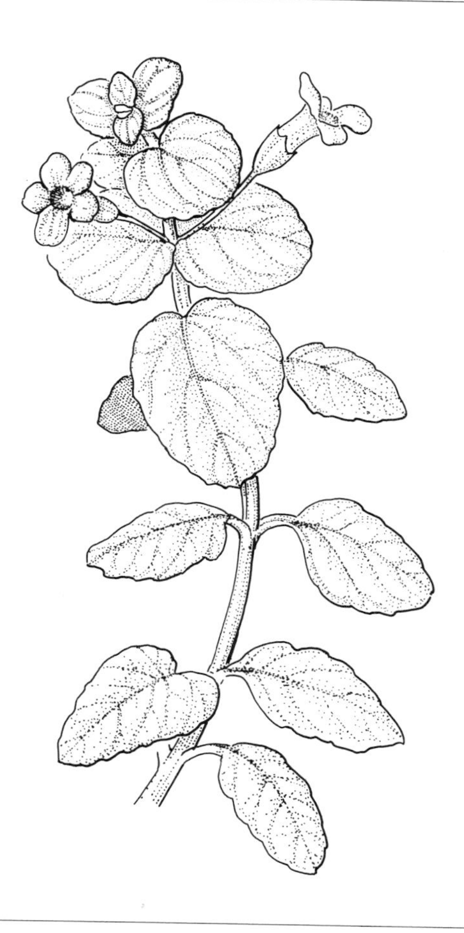

FIGURE 11.5. ***Mimulus glabratus.*** **Pen-and-ink drawing by John Myers (1955–). [Missouri Botanical Garden.]**

tween populations is reinforced to such a degree by the combination of isolation via different courtship patterns, more complex individual development, strongly developed differences between the sexes and chromosomal heterozygosity in one sex, plus limited and often short life spans, that it provides almost universally effective barriers of genetic discontinuity, which can easily be made to coincide with species boundaries.

Conversely, in plants numerous conditions combine to eliminate in many genera the sharp contrast between presence versus absence of gene exchanges among populations that serves so well to delimit animal species. These conditions include the presence of both biparental and uniparental reproduction, which may alternate within the same pair of individuals depending on environmental conditions, random cross-pollination by wind or by unspecialized insects, simple developmental patterns that often allow the formation and establishment of hybrids in nature, great differences in the degree of fertility of F_1 hybrids, and not infrequent vegetative persistence of a single individual during hundreds of years. Populations belonging to closely related plant species possess an entire spectrum of intermediate degrees of gene exchange that may vary greatly depending on different environmental conditions. Plant species are not bounded by any single limiting condition, as B. D. Mishler (1985) has pointed out. Nature forces botanists to adopt a pluralistic species concept.

The need for a pluralistic, but still biological, species concept has recently been recognized by the zoologist A. R. Templeton (1989). His cohesion species concept is similar to those discussed above. It can be applied more widely to different groups of organisms than can a biological concept that is based entirely on the presence versus absence of reproductive isolation and gene flow.

A broad definition of species that is in accord with these principles is as follows. Species are basic units of systematics and evolution. They consist of systems of populations that resemble each other in morphological, ecological, and genetical properties. The populations are held together by various cohesive forces, principally by gene flow but sometimes by partial autogamy, and normally by similarity due to common descent as well as by similar, complex adaptive syndromes that elicit parallel responses to environmental influences. Species

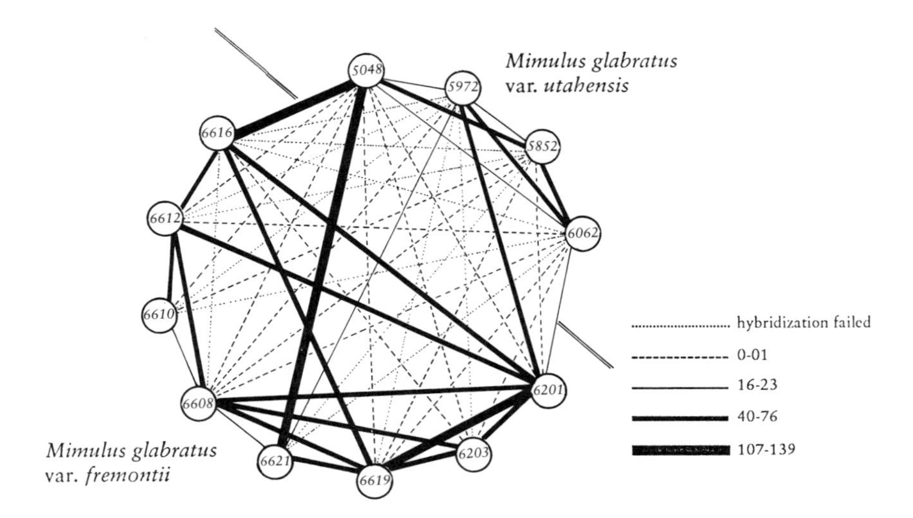

FIGURE 11.6. Artificial hybridization between populations of *Mimulus glabratus* var. *utahensis* (from the Great Basin; populations 5048, 5972, 5852, 6062) and *M. glabratus* var. *fremontii* (Southern Rocky Mountains and Sierra Madre Oriental populations) of the *M. glabratus* complex, which consists of short-lived herbaceous perennials or annuals. Thin lines indicate crosses that produced seeds unable to germinate, weak seedlings, or completely sterile hybrids. Values represent percentages of the expected seed sets. Note the occasional barriers (failed hybridization) existing between populations belonging to the same taxonomic variety, and the high degree of isolation between the two varieties. [From W. Tai and R. K. Vickery Jr. 1970. Reprinted with permission.]

boundaries may be sharply defined to produce complete genetical and physiological isolation from all other species, or they may be locally and temporarily weakened by partial breakdown of interspecific barriers.

Subspecies and Semispecies

Botanists must clearly pay attention to infraspecific categories. Moreover, these must include entities that are sympatric as well as those that are allopatric. As indicated earlier, sympatric population systems that are incompletely isolated from each other are of two kinds: (1) groups within which self-pollination predominates, as in the *Vicia sativa* complex, and (2) groups within which the formation of hybrid swarms depends on local ecological conditions, as in woody genera and in some long-lived perennial herbs. A model of a polytypic species within which allopatric subspecies can be recognized is *Potentilla glandulosa,* as classified by J. Clausen, D. D. Keck, and W. M. Hiesey (1940). Allopatric infraspecific categories are usually designated as subspecies. In local floras (see discussion of the *Vicia sativa* complex), some authors recognize as separate species, sympatric populations that in many regions keep distinct from each other but that elsewhere form localized hybrid swarms. Other authors designate them as "varieties." Examples like *Quercus douglasii* and *Q. dumosa* could be regarded as semispecies, or examples of arrested speciation, but for purposes of a flora they are assigned binomials, as are full species.

Recombinational Species

The example of *Vicia sativa* demonstrates the independence of segregation between morphological characteristics and the elements that compose barriers of reproductive isolation, chiefly chromosomal differences. Because of this independence, partially sterile interspecific hybrids can give rise in later generations to fully fertile, true breeding descendants that are partially isolated from both of their parental species. This course of events has been experimentally documented in *Nicotiana* (H. H. Smith and K. Daly 1959) and is responsible for the process designated as recombinational speciation (V. Grant 1981). Plant taxonomists who have used experimental methods have long recognized that at the level of population systems and species, phylogeny is reticulate. This should be kept in mind in cladistic treatments, especially with groups that include polyploid complexes.

The Relevance of Polyploidy and Apomixis to Taxonomy

One of the most significant factors that affects species patterns and speciation among higher plants is polyploidy. Its frequency has been variously estimated, depending on the criteria used to determine whether or not the ancestors of a particular species have undergone chromosome doubling at some time during their evolutionary history. A conservative estimate is that at least 30% to 40% of modern species are involved. Some of this doubling has accompanied the origin of the genus to which species belong. In other examples, entire subfamilies have originated via polyploidy, including Rosaceae subfamily Maloideae (basic number $x = 17$), Oleaceae subfamily Oleoideae ($x = 23$), Salicaceae ($x = 19$), and Magnoliaceae ($x = 19$). The origin of these groups has most probably involved reticulate evolution at the level of higher categories. Only recognizable polyploid series within genera are considered herein. The percentage of polyploidy among North American species varies over the entire spectrum from 0% to 98–100% (table 11.1).

All aspects of polyploidy have been carefully dis-

TABLE 11.1 Approximate Percentages of Polyploid Species in the Larger, More Widespread Genera of the North American Flora[1]

Percent Polyploid Species	Poaceae[2]	Asteraceae	Ranunculaceae	Rosaceae	Scrophulariaceae	Fabaceae	Other
90–100	*Agrostis* *Elymus*[3] *Poa*	*Senecio*	*Thalictrum*				*Iris* *Viola*
70–89	*Festuca*			*Crataegus*[4] *Potentilla*			*Saxifraga* *Silene*[5]
50–69	*Eragrostis* *Muhlenbergia* *Paspalum*		*Ranunculus*	*Rosa*	*Veronica*		*Plantago* *Polygonum* *Salix*
30–49	*Bromus* *Panicum*	*Artemisia*	*Anemone*	*Rubus*[4]	*Castilleja*		*Galium* *Opuntia* *Vaccinium*
10–29		*Aster* *Erigeron* *Haplopappus* *Helianthus* *Solidago*	*Delphinium*		*Penstemon*	*Lupinus*	*Allium* *Arabis* *Lomatium* *Phacelia* *Phlox*
0–9	*Melica*	*Baccharis* *Brickellia*	*Clematis*		*Pedicularis*	*Lotus* *Trifolium*	*Asclepias* *Berberis* *Hypericum* *Lilium* *Lonicera* *Populus* *Quercus* *Viburnum*

[1] The Poaceae tribe Andropogoneae and the genus *Oenothera* were omitted because of difficulty in coordinating published chromosome numbers with genera and species currently recognized.

[2] Family names are used only when two or more genera, as recognized in all manuals, are included in the table.

[3] The genus is recognized in a broad sense, including many species listed elsewhere under *Agropyron, Sitanion, Hystrix,* and various newly adopted names.

[4] The percentages assigned to the genera *Crataegus* and *Rubus* are somewhat arbitrary due to the difficulty of delimiting species among the apomicts of these genera.

[5] The high frequency of polyploids applies only to North America species. The great majority of Old World species, including some introductions into North America, are diploids.

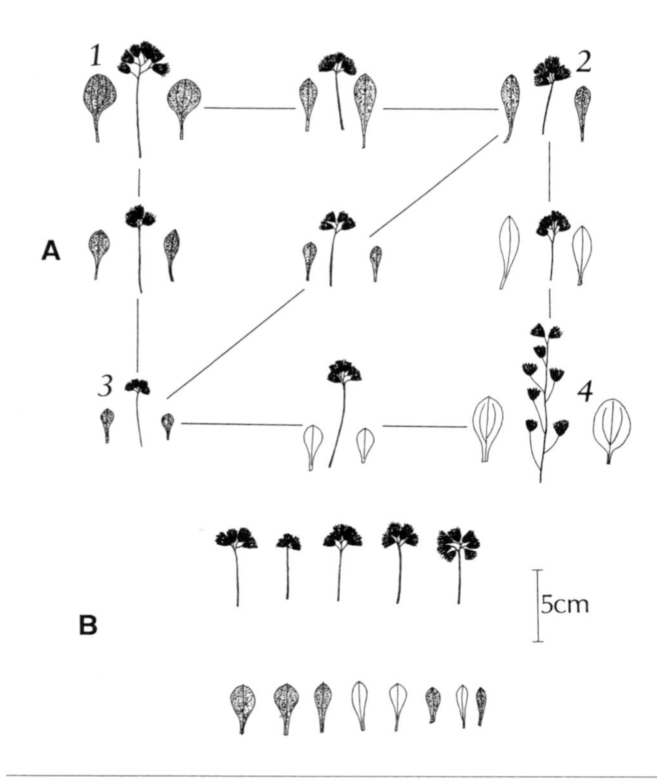

FIGURE 11.7. The hybrid origin of apomicts belonging to the polyploid complex *Antennaria howellii* (=*neodioica*) in North America. Adaxial leaf pubescence is shown by stippling. (A) Capitulescences and leaves of diploid species, 1–*A. plantaginifolia,* 2–*A. neglecta,* 3–*A. virginica,* and 4–*A. racemosa,* plus their artificial diploid hybrids. (B) Similar drawings of polyploid apomicts assigned to *A. howellii* (=*neodioica*). [From R. J. Bayer 1985. Reprinted with permission from Springer-Verlag.]

cussed in the volume edited by W. H. Lewis (1980). Additional information comes from studies of genetic diversity based on electrophoretic marker gene loci on allopolyploid *Tragopogon* (M. L. Roose and L. D. Gottlieb 1976), on genera of Saxifragaceae (D. E. Soltis and P. S. Soltis 1989), and on autopolyploid *Dactylis* (R. Lumaret 1988). For refinements, particularly with respect to ecology and distribution, see G. L. Stebbins (1984, 1985, 1986) and G. L. Stebbins and J. Dawe (1987).

The existence of polyploidy presents two important difficulties in classifying species and infraspecific taxa along traditional lines. First, it promotes reticulate evolution, creating situations in which systematists who would not know the chromosome numbers of the entities with which they are dealing would be likely to regard a common intermediate polyploid species as ancestral to a group. The actual diploid ancestors, which are less common and often more localized, would be considered as derivative. Second, polyploidy usually occurs in association with hybridization, genetic segregation, and natural selection. Hereditary differences within a polyploid complex involve not only differences in chromosome number, but also in structural rearrangements of segments within individual chromosomes and allelic differences at many gene loci. Some differences affect the diagnostic characters used to distinguish between species; others do not, although they may affect considerably the adaptation and distribution of the populations.

Consequently, the classification of populations in polyploid complexes deals with highly complex genetic and cytogenetic segregations, only a small part of which can be followed by the conventional methods of morphological taxonomy. In order to make these difficult problems accessible to those who are not cytogeneticists, some kind of simplification is necessary. Nevertheless, one cannot extrapolate from simplified treatments to determine actual relationships. With polyploid complexes, a case can be made for keeping the working taxonomy that is useful in floras separate from the analytical cytogenetic and biochemical taxonomy that is necessary for understanding evolutionary relationships.

The usual separation of polyploids into two categories, auto- and allopolyploids, must be regarded in the light of this complexity. The terms were originally defined on the basis of chromosomal differences (H. Kihara and T. Ono 1926). Allopolyploids contain two or more different genomes, defined as sets of chromosomes different enough from each other to prevent regular pairing at meiosis in an F_1 hybrid. Kihara and Ono believed that the presence versus absence of such pairing is an absolute difference, usually correlated with species boundaries. Many cytologically intermediate situations were found later. Depending on the parental combination, different interspecific hybrids present a

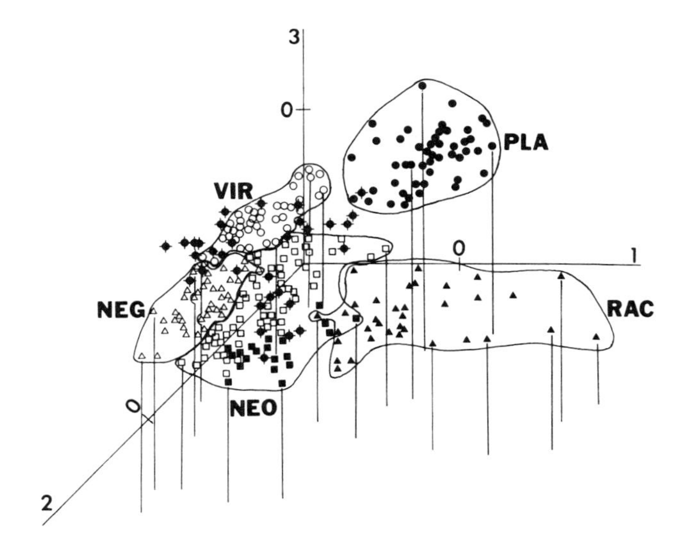

FIGURE 11.8. Scope of variation as represented by multivariate analysis of individual specimens of four diploid species of *Antennaria*, some interspecific hybrids, and the polyploid-apomictic *A. howellii* complex. The diploids are: ⟨△⟩ *A. neglecta* (NEG), ⟨●⟩ *A. plantaginifolia* (PLA), ⟨▲⟩ *A. racemosa* (RAC), ⟨○⟩ *A. virginica* (VIR); ⟨●⟩ interspecific hybrids; ⟨□⟩ and ⟨■⟩ *A. neodioica* complex (NEO). Parentages of hybrids are noted in figure 11.7. [From R. J. Bayer 1985. Reprinted with permission from Springer-Verlag.]

spectrum of conditions between perfect and highly irregular pairing. The degree of regularity can be increased or decreased by the action of different genes.

Furthermore, chromosome pairing in F_1 hybrids is poorly correlated with species distinctness as determined by taxonomic methods. Consequently, a common taxonomic usage—to designate as autopolyploids those entities that fall within the range of morphological variation delimited as a species boundary by taxonomists, and as allopolyploids those which do not—bears little or no relationship to the categories as originally defined and named by Kihara and Ono. Polyploid complexes will be better understood if these terms are either discarded or assigned secondary importance, particularly by taxonomists who have not made careful cytogenetic investigations.

As D. E. Soltis and P. S. Soltis (1989) and R. Lumaret (1988) have shown by their electrophoretic-genetic investigations, all successful polyploids, even those that are strictly autopolyploid, are genetically more heterozygous than their diploid relatives. Relatively homozygous diploid genotypes, when subjected to artificial somatic doubling, produce tetraploids that, either immediately or in the long run, are adaptively less fit than their diploid progenitors (G. L. Stebbins 1985). Consequently, the evolutionary, genetic, and morphological structure of most, if not all, polyploid complexes is best expressed by the pillar model.

Among the complexes at least partially represented in North America, the genus *Dactylis* (Poaceae) and *Epilobium* subgenus *Zauschneria* (Onagraceae) are based on autopolyploidy (G. L. Stebbins 1971), *Antennaria* (Asteraceae) (R. J. Bayer 1985, figs. 11.7, 11.8; 1985b) is based on intermediate or segmental allopolyploidy, and the complex of *Clarkia purpurea* (Onagraceae) is based on genomic allopolyploidy (H. Lewis and M. E. Lewis 1955). Other polyploid complexes may be of a mixed nature.

Many polyploids have achieved great success, but that is by no means universal. Table 11.2 shows that exceptions exist to any rule that might propose a general advantage for polyploidy over diploidy, even when these conditions are present within the same genus. The same conclusion can be drawn from the more complete data compiled by G. L. Stebbins and J. Dawe (1987) from *Flora Europaea* (T. G. Tutin et al. 1964–1980).

Polyploidy in itself does not increase the resistance of plants to cold, drought, and other forms of stress (G. L. Stebbins 1980, 1984, 1985, 1986). The most cogent statement on this point is that the highest levels of polyploidy among cultivated plants are the sugar canes *(Saccharum)* native to the moist tropical lowlands of New Guinea; among woody species in genus *Clerodendrum,* native to the forests of tropical Africa; and among ferns in species of *Ophioglossum,* native to southern India. The correlation between percentage of polyploidy and latitude in European floras on which the hypothesis was first based does not hold for Pacific North America (G. L. Stebbins 1984, 1986; J. G. Packer 1969). Furthermore, better knowledge of the principal genera involved, such as *Calamagrostis, Draba, Poa, Potentilla,* and *Saxifraga,* has revealed the presence in each of these genera of diploid species or cytotypes at high latitudes, or at high elevations in mountain areas. The arid regions of North America are notable for their low percentages of polyploids.

Lists have been compiled of species having very high chromosome numbers ($2n = 96$ or higher) in each of

TABLE 11.2. Widespread Diploid Species in Genera that in Northern America Contain Many Polyploids

Genus	*Species*	*Distribution*
Antennaria	*A. neglecta* (s.s.)	Quebec–Colorado–Virginia–Missouri
Crepis	*C. acuminata*	Northwestern United States
	C. runcinata	Northwestern United States
Silene	*S. acaulis*	Circumpolar, alpine
Triticeae, perennial	"Agropyron" spicatum	Northwestern North America
Bromus	*B. ciliatus*	Northern North America
	B. purgans	Eastern and Central North America
Festuca	*F. elatior*	Northern North America (introduced)
	F. ovina	Circumpolar
Poa	*P. trivialis*	Northern North America, introduced, Europe
Thalictrum	*T. alpinum*	Circumpolar
Crataegus	*C. margaretta*	Ontario–Iowa–Virginia–Missouri
	C. tomentosa	New York–Missouri–Florida–Texas
Potentilla	*P. glandulosa*	Western North America
	P. fruticosa	Circumpolar: boreal, subalpine
Rubus, subg. *Eubatus*	*R. allegheniensis*	Nova Scotia–Ontario–North Carolina
Veronica	*V. serpyllifolia-humifusa*	Northern North America, Eurasia
Viola	*V. pubescens*	Maine–Ontario–Maryland–Kansas

the floristic provinces of North America as defined by A. Cronquist (1982). Interpretation of these results is somewhat complex and will be presented in a subsequent publication. Briefly, the numbers of such species do not differ significantly among the provinces that have the greatest area and the maximum recent history of advances and retreats of floras, i.e., Arctic, Canadian, and Appalachian. They provide support for the secondary contact hypothesis: successful polyploids are usually formed after reunion or secondary contact between previously separated and ecologically divergent races, subspecies, or species.

In a few genera, polyploidy is associated with seed production by asexual means, or apomixis. The principal features of this process, including the various ways by which the events of meiosis and fertilization can be circumvented, have been well known for many years (G. L. Stebbins 1950). Recent research has done no more than refine our knowledge and provide additional examples. In the North American flora, variation patterns dominated by apomixis are best known in *Antennaria, Crepis* sect. *Psilochaenia, Crataegus,* some species of *Erigeron, Poa, Potentilla, Rubus,* and *Taraxacum* (both introduced weeds and native species of the western Cordillera and arctic regions). In the genus *Hieracium,* the largest and most complex series of apomicts known, several of the species introduced from Europe are apomictic, but so far as known, the endemic North American species of the subgenus *Stenotheca* are all diploid and sexual.

An early model for the treatment of agamic complexes (E. B. Babcock and G. L. Stebbins 1938) has been widely accepted and, with revision based on new knowledge, is still valid. It has been recently applied to the two groups of *Antennaria* species found in eastern North America. With the aid of experimental taxonomy (interspecific hybridization and detailed analysis of F_1 hybrids), biosystematic techniques (quantitative determination of the extent of morphological variation via multivariate analysis and cytological studies of meiosis in species and hybrids), and electrophoretic investigations of enzyme variation patterns, the model has yielded a reasonably complete picture of species relationships (R. J. Bayer 1985, 1985b). The absence of gene exchange between apomicts renders impossible decisions about the delimitation of species and subspecies based on presence versus absence of gene flow, so that species boundaries may become more subjective.

Guidelines can be obtained from studying species limits in related groups that reproduce sexually. Such analogies originally suggested that if the morphological pattern of a group of apomicts indicated that they were derived from hybridization between two sexual species that are similar morphologically, most or all of the apomictic clones could be admitted within the limits of one or the other of the parental species. If, on the other hand, the putative parental species were very different from each other, or if a series of similar apomicts appeared to have been derived from hybridization involving three or more different diploid species, they would be best understood by grouping them into highly variable, collective "agamospecies" bearing a different binomial from any of the sexual ancestors. This alternative treatment was adopted in both *Crepis* and *Antennaria.*

Agamic complexes can be treated taxonomically in two other ways. One method, adopted extensively in Europe for *Taraxacum, Hieracium, Rubus,* and other genera, is to recognize each apomictic line as a separate species. Aside from the confusion resulting from the enormous number of so-called "species" generated by this treatment, it has one basic defect with respect to both biosystematics and taxonomic philosophy. In all the genera mentioned above, clusters of obligate apomicts exist sympatrically, at least occasionally, with highly heterozygous and variable populations belonging to a related, fully sexual species. If two such clusters or populations are compared with respect to any group of characters, morphological, cytological, or biochemical, each individual of the sexual population will correspond to a large degree with a clone belonging to the apomictic cluster.

Apomictic clones can be recognized as species only when they occur beyond the range of the related sexual species. The confusion produced by giving the same taxonomic rank, in the same section of the same genus, on the one hand to a single genotype, and on the other to an entire population system, produces a confusing taxonomic system.

An alternative treatment might be to split up the sexual species to make each of their "species" units similar to the apomicts, but this would cause even more confusion. At the other extreme, one might argue for recognizing as a single taxonomic species an entire agamic complex. In addition to producing taxonomic confusion of a different sort, such a practice would violate one of the basic principles of modern taxonomy. Because the superstructure of apomictic clones that form the bulk of the complex are "crystallized hybrids" that contain genes derived from several different sexual parental species, some of which may have been isolated from each other for millions of years, the entire complex is, by its very nature, highly polyphyletic. To treat such a complex as a single basic unit, one species, violates not only current taxonomic practices but also basic biological philosophy.

When dealing with apomictic complexes, the taxonomist faces a situation that nature has evolved, fortunately in only relatively few instances, in defiance of the rules that taxonomists would like to follow. Anybody who is dealing with these situations must recognize this fact and solve the problem in a manner that will be most useful to fellow scientists.

New Ideas on the Treatment of Genera

The delimitation of genera has always raised greater problems than that of species. Some evolutionary taxonomists, particularly zoologists, have maintained that species are objective entities that the taxonomist must learn to recognize on the basis of boundaries that become clearly defined when enough facts are known. Regarded by many taxonomists as erroneous even for species, this statement has never been maintained with respect to genera, except by a few taxonomists who regard certain attributes as of overriding importance.

Nevertheless, to many biologists who are not taxonomists, genera are regarded as more important than species. No species of higher plant or animal is as familiar to experimental biologists as are generic names such as *Drosophila, Mus, Nicotiana, Lycopersicon, Triticum, Saccharomyces,* or *Neurospora.* Consequently, for general biological purposes such as ecology, physiology, morphology, and evolution, exchange of information is facilitated if genera are delimited to be as nearly equivalent to each other as possible, and to retain their scientific names as long as possible. These considerations should be balanced against the need to change names of genera when new information is obtained about their phylogeny.

In contemporary angiosperm taxonomy, two trends have developed that must be carefully watched. One is the splitting of genera on the basis of microscopic details, as has been done in the Asteraceae tribe Eupatorieae (e.g., R. M. King and H. Robinson 1970, 1987). While microscopic characters should by no means be overlooked, their use to establish new genera that are not distinct on the basis of any other characteristics should be discouraged.

The other trend is to base genera on genomic composition (M. E. Barkworth and D. R. Dewey 1985): taxa with different genomic constitutions are placed in different genera. A familiar example is found within genus *Nicotiana.* In this example, S and T are genomes, each containing a gametic set of 12 chromosomes that differ from each other so much that they are unable to pair in the diploid hybrid (ST). If the rule stated above is adopted, cultivated tobacco, *N. tabacum* (SSTT), would be placed in a genus by itself, separated from its diploid ancestors, *N. sylvestris* (SS) and the *N. tomentosiformis* group (TT).

Such a proposal has three serious difficulties. First, the decision as to whether two genomes are similar or different is hard to make in many groups because intermediate conditions, partial differences, exist. For instance, the three genomes contained in bread wheat were once regarded as different from each other, so that under the Barkworth-Dewey proposal, the three wheats, *aestivum* or hexaploid, *dicoccum* or tetraploid, and einkorn or diploid, would have had to be placed in different genera. R. Riley and V. Chapman (1958), however, have identified genes that prevent partly homologous chromosomes from pairing with each other, so that the wheat genomes can now be regarded as similar. The extent to which genes having this effect exist in other plant groups, including the Triticeae, is unknown.

Second, the amount of labor necessary for the cytogenetic delimitation of genomes is great. If the proposal is adopted, different standards of generic limits would exist for those groups that have been investigated for this purpose and those that have not. For instance, if the genomic composition of a neighboring genus, *Bromus,* were fully known, and if *Bromus* were treated according to standards proposed for subtribe Triticinae, it would be split into at least nine genera (G. L. Stebbins 1981). A similar fate would befall *Festuca, Poa, Panicum, Agrostis,* and other large grass genera should they be analyzed in the same manner.

Third, if the proposal were adopted, then every major cytogenetic investigation of a genus of angiosperms would result in several new and unfamiliar generic names. Beginning students of taxonomy are told that they must adopt scientific names because they have greater stability than common names. If they were then told that names should be changed whenever a comprehensive piece of cytogenetic research was completed, they would be justified in being somewhat skeptical of the first argument.

Nevertheless, a more moderate use of cytogenetic or other phylogenetic information is possible. For this purpose, categories of intermediate rank are recognized. Both the subgenus and the section are commonly used, and could be applied to genomic differences without upsetting the apple cart.

12. Pteridophytes

Warren H. Wagner Jr.
Alan R. Smith

General Background

Pteridophytes, the ferns and so-called fern allies, comprise about 3% of the vascular plant species of continental North America north of Mexico. This assemblage of organisms includes very diverse elements such as aquatic quillworts (*Isoëtes*; Lycopodiophyta), desert cliff brakes (*Pellaea*; Polypodiophyta), nearly leafless whisk-ferns (*Psilotum*; Psilotophyta), and weedy horsetails (*Equisetum*; Equisetophyta). These groups are so divergent that they are placed in four different divisions (A. Cronquist et al. 1966).

What is the justification for grouping the pteridophytes together? Common features include: (1) the production of spores, the principal dispersal units; (2) the germination of spores and subsequent development to produce gametophytes that exist independently of the spore-producing plants; and (3) the fertilization of eggs by flagellate, swimming sperms. In addition, pteridophytes share a negative feature—the absence of seeds. The seedless condition is a primitive one that characterized all early land plants.

Most pteridophytes are basically herbs, some as small as 2–3 mm, and only a minority reach heights of more than 2–3 m. Only tree ferns, such as many Cyatheaceae, have tall trunks, and Cyatheaceae are essentially absent, except in cultivation, outside of the tropics. Even in the most tropical areas of the United States (e.g., southern Florida), there are no native tree ferns. Therefore, the tallest pteridophytes in North America north of Mexico are the climbing ferns *(Lygodium);* their height, however, is not accomplished by stems but rather by extremely long leaves that clamber up other plants. Species in the flora have twining leaves that may reach 3–7 m.

With the exception of certain grapeferns *(Botrychium)* and *Isoëtes,* no modern pteridophyte displays true secondary growth: the tissues are all primary, derived from a terminal meristem. Pteridophytes have no cambium, and they lack cork and secondary vascular tissues.

Pteridophytes are noted for retaining many primitive features of the earliest land plants. Some of the early pteridophytes were trees to perhaps 35 m tall with abundant secondary wood (e.g., the progymnosperms and certain lycopods of the Devonian [405–345 million years ago] and the Carboniferous [345–280 million years ago]). The club-moss (Lycopodiaceae), spike-moss (Selaginellaceae), and quillwort (Isoëtaceae) lines of evolution have been distinct from all other vascular plants since approximately the early Devonian. Whisk-ferns *(Psilotum)* and horsetails *(Equisetum)* are so distinctive that we are still in a quandary as to their relationships.

Evidence from the fossil record suggests that the earliest vascular land plants (those possessing xylem and phloem, the specialized water- and food-conducting tissues) had a life cycle similar to modern pteridophytes. These early examples are known today from fossil imprints and petrifactions, especially in Devonian depos-

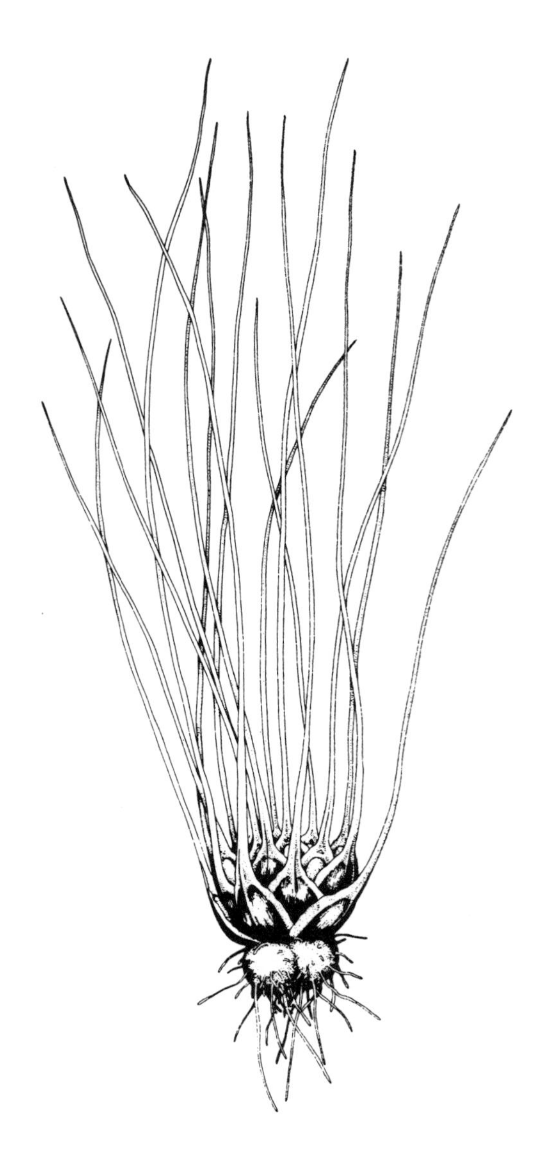

Isoëtes melanopoda. [From W. C. Taylor. 1984. Arkansas Ferns and Fern Allies. Milwaukee. P. 125, by P. W. Nelson. Reprinted with permission.]

its of eastern Canada and the northeastern United States, where some of the world's best preserved fossils of these ancient plants have been discovered. Remarkably detailed features are still preserved, including internal tissues, spore cases, and spores.

Some pteridophytes, such as *Psilotum,* seem nearly to have lost their ability to speciate; many others, such as *Thelypteris,* are apparently speciating actively today. Many of the modern groups show great diversity in form, life cycle, and habitat preference. We discuss some details of this diversity, especially as they pertain to an understanding of the taxonomic treatments in this flora, in the sections that follow.

Life Cycle

The pteridophyte life cycle is characterized by having two separate free-living plants, gametophytes and sporophytes, interconnected by stages of the sexual process. This phenomenon is referred to as alternation of generations. Ordinarily, the more conspicuous and dominant plant is the sporophyte (diploid and $2n$), which is usually perennial and lives for an indefinite period. The gametophytes (haploid and n) tend to be inconspicuous (usually considerably less than 1 cm in the largest dimension) and short-lived. When fertilization occurs and the new embryonic sporophyte forms, the

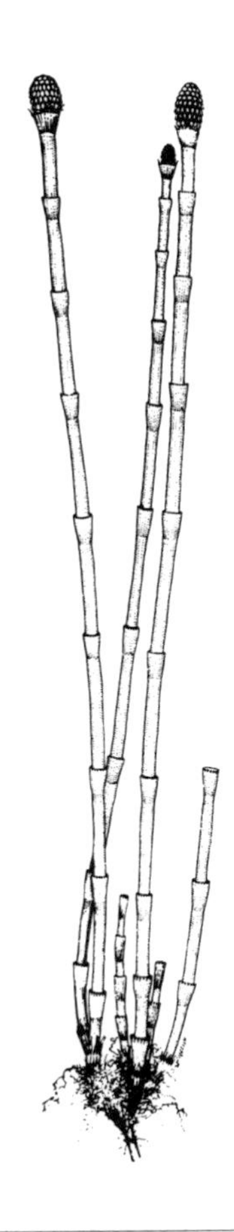

Equisetum laevigatum. **[From W. C. Taylor. 1984. Arkansas Ferns and Fern Allies. Milwaukee. P. 119, by P. W. Nelson. Reprinted with permission.]**

gametophyte dies. Gametophytes are occasionally the dominant or only generation present in a given species in certain temperate outliers of tropical species (see Modifications of the Life Cycle later in this chapter).

Gametophytes of closely related species and genera tend to be much alike and have been reported for only a small percentage of the species of North American pteridophytes. They are also difficult to detect because of their occurrence in small niches, and some grow underground. For these reasons, it is primarily the sporophytes that are described in this flora.

Organs—Roots, Stems, and Leaves

Sporophytes are made up usually of three organ systems: roots, stems, and leaves (fig. 12.1). Roots of pteridophytes have been poorly surveyed, and our knowledge of them is limited. They tend to arise along the stem, commonly near leaf bases. Most pteridophytes have very narrow (less than 0.5 mm thick) and wiry roots, but roots may be thicker (more than 1 mm) and fleshy. The roots may form a dense mass, as in *Osmunda*, that may be cut up and used as a substrate ("osmundine") for growing orchids or other epiphytes, or roots may be scattered and few, as in certain lycopods. Delicate root hairs may be abundant (Vittariaceae), sparse (most ferns), or totally absent (Ophioglossaceae). Most root hairs occur behind the root tips and die off, but they may be persistent in some groups.

The stems of pteridophytes are mainly true rhizomes, i.e., stems that run horizontally at or just beneath the surface of the ground. The growing tips, however, may be ascending or erect. Some ferns have upright stems like tiny tree trunks (caudices) and are unable to form colonies except for occasional upright branchings, as in *Botrychium*. Rarely a pteridophyte may have both upright caudices and horizontal rhizomes, these strongly differentiated as in *Matteuccia*. The rhizomatous habit of many pteridophytes results in extensive vegetative reproduction. Often a large continuous colony is actually a clone, and in terrestrial species, the individual leaves or plants that appear aboveground are from the

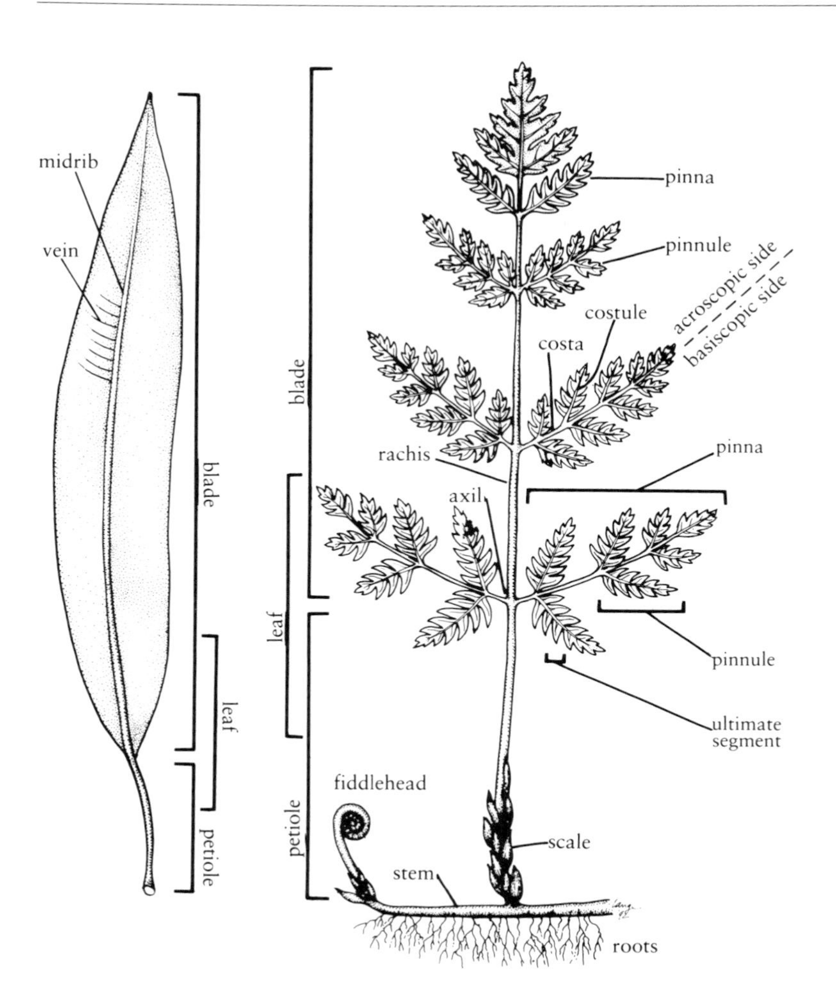

FIGURE 12.1. General morphologic terms describing ferns.

rhizome system of a single original plant. In the jack pine *(Pinus banksiana)* forest of northern Michigan and Wisconsin, the rhizomatous brackens are the dominant understory, covering thousands of square kilometers. Some clones may be hundreds or even thousands of years old and cover a hectare or more, but because they lack secondary growth, their age is difficult to ascertain.

The clonal nature of growth is one of the main reasons why some sterile hybrid ferns and fern allies are capable of becoming persistent and sometimes common elements in the vegetation. Although most ferns are capable of propagating by rhizomes, others spread from leaves or roots (see below). The majority of individual pteridophyte plants encountered in forests, fields, fens, and along pond edges are vegetatively produced. In only a few terrestrial species, e.g., most moonworts and grapeferns, are individual plants formed only by sexual reproduction. Not surprisingly, the majority of plants that occur in narrow crevices of exposed rock cliffs also tend not to form clones. Even if they have the ability, the barren rock surfaces separating the crevices tend to keep the plants compact, small, and isolated.

Anatomically the stems of pteridophytes are simpler than in most seed plants because they are made up only of primary tissues. Complex patterns produced by secondary production of new xylem and phloem are absent. Even the primary tissues, however, may have elaborate arrangements, such as platelike systems of xylem (the plectosteles of certain Lycopodiaceae) or tubes of netted xylem-phloem bundles (the dictyosteles common in leptosporangiate ferns). Cross sections of stems reveal many patterns. These range from solid cores of xylem (protosteles) as in certain lycopods, to nearly closed tubes of xylem (siphonosteles) as in *Dennstaedtia,* to complex open tubular nets of xylem (dictyosteles) as in many ferns. Externally, the stems of most ferns tend to be covered with hairs or scales (see be-

Psilotum nudum. [From I. W. Knobloch and D. S. Correll. 1962. Ferns and Fern Allies of Chihuahua, Mexico. Renner, Tex. Plate 1, fig. 1, by P. Horning.]

low), especially near the growing tips where indument protects the easily injured cell initials (meristem).

Leaves of pteridophytes are extremely diverse, ranging from unvascularized tiny projections (enations) as in *Psilotum,* to small, simple, needlelike or scalelike leaves with only a single, central vein (microphylls) as in the lycopods, to rushlike microphylls up to 50 cm long in *Isoëtes,* and to the large complex leaves with an intricate pattern of many veins found in most ferns (megaphylls). In ferns the leaves are often referred to as fronds. These develop in a distinctive manner with the soft growing tip (meristem) rolled up in the center of the crozier (fiddlehead). The crozier is produced by a process of growth and unfolding, and the characteristic pattern thus formed is known as circinate vernation.

Leaf development that is initiated at the tip is called acropetal, and this development is characteristic of most modern ferns. Leaves of a few genera have another kind of development. For example, *Ophioglossum* leaves grow by intercalation and are conduplicate rather than circinate. Leaves of seed plants, in contrast, tend to develop by overall growth and expansion, i.e., by intercalary differentiation and enlargement.

The parts of a leaf include the petiole (stipe) and the blade (lamina). Petioles of ferns are diverse and offer many taxonomically valuable characteristics. There may be a single round stele or vascular trace, or the trace may be U-shaped or Ω-shaped, or there may be 2–10 or more separate traces. The patterns are characteristic of genera and sometimes families. For example, the petiole trace of most Aspleniaceae is X-shaped in cross section.

The main axis of the blade is called the midrib (rachis). If the blade of a fern is divided into leaflets, these are called pinnae, and if the leaflets are divided again into subleaflets, the subdivisions are called pinnules. To be defined as a pinna or pinnule, a leaf subdivision must be stalked or at least obviously narrowed at the base and not connected to adjacent pinnae or pinnules by greenish blade tissue; leaf subdivisions that are not stalked or narrowed are called segments. If a blade has no divisions it is termed simple; if it is once-divided only, it is regarded as 1-pinnate (fig. 12.2). If the pin-

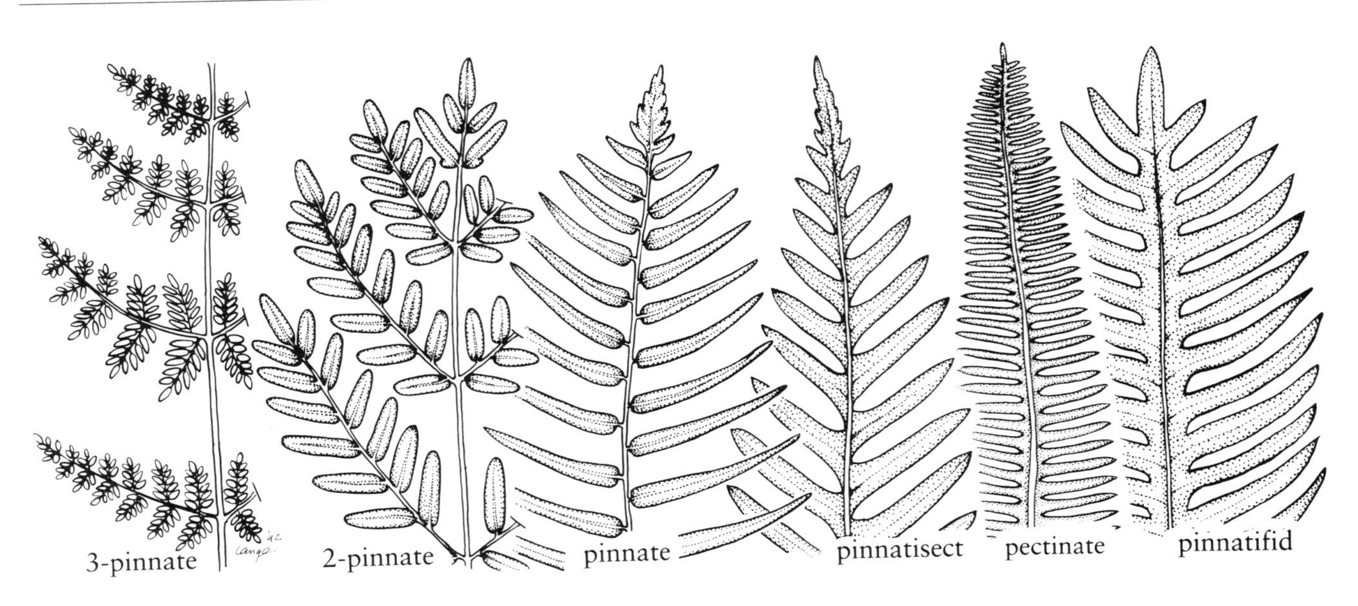

FIGURE 12.2. Leaf division in ferns.

nae are divided further into pinnules, the blade is 2-pinnate. A very finely dissected blade may be 3 (or more)-pinnate (decompound).

Lobed blades are often mistakenly termed pinnate although, in fact, the segments are not narrowed at all at their bases. If the sinuses in a lobed leaf reach only part way to the rachis, the leaf is pinnatifid. If, however, the sinuses extend all the way to the rachis, the blade is considered pinnatisect. Terms referring to the dissection of the blade, along with terms pertaining to blade outline and relative petiole length, are used extensively in describing and identifying ferns.

Indument—Hairs, Scales, and Glands

In ferns, outgrowths of the epidermis of stems and leaves are common and varied, and these features are very important in description and identification. Simple outgrowths made up of a single cell or a chain of several cells are generally called hairs or trichomes and are regarded as the most primitive kind of outgrowth. Some primitive fern groups have only hairs (e.g., Ophioglossaceae, Schizaeaceae sensu stricto, *Dennstaedtia* and certain of its relatives). Trichomes that have two or three parallel rows of cells at the base and a single file of cells at the tip are called bristles (setae). More elaborate outgrowths that form flat plates of 3–20 or more rows of cells are scales (paleae). Scales may be attached basally or attached centrally on a small stalk; the latter are described as peltate. All transitional forms between hairs and scales may be found in some species, even on the same plant.

One of the more distinctive features of many hairs and scales is the presence of enlarged and rounded terminal secretory cells. Glandular hairs, sometimes called glands, may characterize particular species. Such glands are not to be confused with nectaries, which are also secretory structures. Nectaries are rare in pteridophytes and have been found in bracken and certain polypodies (S. Koptur et. al. 1982). The presence of glands in *Dryopteris intermedia* quickly distinguishes it from its look-alike, *D. carthusiana.* Likewise, *Gymnocarpium robertianum* is glandular, and the similar *G. dryopteris* is without glands or nearly so. Glandular cells may occur at tips of hairs or scales. When the leaf of a given species or variety of fern is heavily glandular, the blade will actually stick to the paper when it is pressed (as with the rare glandular variety of *Osmunda cinnamomea*).

Some pteridophytes lack trichomes on stem and leaf surfaces. Most lycopods are hairless, but *Palhinhaea* has hairy stems. Hairlike structures are absent in most other fern allies, although lateral projections along the leaf margins, often called teeth or cilia, may resemble multiseriate hairs *(Lycopodiella, Selaginella). Lycopodium clavatum* has terminal leaf hairs.

Gametophytes of most pteridophytes tend to have distinctive brownish or colorless hairs (rhizoids) that apparently serve to collect water and to anchor the plant. They may also have uniseriate hairs and glands on the prothallial margins and surfaces. Trichomes sometimes occur among or on the gametangia (as in *Huperzia*) or sporangia (as in *Polypodium*) and are then called paraphyses.

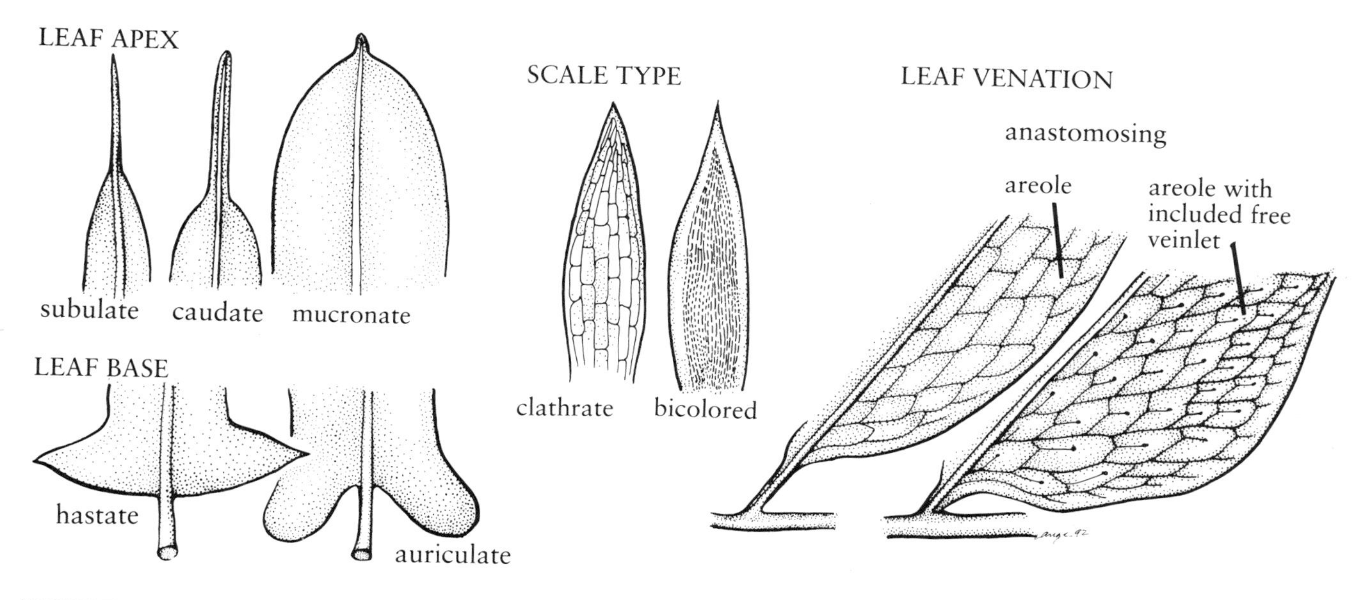

FIGURE 12.3. **Leaf apices, bases, scales, and venation.**

Veins—Free, Dichotomous, or Reticulate

The venation of ferns provides a rich source of information for identification (W. H. Wagner Jr. 1979). Vein patterns may best be seen by holding the blade up to the light, unless the tissues are very thick. Venation in dried, pressed specimens may be seen by clearing them temporarily with a drop of 95% ethanol.

Many ferns, especially primitive ones, have pinnately arranged free veins, with the major veins running out from the rachis laterally and the minor veins repeating the pattern. In leaves in which veins come together to form a network of loops or meshes (veins anastomosing; venation reticulate), the area within a loop is an areole (fig. 12.3).

More advanced ferns often have various types of reticulate venation. Sometimes, as in *Thelypteris,* the anastomosed veins are found only below the sinuses between segments of the pinnae. In other cases, the basal veinlets connect to form a chain of areoles along the pinna midribs (as in *Woodwardia*). Vein connections are found throughout the blade in *Asplenium rhizophyllum* and *Onoclea sensibilis.* Rarely in North American ferns (but commonly in tropical ferns), the spaces within the areoles are occupied by single, free, unbranched or branched veinlets (fig. 12.3) called included veinlets (as in *Ophioglossum*).

Rarely, the venation of a pinna or pinnule may be entirely dichotomous with the veinlets repeatedly forked from the base to the tip of the segments. This condition is illustrated by certain species of *Adiantum* and of *Botrychium.* One of the special features of truly dichotomous venation is that there is no midrib. The small, juvenile or first-formed leaves of many ferns have dichotomous venation, but distinct midribs are generally found in more adult, larger leaves.

Sporangia

The spore-producing organ of pteridophytes is the sporangium. With few exceptions, sporangia are borne on leaves or modified leaves (in *Psilotum* they are produced at the tips of short lateral branchlets). When the leaves that bear sporangia are like those that are only photosynthetic, the leaves are described as monomorphic. If, however, the sporangia are borne on leaves or leaflets that are strongly modified and different from the photosynthetic (or nutritive) leaves, the leaves are said to be dimorphic. In plants that show some divergence of fertile and sterile leaves, the fertile ones are often called sporophylls, and the sterile leaves are trophophylls (W. H. Wagner Jr. and F. S. Wagner 1977). Sporophylls tend to be taller or longer-petioled than trophophylls, and if they are photosynthetic at all, it is only in early stages of development. They usually have little or no laminar tissue. Also, sporophylls tend to be short-lived, lasting only long enough to produce and discharge the spores.

Sporangia of all fern allies and the primitive ferns are basically like those of the pollen-producing microsporangia of gymnosperms or the anthers of angiosperms in having thick walls with a number of cell layers. The sporangia open usually by a transverse split and produce hundreds or thousands of spores. This type of sporangium, which is most like the original vascular

Selaginella eclipes. **[From W. C. Taylor. 1984. Arkansas Ferns and Fern Allies. Milwaukee. P. 213, by P. W. Nelson. Reprinted with permission.]**

plant sporangia of the Lower Devonian, is referred to as a eusporangium, i.e., true or typical sporangium, the one found in practically all vascular plants.

Most ferns, however, have drastically modified and reduced sporangia that are so simple, they appear to be little more than elaborate trichomes bearing a tiny round spore case at the top. This distinctive type is called the leptosporangium (from the Greek *leptos,* slender or small). The spore case itself is few-celled, and the outer wall is only one cell thick at maturity.

The number of spores produced in leptosporangia is usually only 128, 64, or 32, most commonly 64. Most leptosporangia have a bow or annulus made up of strongly modified, thickened cells. The position and extent of the annulus are important characteristics that define a number of fern families, particularly the primitive leptosporangiate ones. The leptosporangia of *Osmunda* have patches of specialized, thickened cells. Those of Schizaeaceae, Lygodiaceae, and Anemiaceae have apical annuli. In Hymenophyllaceae, the annulus is oblique and not interrupted by the stalk. In most other leptosporangiate ferns, the annulus runs lengthwise from the base of the sporecase (where it abuts the tip of the stalk) over the distal end of the sporangium, thus dividing the sporangium into two equal halves. The stalk itself may be made up of 5–6 rows of cells, 2–3 rows (most ferns), or rarely only 1 row (as in Grammitidaceae and Aspleniaceae). The comparative morphology of leptosporangia deserves a great deal more attention than it has received, and it is a likely source of many additional taxonomic characters (K. A. Wilson 1959).

The sporangia are variously arranged, and the particular arrangement is a useful character in recognizing taxa. In *Psilotum* they are borne in fused clusters of three on short lateral branches; in all lycopods and quillworts, they are solitary and borne in or just above the axils of microphylls; and in the horsetails and scouring rushes, 5–7 pendent sporangia are borne on peltate sporophylls that are aggregated in cones. In ferns, sporangia are on leaves or leaf parts, in various arrangements, solitary to many, often in clusters (sori) that are linear, oval, or round; in some species the sporangia completely cover the abaxial surface (sporangia acrostichoid, as in *Acrostichum*). Sori are often protected by flaps formed by the leaf margin (false indusia) or specialized scales or cuplike coverings (true indusia) that are separately produced from the area on which the sorus is borne (receptacle). It is important, especially in ferns, to obtain fertile leaves that bear sporangia because this usually greatly simplifies identification, especially at the generic level.

Spores

Spores have been extensively employed in discriminating taxa of pteridophytes at all levels (A. F. Tryon and B. Lugardon 1991). Sometimes almost identical spore types appear in unrelated taxa, but generally there are good correlations of the spore type with taxonomy (A. F. Tryon 1986). The functions and adaptive significance (if any) of the different spore types are largely unknown. It is known that green spores (containing chloroplasts) are common in epiphytes growing in tropical rainforests (Hymenophyllaceae, Grammitidaceae) and in temperate genera that release their spores early in the spring (*Onoclea, Matteuccia,* Osmundaceae).

Spores of most modern pteridophytes are all of one type, and the taxa are said to be homosporous. Individual spores are mostly 20–60 μm in length or diameter and are almost invisible without a microscope. En masse spores may be white, yellow, orange, green, brown, or nearly black. A number of pteridophyte orders have evolved the condition known as heterospory, in which two types of spores are produced by a given species: small spores mostly 20–30 μm long (called male) and large spores 200–700 μm in diameter (called female). Some of the life cycle modifications associated with heterospory will be discussed in a following section.

Spore shape and symmetry also vary. Spores may be tetrahedral or nearly globose, and the scar (laesura) on the inner (proximal) surface (toward the center of the original tetrad) may be triradiate; or the spores may be more or less reniform or bean-shaped and bilateral, with a single straight laesura. The type with the triradiate laesura is called trilete, and the one with the linear laesura, monolete.

Spores also differ greatly in the development of the wall layers. The basic layers are the endospore and the exospore, and they differ in their development and structure from the wall of pollen grains in seed plants (B. Lugardon 1978). An outer tapetal deposit, known as the perispore or perine, may also be evident under the light microscope as a massive buildup of ridges, wings, or warts. B. Lugardon (1971) determined that practically all ferns have at least a rudimentary perispore.

In nature, spores are distributed largely by air convection, and it is believed that long-distance dispersal by wind is far more common in pteridophytes than in seed plants. By and large, pteridophytes tend to have wider ranges than seed plants (A. R. Smith 1972). Most spores are easy to germinate on soil, flowerpot chips, and even tap water. Spores sometimes germinate after decades, but green spores remain viable only a week or so after their release from the sporangium. Spores of pteridophytes with underground gametophytes associated with fungi germinate only in the dark (D. Whittier 1972).

Germination of spores generally occurs on disturbed and exposed areas of moist soil, in rock crevices, or on rotting logs, often among bryophytes. The best places to look for young gametophytes are on shaded soil banks, at edges of streams, and on mossy bark at bases of trees. Under favorable conditions gametophytes appear by the thousands, especially when the spore parents are nearby. More common perhaps is the appearance of a single plant or a few isolated ones in a tiny microhabitat, but such isolated and sporadic specimens are rarely observed.

Gametophytes

Sexual fusion in pteridophytes occurs on inconspicuous, free-living plants known as gametophytes. These are often associated with similar-appearing but much more common mosses and liverworts. Most gametophytes are green and surficial (surface borne), but some are nongreen and subterranean. Surficial photosynthetic prothalli are generally flat and have clear-cut upper and lower surfaces. The upper surface in *Equisetum* and *Lycopodiella* is provided with projecting lobes or flattened processes. The under surface, i.e., the side next to the substrate, bears specialized hairs (rhizoids) that presumably function for anchorage and water uptake. Subterranean gametophytes tend to be much fleshier, either cylindric or thickly wafer-shaped, and yellowish to brownish. They are extremely difficult to find in nature.

Both surficial and subterranean gametophytes depend on free water through which the sperm must swim to reach the egg and achieve fertilization. The male gametes (sperms) are provided with special organelles equipped with cilia that propel the gametes by their motion. There may be two cilia per sperm (Lycopodiaceae, Selaginellaceae) or dozens (all other modern pteridophytes). Sperms are produced in specialized cases (antheridia) that are either sunken in the gametophyte or protrude from it. The antheridia release the sperms through a pore. Release depends on maturity of the sperms and the presence of water. The female gamete (egg) is located in a bottlelike organ, the archegonium. Sperms swim to the opening at the top of the neck of the archegonium. The neck provides a passageway to the enlarged base (venter), where the egg is located and where fertilization takes place.

Both antheridia and archegonia may be present on an individual gametophyte, especially if it is growing singly in a culture dish (E. J. Klekowski Jr. 1969). Thus it is possible to have intragametophytic selfing, the fertilization of an egg by a sperm from the same gametophyte. In nature, however, it is believed that the tendency, at least in ferns and *Equisetum,* is for gametophytes to pass sequentially through all-archegonial or all-antheridial stages before becoming bisexual. Female gametophytes release soluble substances (antheridiogens) that stimulate nearby gametophytes to develop antheridia only. This tends to promote cross-fertilization between gametophytes of the same or closely related species. It may help explain the well-known predilection for interspecific hybridization to occur, even in species with subterranean gametophytes (W. H. Wagner Jr. et al. 1985).

Embryos and Young Sporophytes

The embryogeny in a number of pteridophytic groups is still unknown or poorly known; what is known has been summarized mainly by D. W. Bierhorst (1971). The embryos of many pteridophytes develop their organs early, and it is often possible to distinguish the first leaf, the stem, and the root at a few-celled stage. Young sporophytes differ almost as much among the major groups as do the adult sporophytes. In some families, e.g., Equisetaceae, Lycopodiaceae, Psilotaceae, the early stages are simple and similar to greatly reduced adult stages. In many ferns, however, the juvenile leaves are often strikingly unlike the mature leaves; the "sporelings" have dichotomously constructed early fronds, the two halves corresponding to the two basal pinnae or vein trusses of the intermediate and mature leaves (W. H. Wagner Jr. 1952).

Chromosomes

Spore mother cells, located inside the developing sporangia, are often used to study the chromosomes of pteridophytes. The spectacular work of Irene Manton of the University of Leeds, whose book *Problems of Cytology and Evolution in the Pteridophyta* (1950) aroused pteridologists around the world, led to numerous profound changes in our taxonomic concepts. By making preparations of spore mother cells undergoing meiosis, she was able to determine the somatic number ($2n$) of chromosomes for individual species and also the base number (x) of chromosomes for each genus she studied. Additionally, she was able to observe the cytological effects of hybridization between species.

From the enormous number of studies that have been reported since 1950, we can make certain generalizations about pteridophyte chromosomes. Homosporous pteridophyte genera have high chromosome base numbers (i.e., the lowest haploid multiple) ranging from about $x = 20$ to $x = 110$. These are among the highest base numbers known in vascular plants. Genetic evidence suggests that homosporous ferns with high base chromosome numbers are diploid (C. H. Haufler and D. E. Soltis 1986), even though they may be derived from ancient polyploids (paleopolyploids). Heterosporous pteridophytes, however, have low chromosome base numbers like those of most seed plants, i.e., $x = 7–11$. This generalization applies across all divisions of pteridophytes.

Polyploidy may be superimposed on the base numbers and, at least among homosporous taxa, very high numbers are sometimes attained (in *Ophioglossum*, n numbers exceed 600). Of particular interest to taxonomists is the discovery that certain base numbers characterize major groups; adiantoid ferns have $x = 29$, 30; dryopteroid ferns, $x = 40–42$; osmundoid ferns, $x = 22$; huperzioid lycopods, $x = 67$, 68; equisetums, $x = 108$; and so on.

By observing meiotic chromosomes, we can usually determine whether a given plant is a hybrid or not. Meiosis in hybrids is generally very irregular, and the spores that result are malformed. Malformation of spores is generally attributed to the genetic imbalance resulting from uneven distribution of chromosomes. Sexual species usually have normal meiosis and spore production, whether they are diploid or polyploid. Even normal species may produce triploid individuals, however, by chance fusion of haploid and diploid gametes. Such individuals are invariably sterile (or apogamous, see below) because the odd number of chromosomes usually prevents a normal reduction division.

Hybridization is common in many groups of pteridophytes and plays an important role in the taxonomy of these plants. In a few cases, species may hybridize and the progeny retain the capacity for normal meiosis and spore formation. This is best illustrated in North America by two genera of club-mosses, *Diphasiastrum* and *Lycopodiella*, in which interspecific hybrids appear to be perfectly normal meiotically and produce well-formed, viable spores. With few exceptions, these club-moss hybrids are always in the minority in mixed populations, and their significance is poorly understood.

Modifications of the Life Cycle

From the life cycle described above there are many deviations. The evolution of heterospory extensively modified the sexual life cycle characteristic of homosporous pteridophytes. Heterosporic life cycles involve two strikingly different gametophytes, the microgametophyte (male) and the megagametophyte (female), both of which develop within the walls of their spores (endosporic development). The male gametophytes develop from small microspores, the female gametophytes from much larger megaspores, which are readily visible to the naked eye. They produce, respectively, the sperms and eggs. Fertilization takes place in water on the ground or in ponds. The heterosporous condition is known in pteridophytes as early as the Devonian and has evolved in a number of independent lines, including the ligulate lycopods (Isoëtaceae and Selaginellaceae), Marsileaceae, Salviniaceae, and Azollaceae. The heterosporous cycle may have been originally an adaptation to strongly seasonal or xeric habitats where moisture was available only for short periods during sporadic or vernal rains. Reproduction by this process is rapid because it is not

Osmunda regalis var. _spectabilis_. [From W. C. Taylor. 1984. Arkansas Ferns and Fern Allies. Milwaukee. P. 171, by P. W. Nelson. Reprinted with permission.]

necessary to grow a whole photosynthetic thallus, as in the homosporous cycle.

An asexual modification of the sexual homosporous life cycle is known as apogamy. The apogamous life cycle is common in leptosporangiate ferns and is especially widespread in the dry parts of western North America, among rock ferns like the cheilanthoids. Many well-known species, such as *Phegopteris connectilis, Pellaea atropurpurea, Pteris cretica,* and *Asplenium resiliens,* show this type of reproduction.

In the most common type of apogamy, there are two generations, but fertilization is not required for production of the sporophyte, which originates directly from the tissues of the gametophyte. The same chromosome number in sporophyte and gametophyte is maintained by doubling of chromosomes by endomitosis (nuclear division without cell division) in the formation of the spore mother cells. Therefore, only half as many spore mother cells develop, but each of these has double the number of chromosomes. These $4n$ spore mother cells undergo meiosis to produce normal-appearing $2n$ spores with the same chromosome number as the parent sporophyte. Generally, apogamy can be recognized because there are half as many spores as normal, and these are larger than for sexual members of the same genus. If the number of sexual spores is normally 64 per sporangium, the apogamous fern will have only 32.

Commonly an apogamous species will also have a sexual form, and the former may originate several times and in several different places from the latter (G. J. Gastony and L. D. Gottlieb 1985). Sexual species usually have an even number of genomes ($2x$, $4x$, $6x$) whereas apogamous ones have an odd number ($3x$, $5x$), but exceptions are frequent (e.g., *Pellaea glabella*).

Apogamous taxa cannot hybridize with each other, but they are often capable of hybridizing with sexual taxa because they can produce antheridia and viable male gametes, even if they cannot form normal archegonia and eggs. Apogamy is inherited by the hybrid, as in *Asplenium heteroresiliens* (*A. heterochroum* [sexual] × *A. resiliens* [apogamous]).

In some pteridophytes one or the other generation is partially or entirely eliminated. Vegetative asexual reproduction is accomplished usually by creeping rhizomes. This is most notable in sterile hybrid ferns, such as *Osmunda* ×*ruggii (O. claytoniana* × *regalis),* which are capable of forming enormous colonies. One can only estimate the age of such large clones because the old rhizome branches die and rot away; they may be well over a thousand years old.

Specialized buds are also produced on one or another of the vegetative organs. Root buds are well known in certain rock-inhabiting Florida aspleniums. Such buds are also found in most species of *Ophioglossum*. Stems may produce minute brood bodies (gemmae), such as those found around the subterranean shoot apices of *Psilotum*. These play a role in colonization, especially in greenhouses, where plants are scattered from pot to pot by soil-borne gemmae. In certain species of prairie-inhabiting moonworts *(Botrychium)*, masses of tiny spherical gemmae develop along the stem. These break off in the soil and become new plants.

Single leaf buds at the tips of long narrow fronds are familiar in a number of ferns, notably *Asplenium palmeri* and the walking fern *A. rhizophyllum*. Buds are also formed along the rachises of a number of dark-petioled spleenworts, such as *A. monanthes*. The common eastern American *A. platyneuron* produces a single bud in the center of the lowermost reduced pinna of large fronds, and a single plant is therefore able to produce clusters of numerous plants over a period of years.

In *Actinostachys pennula*, all new leaves are produced from the base of preexisting leaves, so that a tough, tangled mass of many dozens of petiole proliferations is produced. There is no main shoot apex at all; each leaf comes from a proliferation of a previous leaf. The most remarkable leaf proliferations in the North American fern flora are the highly specialized bulblets of *Cystopteris bulbifera*. Nearly spherical plantlets are abscised from the distal part of the rachis and pinna costae and fall or roll to positions where they germinate and form large colonies. The complex and somewhat samaralike gemmae of the typical firmosses *(Huperzia)* are notable for spreading widely not only the normal sexual phase of the species, but their sterile, misshapen-spored hybrids as well.

Perhaps the most surprising examples of vegetative propagation are those found in primarily tropical genera that have apparently spread and remained in the temperate zones by gametophytic gemmae. Members of three families—Hymenophyllaceae, Grammitidaceae, and Vittariaceae—occur in the Appalachian region as far north as New England, but they often appear only as gametophytes that may produce extensive colonies from the minute, one- or few-celled propagules (D. R. Farrar 1967, 1985). The gametophytes themselves are simple algalike filaments or ribbons, easily confused with green algae or liverworts. These gametophytes have few or no gametangia, and they are apparently adapted to growing only in shaded, humid, bryophytic habitats, mainly in rock crevices and grottoes. They normally produce no sporophytes at all and are the only representatives of their respective families in the floras of a number of states. In this flora we key them out alongside typical sporophytic members of their genera.

Habitats

The optimal habitat for most North American pteridophytes is rich, moist, mesic forest. Ferns and fern allies reach their greatest abundance and diversity in such habitats. Many genera are common, e.g., *Huperzia, Lycopodium, Equisetum, Osmunda, Dryopteris, Polystichum, Thelypteris,* and *Athyrium*. The plants average 3–10 times as tall as those of rocky exposed habitats, and many of them are capable of producing extensive clones. It is common for more than one species in a genus to cooccur in favorable habitats, and it is in such instances that we can often gain great insight into taxonomic relationships (W. H. Wagner Jr. and F. S. Wagner 1983).

Because of extensive farming in the past, as well as fires and lumbering, much of the continent is spotted with areas of disturbance that are in the process of returning to a climactic and stable state. Such successional vegetational formations provide habitats for species dependent on disturbance. Probably the most conspicuous pteridophyte of this type is *Pteridium aquilinum*, which is widespread in fields and second-growth open forests. It forms expansive clones that may produce the dominant vegetation in, for example, *Pinus banksiana–Populus tremuloides* associations in the upper Great Lakes region.

Among other second-growth specialists are many clubmosses *(Lycopodium* and *Diphasiastrum), Ophioglossum,* and *Botrychium. Asplenium platyneuron* may form large colonies on old farm sites and in previously logged areas. Another widespread successional plant in eastern North America is *Dennstaedtia punctilobula*, found commonly in pastures and old fields. Roadside borders and banks provide habitats for such successional species, as do cultivated areas. The pteridophytes that occur in these habitats are not necessarily weedy. Indeed, one of our rarest southeastern pteridophytes, *Botrychium lunarioides*, occurs almost exclusively in weedy and grassy roadsides and banks, lawns, and cemeteries. The last is the normal habitat for a number of our minute tropical adder's-tongues in the southeastern United States. A typical cemetery in Louisiana might yield as many as four species of *Ophioglossum* and two species of *Botrychium*.

The naturalist Edgar T. Wherry (1885–1982) stimulated much interest in the importance of soil and ecology in the biology of pteridophytes, and his influence

Ophioglossum crotalophoroides. **[From W. C. Taylor. 1984. Arkansas Ferns and Fern Allies. Milwaukee. P. 159, by P. W. Nelson. Reprinted with permission.]**

was profound. Wherry was first drawn to rock-inhabiting ferns because of his interest in geology and mineralogy. He pointed out that acidic rocks like granites, quartzites, and sandstones have their own suites of pteridophyte species (especially huperzias and certain aspleniums). Likewise, calcareous rocks like certain shales, limestones, and dolomites harbor sets of characteristic species, particularly other aspleniums, pellaeas, and notholaenas. It is interesting that ferns in calcareous habitats are commonly bluish or whitish in aspect. A few of the western North American rock ferns grow primarily on serpentine rocks (e.g., *Polystichum lemmonii* and *Aspidotis densa*).

Ferns of rocky habitats, such as cliffs, walls of sinkholes, boulders, and talus slopes, tend to be more or less xeromorphic and able to withstand considerable drying. They are usually small and tufted, and the rhizome is short, often appearing to be erect. The laminar tissue is mostly leathery. The apogamous life cycle occurs widely among rock ferns, and both sexual diploids and apogamous triploids may be found together on the same rocks. The major North American genera of rock-growing pteridophytes are *Huperzia*, *Selaginella* (*S. rupestris* group), *Polypodium, Cheilanthes, Argyrochosma, Notholaena, Pellaea, Polystichum, Asplenium, Gymnocarpium, Cystopteris,* and *Woodsia*. One species of *Dryopteris,* a normally soil-growing genus, is *D. fragrans,* which is restricted to various rock types in northern North America.

Wetland habitats for pteridophytes are varied and include such diverse types as lakes and ponds, fens, bogs, conifer swamps, and hardwood swamps. In the southern United States, *Azolla* and *Salvinia* are common in ponds and canals. In the South and Southwest, *Marsilea* and *Pilularia* are frequent along the banks of ponds and streams. Throughout North America various *Isoëtes* grow ephemerally in temporary ponds on prairies and rock outcrops or continually in permanent marshes, lakes, and slow-moving streams.

Marshes, fens, bogs, and swamps offer ideal sites for many genera. Open grassy marshes provide the habitat for the abundant *Thelypteris palustris* in eastern North America. Acid bogs in the Great Lakes area and along the eastern coastal plain are habitats for *Lycopodiella* and *Pseudolycopodiella,* which occur also in burned-over, damp, grassy and sandy areas, together with sundews *(Drosera)* and pitcher-plants *(Sarracenia)*. At various places from Delaware to Newfoundland, the

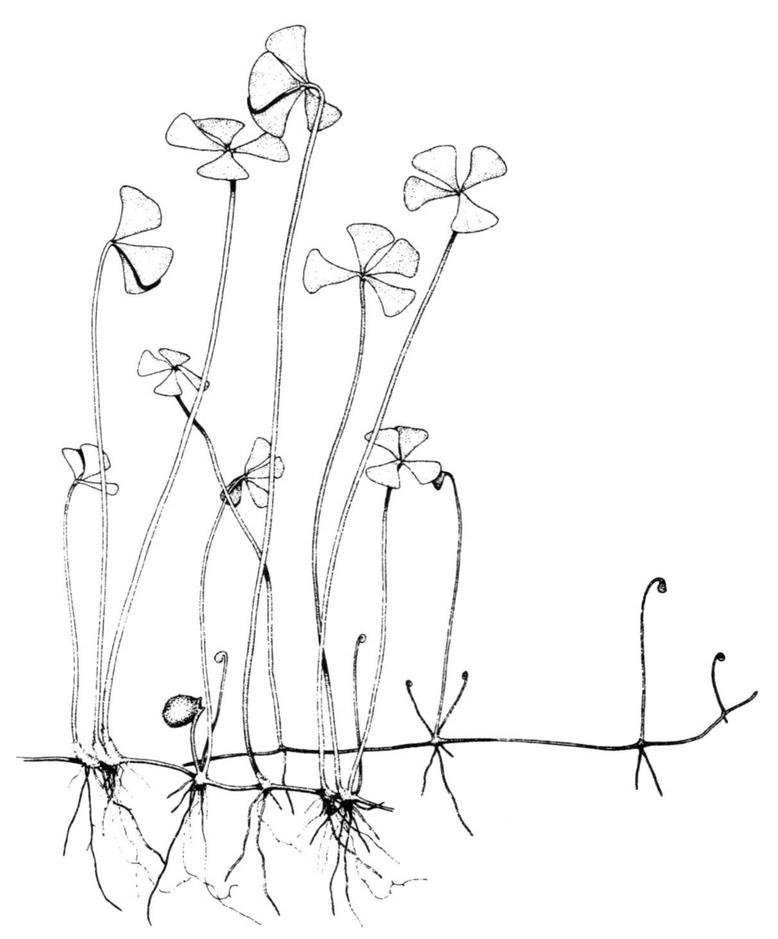

Marsilea vestita. **[From W. C. Taylor. 1984. Arkansas Ferns and Fern Allies. Milwaukee. P. 147, by P. W. Nelson. Reprinted with permission.]**

famous curly-grass fern *(Schizaea pusilla)* may be found in such acidic sites.

Deep hardwood and conifer swamps in eastern North America generally support species of *Equisetum, Osmunda, Dryopteris, Athyrium, Woodwardia,* and *Thelypteris.* The cooccurrence in swamps of black ash *(Fraxinus nigra),* red maple *(Acer rubrum),* and yellow birch *(Betula alleghaniensis),* as well as arbor vitae *(Thuja occidentalis),* is an excellent sign of potential pteridophytic diversity. Oddly, a number of normally wetland species may occasionally appear on rock cliffs, along ledges and near their bases. These include such unexpected occurrences on rock as *Osmunda regalis, Athyrium filix-femina, Cystopteris bulbifera, Woodwardia areolata,* and *Thelypteris simulata.*

Probably the best example of a truly xerophytic pteridophyte is *Selaginella densa,* which occurs as an inconspicuous, dull gray crust on the deserts and plains of an enormous section of the continent from Manitoba and British Columbia to California and Texas. In terms of total biomass, it may actually constitute our most abundant pteridophyte. In general, in desert regions pteridophytes grow in the protection of shaded fissures and ledges of rock cliffs and boulders. Compared to seed plants, pteridophytes are generally rare in xeric localities, probably a result of their need for water to achieve fertilization. As indicated above, some of these plants manage to exist in desert regions by remaining dormant, making appearances only in brief wet periods.

Epiphytic ferns and fern allies are sporadic, and there are only a few in the flora. The best known in western North America are *Polypodium scouleri* and *Selaginella oregana,* both occurring near the Pacific Coast. In the Southeast, *Pleopeltis polypodioides* forms huge masses on trunks, boughs, and in crotches of trees. Mosslike filmy ferns, especially *Trichomanes petersii,* produce carpets on damp trunks near the ground in shaded forests. In peninsular Florida, a number of epiphytes belonging to species common in the Antilles are

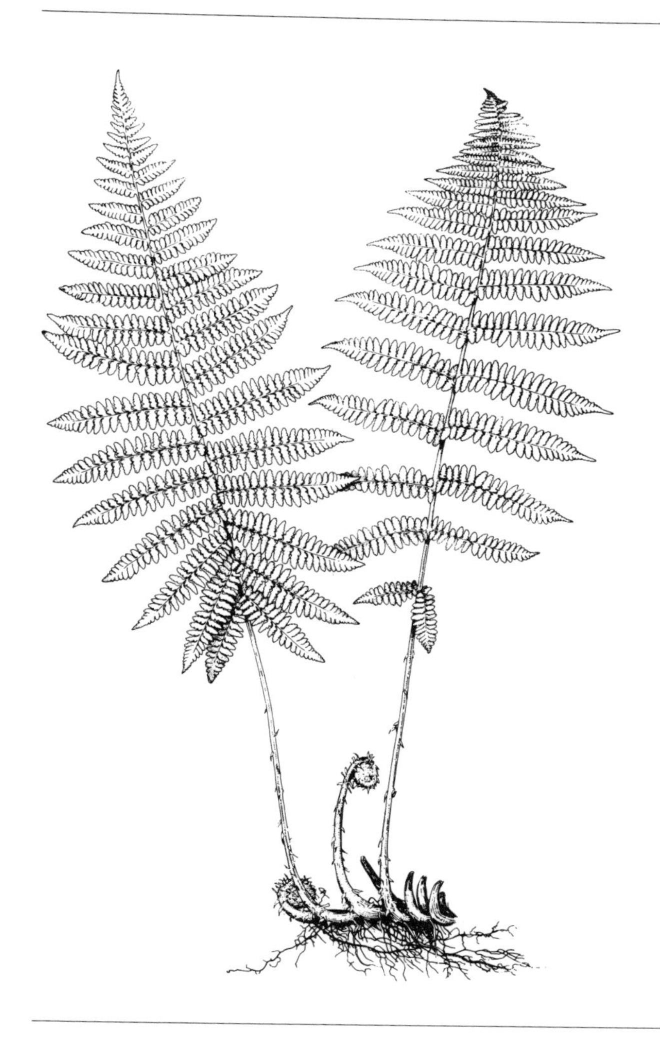

Athyrium filix-femina **subsp.** ***asplenioides.*** **[From W. C. Taylor. 1984. Arkansas Ferns and Fern Allies. Milwaukee. P. 57, by P. W. Nelson. Reprinted with permission.]**

found. The most unusual of these is *Cheiroglossa palmata,* which hangs from the leaf bases of palmettos *(Sabal).*

Some epiphytes occur also on rocks, as illustrated by the northern occurrences of *Trichomanes petersii* and *Pleopeltis polypodioides* on shaded rock faces and crevices. The reverse also occurs. Normally rock-inhabiting ferns such as *Polypodium virginianum* and *Asplenium rhizophyllum* may also be found occasionally as epiphytes in swamps. Both of the common terrestrial spinulose woodferns, *Dryopteris carthusiana* and *D. intermedia,* are occasionally found up to 4 m from the ground, growing in tree crotches. .

Geography

The pteridophyte flora of North America north of Mexico is much larger (441 species) than that of Europe (about 175 species). But compared with that of eastern Asia (with an estimated 2000 species in China alone), it is very much smaller, perhaps only one-fourth or one-fifth as large. Eastern Asia is much more diverse topographically and climatically than North America, and it probably has the richest temperate-subtropical fern flora in the world. Most of North America is temperate or arctic, and much of the western part is arid.

The mountain systems of North America are simpler than those of China. Wet tropical or subtropical mountains harbor the bulk of the pteridophytes of the globe, and such mountains are absent in North America. One mountain in Borneo, Mt. Kinabalu, may have more species of pteridophytes than all of North America. The tiny Central American country of Costa Rica, with an area slightly less than half the size of Virginia, has nearly 1000 species of pteridophytes, more than double that of the entire flora area. Even the few tiny, mountainous, tropical islands of Hawaii boast a diversity that is 40% of that of the entirety of North America north of Mexico.

To discuss pteridophyte distribution, the continent

may be subdivided into floristic provinces: Arctic and subarctic (Tundra and Northern Conifer Forest of H. A. Gleason and A. Cronquist 1964, as given in D. B. Lellinger 1985, fig. 3); eastern North America (Eastern Deciduous Forest, extended to include Newfoundland as a continuum); Coastal Plain (same as Gleason and Cronquist); Peninsular Florida (West Indian Province); Great Plains and Prairie (Grassland); Rocky Mountains–Cascade Mountains–Sierra Nevada (Cordilleran Forest); West Coast (Californian, plus the coastal area north to Alaska); and Sonoran (western Texas to Mojave Desert).

Most of the arctic species are circumboreal, and the number of species and their endemism is low (R. M. Tryon 1969). Eastern North America is very rich in species, especially in the Appalachian Mountain system. Some of these species, or at least closely related sister species, are found in Japan and China (M. Kato and K. Iwatsuki 1983; K. H. Shing 1988). Peninsular Florida is a very distinctive area of the continent floristically, but its pteridophytes are mostly at the edge of the range of widespread tropical American species (R. M. Tryon 1969). The Great Plains and Prairie have local areas of interest, usually small isolated elevated regions like the Cypress Hills (Alberta-Saskatchewan) and Black Hills (South Dakota), but otherwise the grasslands mostly are relatively uninteresting pteridologically.

In the Rocky Mountains, Cascade Range, and Sierra Nevada, pteridophytes are often rare and localized, and they occur at progressively higher elevations southward. The same species that occur nearly at sea level in northern Canada and Alaska may turn up at 3500 m at the southern ends of their ranges. Isolated mountain ranges in southern California and Arizona contain many primarily Mexican taxa. The West Coast region lies on the western slopes of the major ranges from California to Alaska and is rich in pteridophytes. Species seem to form overlapping south-north ranges, one species extending so far and overlapping another in the same genus. The diversity is greatest in California and decreases in Canada and Alaska where subarctic and arctic floras abut. Some of the most luxuriant pteridophyte growth in North America occurs as spectacular assemblages near the Pacific Coast, including prominent species of *Polystichum, Dryopteris,* and *Athyrium,* but the number of species is relatively low. It is sometimes difficult to draw a sharp line between the West Coast and the Cordilleran phytogeographic regions.

The demarcation of well-defined pteridophyte floristic areas is confounded by the presence of numerous disjunctions. Disjunctions of several hundred kilometers are common, and some disjunctions are much greater (W. H. Wagner Jr. 1972). A species that is common and dominant in one area may reappear as a highly localized outlier in another. A number of species have their centers in the west or southwest but have sporadic and rare disjunctions in the east; examples are *Pellaea wrightiana* (disjunct in North Carolina) and *Astrolepis sinuata* (disjunct in Georgia). *Asplenium septentrionale* is centered in the Rocky Mountains with outlying populations in West Virginia.

A spectacular series of west-east disjunctions is seen in the scattered isolated populations in the region from Lake Superior to northeastern Canada, including such predominantly western taxa as *Botrychium hesperium, Aspidotis densa, Cryptogramma acrostichoides, Polystichum scopulinum,* and *Thelypteris quelpaertensis.* East to west disjunctions include such extreme examples as *Cystopteris bulbifera* and *Asplenium platyneuron,* both very widespread and frequent species in the east that reappear as isolated populations in Arizona.

Disjunctions also occur north to south in both the eastern and western mountains, where low elevation northern species climb higher and higher southward and end up very widely scattered on the topmost peaks. Since 1940, for example, plants of the three well-known northeastern moonworts, *Botrychium lanceolatum* subsp. *angustisegmentum, B. matricariifolium,* and *B. simplex,* have been reported farther south from Pennsylvania (where the species are frequent to common) down to southern North Carolina along the Blue Ridge Mountains (where the species are rare and localized at elevations above 1200 m). In the central Rockies, *B. hesperium* is known from numerous localities, but farther south, in Arizona, records for it exist only from Mount Baldy, at a very high elevation.

Although far fewer than in seed plants, introductions of exotic species of pteridophytes by human commerce have taken place. Naturalized alien taxa are found mainly in southeastern North America, but there are fewer than 30 species altogether. A few species, such as *Marsilea quadrifolia,* have spread in temperate areas of the East. Among the most prominent of the southern introductions are *Thelypteris dentata,* which is particularly common in and around greenhouses, and *Macrothelypteris torresiana,* now widespread in the Southeast. *Cyrtomium falcatum* is apparently spreading slowly in southern California and in the Gulf states. Of special interest is *Pteris vittata,* presumably from eastern Asia, which has established itself in Florida, where it has encountered the native analog, *P. bahamensis,* and crossed with it to form colonies of *P.* × *delchampsii,* a vigorous nothospecies, especially in the vicinity of Miami.

Between 20% and 25% of all of the reproductively competent, native pteridophyte species (including all sexual, apogamous, and gemmiferous taxa) are endemic to North America north of Mexico. The province of highest endemism is also the region of the greatest overall diversity, namely the Eastern Deciduous Forest. It contains approximately 50 endemics, twice the number of the West Coast Province and five times that of the southeastern coastal plain. As rich as the Florida hammocks are for ferns, most of the species there are nonendemic, and they are merely outliers of the Antillean flora.

Classification—The Higher Ranks

Classification and phylogeny of the approximately 10,000 species of pteridophytes are controversial because many of the differences are profound, and evolutionary lines have apparently been separate since the time of the earliest land vegetation. The problems of classification involve such questions as how much evolution has occurred, how extensive are the gaps in today's patterns of diversity, and where do connections occur. As to the last, intensive search has failed to reveal convincing connections between the lycopods, the horsetails, the whisk-ferns, and the true ferns. Highly speculative extrapolations have been made, but they are tenuous. Each of these assemblages possesses an ensemble of very distinctive characteristics. As a consequence, these groups have been accorded divisional status by many, including A. Cronquist et al. (1966; see table 12.1), whose classification we adopt. As a comparison, the approximately 250,000 species of flowering plants occupy a single division, Magnoliophyta.

The most evolutionarily advanced pteridophyte taxa are the most difficult to fix in the system because so many characters are involved, and apparently multiple parallelisms and convergences exist. During the past half century more attention has been given to the familial rank. The traditional family Polypodiaceae, which historically encompassed most of the leptosporangiate ferns, has been subdivided into many widely accepted families.

Among the genera that are still widely questioned as to family affinities is *Ceratopteris*. Is it a member of the Pteridaceae, perhaps a subfamily of it, or does it constitute a separate monotypic family, Parkeriaceae (as treated in this flora)? Some taxonomists (e.g., R. E. G. Pichi-Sermolli 1977) also segregate other genera in the flora of North America, e.g., *Botrychium* (Botrychiaceae), *Cryptogramma* (Cryptogrammaceae), *Woodsia* (Woodsiaceae), and *Lomariopsis* (Lomariopsidaceae) (R. E. G. Pichi-Sermolli 1977). Even *Adiantum*, treated here along with numerous other genera of cheilanthoid affinities, is recognized as being rather isolated and is sometimes regarded as the sole genus in Adiantaceae.

Much of the disagreement about family definition involves the group that is most numerous in the world

TABLE 12.1. Higher Classification (Divisions and Families) of Pteridophytes

Division	*Family*	*No. Genera in FNA*	*No. Species in FNA*
Psilotophyta	Psilotaceae	1	1
Lycopodiophyta	Lycopodiaceae	7	27
	Selaginellaceae	1	34 (39)
	Isoëtaceae	1	24
Equisetophyta	Equisetaceae	1	10 (11)
Polypodiophyta	Ophioglossaceae	3	37 (1)
	Osmundaceae	1	3
	Gleicheniaceae	1	1
	Schizaeaceae	2	2
	Lygodiaceae	1	1 (3)
	Anemiaceae	1	3
	Parkeriaceae	1	2 (3)
	Pteridaceae	13	86 (91)
	Vittariaceae	1	3
	Hymenophyllaceae	2	11
	Dennstaedtiaceae	4	6
	Thelypteridaceae	2 (3)	22 (24)
	Blechnaceae	2	6
	Aspleniaceae	1	31
	Dryopteridaceae	16 (18)	74 (80)
	Grammitidaceae	1	1
	Polypodiaceae	7	25
	Marsileaceae	2 (3)	6 (7–8)
	Salviniaceae	1	1
	Azollaceae	1	3

Numbers for genera and species are for native taxa; numbers in parentheses reflect the totals including naturalized taxa, if any. Divisions are according to A. Cronquist et al. 1966; families and their sequence are as in this flora.

today, the Dryopteridaceae sensu lato, ferns that have basically reniform or peltate indusia and well-developed perispores. Especially controversial are the *Thelypteris* group and the *Asplenium* group, both of which have been raised to family status on the basis of petiole anatomy, trichomes, chromosome number, and a general "coherence" in each, as well as an absence of genera linking them to other families. If we compare, on the basis of the actual number of differences, the elements formerly placed in the family Schizaeaceae sensu lato, namely the *Anemia, Lygodium,* and *Schizaea* groups, we find that the differences among them are so numerous and striking that each should constitute a different family. The same is true also of the *Huperzia, Lycopodium,* and *Lycopodiella* groups of the traditional Lycopodiaceae.

Table 12.1 represents our classification of the pteridophytes, an attempt to bring together somewhat divergent classifications into a single, relatively consistent treatment. For other viewpoints, the reader should consult J. A. Crabbe et al. (1975), K. U. Kramer and P. S. Green (1990), R. E. G. Pichi-Sermolli (1977), and R. M. Tryon and A. F. Tryon (1982).

Classification—Genus to Variety

Historically, many fern genera have been established on the basis of characteristics (e.g., dimorphism, venation, and blade dissection) that we now regard as insufficient. A good example is in the spleenworts, where *Phyllitis, Camptosorus,* and *Ceterach* have traditionally been separated from *Asplenium* but are now reunited with it. It has been demonstrated that these segregate "genera" are very similar to *Asplenium* anatomically and cytologically, and numerous hybrids among them have been found.

Other genera have been subdivided as we learn more about their morphology and tropical relatives. Such an example involves *Thelypteris,* in the past merged with *Dryopteris.* In this case, anatomy and cytology, plus the total absence of hybrids, have led to recognition of separate genera. They differ in so many characters that most workers place them in separate families, as we do here.

In *Flora of North America,* we have attempted to maintain a more or less equal level of comparability among the genera. Some of the changes we have accepted (e.g., the generic separation of elements usually included within *Athyrium* s.l.) are still relatively novel, and in the future they may prove, at least in part, to be untenable. There have been numerous attempts to segregate genera in the past that have had to be reversed. *Asplenium* and its segregate genera, discussed above, is a good example, as is *Woodwardia,* from which *Anchistea* and *Lorinseria* once were separated. The family Thelypteridaceae has been variously treated in floras and monographs to comprise a single genus or 11 genera in North America. We have opted for an intermediate course, recognizing three genera in the flora. In some groups, e.g., the adiantoid and the dryopteroid ferns in particular, we have possibly accepted too many segregates, while in others, e.g., selaginellas and botrychiums, we have, perhaps, accepted too few. Only further research, plus quantitative evaluations of comparability, can be expected to resolve inconsistencies that may exist.

The most profound change in the taxonomy of North American pteridophytes involved the species concept and took place during the past 30 years, inspired in large part by the use of new evolutionarily informative characters and more objective approaches to taxonomy. This change can best be illustrated for eastern North America by comparing the treatments in the most recent edition of *Gray's Manual of Botany* (M. L. Fernald 1950) with the treatments given here. The former may be referred to as the "variety/form" school because of the emphasis on those ranks, and the latter the "species/hybrid" school. The former founded species primarily on relative similarity of morphology; the latter likewise employs morphology but also stresses anatomy, cytology (especially chromosomes), spore viability, breeding characteristics, and syntopic cooccurrence. The varietal category as used prior to 1950 included a wide conglomeration of elements that we now refer to, in our recent taxonomy, as either distinct species, hybrids, trivial forms, or geographical subspecies (W. H. Wagner Jr. 1960).

The category of variety was also used by authors of the variety/form school when they were in doubt as to whether the entity was a distinct species or not; its use tended to be a "trial balloon" without necessarily committing the author to recognizing the entity as a species. *Dryopteris intermedia* was made a variety of *D. carthusiana* (as *D. spinulosa*), to which it bore a certain resemblance. *Diphasiastrum digitatum* (as *Lycopodium*) was made a "variety" of *D. complanatum.* The sterile hybrid, *Equisetum* ×*ferrissii (E. laevigatum* × *hyemale)* was construed as a "variety" of one of the parents, *E. hyemale.* The fertile hybrid, *Dryopteris clintoniana (D. cristata* × *goldiana)* was treated as a "variety" of *D. cristata. Botrychium oneidense* was made a "form" of *B. dissectum,* and *B. minganense* a "form" of *B. lunaria.* Numerous other examples could be cited.

The *Gray's Manual* treatment in 1950 included 120 named species, 159 additional varieties and forms, and 12 named and recognized hybrids. Forty years later, those ratios have been drastically changed, as seen in the treatment herein. The number of species in eastern North America has been increased by approximately 40%, the number of varieties and forms has been reduced by 90%, and the number of named hybrids has been increased 400%, obviously a dramatic change in the systematics of pteridophytes. Pioneers in the presentation of the current taxonomy have been J. T. Mickel (1979) and D. B. Lellinger (1985), who brought together much of the modern monographic literature in their manuals of North American pteridophytes. More recently, W. J. Cody and D. M. Britton (1989) have provided a similar manual for Canadian pteridophytes. All of the problems have not been solved, however. The relationships within the *Athyrium filix-femina* complex are still not fully understood. Whether the mainly tropical *Pteridium aquilinum* var. *caudatum* is conspecific with the temperate northern hemisphere var. *aquilinum* is still debated.

It is especially gratifying to note that recent floristic accounts have, for the most part, been abundantly supported by new comparative characters such as chromosomes, spore wall structure, phenolic compounds, isozymes, and especially, closer observations of the behavior of populations in the wild. The role of hybridity in forming nothospecies (hybrid species) and secondary species is now better understood. Geographical segregates of species are treated as subspecies if the differences are numerous and as varieties if the differences are few, although there is, as would be expected, some arbitrariness in this regard.

Some authors occasionally use one or the other of these infraspecific categories to accommodate different cytological forms, such as diploid and tetraploid populations that differ hardly or not at all morphologically. These "subspecies" or "varieties" may coexist syntopically without successful interbreeding (e.g., *Asplenium trichomanes* subsp. *trichomanes*, $2n = 72$, and subsp. *quadrivalens*, $2n = 144$). The category of form is usually applied to variants that differ in a single character (e.g., glandular form of *Dryopteris expansa*, dwarf form of *Botrychium multifidum*, forked form of *Pellaea atropurpurea*, or sterile-fertile frond intermediates of *Onoclea sensibilis*). It is now more clearly understood that there are few scientific reasons for recognizing such plants taxonomically. Horticultural or other reasons may exist for doing so.

To summarize the changes in classification at the generic rank and below, we provide the following figures. Approximately 75 species in the flora have undergone a name change as compared with the most recent comprehensive treatment of Canadian and United States pteridophytes by D. B. Lellinger (1985). This is, in part, a reflection of our changing generic concepts in ferns and allies, particularly in the lycopodioid and cheilanthoid assemblages. In addition, 58 taxa have been added to the flora since Lellinger's work. Of the additions, 29 are newly described taxa, mostly species. The total number of these recent changes approaches 30% of the pteridophyte flora!

The Flora of North America project itself has provided the impetus for many of these changes. Projects already underway when FNA began were accelerated so that the work could be incorporated into the flora. Recent studies have focused on taxonomically difficult genera in which allopolyploid and cryptic speciation (C. A. Paris et al. 1989) have occurred, e.g., *Isoëtes*, *Botrychium*, *Pellaea*, *Woodsia*, and *Polypodium*. To resolve problems in these genera and others, taxonomists have used cytotaxonomic and electrophoretic methods as well as more rigorous morphometric comparisons.

We view the changes described herein as a natural progression and growth in our understanding of complex relationships, changes made possible by new techniques and a more consistent application of older methods. As a result of these new studies, we believe that systematic pteridology is now a more viable and exciting area of endeavor than at any time in its history.

Interspecific Hybrids

There are approximately 100 species of hybrid origin in the flora, over 20% of the total. The role of interspecific hybridization in the diversity of North American ferns was grossly underestimated in the first half of the twentieth century. An early pioneer in documenting hybridization was Margaret Slosson (1902), who resolved the origin of *Asplenium ebenoides* by bringing together gametophytes of *A. platyneuron* and *A. rhizophyllum*, work later confirmed by K. S. Walter et al. (1982). Subsequently, R. C. Benedict, who was interested in fern genetics, especially in *Nephrolepis*, and E. T. Wherry, who was interested in fern ecology, especially in *Asplenium*, directed professional and amateur attention to reticulate fern evolution, culminating in major reinterpretations of a number of genera. The advent of chromosome studies (I. Manton 1950) refined the experimental methods, and isozyme analysis has substantiated hybrid origins in numerous plants. It is now widely accepted that the study of nothospeciation, i.e., the origin of taxa by hybridization, is an essential pursuit in pteridology.

Only closely related species can be expected to hybridize, normally only members of the same genus or subgenus. In North America, hybridization is frequent between members of *Equisetum* subg. *Hippochaete*, *Huperzia*, *Diphasiastrum*, *Lycopodiella*, *Botrychium* subg. *Botrychium*, *Polypodium*, *Cheilanthes*, *Pellaea*, *Dryopteris*, *Polystichum*, and *Asplenium*, but it occurs in many other genera as well. Where extensive interspecific networks have been generated by hybridization in nature, we give diagrams summarizing the relationships in the generic treatments.

Some hybrid combinations are extremely rare and represented by only solitary or few individuals in a given locality. Examples are *Dryopteris carthusiana* × *goldiana* and *Asplenium platyneuron* × *ruta-muraria* (= *A.* ×*morganii*). Other hybrids are common, e.g., *Di-*

phasiastrum sitchense × *tristachyum* (= *D.* ×*sabinifolium*), *Equisetum hyemale* × *laevigatum* (= *E.* ×*ferrissii*), and *Dryopteris carthusiana* × *intermedia* (= *D.* ×*triploidea*). Common and/or conspicuous hybrids are often designated with a name in binomial form (rather than by the hybrid formula name).

Both sterile and fertile hybrids may display vigorous growth (and make excellent garden or greenhouse plants). Fertile hybrids produce normal spores and pass through a normal alternation of generations. They are capable of building up large populations either by vegetative or sexual means as discussed previously. If one wishes to distinguish sterile and fertile hybrids with the same parentage, the following convention (proposed by C. R. Werth and W. H. Wagner Jr. 1990 as an amendment of the *International Code of Botanical Nomenclature*) can be used. Hybrids that are sterile (sexually incompetent) are indicated by the × sign (*Asplenium* ×*bradleyi,* 2*x* form); if fertile, the times sign is placed in square brackets (*A.* [×]*bradleyi,* 4*x* sexual; *A.* [×]*heteroresiliens,* 5*x,* apogamous). Many hybrids, such as *Dryopteris goldiana* × *ludoviciana* (= *D.* [×]*celsa* or *Asplenium montanum* × *rhizophyllum* (= *A.* [×]*pinnatifidum*), were designated as normal species long before their hybrid origin was suspected.

Hybrid species (nothospecies) can be placed in two categories, depending on their reproductive system. Most often, homoploid hybrids (with the same chromosome number as the putative parents) are sterile. The malformed spores can be readily observed, even from herbarium sheets (W. H. Wagner Jr. et al. 1986). The spores are variable in size and shape, many of them obviously collapsed or otherwise distorted. Usually only a few out of thousands of such spores will germinate in culture.

Homoploid fertile hybrids are rare, and they are known in North America only in two genera of Lycopodiaceae, *Diphasiastrum* and *Lycopodiella.* In these hybrids, the chromosome behavior at meiosis appears to be entirely normal, as are the resultant spores. Why these homoploid fertile hybrids do not simply swamp the parental species is unknown; the intermediates are distinctive, as a rule, and they are readily separated from their parents. Backcrossing has not been observed, and thus introgression has not been demonstrated, even though many of the hybrids, especially certain ones in *Diphasiastrum,* are frequent or even common.

In some predominantly sterile hybrids, large unreduced spores (with double the chromosome complement) are formed; these are theoretically capable of germinating to produce gametophytes. It is hypothesized that such gametophytes may be able to generate new sporophytes directly by apogamy (V. M. Morzenti 1962). This may explain why such "sterile" hybrids as *Asplenium* ×*trudellii (A. montanum* × *pinnatifidum)* and *Dryopteris* ×*boottii (D. cristata* × *intermedia)* are so common in nature.

The study of fern hybrids raises an unanswered question: Why have some sterile hybrids undergone chromosome doubling to restore fertility (allopolyploidy) while others have not? A number of sterile hybrids have never been known to double, even when they are very common, as are the numerous hybrids involving *Dryopteris marginalis.* Why has *Asplenium* ×*ebenoides (A. platyneuron* × *rhizophyllum)* doubled only at one locality in Alabama, whereas *A.* [×]*bradleyi (A. montanum* × *platyneuron*) has practically always doubled? We still don't know.

13. Gymnosperms

James E. Eckenwalder

The classification of gymnosperms, like that of the flowering plants, is based primarily on structures of the sporophyte associated with reproduction because the gametophyte phases of seed plants are wholly included within sporophytic reproductive structures. The major groups of living gymnosperms, recognized by different taxonomists at ranks ranging from division through class and order to family, are distinguished by fundamental differences in the organization of their reproductive structures. Plants of three of the four extant top-rank gymnosperm taxa, here treated as divisions Cycadophyta, Coniferophyta, and Gnetophyta, grow naturally in North America. The fourth division, Ginkgophyta, was worldwide in the Mesozoic and Tertiary but now has a single extant species, *Ginkgo biloba,* the maidenhair tree. It is cultivated in North America and other temperate areas, but it is apparently no longer found in natural stands (S. Y. Hsu 1980).

The indigenous North American orders all possess pollen cones separate from seed cones but differ in whether the cones of one or both sexes are simple or compound. A simple cone consists of spore leaves inserted directly on a main axis. Both the pollen cones and the seed cones of the Cycadophyta, but only the pollen cones of the Coniferophyta, are simple. Compound cones have bracts along the main axis that subtend fertile dwarf shoots. In the Gnetophyta, both the pollen and the seed cones are compound. The seed cones of the Coniferophyta are also compound, but their fertile dwarf shoots have been transformed into seed-bearing cone scales that deceptively resemble spore leaves. The work of the Swedish paleobotanist R. Florin (1951) that elucidates the origin of the conifer cone scale is one of the great detective stories in plant morphology.

Florin also argued that the solitary terminal seeds of the Taxaceae had a separate evolutionary history from the compound seed cones of the conifers, and he advocated excluding Taxaceae from the conifers. Recent research weakens some of the force of his arguments (R. A. Price 1990), and some authors see the terminal ovule as the result of reduction, a process analogous to that operating within the undoubtedly coniferous genus *Juniperus* (cf. P. B. Tomlinson et al. 1989).

Taxonomic controversies are common at all ranks in gymnosperms, and thus the classification adopted here, a modification of the standard classification of R. K. F. Pilger (1926), is far from universally accepted. Many past disputes have been finally laid to rest, but new ones continue to emerge. Disagreements include the assignment of ranks for accepted taxa, as with the rank of the taxa here treated as divisions, and extend to substantial differences in circumscription of taxa, as with inclusion or exclusion of Taxaceae in the conifers.

Taxonomic disagreements concerning familial and generic relationships in Cycodophyta and Gnetophyta hardly affect our view of North American taxa because we have only one genus of each, *Zamia* and *Ephedra.* Although controversies surrounded the separation of these genera from Cycadaceae and Gnetaceae, respec-

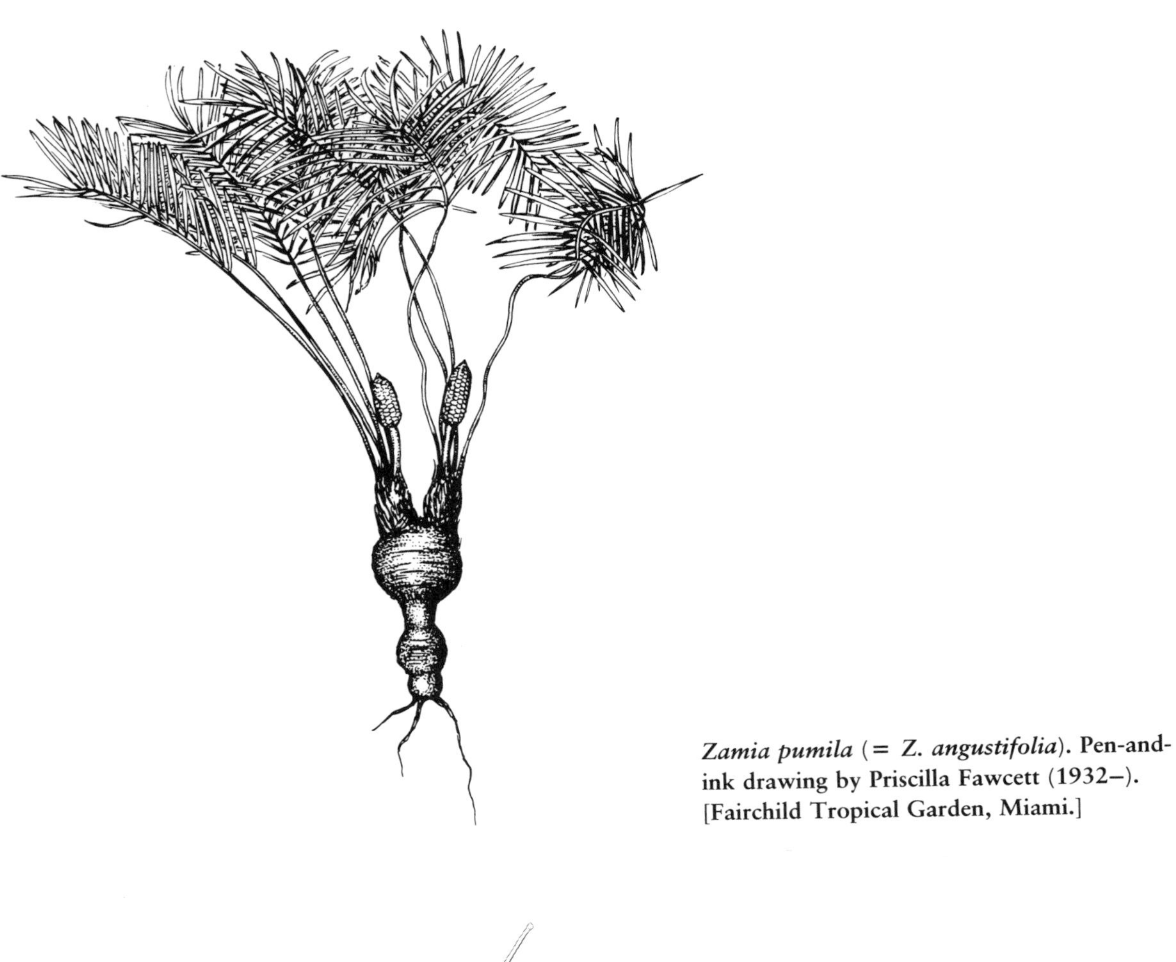

Zamia pumila (= *Z. angustifolia*). Pen-and-ink drawing by Priscilla Fawcett (1932–). [Fairchild Tropical Garden, Miami.]

Ephedra antisyphilitica. [From S. Watson. 1871. United States Geological Exploration of the Fortieth Parallel.... (Vol. 5) Botany. Washington. Plate 39, by J. H. Emerton.]

Thuja plicata. [From W. L. Jepson. 1909. The Trees of California. San Francisco. Fig. 69, by Mary H. Swift.]

tively, most gymnosperm taxonomists now seem comfortable with including them in the segregate families Zamiaceae (Cycadophyta) and Ephedraceae (Gnetophyta). Even so, disputes continue over the number of species in each genus, more so with respect to *Zamia* (J. E. Eckenwalder 1980b) than *Ephedra* (L. D. Benson 1943), although some taxa of the latter have been treated either as species or varieties. Here, we favor a broad specific circumscription.

Disagreements exist on the circumscriptions of families, genera, and species in the Coniferophyta, in addition to the issue of segregation of Taxaceae. Both Pinaceae and Taxaceae are firmly established families in the literature, but the Cupressaceae, as used here, include genera *(Sequoia, Sequoiadendron,* and *Taxodium)* that were treated in the segregate family Taxodiaceae in almost all previous North American floras. In contrast to the other widely accepted conifer families, which differ in both reproductive and vegetative traits, Cupressaceae and Taxodiaceae overlapped in all traits except phyllotaxis. The discovery of the Chinese *Metasequoia,* an apparent close relative of *Sequoia* with decussate phyllotaxis, which is characteristic of most cupressoids, eliminated even that difference. A suggestion that these two families be merged under the name Cupressaceae (J. E. Eckenwalder 1976) has gradually gained acceptance (J. A. Hart and R. A. Price 1990) as new lines of evidence (e.g., the immunological comparisons made by R. A. Price and J. M. Lowenstein 1989) and new kinds of analyses (e.g., the cladistic analysis by J. A. Hart 1987) seem to confirm the artificiality of the traditional familial separation.

The arrangement of genera within the combined family is poorly established, and neither the scheme by R. K. F. Pilger (1926) nor that by Li H. L. (1953) is satisfactory. The arrangement adopted here is a compromise for North American genera, pending a thorough analysis of the family worldwide. The taxodioid genera are placed first because they display the greatest proportion of presumably primitive characteristics among the North American genera. In contrast to Pilger's scheme, ours segregates *Sequoiadendron* from *Sequoia,* as is now almost universally accepted following J. T. Buchholz (1939). *Cupressus* most closely resembles the three taxodioid genera among the cupressoids, *Chamaecyparis* is close to *Cupressus,* and *Thuja* and *Calocedrus* seem

Pinus elliottii. **[From G. Engelmann. 1880. Revision of the genus *Pinus,* and description of *Pinus Elliottii.* Trans. Acad. Sci. St. Louis 4: 161–190. Plate 1, by Paulus Roetter.]**

to display increasing asymmetry of the cones, while sharing the flattened branchlet sprays of *Chamaecyparis.*

Following Li H. L. (1953), as corrected by R. Florin (1956), we segregate *Calocedrus* from *Libocedrus* (which is thus restricted to Southern Hemisphere species), and this too is generally unquestioned today. *Juniperus* appears to be an independent derivative of a *Cupressus*-like ancestor, with cones specialized for animal dispersal. The naturalized Australian genus *Callitris* stands apart from the native North American cupressoids, and it is placed last here because of its apparent divergence from *Sequoia, Sequoiadendron,* and *Taxodium,* whereas Pilger made it first in his sequence of Cupressaceae in the strict sense.

At the species level, most genera of Cupressaceae are relatively unproblematic, each having 1–3 North American species, but species concepts in *Taxodium* and *Cupressus* are particularly controversial, and here we are conservative in both cases. Furthermore, we do not accept some recently proposed, but not widely accepted, segregate species in *Chamaecyparis* and *Juniperus.*

Generic arrangements in the Pinaceae are also debated. There are two major schemes, those by P. van Tieghem (1891) and by F. Vierhapper (1910). Recent evidence (cf. R. A. Price et al. 1987, on immunological distances; M. P. Frankis 1988, on seed characters) has favored van Tieghem's arrangement (R. A. Price 1989), and his view is adopted here. In this arrangement, *Abies* and *Tsuga* are the sole North American representatives of subfamily Abietoideae. A recent suggestion by C. N. Page (1988) to take up the genus *Hesperopeuce* for *Tsuga mertensiana* is not adopted here, pending further

Taxus floridana. [From C. S. Sargent. 1890. The Silva of North America .... 14 vols. Boston and New York. (Facsim. ed. 1947.) Vol. 10, plate 515, by C. E. Faxon.]

evidence and discussion. *Pseudotsuga,* with a north Pacific distribution, and *Larix, Picea,* and *Pinus,* all circumboreal, are the North American genera of the Pinoideae. *Pseudotsuga* and *Larix* are closely linked by aspects of pollination biology (J. Doyle and M. O'Leary 1935), among other features. *Picea* and *Pinus* are relatively isolated from the former and from each other.

Species delimitation in Pinaceae has been relatively stable, perhaps because the economic importance of these plants led to early exploitation and taxonomic study throughout the continent. There have been two major exceptions to this near universal agreement. In a treatment rejected by every subsequent North American author, F. Flous (1937) recognized 12 species of Douglas-fir where others see only *Pseudotsuga menziesii.* More recently D. K. Bailey (1970, 1987; Bailey and F. G. Hawksworth 1979) has proposed several revisions in the taxonomy of southwestern *Pinus,* only some of which are adopted here, the others not yet widely accepted. Although species delimitations within the North American Pinaceae have remained largely uncontroversial, there have been numerous nomenclatural disputes, most of which have been settled by the research leading to the U.S. Forest Service's most recent checklist (E. L. Little Jr. 1979).

Taxaceae are so poorly represented in North America, and the species so distantly related to one another, that no controversy has arisen regarding the circumscriptions of the five taxa recognized by R. K. F. Pilger (1903). We follow most contemporary authors in treating our three representatives of *Taxus* as endemic species rather than subspecies of a circumboreal *T. baccata.* The two endemic species of the presumably more primitive *Torreya* were also so recognized by Pilger.

14. A Commentary on the General System of Classification of Flowering Plants

Arthur Cronquist *(deceased)*

Introduction

For the past century or more, major schemes of classification of flowering plants have attempted to portray the ensemble of similarities and differences among plants in an evolutionary context. Authors have implicitly or explicitly considered their schemes to be phylogenetic, and indeed C. E. Bessey (1915) labeled his epochal treatment *The Phylogenetic Taxonomy of Flowering Plants*. Phylogenetic in this traditional sense means simply that the scheme is compatible with presumed evolutionary relationships.

The much narrower and more rigid concept of "phylogenetic" recently promoted by cladistic theorists requires that any proper taxonomic group must include all the descendants of the nearest common ancestor of the group. A. Cronquist (1987) and others have discussed some of the problems of applying this concept. No general system of classification of flowering plants has yet been produced on cladistic principles.

The well-known and widely used system of Adolf Engler (H. G. A. Engler and K. Prantl 1887–1915) was considered by its author(s) to be phylogenetic at least in a general sense. That view, although defensible at the time, is no longer tenable. The system and its concepts do not meet the test of providing for all the evidence. The great weakness of the Englerian system, from a current point of view, is that it does not distinguish adequately between primitive simplicity and simplicity by reduction. Inasmuch as most students of the subject now agree that floral reduction has been a pervasive (though not exclusive) trend within the angiosperms, the system must be extensively recast.

The search for general agreement and a set of principles that will permit everything to fall into place has led taxonomists to revive, modify, and expand the concepts of floral evolution first presented in embryonic form by A. P. de Candolle in 1813. The system of G. Bentham and J. D. Hooker in their *Genera Plantarum* (1862–1883) was a lineal descendant of that of de Candolle. Although post-Darwinian in publication, it was pre-Darwinian in concept, and its authors never claimed anything else. It was, however, an important historical link in the progression from de Candolle's "natural" system to the avowedly phylogenetic system of C. E. Bessey. The oft-noted insertion of the gymnosperms between the monocotyledons and dicotyledons in the Bentham and Hooker system is a curious anachronism that does not seriously affect their treatment of the angiosperms.

This commentary draws heavily on my 1981 and 1988 books, cited in the bibliography. The New York Botanical Garden, which holds the copyright, has permitted some duplication of phraseology. The phylogenetic diagrams, originally published in the 1981 book, are specifically released from copyright.

Charles Edwin Bessey, 1845–1915. [Courtesy of the Special Collections, Library of The New York Botanical Garden, Bronx, New York.]

The now widely accepted strobilar hypothesis of floral evolution dates from C. E. Bessey (1897), who saw "the reproductive strobilus in the form of a flower, in which the sterile leaves are well set off from those which bear the spores." Bessey's interpretation was a particular application and modification of F. O. Bower's (1894) theory of the strobilus in archegoniate plants. This interpretation is perfectly compatible in principle with A. P. de Candolle's pre-Darwinian views, with the addition that the relationships among the taxa are considered to be evolutionary rather than purely conceptual.

Authors since C. E. Bessey's time, including among others J. Hutchinson (1926–1934 et seq.), K. R. Sporne (1949 et seq.), A. L. Takhtajan (1959 et seq.), R. F. Thorne (1963 et seq.), and A. Cronquist (1968 et seq.), have built on his concepts of what we would now call polarity in the morphology of angiosperms. (Takhtajan's publications in this field actually began in 1942, but they did not begin to have a major impact on botanical thinking until the 1959 book here cited). Some differences of opinion about polarity remain, but all of the widely respected systems of classification proposed in the last several decades fit into the de Candolle-Bentham and Hooker-Bessey tradition. (Takhtajan told me that he took his inspiration more nearly from H. Hallier [1901 et seq.] than from Bessey, but his views on polarity are in the Besseyan mainstream, which he himself has helped to channel.) Considerable blocks from the Englerian system have remained virtually intact in the newer systems, but they are rearranged *inter se* to reflect more recent concepts of polarity.

The general system of classification of angiosperms used in the Flora of North America is the "integrated system" of A. Cronquist (1981, slightly modified in 1988). Among other modern systems, the integrated system is most similar to those of A. L. Takhtajan, as presented in progressively modified versions over several decades, most recently in 1986 and 1987. Although these two sets of systems are conceptually similar, Takhtajan recognizes a significantly larger number of families, orders, and subclasses than Cronquist. Furthermore, some families, such as the Euphorbiaceae and Urticaceae, appear in very different places in the works of these two authors. Differences in the rank at which taxa are received (and thus in the number recognized at a given level) are inherently unresolvable except by fiat. Differences in the position of various taxa in different systems often reflect the difficulty of distinguishing convergence or parallelism from synapomorphy.

Armen L. Takhtajan, 1910– (left) and A. Cronquist (1919–1992). [Courtesy of the author.]

Robert Folger Thorne, 1920–. [Hunt Institute, Carnegie Mellon University, Pittsburgh.]

Molecular techniques may eventually help to resolve some of these problems, but the time (1992) is not quite yet.

Dicots and Monocots

It has been recognized for more than a century that the angiosperms form a natural group that consists of two subgroups. These two subgroups have usually been called dicotyledons and monocotyledons (or equivalent names with Latinized endings), from the most nearly constant of the several differences between them. As formal taxa they are here considered to be classes, called Magnoliopsida and Liliopsida. Less formally they may be called dicots and monocots.

Both groups occur in a wide variety of habitats, but the dicotyledons are the more diverse in habit. About half of all the species of dicots are more or less woody-stemmed, and many of them are definitely trees, usually with a deliquescently branched trunk. The monocots, in contrast, are predominantly herbaceous. Fewer than 10% of all monocots are woody, and most of these belong to the single large family Arecaceae (Palmae), which has only a few species in the area of our flora. Woody monocots usually have an unbranched (or sparingly branched) stem with a terminal crown of leaves, a habit that is rare among dicots. The difference in habit is partly a reflection of the complete absence of typical cambium in monocots, in contrast to its usual presence in dicots.

Differences exist in the underground as well as the aerial parts of dicots and monocots. In monocots the primary root soon aborts, and the mature root system is wholly adventitious. Many dicots likewise have an adventitious root system, but a primary root system, derived from the radicle, is more common. The adventitious, fibrous root system of monocots is a consequence of the absence of cambium. Having no adequate means of secondary thickening, individual roots cannot persist, enlarge, and ramify. The roots of some monocots do manage to penetrate deeply into the soil, but the largest single family, the Orchidaceae, is shallow-rooted and mycorrhizal, and the next largest family, the Poaceae, tends to exploit mainly the upper part of the soil, often forming a dense turf. Another large family, the Liliaceae, often has contractile roots that pull the bulb progressively deeper into the soil with the passing years. Creeping rhizomes, which may penetrate to any depth, serve as rootstocks for many of the Poaceae, Cyperaceae, Liliaceae, and other monocots.

All differences between dicots and monocots are subject to overlap or exception. The most nearly constant difference is the number of cotyledons, but some dicots have only one cotyledon, and some members of both groups have an undifferentiated embryo without cotyledons. The several differences between dicots and monocots are summarized in table 14.1.

It is widely agreed that the monocots are derived from primitive dicots, and that therefore the monocots must follow rather than precede the dicots in any proper linear sequence. The solitary cotyledon, parallel-veined leaves, absence of a cambium, dissected stele, and adventitious root system of monocots are all regarded as apomorphic characters within the angiosperms, and any plant that was plesiomorphic (i.e., more primitive than the monocots) in these several respects would certainly be a dicot. Monocots are more primitive than the bulk of the dicots in having mostly 1-aperturate pollen, but several of the more archaic families of dicots also have 1-aperturate pollen.

Despite the individual failure of all the characters used to separate monocots from dicots, there is seldom any problem in assigning a particular family or order to the one group or the other. Only the Nymphaeales (here, as customarily, assigned to the dicots) excite any continuing controversy in this regard (R. W. Haines and K. A. Lye 1975). Both Takhtajan and I regard Nymphaeales as the probable sister group of the monocots.

TABLE 14.1. Main Differences between Dicots and Monocots

Dicots	*Monocots*
Cotyledons 2 (seldom 1, 3, or 4, or the embryo seldom undifferentiated)	Cotyledon 1 (or the embryo sometimes undifferentiated)
Leaves mostly net-veined	Leaves mostly parallel-veined
Intrafascicular cambium usually present	Intrafascicular cambium lacking; usually no cambium of any sort
Vascular bundles of the stem usually borne in a ring that encloses a pith	Vascular bundles of the stem generally scattered, or in 2 or more rings
Floral parts, when of definite number, typically borne in sets of 5, less often 4, seldom 3 (carpels often fewer)	Floral parts, when of definite number, typically borne in sets of 3, seldom 4, almost never 5 (carpels often fewer)
Pollen typically 3-aperturate, or of 3-aperturate–derived type, except in a few of the more archaic families	Pollen 1-aperturate or of 1-aperturate–derived type
Mature root system either primary or adventitious, or both	Mature root system wholly adventitious

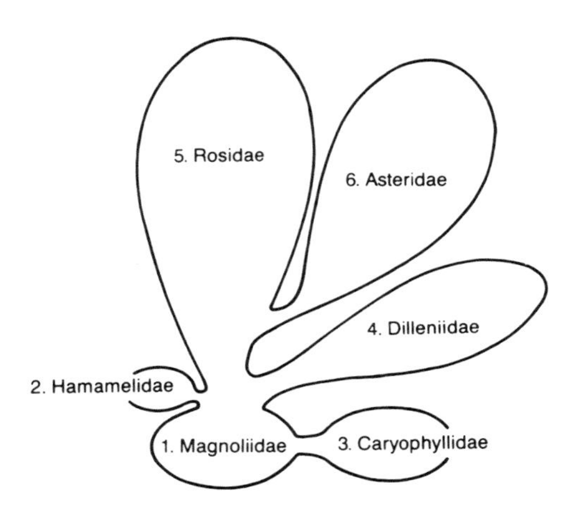

FIGURE 14.1. Putative relationships among the subclasses of dicotyledons. The size of the balloons is roughly proportional to the number of species.

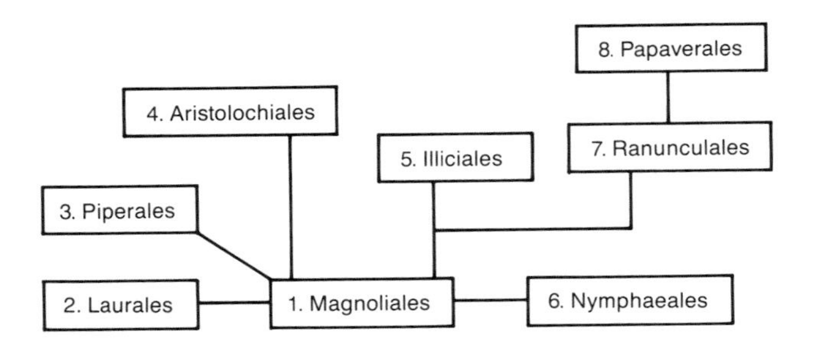

FIGURE 14.2. Putative relationships among the orders of Magnoliidae.

The Subclasses of Dicots

The dicots are here considered to consist of six subclasses, Magnoliidae, Hamamelidae, Caryophyllidae, Dilleniidae, Rosidae, and Asteridae (fig. 14.1). These are groups that cohere on the basis of all our information but that cannot precisely be defined phenetically. They can be characterized only in generalities and in terms of "critical tendencies" (a term suggested by H. F. Wernham 1912). They are polythetic groups, established by the accretion of apparently related members. Our minds demand some sort of organization of the many orders of dicots into a smaller number of affinity groups, even though the resulting groups are poorly characterized.

Magnoliidae

In general, it may be said that the Magnoliidae are the basal complex from which all other angiosperms have been derived. The Magnoliidae consist of nearly all of what has been informally called the Ranalian complex, plus a few other families. In cladistic terms the subclass is a prime example of a paraphyletic group, that is, a group that does not contain all the descendants from the immediate common ancestor.

The Magnoliales may be briefly characterized as those families of Magnoliidae that have ethereal oil cells, 1-aperturate (or sometimes inaperturate or 2-aperturate) pollen, and usually hypogynous flowers with a well-developed perianth of usually separate tepals. They are all woody plants. All of the putatively primitive features of angiosperms occur in one or another family of Magnoliales, but some of these features occur in other orders as well.

Based on the comparative morphology of modern species and on early Cretaceous fossil pollen and leaves, the Magnoliales must be considered the most archaic order of flowering plants (fig. 14.2). The fossil record clearly shows that the earliest angiosperm pollen grains were monosulcate and that simple, entire leaves with relatively poorly organized pinnate venation are the ancestral archetype. Among modern angiosperms, only the Magnoliales can accommodate this combination of characteristics.

No one family of Magnoliales can be considered ancestral to the rest of the order. Each family presents its

Nelumbo lutea. [From C. F. Reed. 1970. Selected Weeds of the United States. Washington. Fig. 84, by Regina O. Hughes.]

own combination of primitive and advanced characteristics.

The Nymphaeales may be briefly characterized as those families of the Magnoliidae that are aquatic herbs without ethereal oil cells and without vessels in the shoot. Most of them have 1-aperturate or inaperturate pollen, but the distinctive Nelumbonaceae have 3-aperturate pollen. Most of the Nymphaeales have long-petiolate leaves with broad, cordate to hastate or peltate, floating blades, but in *Ceratophyllum* the leaves are all slender and submersed, and in *Nelumbo* many of the leaves are emergent.

Unlike most orders of angiosperms, the Nymphaeales have an obvious ecological niche. They inhabit still waters, and many of their characteristics reflect adaptation to this habitat. They do not occupy this habitat to the exclusion of other groups, however. *Nymphoides* (Menyanthaceae) is very nymphaeaceous in aspect and habitat, and *Myriophyllum* (Haloragaceae) likewise recalls *Ceratophyllum.*

Based on leaf fossils, it appears that the Nymphaeales have a long history, going all the way back to the Albian stage of the Lower Cretaceous. The fairly numerous and varied Albian fossils of this sort are coming to be called nymphaeaphylls. The monosulcate pollen of many of the modern Nymphaeaceae and Cabombaceae bespeaks an ancestry among the more primitive dicotyledons, but it does not specify the time of origin. Nymphaealean pollen is in fact not very distinctive, and it can be traced with some certainty only to the uppermost Cretaceous.

It is here considered that the modern Nymphaeales all descend from a group of primitive dicotyledons that took to an aquatic habitat and became herbaceous very early in the history of the angiosperms. A subsequent early dichotomy resulted in a line leading to the modern Nelumbonaceae and a line leading eventually to the other four families. The modern families of the order thus represent a series of isolated endlines, comparable on a smaller scale to the series of isolated endlines that comprise the families of the modern orders Magnoliales and Laurales.

The Ranunculales and Papaverales form a pair of closely related orders that stand somewhat apart from the remainder of the Magnoliidae. The mostly apocarpous flowers of the Ranunculales and the isoquinoline alkaloids of both orders tie the pair to the Magnoliidae, but the mostly 3-aperturate pollen separates them from other Magnoliidae except Illiciales and *Nelumbo*, and the absence of ethereal oil cells separates them from other orders except the Nymphaeales. The scanty fossil record does not carry the pair back beyond the Tertiary. Conceivably these two orders are of relatively recent origin, despite their retention of some archaic features.

Hamamelidae

The Hamamelidae are a group of mostly wind-pollinated families with reduced, usually apetalous flowers that are often borne in catkins. This subclass consists chiefly of the core of the traditional Amentiferae, after some unrelated families such as the Salicaceae have been excluded. Plants with highly reduced flowers also occur in each of the other subclasses of dicots, but the Hamamelidae are an ancient major group with reduced flowers.

The Hamamelidales are morphologically central to their subclass. Except for the highly archaic Trochodendrales (two monotypic families), all the other orders appear to tie back directly or indirectly to the Hamamelidales (fig. 14.3).

The Hamamelidae can be traced back through the platanoid line (Hamamelidales) to near the middle of the Albian (final) stage of the Lower Cretaceous period. Members of the orders Juglandales, Myricales, Fagales, and Casuarinales have a distinctive sort of pollen that appears to take its evolutionary origin in the *Normapolles* complex of Middle Cenomanian (early Upper Cretaceous) time. The Urticales also have a long fossil record, with pollen dating from the Turonian, some 90 M.Y.B.P. (the next stage above the Cenomanian).

The Urticales have traditionally been associated with some of the other orders here referred to subclass Hamamelidae. The reduced flowers provide the obvious initial basis for such a treatment, but vegetative anatomy (O. Tippo 1938; E. M. Sweitzer 1971) and leaf venation (J. A. Wolfe 1973) have also been adduced to support the association. Even the pollen is said to resemble that of the *Normapolles* group of early Upper Cretaceous fossils. If the virtually apocarpous archaic genus *Barbeya* is, as by most authors, associated in some way with Urticales, it further strengthens the connection of the latter to Hamamelidae rather than to any other subclass.

A longstanding school of thought, exemplified most recently by R. F. Thorne (1973), C. C. Berg (1977), and A. L. Takhtajan (1987), holds that the Urticales are allied to the Malvales rather than to the Hamamelidales. The stratified phloem and presence of mucilaginous cells and ducts in both orders have played a large role in this assessment of relationship. Thorne has tried to mitigate the problem of *Barbeya* by excluding it from Urticales and considering it to be *incertae sedis*.

Caryophyllidae

The Caryophyllidae consist of the large order Caryophyllales plus two smaller orders (Polygonales and Plumbaginales) that are customarily associated with it (fig. 14.4). Each of the three orders is well marked, but no one distinctive feature marks the subclass. The simplest way to characterize the group is to say that it consists of those dicotyledons that have bitegmic, crassinucellar ovules and either have betalains instead of anthocyanins or have free central or basal placentation in a compound ovary. Furthermore, most species are herbaceous, and woody species usually have anomalous secondary growth or otherwise anomalous stem structure. The stamens, when numerous, originate in centrifugal sequence, and the pollen grains are usually 3-nucleate. The food storage tissue of the seed is typically starchy, and very often has clustered starch grains.

The fossil record as presently interpreted carries the Caryophyllidae back only to the Maastrichtian epoch of the latest Upper Cretaceous, some 70 M.Y.B.P. (pollen of Amaranthaceae or Chenopodiaceae). The relatively short fossil history, as contrasted to the Magnoliidae, Hamamelidae, and Rosidae, is consonant with the primitively herbaceous habit of Caryophyllidae. Aside from the ancestors of the Nymphaeales, dicotyledonous herbs apparently played only a negligible role in the vegetation of the Cretaceous.

The ancestry of Caryophyllidae may lie in or near Ranunculaceae. In the absence of known fossil connections, it may be supposed that the common ancestor of the Caryophyllidae was an herb with hypogynous flowers and separate carpels, without petals. The number of potentially ancestral groups is thus immediately limited. The possibility that members of the Ranunculaceae may be at least collateral ancestors is bolstered by the fact that some of them have pollen very much like that of many Caryophyllales. The floral trimery of the Polygonaceae also has ample precedent in the Ranunculaceae. The evolutionary significance of the centrifugal androecium in Caryophyllales can scarcely be evaluated until a satisfactory general interpretation of the origin of centrifugality is achieved. Further speculation about the ancestry of Caryophyllidae is hampered by the uncertainty about the affinity of Polygonales and Plumbaginales to Caryophyllales.

Chemical and ultrastructural studies of the past two or three decades have led to a consensus as to the contents and characterization of the Caryophyllales, now one of the best defined major orders of angiosperms. All investigated members of the order have a character-

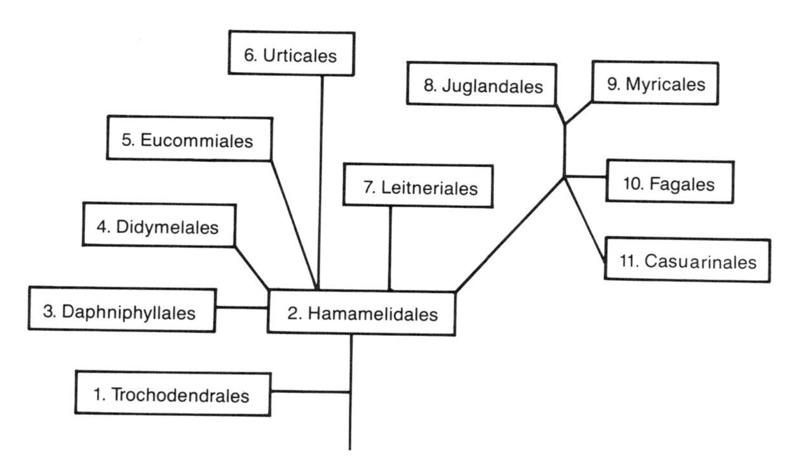

FIGURE 14.3. Putative relationships among the orders of Hamamelidae.

istic type of sieve-tube plastid that is unknown in other angiosperms (H.-D. Behnke 1976). The plastid contains a set of bundles of proteinaceous filaments that collectively form a subperipheral ring. Often there is a larger central protein crystalloid, which may be either globular or polyhedral. Sieve-tube plastids with proteinaceous inclusions occur in some other dicotyledons (notably some members of the Magnoliidae and Fabaceae), but in those groups the inclusions do not form a ring of filaments.

Ten of the twelve families of Caryophyllales consistently produce betalains but lack anthocyanins. Betalains are otherwise unknown in angiosperms, although they have been found in some Basidiomycetes. The absence of betalains and presence of anthocyanins in the Caryophyllaceae and Molluginaceae led some botanists a few years ago to propose a fragmentation of the order, but this difference is now generally acknowledged to be of no more than subordinal importance.

Dilleniidae

The weakest distinction among the subclasses of dicotyledons is that between the Dilleniidae and Rosidae. Both groups are more advanced than the Magnoliidae in one or another respect, but less advanced than the Asteridae. The two taxa are kept apart as subclasses because each seems to constitute a natural group separately derived from the ancestral Magnoliidae rather than because of any definitive distinguishing characteristics. The same sorts of evolutionary advances have occurred in both groups, but with different frequencies. Despite the lack of solid distinguishing criteria, it is conceptually more useful to hold the two as separate subclasses than to combine them into one or to abandon any attempt at the organization of the Magnoliopsida into subclasses.

The Dilleniidae cannot be fully characterized morphologically. Except for the rather small (400 species) order Dilleniales, the vast majority of the Dilleniidae are sharply set off from characteristic members of Magnoliidae by being syncarpous. With few exceptions, the species of Dilleniidae with numerous stamens have the stamens initiated in centrifugal sequence. In this respect, they differ from Rosidae, in which the species with numerous stamens usually have a centripetal sequence of development.

More than a third of the species of Dilleniidae have parietal placentation, in contrast to the relative rarity of this type in the Rosidae. (Application of the term "parietal" is here restricted to compound ovaries.) Another third of the species (not the same third) are sympetalous, but only a very few of these (e.g., *Diapensia*) have isomerous, epipetalous stamens alternate with the corolla lobes and also unitegmic, tenuinucellar ovules as in Asteridae. Sympetaly is rare in the Rosidae. Ovules in the Dilleniidae as a whole are bitegmic or less often unitegmic, with various transitional types, and they range from crassinucellar to tenuinucellar. Often they are bitegmic and tenuinucellar, a combination rare outside this group.

Compound leaves with distinct, articulated leaflets are much more common in the Rosidae than in the Dilleniidae. Uniovulate or biovulate locules are much less common in the Dilleniidae than in the Rosidae, but they are well represented in the Malvales, whose position in the Dilleniidae is well established. Not many of the Dilleniidae have a typical nectary disk of the sort so common in the Rosidae, but other types of nectaries are common.

It seems clear that the Dilleniidae take their origin in the Magnoliidae. The apocarpous order Dilleniales, especially the family Dilleniaceae, forms a connecting link between the two subclasses. If the rest of the Dilleni-

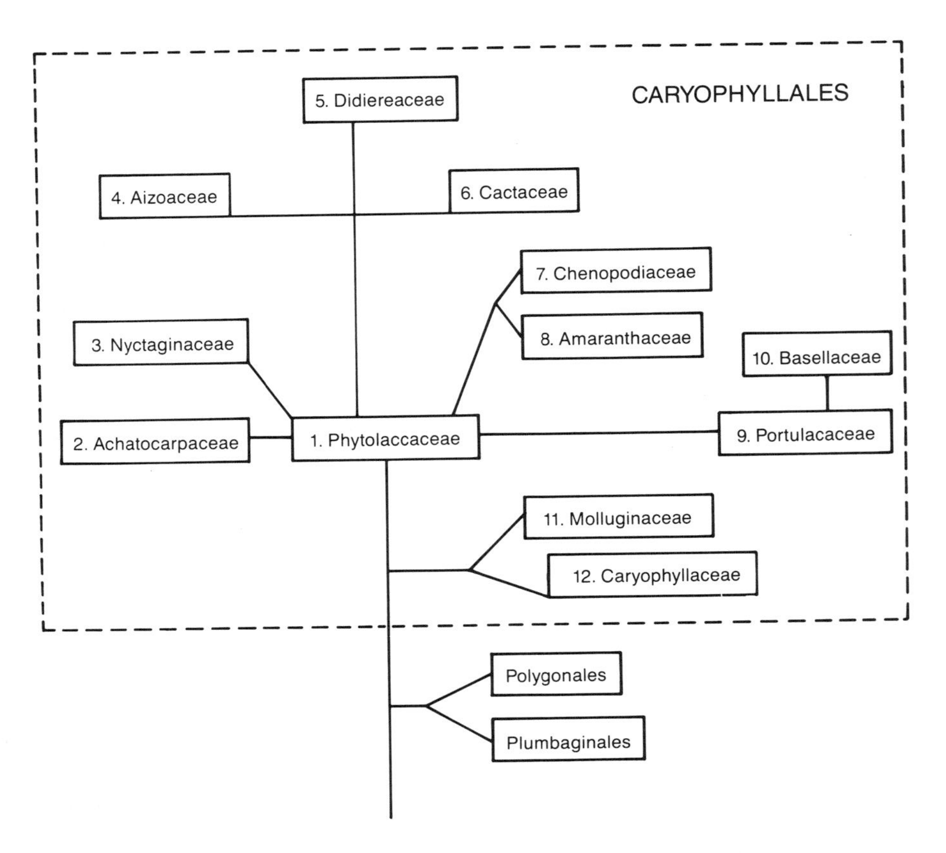

FIGURE 14.4. Putative relationships among the families and orders of Caryophyllidae.

idae did not exist, the Dilleniales could easily be accommodated as a peripheral order of Magnoliidae. On the other hand, the Dilleniales do not closely resemble any one family of Magnoliidae, and they are sharply set off by a series of chemical features.

Unlike the vast majority of the Magnoliidae, the Dilleniaceae have ellagic acid, proanthocyanins, and raphides. The ethereal oil cells so characteristic of the woody Magnoliidae are absent from the Dilleniaceae. The Dilleniaceae are almost without alkaloids, and the few that have been found in some species have nothing to do with the characteristic benzyl-isoquinoline alkaloids of Magnoliidae.

Furthermore, there is a still unresolved controversy about the ancestry of the multistaminate, centrifugal androecium such as that in the Dilleniidae. If, as some authors maintain, this kind of androecium is necessarily derived from an ordinary, cyclic, oligostemonous androecium by secondary increase in number of stamens, then the gap between the Dilleniaceae and Magnoliidae is further widened.

The centrifugal androecium and frequently campylotropous or amphitropous ovules of the Dilleniidae recall the Caryophyllidae, but the latter have their own set of specialized features not found in the Dilleniaceae and other Dilleniidae. Any evolutionary relationship between the Dilleniaceae and Caryophyllidae must be rather remote in time.

Pollen that appears to represent the Dilleniidae dates from about the beginning of the Upper Cretaceous, but this early pollen is not clearly referable to an order. Otherwise the fossil record as presently understood gives no clear indication of the origin of the group. If my hypothesis of chemical evolution (A. Cronquist 1977b, elaborated in 1988) is correct, the Dilleniidae probably originated about the same time as the Hamamelidae and the Rosidae. The Hamamelidae and the more archaic members of the Dilleniidae and Rosidae characteristically rely heavily on hydrolyzable tannins as defensive weapons. The more advanced members of the latter two groups have largely discarded tannins in favor of more recently evolved defenses.

In contrast to their evident separation from the Magnoliidae and Caryophyllidae, the Dilleniaceae are obviously allied to such syncarpous families as the Actinidiaceae and Theaceae. For purposes of conceptual

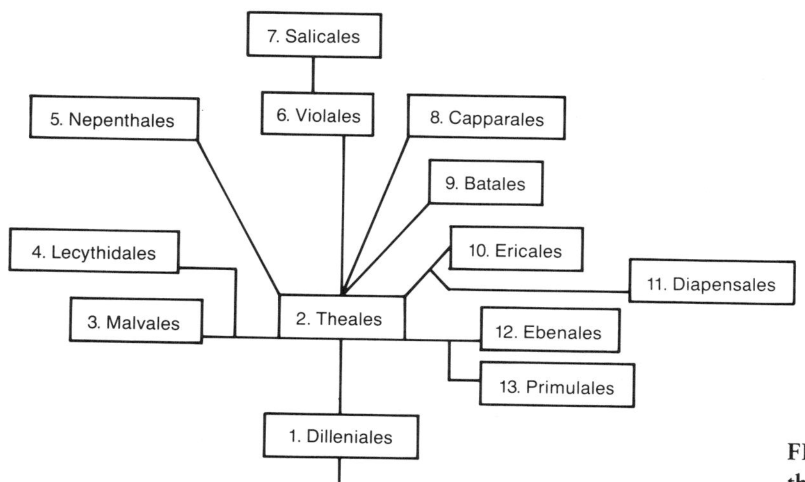

FIGURE 14.5. Putative relationships among the orders of Dilleniidae.

organization, however, it is useful to put the largely apocarpous family Dilleniaceae in a separate order from the largely syncarpous Theales. The Dilleniaceae appear to be the modern remnants of the group that gave rise not only to Theales but also to nearly all the rest of the subclass Dilleniidae. The Paeoniaceae, the only other family of the Dilleniales, stand somewhat apart and do not appear to be in the main line of evolution of the subclass.

Theales are the central group of Dilleniidae, from which all the other orders except the Dilleniales appear to have evolved. The Malvales, Lecythidales, Violales, Capparales, and Nepenthales all appear to have arisen from a common complex in the Theales. The Salicales are an amentiferous offshoot from the Violales, and the Batales appear to be related to the Capparales. The remaining four orders take their origin in a different part (or parts) of the Theales. The Ericales are evidently allied to the Actinidiaceae (Theales), and the Diapensiales appear to be allied to the Ericales. The Ebenales and Primulales are somewhat more remote, but they may be allied to each other and to a lesser extent to Ericales. Only the Theales provide a reasonably likely origin for these groups.

The concepts of relationships within Dilleniidae that are here expounded may be conveniently expressed in the phylogenetic diagram (fig. 14.5).

Rosidae

The 18 orders that make up the Rosidae evidently cohere as a natural group. Only the Euphorbiales and Rafflesiales are obviously debatable, the latter because their morphological reduction in association with parasitism makes their affinities hard to establish. The position of the Euphorbiaceae has long been uncertain. Some authors have associated the Euphorbiaceae with Malvales, others with orders here assigned to Rosidae, and some have used the apparently dual affinity of the Euphorbiaceae to cast doubt on the rosid-dilleniid distinction.

The Buxaceae and Simmondsiaceae, often (as here) associated with the Euphorbiaceae, may not properly belong there. Takhtajan may well be right in putting these two small families into the Hamamelidae in the more recent versions of his system. For the Buxaceae alone, that would present no great problem. Unfortunately the foliaceous-accrescent sepals of *Simmondsia* are quite out of harmony with the Hamamelidae, yet the affinity of *Simmondsia* (even as a separate family Simmondsiaceae) to the Buxaceae is generally acknowledged. The position of the several families of Euphorbiales as here constituted would be a suitable subject for molecular techniques, especially because of the wide diversity of recent opinions based on phenetic data.

The Rosales as here constituted form an exceedingly diverse order, standing at the evolutionary base of their subclass. In effect, they are what is left after all the more advanced, specialized orders of Rosidae have been delimited (fig. 14.6). Aside from internal phloem and a parasitic or highly modified aquatic habit, most of the features that mark the more advanced orders of Rosidae (and indeed even some of the features of Asteridae) can be found individually within the order Rosales, but in the Rosales these features do not occur in the combinations that mark the more advanced groups.

Two characteristics that are very common in the Rosales are much less common among other orders of the

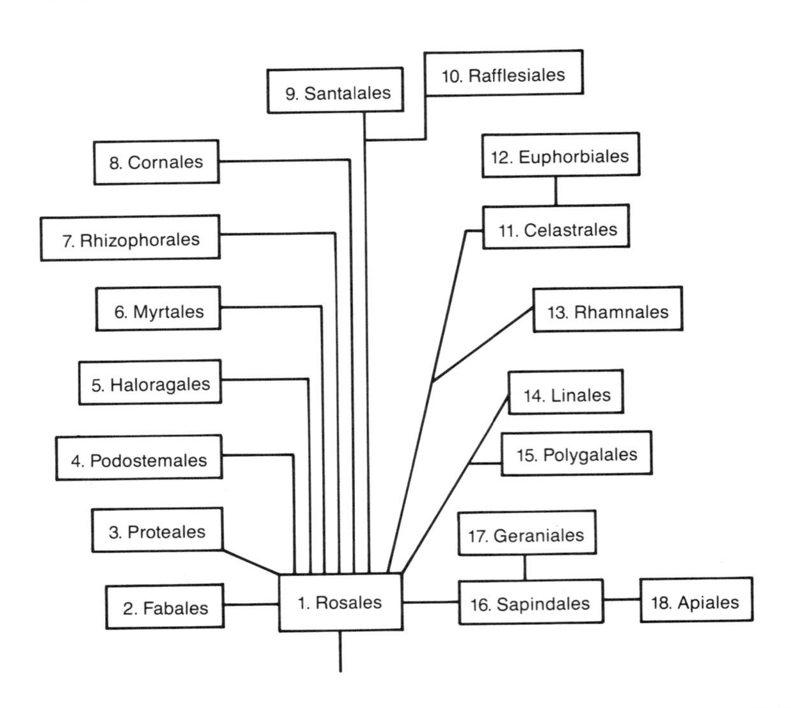

FIGURE 14.6. Putative relationships among the orders of Rosidae. Here as elsewhere the length of the lines reflects the exigencies of two-dimensional representation.

subclass. These are a polymerous androecium and an apocarpous gynoecium (although the gynoecium is monocarpous in the large order Fabales). Furthermore, a great many of the Rosales have more or less numerous ovules per carpel, and relatively few have only one or two. In contrast, many of the other orders of Rosidae have only one or two ovules per carpel, and most of those that have numerous ovules are well differentiated from the Rosales in other respects.

The order Fabales, as here treated, consists of three closely allied families, the Fabaceae, Caesalpiniaceae, and Mimosaceae. The existence of three major groups (here called families) that collectively constitute a larger group (here called an order) is widely admitted. The rank of the groups is in dispute, however. Many authors have preferred to recognize a single family Leguminosae (or Fabaceae sensu lato), with three subfamilies. The broadly defined family is then often included in Rosales. There is no objective right or wrong here. I prefer the treatment here presented as being more in harmony with the customary definitions of families of angiosperms. The legumes, whether treated as one family or three, form such a coherent group with such an abundance of genera and species that they would appear to dominate any other order to which they might be assigned. At some 18,000 species, they are second only to Asterales among dicotyledonous orders. Morphologically, the Fabales form one of the better-defined orders of dicotyledons. The combination of compound, stipulate leaves and flowers with a single carpel will distinguish the vast majority of Fabales from virtually all other groups.

The closest linkage of the Fabales to Rosales lies in the fairly small (300–400 species), mainly tropical family Connaraceae, which is not represented in our flora. The Connaraceae lie in the nebulous area where the Rosales, Fabales, and Sapindales join at their evolutionary base.

The Myrtales, with some 14 families and more than 9000 species, are one of the better defined large orders of dicotyledons. In addition to the mostly strongly perigynous to epigynous flowers and the overlapping similarities among the constituent families, the order is marked by two otherwise uncommon anatomical features: internal phloem and vestured pits in the vessel segments.

The most discordant family of the Myrtales, and the only one that still provokes controversy about its possible inclusion in the order, is the Thymelaeaceae. The Thymelaeaceae are marked by their usually pseudomonomerous ovary, often unusual pollen, and a distinctive set of secondary metabolites, commonly including the simple coumarin daphnin (or allied compounds). Furthermore, some few members of the family are unusual in Myrtales in having essentially hypogynous flowers. More ordinary kinds of pollen and gynoecia, with transitional types, also occur in the family, however. Thus it is unnecessary to seek placement of the Thymelaeaceae in another order. Indeed the internal phloem, vestured pits, and strongly perigynous,

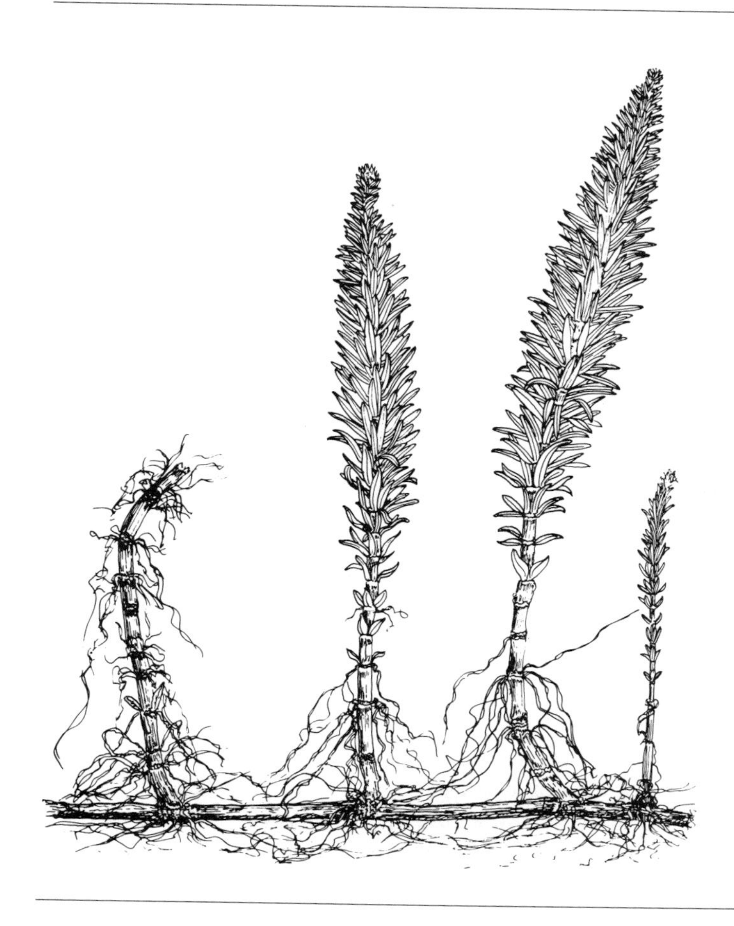

Hippuris vulgaris. [From D. S. and H. B. Correll. 1972. Aquatic and Wetland Plants of Southwestern United States. Washington. Fig. 572, by Vivien Frazier.]

polypetalous to apetalous flowers of characteristic members of the family would be out of harmony with any other order that might be suggested as a haven for it.

Furthermore, the characteristic obturator of the Thymelaeaceae, though not identical in detail, might be compared with the obturator of the Combretaceae, and the glandular-punctate leaves of some Thymealeaceae recall those of the Myrtaceae. It would, of course, be possible to recognize an order Thymealeales to provide for this one family, as some authors have done, but the segregate order would still stand alongside Myrtales. We should also note that P. G. Martin and J. M. Dowd (1986) linked the Thymelaeaceae to the Myrtales on the basis of the sequence of amino acids in the terminal 40 residues of the smaller subunit of ribulose biphosphate carboxylase.

The Sapindales, with 15 families and about 5400 species, form a well-characterized natural group. Only two families are really peripheral. The Staphyleaceae connect the Sapindales to the ancestral Rosales, in the vicinity of the Cunoniaceae, and the Zygophyllaceae are suggestive of the Geraniales. These two families also differ from the bulk of the order in often having more than two ovules per locule.

The features common to most members of the Sapindales, which make it useful to distinguish them as a group from the Rosales, are the compound or cleft leaves, haplostemonous or diplostemonous androecium, well-developed nectary disk, and syncarpous ovary with a limited number of ovules (usually only one or two) in each locule. All of these features can be found individually in the Rosales, but mostly not in combination.

Asteridae

The Asteridae are the best characterized subclass of dicotyledons, marked by their sympetalous flowers, in which the stamens are isomerous and alternate with the corolla lobes, or fewer than the corolla lobes. Much less than 1% of the species of Asteridae fail this test, and probably no more than 1% of the species that do meet the test do not belong to Asteridae. The tenuinucellar ovule with a massive single integument is a further marker of the group, but there are more exceptions to this feature, both within and without the Asteridae. The Callitrichales are the most aberrant order in the subclass. Except for the vestigial calyx in *Hippuris,* the Callitrichales lack a perianth entirely.

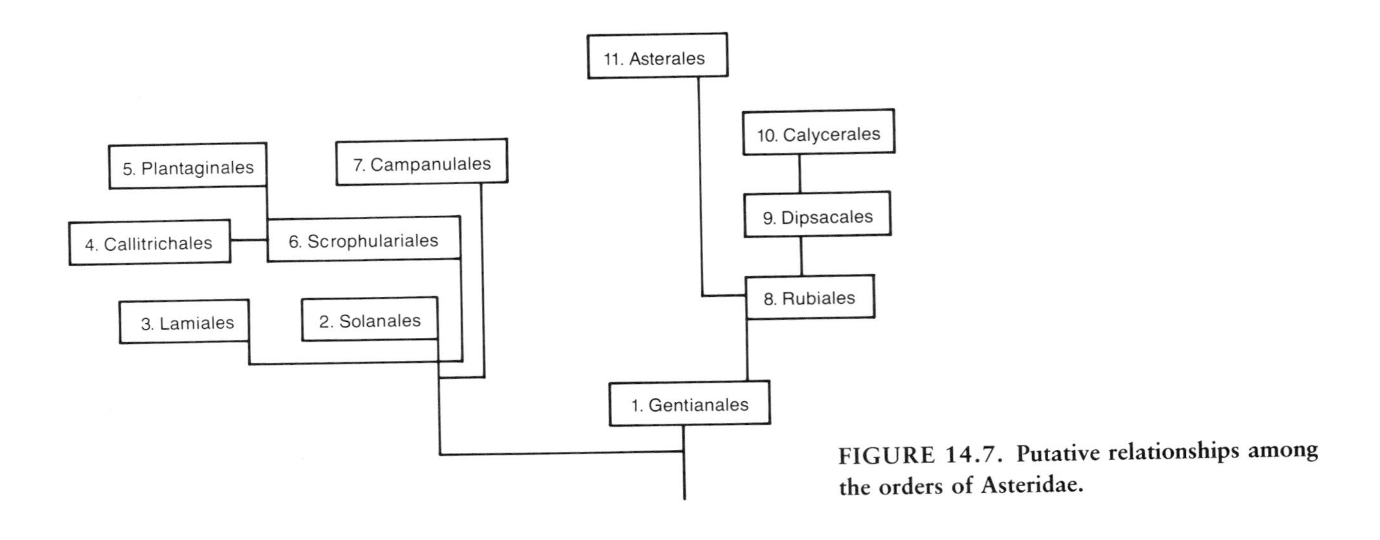

FIGURE 14.7. Putative relationships among the orders of Asteridae.

Chemically, the Asteridae are noteworthy for the frequent occurrence of iridoid compounds, the usual absence of ellagic acid and proanthocyanins, and the apparently complete absence of betalains, benzyl-isoquinoline alkaloids, and mustard oils. The absence of betalains and benzyl-isoquinoline alkaloids tends to set Asteridae off from most Caryophyllidae and Magnoliidae, respectively, and the usual absence of ellagic acid and proanthocyanins tends to set them off from Rosidae, Dilleniidae, and Hamamelidae. The distinction is far from absolute, however, because these substances are missing from many members of the subclasses they are supposed to characterize.

Ancestry of the Asteridae very probably lies in the order Rosales sensu latissimo. Iridoid compounds, a sympetalous corolla, stamens isomerous and alternate with the petals, a compound pistil with numerous ovules on axile placentas, unitegmic ovules, and tenuinucellar ovules occur in this order, but not in combination. The nectary disk of many Asteridae also finds a ready precedent in Rosales, as do the simple, stipulate, opposite leaves of the more archaic members of the group. The combination of these features into a functional whole marks the transition from the ancestral Rosales to the first members of the Asteridae. All the other orders of Rosidae are already too advanced to serve as plausible ancestors of the Asteridae.

The Asteridae make a relatively late entry into the fossil record. There is no reason to suppose that the group originated before the beginning of the Tertiary period, and its members began to play a prominent role in the flora of the world only during the Oligocene.

The Asteridae are the most advanced subclass of dicotyledons, and possibly the most recently evolved; only the Caryophyllidae may be more recent. More than any other subclass, they exploit specialized pollinators and specialized means of presenting the pollen. It seems likely that the rise of the Asteridae is correlated with the evolution of insects capable of recognizing complex floral patterns.

The Asteridae form such a coherent group that delimitation of the orders becomes problematical. The Gentianales appear to stand near the base of the subclass (fig. 14.7). The Rubiales link to both the Gentianales and the Dipsacales. The Menyanthaceae and Buddlejaceae, often included in Gentianales, are here referred to Solanales and Scrophulariales, respectively.

The largest order of Asteridae, in terms of number of families, is Scrophulariales. The Scrophulariaceae are the largest family of the order and are also central to it. Four of the other families (Acanthaceae, Bignoniaceae, Globulariaceae, and Orobanchaceae) are connected to the Scrophulariaceae by genera or groups of genera that have by different authors been referred to the central or the peripheral family. Although the seven remaining families of the order are more sharply delimited, most of them may logically be considered to be specialized derivatives of the Scrophulariaceae.

The order Asterales consists of the single worldwide family Asteraceae (Compositae), with perhaps as many as 20,000 species. The Asteraceae are one of the more successful families of flowering plants, represented by numerous genera, species, and individuals. Not many of them are forest trees, and only a few are aquatic, but otherwise they exploit most of the obvious kinds of ecological opportunities open to angiosperms.

Affinities of the Asteraceae have been vigorously but inconclusively debated. Traditionally they have been thought to be allied to the Campanulaceae, and often even included in the same order. I have argued (1977, 1981) that patterns of relationship within the family suggest that the tribe Heliantheae is the basal group, and that their immediate ancestors must have been woody plants with opposite leaves and a cymose inflorescence, as in the Rubiaceae. Ongoing studies of chloroplast DNA in the Asteraceae and some other families by R. K. Jansen and J. D. Palmer (e.g., 1988) and their associates suggest that *Barnadesia* and its immediate allies in the tribe Mutisieae may be the sister group of all the other Asteraceae. The consequences of this view to the taxonomic and phylogenetic interpretation of the family are only beginning to be explored.

Monocots

The differences between monocots and dicots and the probable origin of the monocots from dicots have already been discussed. Some further considerations are in order here.

The fossil pollen record suggests that the origin of monocots from primitive dicots in Aptian-Albian times was the first significant dichotomy in the evolutionary diversification of the angiosperms. The wide variety of monocotyledonous leaves found in the late Lower Cretaceous and through the Upper Cretaceous attests to the continuing diversification of the group during this time, but most of these leaves cannot be referred with any certainty to modern groups.

The first modern family of monocots to be clearly represented in the fossil record is the Arecaceae (subclass Arecidae), near the base of the Santonian (or perhaps the Conacian) epoch, but palms are surely not primitive monocots. Their large, distinctive, readily fossilized leaves merely make the group easy to recognize from its inception. Pollen that probably represents Cyperales or Restionales (subclass Commelinidae) appears in the late Upper Cretaceous, probably before the Maastrichtian, and the distinctive leaves of Zingiberales show up in the Maastrichtian. Pollen thought to represent *Pandanus* occurs in Maastrichtian deposits in North America.

The Alismatidae and Liliidae are not certainly recognizable before the Tertiary, but some of the miscellaneous Upper Cretaceous fossil monocot leaves might well belong to one of these groups. We can be reasonably sure that the palms and the Zingiberales did not arise long before the first appearance of their characteristic leaves in the fossil record, but we cannot be so confident that other large groups did not long antedate their first identifiable fossils. Thus the fossil record, as presently understood, is compatible with any of several different views about the Cretaceous diversification of monocots. The principal constraint is the recognition of a very early dichotomy between monocots and dicots.

The dicots that gave rise to the monocots may have had apocarpous flowers with a fairly ordinary (not highly specialized) perianth, and with 1-aperturate pollen. They must have been herbs without a very active cambium, and they presumably had laminar placentation. The only modern group of dicots that meets these specifications are the Nymphaeales.

It is not here suggested that the Nymphaeales are directly ancestral to the monocots as a whole, but rather that the premonocotyledonous dicots were probably something like the modern Nymphaleales. As noted, an aquatic group of angiosperms with leaves much like those of the modern Nymphaeales was already proliferating in the Albian epoch of the Lower Cretaceous. The modern Nymphaeales are aquatic, mostly lack vessels, and show tendencies toward the fusion of two cotyledons into one.

An interesting difference between monocots and dicots is that whereas in dicots the vessels appear first, phyletically, in the secondary wood of the stem and spread to other tissues and organs, in monocots they appear first in the roots. This fact led V. I. Cheadle (1953) and others to suppose that vessels originated independently in the two classes. They therefore consider that the evolutionary divergence of the two classes preceded the origin of vessels. As I have argued elsewhere (1988), it is at least equally plausible that in the ancestral premonocots, as in their probable relatives the Nymphaeales, vessels were phyletically lost in association with the aquatic habitat. Loss of the cambium eliminated at one stroke all vessels that had not worked their way, phyletically, into the primary tissues. Subsequent evolution of vessels in the monocots had to begin essentially *de novo,* in association with the return to a terrestrial habitat.

The typical parallel-veined leaf of monocots is here considered to be a modified, bladeless petiole. This morphological interpretation for the leaves of *Sagittaria* was proposed a century and a half ago by A. P. de Candolle (1827, vol. 1, p. 286). It was further elaborated in evolutionary terms and applied to monocots as a whole by A. Arber in 1925. It is the only hypothesis known to me that permits all the information about monocots to fall into place and make sense. Even the ontogeny of the typical monocot leaf is highly compatible with the petiolar hypothesis. The blade typically develops from a portion of the leaf primordium some-

Sagittaria latifolia. **[From C. F. Reed. 1970. Selected Weeds of the United States. Washington. Fig. 12, by Regina O. Hughes.]**

what behind the tip and matures basipetally; the primordial tip is inactive or produces only a terminal point or small appendage on the blade.

Terrestrial monocots with a well-defined, net-veined leaf blade are here considered to be derived from ancestors with narrow, parallel-veined leaves lacking a well-defined blade. All transitional stages can be seen in several families. An attempt to read the system the other way (R. M. T. Dahlgren et al. 1985) means that we must start with broad, more or less net-veined leaves in diverse groups of monocots having little to do with each other, and have all these converge in both floral and vegetative characters into a hopelessly polyphyletic core of typical monocots.

Three principal ways exist by which the typical monocot leaf can become broad and more or less net-veined. One way is to spread the main veins farther apart near the middle of the blade and amplify the cross-connections among them. Subsequently the main veins can fade out before reaching the leaf tip, so that a more or less palmate venation is established. *Alisma, Sagittaria, Dioscorea, Smilax, Trillium,* and many aroids exemplify this type of change. A second way is for each of the many closely set parallel veins to diverge in turn toward the margin, the outermost veins first, those nearest the midrib last. The result is a pinnately veined leaf with numerous closely parallel primary lateral veins. Members of Zingiberales reflect this sort of change. A third way, known only in the palms and Cyclanthales, differs from the second way in the intercalation of new tissue between the lateral veins during the early growth of the leaf. The plicate structure of palm and cyclanth leaves reflects this ontogeny.

If the interpretation here presented is correct, the aquatic ancestry of the monocots has had a profound effect on the subsequent evolutionary history of the group. As aquatic herbs, early monocots were preadapted to evolve terrestrial herbaceous forms, filling a niche (or set of niches) not then effectively occupied by dicots. Nevertheless, the evolution of an effective water-conducting system (with well-developed vessels), of expanded, net-veined leaves, of a branching, arborescent

Alisma triviale. **[From D. S. and H. B. Correll. 1972. Aquatic and Wetland Plants of Southwestern United States. Washington. Fig. 55, by Vivien Frazier.]**

habit, and of a means of secondary thickening has not been easy for them. No monocot has evolved a coherent syndrome of these features that would permit a broad-scale evolutionary challenge to woody dicotyledons. Even among those monocots that have evolved a broad, more or less net-veined blade, traces of the ancestral parallel-veined pattern usually persist. We have noted that aroids, palms, and the Zingiberales have taken three essentially different routes in the evolutionary expansion of leaves, and the difference is reflected in the mature morphology.

Although the monocots are considered to have an aquatic ancestry, the situation is not simple. It appears that terrestrial monocots, derived from aquatic early monocots, have themselves repeatedly given rise to groups that have returned to the water. Among the modern Alismatidae, there appears to be a progressive adaptation to an aquatic and eventually marine habitat. In the subclass Arecidae, the mainly terrestrial family Araceae has some secondarily aquatic forms that point toward the thalloid, aquatic family Lemnaceae. In a third subclass, the Commelinidae, such aquatic families as the Mayacaceae, Sparganiaceae, and Typhaceae appear to be derived from terrestrial ancestors within the group. The aquatic habit of the Pontederiaceae (Liliidae) may likewise be secondary.

The nature of the single cotyledon in monocots has occasioned much study and controversy. Like the foliage leaves, the cotyledon often has a basal sheath surmounted on one side by a limb that may or may not be divided into blade and petiole. Typically, the sheath is closed and tubular, at least near the base. The vascular supply typically consists of two near-median bundles, as in the individual cotyledons of dicots. This implies that the monocot cotyledon is equivalent to a single leaf and is not a double structure as has sometimes been supposed. The sheathing base of the cotyledon is thus left unexplained, except that it is comparable to the sheathing base of a foliage leaf.

Drawing on evidence from living members of the Nymphaeales (dicots), I have suggested an alternative interpretation (1968, 1981, 1988): two ancestral coty-

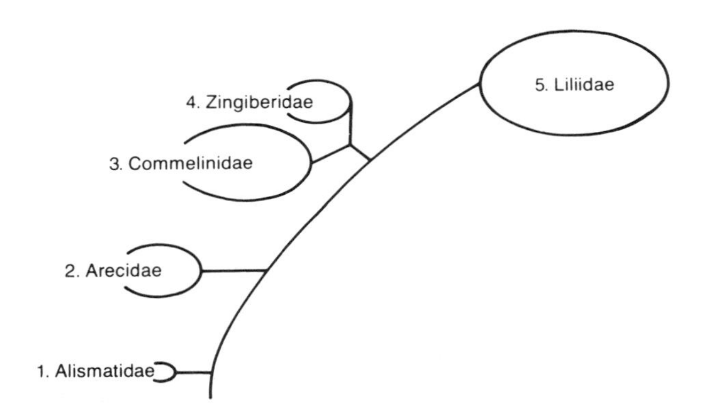

FIGURE 14.8. Putative relationships among the subclasses of monocotyledons.

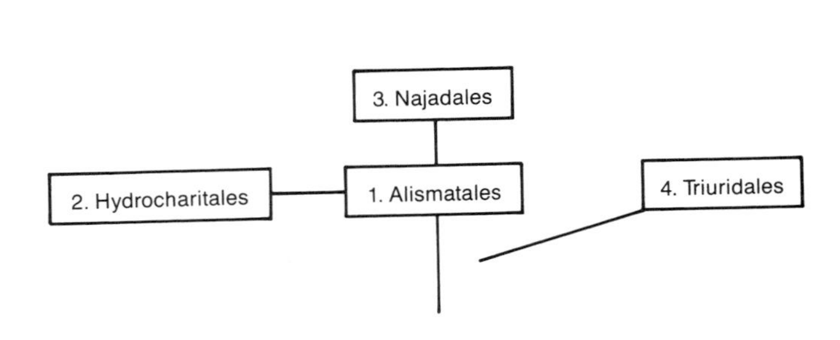

FIGURE 14.9. Putative relationships among the orders of Alismatidae.

ledons have become connate by their margins toward the base, forming a bilobed, basally tubular, compound cotyledon. One of the lobes has subsequently been reduced and lost, and its vascular supply suppressed, so that the embryo has, in effect, a single cotyledon with a sheathing, tubular base. I have further suggested that this modified cotyledonary structure has so firmly impressed itself on the growth pattern of the embryo that subsequent leaves are also built on the same plan. The sheathing base of monocot leaves is therefore a reflection of cotyledonary structure, rather than the reverse.

Regardless of the morphological nature of the single cotyledon, throughout the Liliopsida it is basically the same organ. Highly modified though the cotyledon may be in such plants as grasses, there is no need to assume that the monocotyledonous condition of the Liliopsida is of more than one origin. A more ample discussion of the nature and possible evolutionary history of the cotyledon in Liliopsida is provided by A. Arber (1925) and A. J. Eames (1961).

The unique septal nectary of many Liliopsida helps to unify the class and also to strengthen the concept that the Alismatidae are near basal. The structure is apparently unknown in the Magnoliopsida. According to W. H. Brown (1938), septal nectaries "occur in the septa between two carpels and represent places where the adjacent walls of the carpels have not fused. They discharge nectar to the outside by means of small openings. They are such complicated structures that they would seem to indicate a relationship between all plants having them."

Septal nectaries are characteristic of those Arecaceae that are nectariferous, and of the Liliales, Bromeliales, and Zingiberales. Not every genus in every family of these orders has septal nectaries, but they are common enough so that their absence is exceptional rather than typical. The Smilacaceae and the tribe Tulipeae of the Liliaceae are among the more notable exceptions. The complex, external nectaries of some of the Zingiberales are evidently derived from septal nectaries (V. S. Rao 1970). Some of the Orchidaceae also have modified septal nectaries. It will be noted that septal nectaries occur in three of the four subclasses of Liliopsida that are typically syncarpous. Most of the Commelinidae lack nectaries entirely.

The antecedents of septal nectaries probably lie in the mostly apocarpous subclass Alismatidae. As W. H. Brown (1938) has pointed out, *Sagittaria* and other Alismatidae have nectaries between the petals and staminodes, and between and around the staminodes and lower carpels. *Alisma*, with a single whorl of separate carpels, has a nectary at the base of the slit between any two adjacent carpels. The palms, which range from apocarpous to syncarpous, have correspondingly alismatoid to septal nectaries. Presumably, a similar change occurred in the line(s) leading to the Zingiberidae and Liliidae.

The Subclasses of Monocots

None of the subclasses of monocots can be considered ancestral to any of the others (fig. 14.8). The Alismatidae, Commelinidae, Zingiberidae, and Liliidae are fairly well characterized, but the Arecidae are a more loosely knit affinity group.

Alistmatidae

The Alismatidae have often been considered to be the most archaic group of Liliopsida. They can scarcely be

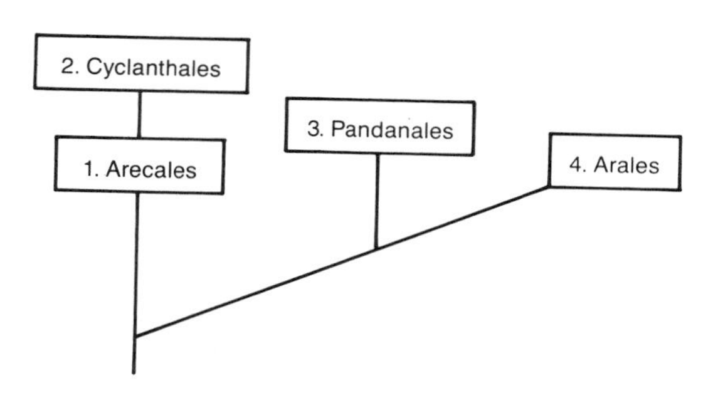

FIGURE 14.10. Putative relationships among the orders of Arecidae.

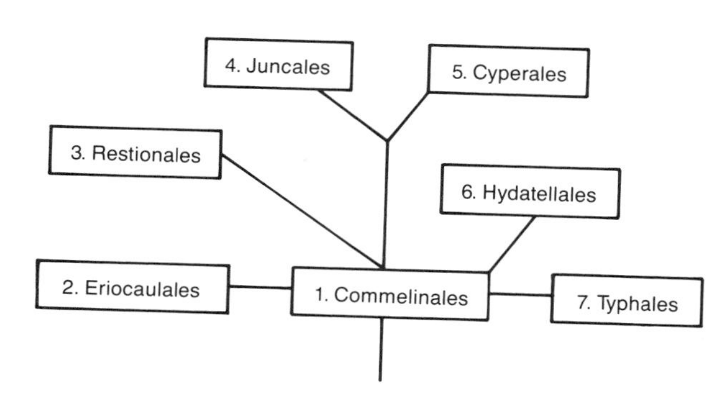

FIGURE 14.11. Putative relationships among the orders of Commelinidae.

on the main line of evolution of the class, however, because a primitive monocot should have binucleate pollen and endospermous seeds. The anomalous small order Triuridales does have endospermous seeds, but the mycotrophic, nongreen habit of this group sets it well apart from any possible mainstream of monocot evolution.

The Alismatidae are here considered to be a near-basal side branch of the monocots, a relictual group that has retained a number of primitive characteristics. The apocarpous gynoecium of most members of the Alismatidae, combined with the mainly 1-aperturate pollen of the Liliopsida as a whole, indicates that any connection of the Liliopsida to the Magnoliopsida must be to the archaic subclass Magnoliidae. It should be noted, however, that the ontogeny of the pleiomerous androecium in the Alismatidae is quite different from that in the Magnoliidae, so that the evolutionary homology can be questioned. The putative relationships among the orders of Alismatidae are given in figure 14.9.

Arecidae

Our two orders of the subclass Arecidae, the Arecales (palms) and Arales (aroids), are only loosely linked and are best discussed separately (fig. 14.10). The palms are the only monocotyledons to combine an arborescent habit, a broad leaf blade, and a well-developed vascular system that has vessels in all vegetative organs. This obviously functional syndrome approaches that of woody dicotyledons, but palms lack an adequate means of secondary growth and a means of expanding the coverage of the crown. Furthermore, palms have never developed the deciduous habit, and with minor exceptions they have not adapted to temperate or cold climates. Thus, their ecological amplitude is limited, as compared with that of woody dicots. They do well in tropical regions that are moist enough to support evergreen tree growth but not moist enough to support a dense forest, and they are also common components of the understory in tropical rainforests.

The Araceae form the principal family of Arales, with the Lemnaceae and Acoraceae as much smaller appendages. The leaves are usually more or less expanded and tend to be net-veined, but they lack the ontogenetic peculiarity previously noted for the palms. Most aroids are herbs of the forest floor or are vines climbing on forest trees. *Pistia,* a free-floating aquatic aroid, has a relatively small spadix. It is seen as pointing the way to the free-floating, thalloid Lemnaceae, which reproduce mainly vegetatively but occasionally produce very small, few-flowered inflorescences. *Acorus,* with ensiform, unifacial leaves, has usually been included in the Araceae, but it also differs from typical Araceae in a number of technical features, including the presence of ethereal oil cells and absence of raphides. Its treatment as a separate family Acoraceae now seems well justified (M. H. Grayum 1987).

Commelinidae

The Commelinidae are a well-marked group characterized by progressive floral reduction, absence of septal nectaries, and eventually the abandonment of insect pollination in favor of anemophily. The Commelinales, standing at the base of the subclass (fig. 14.11), have entomophilous flowers of normal appearance but without nectar or nectaries. The other orders have small flowers of more or less reduced structure, but *Eriocaulon* (Eriocaulales) has reverted to entomophily and has

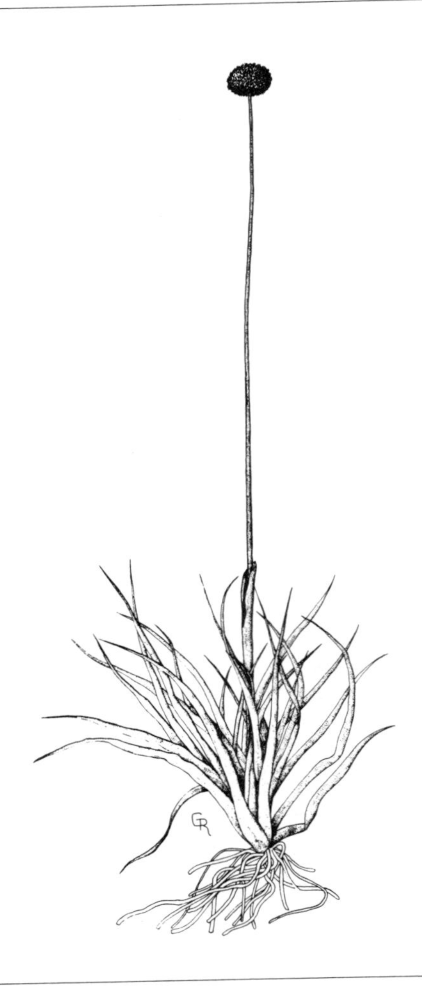

Eriocaulon compressum. [From R. K. Godfrey and J. W. Wooten. 1979. Aquatic and Wetland Plants of Southestern United States: Monocotyledons. Athens, Ga. Fig. 293, by Grady Reinert. Reprinted with permission from the University of Georgia Press.]

nectariferous glands just within the tip of the tiny petals. The Typhales (Typhaceae and the closely related Sparganiaceae) have been variously disposed in different schemes. They are here included in the Commelinidae because of their floral reduction associated with anemophily, their paracytic stomates, the presence of vessels in all vegetative organs, their mealy, starch-bearing endosperm, and the bound ferulic acid in their cell walls. All of these features are perfectly fine for the Commelinidae, but collectively they are difficult to reconcile with any other subclass.

Although the Cyperaceae and Poaceae have traditionally been associated in older systems of classification, in recent years several authors have taken each family to represent a separate, unifamilial order, or have associated the Cyperaceae with the Juncaceae and the Poaceae with the Restionales. Neither of these alternatives is really wrong, but I am not convinced that the change is yet necessary. When the taxonomic distribution of diffuse centromeres in this group of families is better known, it may be time to reconsider the ordinal alignments.

Zingiberidae

The two sharply defined orders (Bromeliales and Zingiberales) that collectively make up the subclass Zingiberidae have not usually been closely associated in past systems. Nevertheless, they would be discordant elements in any other subclass, and they have certain features in common that collectively distinguish them from other subclasses. They resemble the Liliidae (and differ from the Commelinidae) in commonly having septal nectaries and in usually having the vessels confined to the roots.

They resemble Commelinidae (and differ from most Liliidae) in their starchy endosperm with compound starch grains, and they further differ from typical Liliidae (and resemble the order Commelinales) in having the sepals well differentiated from the petals, often green and herbaceous in texture. They differ from both the Liliidae and the Commelinidae in that the number of subsidiary cells around the stomates is usually four or more.

Liliidae

The Liliidae characteristically (although not always) have showy flowers, with the tepals all petaloid, and they have intensively exploited insect pollination. With few exceptions, they have not taken the path of floral reduction, and none of them has a spadix. Although some few Liliidae are arborescent, some have broad, net-veined leaves, and some have vessels throughout the shoot as well as in the root, none of them has coordinated these features into a working system that could present a competitive challenge to woody dicotyledons.

I take the narrow, parallel-veined leaf to be primitive within the Liliales. Broader, more or less net-veined types have evolved repeatedly but still show traces of the ancestral condition. Often they have several main veins that arch out from the base and converge toward the tip. It is not difficult to envisage the evolution of this type of venation by increase in the width of the blade, without any increase in the number of main veins. Concomitantly, the connecting cross-veins are elaborated to vascularize the expanded space between the main veins.

I take the primitive type of endosperm in the Liliales to be fleshy or cartilaginous, with food reserves of protein, oil, and possibly some starch, but not hemicellulose. This type is readily compatible with the endosperm of archaic dicotyledons. Both strongly starchy endosperm and very hard endosperm with reserves of hemicellulose in the thickened cell walls appear to be advanced features.

Although the Dioscoreaceae and the geographically extralimital family Taccaceae do not appear to be primitive in other respects, they have a more primitive embryo than other families of the Liliales. The embryo is slipper-shaped or obliquely ovate, the cotyledon is evidently lateral, and the plumule is more or less distinctly terminal. In the other families, the embryo is mostly barrel-shaped or ellipsoid to ovoid or cylindric, with a terminal cotyledon and a tiny, lateral, often scarcely distinguishable plumule that may be sunken into a small pocket.

Despite the embryo, the Dioscoreaceae are a poor candidate for a position near the base of the monocotyledons, as recently proposed by a few authors. They have a compound, inferior ovary with septal nectaries, and a horny endosperm with hemicellulose deposited in the thickened cell walls. An extraordinary reversal of current views on polarity would be required to consider these as plesiomorphic features in monocotyledons in general. Furthermore, if the views here presented about evolution of leaf form and venation in the monocots are correct, the broad, more or less net-veined leaves of the Dioscoreaceae must remove them even farther from the ancestry of the class.

The delimitation of the family Liliaceae is much disputed. In the traditional Englerian system most of the lilioid monocots with a superior ovary were put into the Liliaceae, and those with an inferior ovary into the Amaryllidaceae. All hands now agree, however, that a cleavage on this basis is unnatural, because the genera with inferior ovaries properly belong in several groups that relate separately to the ancestral group with a superior ovary.

There is enough diversity within the traditional Liliaceae plus Amaryllidaceae to provide for several families, but the problem is how to go about the dismemberment. The separation of *Smilax* and its allies as a distinct family (or cluster of smaller families) now seems to be generally accepted. Most authors now also accept the Agavaceae, but they are not agreed as to its delimitation. If the Agavaceae are defined cytologically (x = 5 large and 25 small chromosomes), then they include about 8 genera of coarse plants with firm, perennial leaves, plus the habitally very different genus *Hosta*. If, on the other hand, the Agavaceae are defined largely on the basis of habit and chemistry, then the number of genera goes up to about 18, and the chromosome number is more variable but the karyotype still comprises a few large chromosomes and many small ones. A consistent treatment then requires the recognition of the mainly African family Aloaceae as another line that has independently undergone the same sort of habital changes as the Agavaceae.

I have chosen to recognize the Agavaceae, Aloaceae, and Smilacaceae as distinct families, and to relegate the remainder of the traditional Liliaceae plus Amaryllidaceae to a single rather amorphous family Liliaceae. I would be happy enough to divide this group into several families, if I could find a reasonable way to do it, but I have not found the way. R. M. T. Dahlgren et al. (1985) distributed the Liliaceae sensu meo into 27 families in 4 orders (Melanthiales, part of the Liliales, part of the Dioscoreales, and part of the Asparagales) with such minimal differences that I have not been able to comprehend the essential nature of the groups. We still await a comprehensive reorganization of the lilies into several families more comparable to other recognized families of angiosperms.

Rolf Martin Teodor Dahlgren, 1932–1987. [Hunt Institute, Carnegie Mellon University, Pittsburgh, photograph by W. H. Hodge.]

The Orchidales differ from the Liliales essentially in their strongly mycotrophic habit, and in their very numerous, tiny seeds with a minute, mostly undifferentiated embryo and no endosperm. The reduction of the embryo is at least in part a consequence of mycotrophy; these two features are also associated in some other groups of angiosperms. The ovary of the Orchidales is always inferior and only seldom has typical septal nectaries, although the ovarian nectaries of some Burmanniaceae and Orchidaceae may well be derived from septal nectaries.

The combination of mycotrophy and numerous tiny seeds offers certain evolutionary opportunities as well as imposing some limitations. The plants are physiologically dependent on their fungal symbionts, sometimes for food, sometimes only for other factors as yet not fully understood, but in any case they can grow only where their fungal symbiont finds suitable conditions. The dustlike seeds of the Orchidaceae are admirably adapted to being carried by the wind and lodging in the bark of trees, and many orchids are epiphytes. The production of many ovules is, of course, of no value if they are not fertilized. One way to increase the likelihood of fertilization is to offer special attractants to a limited set of pollinators and to have the pollen grains stick together in masses so that many are transported at once. This strategy puts the plants in thrall to their pollinators, but it opens the door to explosive speciation.

Only the Orchidaceae have efficiently exploited the evolutionary possibilities of the order. The floral characteristics that distinguish the Orchidaceae from their immediate allies clearly reflect progressive specialization for the massive transfer of pollen by specific pollinators.

The Orchidales are evidently derived from the Liliales as here defined. All the characteristics in which the Orchidales differ from the Liliales are apomorphies. Within the Liliales, only the epigynous segment of the Liliaceae has the characteristics from which the Orchidaceae, Burmanniaceae, and Corsiaceae (extraterritorial to our flora) might well have arisen. Although the Orchidaceae never have more than three stamens, two of these are considered to come from the ancestral inner cycle, and one from the outer. Thus a hexandrous ancestry seems likely.

Epilogue

Nothing can be more certain than that future studies will lead to changes in the system here discussed. I can hope that considerable segments of it are now stable, but only time will tell. Just as we stand on the shoulders of our predecessors, future taxonomists will stand on ours.

15. Flowering Plant Families: An Overview

James L. Reveal

One of the purposes of the Flora of North America project is to present a series of established names at all ranks for extant vascular plants north of Mexico, and thus by its very nature, this work will be a standard by which all future modifications will be judged. The problems are more acute at the generic and specific ranks, but major changes will undoubtedly occur also at the ranks of family and order during the useful life of this and the following volumes.

This is not an indictment of modern systems of classification, especially that proposed by Cronquist and adopted here (A. Cronquist 1981, 1988). Rather, the state of our knowledge regarding the circumscriptions of vascular plant taxa at higher ranks will continue to undergo revision, as a result of improvements suggested by an ever-expanding group of biologists who are using new techniques to address questions of relationship.

The longstanding difficulty that these researchers will encounter, and one that the present work is certain to promote, is the problem of acceptance of any change in the established norm. The wide acceptance of the de Candolle and Lindley systems during the first half of the nineteenth century certainly slowed the adoption of changes suggested by Bentham and Hooker in the 1860s. In contrast, the changes proposed by Engler and Prantl in the last two decades of the 1800s were readily adopted because, by then, the need for change was recognized. Moreover, the detailed studies by Engler and Prantl and their associates justified acceptance of the proposed modifications. The Engler and Prantl model remained the norm until the late 1950s when Cronquist and Takhtajan provided a series of innovative changes that were accepted because, once again, a need for substantial change was recognized, and sound rationales were presented for that change. During each era, including the present, other views were presented (Airy Shaw in J. C. Willis 1973; R. M. T. Dahlgren 1983; R. M. T. Dahlgren et al. 1985; G. Dahlgren 1989, 1989b; A. Goldberg 1986, 1989; V. H. Heywood 1978; J. Hutchinson 1959, 1973; D. J. Mabberley 1987; H. Melchior 1964; E. Rouleau 1981; G. L. Stebbins 1974; A. L. Takhtajan 1987; R. F. Thorne 1983, 1992, 1992b), but at some point, one system began to dominate.

With this in mind, the following commentary is offered. It is a reminder to the reader that there are differing views as to circumscriptions and relationships. Finally, whereas classifications at the lower ranks can be evaluated readily in the herbarium or in the field, relationships at the family and ordinal ranks are more difficult to demonstrate simply because of the time required to gain sufficient knowledge.

Cronquist (1981, 1988) used 3 major superfamilial ranks within Magnoliophyta, the flowering plants (appendix 15.1): class, subclass, and order. In his system, the dicotyledons were placed in Magnoliopsida, the monocotyledons in Liliopsida. Within Magnoliopsida, he distinguished 6 subclasses and 64 orders; in Liliopsida, 5 subclasses and 19 orders. A total of 389 families were accepted, 323 of them dicots and 66 monocots.

Magnolia glauca. **Lithograph by Isaac Sprague (1811–1895). [Hunt Institute, Carnegie Mellon University, Pittsburgh.]**

A. L. Takhtajan (1987) used 4 major superfamilial ranks: class, subclass, superorder, and order. Eight subclasses of dicotyledons were distinguished and 4 among the monocots. Within the dicots, Takhtajan accepted 36 superorders, 128 orders, and 429 families; within the monocots, he had 16 superorders, 36 orders, and 104 families. Recently Takhtajan (pers. comm. 1993) has accepted 450 dicot families arranged among 145 orders, 45 superorders, and 10 subclasses. R. F. Thorne (1992) currently recognizes 437 families (351 dicots) distributed among 28 superorders (19 dicots) and 71 orders (52 dicots). (See appendix 15.2 for a concordance of the families recognized by these three authors.)

The first subclass in the Cronquist system is Magnoliidae. Takhtajan divided this into two subclasses, Magnoliidae and Ranunculidae, and the first three superorders of the Thorne system basically encompass the same series of families. These three authors arranged the orders and families in differing linear sequences, although the overall relationships among the families are remarkably similar, with only a few exceptions (e.g., Takhtajan and Thorne placed Coriariaceae and Sabiaceae with the rosoids).

As to circumscriptions of the North American families included in Cronquist's Magnoliidae, some differences of opinion exist. Cronquist defined Ranunculaceae to include Glaucidiaceae and Hydrastidaceae, which both Takhtajan and Thorne recognized as separate. Cronquist and Takhtajan maintain Fumariaceae, which Thorne put in Papaveraceae.

Hamamelididae (spelled by Cronquist as "Hamamelidae") were treated by both Takhtajan and Thorne in two parts. Takhtajan defined Hamamelididae to include such North American families as Hamamelidaceae (plus Altingiaceae, which Cronquist retained in Hamamelidaceae), Platanaceae, Fagaceae, Betulaceae, Myricaceae, Juglandaceae, and Casuarinaceae. Takhtajan placed Ulmaceae, Moraceae, Cannabaceae, and Urticaceae in Dilleniidae. Takhtajan and Thorne included among their hamamelids two families that Cronquist placed in Rosidae: Buxaceae and Simmondsiaceae.

The placement of Leitneriaceae is uncertain. Cronquist included them within their own order in Hamamelididae, between his Urticales and Juglandales. Takhtajan placed the order in Rosidae next to Rutales, and Thorne included the family within Rutales. Most of the hamamelid families Cronquist recognized are circumscribed similarly by other recent workers.

Caryophyllidae, with the exception of Polygonaceae

and Plumbaginaceae, are a tightly knit group, uniformly circumscribed by most recent workers. There is some disagreement over the circumscriptions of the various families.

The placement of Plumbaginaceae has been either near Caryophyllales or Primulales. Cronquist and Takhtajan accepted the former position, Thorne the latter. The isolated nature of Polygonaceae has not been questioned, and their association with Caryophyllales is based more on tradition than on firm evidence.

Cronquist's Dilleniidae, with 77 families, are the second largest subclass of his Magnoliopsida, and one whose representatives in our flora tend to be highly specialized. The families of the subclass are more narrowly defined by Takhtajan (who accepted 153), and they were somewhat more scattered among the rosoid families by Thorne. The overall arrangement of the families, however, is fairly similar.

The traditional inclusion of Paeoniaceae in Ranunculaceae is no longer accepted, with Cronquist referring the family to Dilleniidae. Takhtajan and Thorne placed Paeoniaceae near Glaucidiaceae (which Cronquist included within Ranunculaceae) in Ranunculidae.

Cronquist placed Symplocaceae in Ebenales, whereas Takhtajan and Thorne retained the family in Theales. Cronquist's placement of Sarraceniaceae, Nephenthaceae, and Droseraceae in a single order, Nepenthales, in Dilleniidae, has been questioned by monographers and others. Nonetheless, there is no firm opinion as to an acceptable alternative. Takhtajan retained Sarraceniaceae in Dilleniidae, Nepenthaceae in Magnoliidae, and Droseraceae in Rosidae. Thorne's placements have been equally diverse.

The Flacourtiaceae of Cronquist and most previous workers are the "garbage pail" of Dilleniidae. They were separated by Takhtajan into Berberidopsidaceae, Aphloiaceae, Kiggelariaceae, and Plagiopteraceae, but only the Plagiopteraceae were removed from the immediate vicinity of Flacourtiaceae.

Cronquist included the Loasaceae, without conviction, in Violales of Dilleniidae. Thorne and Takhtajan placed the family in its own superorder near Solanaceae in Asteridae. All three concurred that the Salicaceae belong to Dilleniidae (or their equivalent) and are not among the amentiferous families as defined by Engler and Prantl. The closeness of Tamaricaceae, Frankeniaceae, and Fouquieriaceae to Salicaceae was also accepted.

A consensus on the definition of Ericaceae has been gradually emerging, with the various families recognized through most of the 1800s now reduced primarily to Ericaceae, with the occasional recognition (as by Cronquist) of Pyrolaceae and Monotropaceae.

Rosidae, as defined by Cronquist, comprise 18 orders and 116 families, many being common in temperate regions although most have tropical distributions. Rosales are the basal element; they are a coherent group with little disagreement about their overall makeup. Opinions vary regarding the circumscriptions of families that Engler and Prantl related to Saxifragaceae, and regarding the relationship of Cornales (Cornaceae in a broad sense) to the woody members of Saxifragaceae. The circumscription of Grossulariaceae by Cronquist included most of the woody saxifrage genera. Other recent authors have restricted, more properly in my opinion, Grossulariaceae to the genus *Ribes* and have distinguished several additional families for North American plants, notably Escalloniaceae, Phyllonomaceae, Iteaceae, and Pterostemonaceae. The apparent similarity of woody saxifrage families with Cornales was noted by both Takhtajan and Thorne.

The circumscription of herbaceous Saxifragaceae is equally unsettled. Cronquist included Lepuropetalaceae, Parnassiaceae, and Penthoraceae in his Saxifragaceae, but others have preferred to recognize these as distinct.

The arguments for and against treating Rosaceae and Fabaceae (Fabales) as several families have been discussed since the early 1800s. Cronquist, Takhtajan, and Thorne all recognize Rosaceae in a broad sense. Cronquist has separated Mimosaceae and Caesalpiniaceae from Fabaceae, but both Takhtajan and Thorne included them in the Fabaceae.

Cronquist placed the Euphorbiaceae in Rosidae, along with Buxaceae and Simmondsiaceae. Takhtajan, however, placed Euphorbiaceae in a terminal position in his Dilleniidae, with Buxaceae and Simmondsiaceae placed in their own orders in Hamamelididae. Thorne retained Simmondsiaceae in Euphorbiales near Urticales, and moved Buxaceae into the equivalent of Hamamelididae.

Cronquist's placement of Rhamnales next to Euphorbiales was accepted by Thorne but rejected by Takhtajan. Both Takhtajan and Thorne placed Elaeagnaceae in an order, Elaeagnales, situated between Santalales and Proteales. Cronquist placed Elaeagnaceae and Proteaceae in Proteales.

The circumscription of Zygophyllaceae was reviewed by Takhtajan, who recognized five families where Cronquist had but one; the additional North American name among the five is Peganaceae. Likewise, Takhtajan recognized five families in what Cronquist defined as Geraniaceae, but all of the segregates are beyond our flora.

The distinction between Araliaceae and Apiaceae has long been regarded as weak, with tradition being as significant as morphological characters in distinguishing the two families. Combining the two was promoted by Thorne (1983), who now suggests (1992, 1992b), as have many Old World workers, that if a distinction is made between Araliaceae and Apiaceae, then Hydrocotylaceae must be recognized as well. Takhtajan (pers. comm. 1993) now concurs.

The majority of the sympetalous dicotyledons were placed in Asteridae by Cronquist, who defined the group as a subclass of 11 orders and 49 families. Takhtajan, on the other hand, recognized two unequal groups, the bulk of the families going to Lamiidae, and Asteridae being restricted mainly to Campanulales and Asterales. In addition, Takhtajan referred Dipsacales to a terminal position within Rosidae.

Cronquist has held to traditional circumscriptions of the 49 families of the Asteridae. The families tend to have large numbers of species, with members found in a variety of both temperate and tropical habitats. Even when the families are more narrowly defined, as done by Takhtajan, they still tend to be large and widely distributed.

The basal group within Asteridae is Gentianales. Takhtajan divided Loganiaceae into three families, including Spigeliaceae, and placed Rubiaceae next to them, a position with which Thorne agreed. The traditional separation of Apocynaceae and Asclepiadaceae was maintained by Cronquist but not accepted by Thorne.

Takhtajan divided Cronquist's Solanaceae into three families, and he recognized two New World tropical families, Sclerophylacaceae and Goetzeaceae. Both Cronquist and Thorne separated Cuscutaceae from Convolvulaceae, but Thorne did not.

The diverse Scrophulariaceae and their allies were variously treated. Thorne and Takhtajan have retained Orobanchaceae within the Scrophulariaceae. Cronquist placed the Paulowniaceae in Bignoniaceae, but Takhtajan and Thorne put them in Scrophulariaceae. Selaginaceae, known only from cultivation in our range, was included in Globulariaceae by Cronquist, in Scrophulariaceae by Takhtajan, and was recognized as distinct by Thorne. Martyniaceae, placed in Pedaliaceae by Cronquist, were accepted as distinct by Takhtajan and Thorne.

Lobeliaceae were distinguished from Campanulaceae by Takhtajan, but treated within the Campanulaceae by both Cronquist and Thorne. In addition, Takhtajan adopted Nemacladaceae for three genera from western North America; he also distinguished Cyphiaceae and Cyphocarpaceae, two families beyond our flora.

Within Asteridae, the Caprifoliaceae and relatives are particularly complex. Thorne has moved Viburnaceae and Sambucaceae into an expanded Adoxaceae.

Notable studies have been done recently on the families in the Liliopsida, with the late Rolf Dahlgren in the forefront (R. M. T. Dahlgren et al. 1985). His premature death slowed progress toward understanding the evolution of the flowering plants.

Dahlgren et al. rejected the concept that the monocots had an aquatic ancestry, and their view was adopted by Thorne but not by Cronquist or Takhtajan. While this view does not alter the circumscription of any family, it does influence what is to be regarded as the basal taxon. Cronquist and Takhtajan considered this to be Alismatidae; Dahlgren and Thorne began with the equivalent of Liliidae.

The tradition of distinguishing monocots and dicots at one of the higher taxonomic ranks, as by Cronquist and Takhtajan at the rank of class, was not retained by Dahlgren and Thorne, who considered the flowering plants to be divided into a series of superorders. True, members of each of their superorders have either monocotyledonous or dicotyledonous features, and no one superorder includes both monocotyledonous and dicotyledonous members. Dahlgren and Thorne emphasized that there is no fundamental difference in these features and that monocotyledonous plants arose from several different dicotyledonous groups.

In the monocots, Cronquist maintained families with broad, traditional definitions, while Takhtajan generally circumscribed families more narrowly. The interrelationships posited for particular groups, ranking aside, are rather similar in the two schemes. Even the groupings proposed by Dahlgren and Thorne generally comprised the same complements of families. But there remains great disagreement regarding the lilioid families.

Cronquist defined Liliales as a taxon of 15 families and nearly 8000 species, with Liliaceae defined to include what Dahlgren considered to represent 4 orders (at least in part) and 27 families! In 1988 Cronquist labeled his Liliaceae "inchoate," and stated that he would divide them into several families if he could "find a reasonable way to do it." Takhtajan did attempt to find a reasonable way, and he distributed elements found in Cronquist's Liliaceae among 5 orders (Liliales, Amaryllidales, Asparagales, Dioscoreales, and Alstroemeriales) containing 23 families. Thorne now has recognized 3 orders (Liliales, Asparagales, and Dioscoreales) with 20 families.

The problems noted with Liliaceae are mirrored in the circumscriptions of other families such as Agavaceae, Aloaceae, and Smilacaceae. For the most part, resolution involves families found outside our area, but one may be noted here. Aloaceae were defined by Cronquist to exclude Asphodelaceae, which he placed in Liliaceae. Dahlgren, Thorne, and Takhtajan considered Aloaceae to be no more than a subfamily or tribe of Asphodelaceae.

Conclusion

Systems of classification attempt to show relationships and to provide a convenient means of expressing them. Such arrangements are difficult to substantiate, but consensus on at least some of them is possible.

The differences of opinion expressed by Cronquist, Takhtajan, and Thorne in their respective classifications are not as great as they might seem. Rarely does one author circumscribe a family to include taxa that another assigns to widely separated places. More frequently, a group of families will be associated by one author, but scattered about by another. In most of these instances, the families involved are highly specialized (e.g., Droseraceae, Nepenthales). Where there are truly complex differences of opinion (e.g., Grossulariaceae and Cornales), there is often no evident way to resolve them. While Cronquist has accepted only 388 families as compared to 533 recognized by Takhtajan, the difference is not so much a question of relationship as one of rank (J. L. Reveal 1993). The vast majority of segregate families accepted by Takhtajan are directly associated in his scheme with the families into which Cronquist has put them.

Revised and more sophisticated taxonomic schemes will certainly enhance our knowledge of plants. One of the strengths of the Linnaean system of botanical nomenclature, however, is its fundamental stability. Thus, despite altered family conceptions and revised family sequences, the names by which plants are called (i.e., genus and species names) nearly always remain the same. For example, it makes no difference to the name *Desmanthus illinoensis,* the Illinois bundleflower, if the plant is treated as a member of the Mimosaceae, or within a semidistinctive subfamily of the Fabaceae, the bean family. Similarly, the nomenclature within the genus *Euphorbia* (the spurges) is not affected whether its family, the Euphorbiaceae, is treated among the Rosidae, as in Cronquist, or among the Dilleniidae, as in Takhtajan.

Acknowledgments

This summary would not have been possible without the assistance of the late Arthur Cronquist, Armen L. Takhtajan, and Robert F. Thorne, who have kindly provided me with their most up-to-date information. In particular, Drs. Takhtajan and Thorne have provided me with information that they have yet to publish, and for permission to use their data I am most grateful. My colleague in the review of family nomenclature, Ruurd D. Hoogland, remains, as always, most helpful. Work on the nomenclature of vascular plant families is supported in part by National Science Foundation Grant BSR-8812816.

APPENDIX 15.1. Families of Flowering Plants and Pertinent Synonymy According to the Cronquist System

A. Magnoliopsida Cronquist, Takhtajan & Zimmermann, 1966.

I. Magnoliidae Novák ex Takhtajan, 1967.
Nelumbonanae Takhtajan ex Reveal, 1992.
Ranunculanae Takhtajan ex Reveal, 1992.

1. Magnoliales Bromhead, 1838.
Annonales Lindley, 1833.
Austrobaileyales Takhtajan ex Reveal, 1992.
Eupomatiales Takhtajan ex Reveal, 1992.
Lactoridales Takhtajan ex Reveal, 1993.
Winterales A. C. Smith ex Reveal, 1993.

1. Winteraceae R. Brown ex Lindley, 1830, nom. cons.
Takhtajaniaceae (J. Leroy) J. Leroy, 1980.

2. Degeneriaceae I. Bailey & A. C. Smith, 1942, nom. cons.

3. Himantandraceae Diels, 1917, nom. cons.

4. Eupomatiaceae Endlicher, 1841, nom. cons.

5. Austrobaileyaceae (Croizat) Croizat, 1943, nom. cons.

6. Magnoliaceae A. L. de Jussieu, 1789, nom. cons.
Liriodendraceae F. Barkley, 1975.

7. Lactoridaceae Engler, 1888, nom. cons.

8. Annonaceae A. L. de Jussieu, 1789, nom. cons.
Hornschuchiaceae J. Agardh, 1858.
Monodoraceae J. Agardh, 1858.

9. Myristicaceae R. Brown, 1810, nom. cons.

10. Canellaceae Martius, 1832, nom. cons.

2. Laurales Perleb, 1826

11. Amborellaceae Pichon, 1948, nom. cons.

12. Trimeniaceae (Perkins & Gilg) Gibbs, 1917, nom. cons.

13. Monimiaceae A. L. de Jussieu, 1809, nom. cons.
Atherospermataceae R. Brown, 1814.
Siparunaceae (A. de Candolle) Schodde, 1970.
Hortoniaceae (Perkins & Gilg) A. C. Smith, 1971.

14. Gomortegaceae Reiche, 1896, nom. cons.

15. Calycanthaceae Lindley, 1819, nom. cons.
Chimonanthaceae Perleb, 1838.

16. Idiospermaceae S. T. Blake, 1972.

17. Lauraceae A. L. de Jussieu, 1789, nom. cons.
Cassythaceae Bartling ex Lindley, 1833, nom. cons.
Perseaceae Horaninow, 1834.

18. Hernandiaceae Blume, 1826, nom. cons.
Gyrocarpaceae Dumortier, 1829.
Illigeraceae Blume, 1833.

3. Piperales Dumortier, 1829.
Chloranthales A. C. Smith ex J.-F. Leroy, 1983.

19. Chloranthaceae Blume, 1827, nom. cons., emend. prop.
Hedyosmaceae Caruel, 1881

20. Saururaceae Richard ex E. Meyer, 1827, nom. cons.

21. Piperaceae C. Agardh, 1824, nom. cons.
Peperomiaceae A. C. Smith, 1981.

4. Aristolochiales Dumortier, 1829.

22. Aristolochiaceae A. L. de Jussieu, 1789, nom. cons.
Asaraceae Ventenat, 1799.

5. Illiciales H. H. Hu ex Cronquist, 1981.

23. Illiciaceae (de Candolle) A. C. Smith, 1947, nom. cons.

24. Schisandraceae Blume, 1830, nom. cons.

6. Nymphaeales Dumortier, 1829.
Nelumbonales Burnett, 1835.
Ceratophyllales Bischoff, 1840.

25. Nelumbonaceae (de Candolle) Dumortier, 1829, nom. cons.

26. Nymphaeaceae Salisbury, 1805, nom. cons.
Euryalaceae J. Agardh, 1858.
Nupharaceae A. Kerner, 1891.

27. Barclayaceae (Endlicher) H. L. Li, 1955.

28. Cabombaceae A. Richard, 1828, nom. cons.
Hydropeltidaceae (de Candolle) Dumortier, 1822.

29. Ceratophyllaceae Gray, 1821, nom. cons.

7. Ranunculales Dumortier, 1829.
Glaucidiales Takhtajan ex Reveal, 1992.

30. Ranunculaceae A. L. de Jussieu, 1789, nom. cons.
Thalictraceae Rafinesque, 1815.
Anemonaceae Vest, 1818.
Helleboraceae Vest, 1818.
Calthaceae Martinov, 1820.
Clematidaceae Martinov, 1820.
Hydrastidaceae Martinov, 1820.
Actaeaceae Rafinesque, 1828.
Nigellaceae J. Agardh, 1858.
Delphiniaceae Brenner, 1886.
Glaucidiaceae Tamura, 1972.

31. Circaeasteraceae Hutchinson, 1926, nom. cons.
Kingdoniaceae A. Foster ex Airy Shaw, 1965.

32. Berberidaceae A. L. de Jussieu, 1789, nom. cons.
Leonticaceae Berchtold & J. S. Presl, 1820.
Podophyllaceae de Candolle, 1821, nom. cons.

APPENDIX 15.1 (Continued).

Diphylleiaceae Schultz-Schultzenstein, 1832.
Nandinaceae Horaninow, 1834.

33. Sargentodoxaceae O. Stapf ex Hutchinson, 1926, nom. cons.
34. Lardizabalaceae Decaisne, 1839, nom. cons.
35. Menispermaceae A. L. de Jussieu, 1789, nom. cons.
    Pseliaceae Rafinesque, 1838.
36. Coriariaceae de Candolle, 1824, nom. cons.
37. Sabiaceae Blume, 1851, nom. cons.
    Wellingtoniaceae Meisner, 1840.
    Meliosmaceae Endlicher, 1841.

8. Papaverales Dumortier, 1829.
38. Papaveraceae A. L. de Jussieu, 1789, nom. cons.
    Chelidoniaceae Martinov, 1820.
    Eschscholziaceae Seringe, 1847.
    Platystemonaceae (W. R. Ernst) A. C. Smith, 1971.
39. Fumariaceae de Candolle, 1821, nom. cons.
    Hypecoaceae (Dumortier) Willkomm & Lange, 1880.
    Pteridophyllaceae (Murbeck) Nakai ex Reveal & Hoogland, 1991.

II. Hamamelididae Takhtajan, 1967.

9. Trochodendrales Takhtajan ex Cronquist, 1981.
40. Tetracentraceae A. C. Smith, 1945, nom. cons.
41. Trochodendraceae Prantl, 1888, nom. cons.

10. Hamamelidales Grisebach, 1854.
    Cercidiphyllales H. H. Hu ex Reveal, 1992.
    Eupteleales H. H. Hu ex Reveal, 1992.
    Myrothamnales Nakai ex Reveal, 1992.
42. Cercidiphyllaceae Engler, 1909, nom. cons.
43. Eupteleaceae K. Wilhelm, 1910, nom. cons.
44. Platanaceae Lestiboudois ex Dumortier, 1829, nom. cons.
45. Hamamelidaceae R. Brown, 1818, nom. cons.
    Fothergillaceae Nuttall, 1818.
    Parrotiaceae Horaninow, 1834.
    Altingiaceae Lindley, 1846, nom. cons.
    Disanthaceae Nakai, 1943.
    Rhodoleiaceae Nakai, 1943.
46. Myrothamnaceae Niedenzu, 1891, nom. cons.

11. Daphniphyllales Pulle ex Cronquist, 1981.
47. Daphniphyllaceae Müller Argoviensis, 1869, nom. cons.

12. Didymelales Takhtajan, 1967.
48. Didymelaceae Leandri, 1937.

13. Eucommiales Němejc ex Cronquist, 1981.
49. Eucommiaceae Engler, 1909, nom. cons.

14. Urticales Dumortier, 1829.
    Barbayales Takhtajan ex Reveal & Takhtajan, 1993.
50. Barbeyaceae Rendle, 1916, nom. cons.
51. Ulmaceae Mirbel, 1815, nom. cons.
    Celtidaceae Link, 1829.
52. Cannabaceae Endlicher, 1837, nom. cons.
    Lupulaceae Link, 1829.
53. Moraceae Link, 1829, nom. cons.
    Artocarpaceae R. Brown, 1818.
    Dorsteniaceae Chevallier, 1827.
    Ficaceae (Dumortier) Dumortier, 1829.
54. Cecropiaceae C. C. Berg, 1978.
55. Urticaceae A. L. de Jussieu, 1789, nom. cons.
56. Physenaceae Takhtajan, 1985.

15. Leitneriales Engler, 1897.
57. Leitneriaceae Bentham, 1880, nom. cons.

16. Juglandales Dumortier, 1829.
    Rhoiptelales Novak ex Reveal, 1992.
58. Rhoipteleaceae Handel-Mazzetti, 1932, nom. cons.
59. Juglandaceae A. Richard ex Kunth, 1824, nom. cons.
    Platycaryaceae Nakai, 1930.
    Pterocaryaceae Nakai, 1930.

17. Myricales Engler, 1897.
60. Myricaceae Blume, 1829, nom. cons., emend. prop.

18. Fagales Engler, 1892.
    Betulales Burnett, 1835.
    Balanopales Engler, 1897.
61. Balanopaceae Bentham, 1880, nom. cons.
62. Ticodendraceae Gómez-Laurito & Gómez P., 1991.
63. Fagaceae Dumortier, 1829, nom. cons.
    Quercaceae Berchtold & J. S. Presl, 1820.
    Castaneaceae Baillon, 1878.
64. Nothofagaceae Kuprianova, 1962.
65. Betulaceae Gray, 1821, nom. cons.
    Corylaceae Mirbel, 1815, nom. cons.
    Carpinaceae Vest, 1818.

19. Casuarinales Lindley, 1833.
66. Casuarinaceae R. Brown, 1814, nom. cons.

III. Caryophyllidae Takhtajan, 1967

20. Caryophyllales Perleb, 1826.
67. Phytolaccaceae R. Brown, 1818, nom. cons.

APPENDIX 15.1 (Continued).

Petiveriaceae C. Agardh, 1824.
Rivinaceae C. Agardh, 1824.
Sarcocaceae Rafinesque, 1837.
Agdestidaceae (Heimerl) Nakai, 1942.
Barbeuiaceae (H. Walter) Nakai, 1942.
Gisekiaceae (Endlicher) Nakai, 1942.
Hilleriaceae Nakai, 1942.
Seguieriaceae Nakai, 1942.
Stegnospermataceae (A. Richard) Nakai, 1942.

68. Achatocarpaceae Heimerl, 1934, nom. cons.

69. Nyctaginaceae A. L. de Jussieu, 1789, nom. cons.
Jalapaceae Batsch, 1802.
Allioniaceae Horaninow, 1834.
Bougainvilleaceae J. Agardh, 1858.
Pisoniaceae J. Agardh, 1858.
Mirabilidaceae W. Oliver, 1936.

70. Aizoaceae F. Rudolphi, 1830, nom. cons.
Ficoideaceae A. L. de Jussieu, 1789.
Carpanthaceae Rafinesque, 1808.
Sesuviaceae Horaninow, 1834.
Mesembryanthemaceae Fenzl, 1836, nom. cons.
Tetragoniaceae Nakai, 1942, nom. cons.

71. Didiereaceae Drake del Castillo, 1903, nom. cons.

72. Cactaceae A. L. de Jussieu, 1789, nom. cons.
Opuntiaceae Martinov, 1820.
Cereaceae de Candolle & Sprengel, 1821.
Leuchtenbergiaceae Salm-Dyck, 1854.

73. Chenopodiaceae Ventenat, 1799, nom. cons.
Atriplicaceae A. L. de Jussieu, 1789.
Salicorniaceae Martinov, 1820.
Corispermaceae Link, 1829.
Betaceae Burnett, 1835.
Salsolaceae Moquin-Tandon, 1849.
Blitaceae Adanson ex T. Post & Kuntze, 1903.
Dysphaniaceae (Pax) Pax, 1927, nom. cons.
Halophytaceae Soriano, 1984.

74. Amaranthaceae A. L. de Jussieu, 1789, nom. cons.
Celosiaceae Martinov, 1820.
Achyranthaceae Rafinesque, 1837.
Gomphrenaceae Rafinesque, 1837.
Deeringiaceae J. Agardh, 1858.

75. Portulacaceae A. L. de Jussieu, 1789, nom. cons.
Montiaceae Rafinesque, 1820.
Hectorellaceae Philipson & Skipworth, 1961.

76. Basellaceae Moquin-Tandon, 1840, nom. cons.
Anrederaceae J. Agardh, 1858.
Ullucaceae Nakai, 1942.

77. Molluginaceae Hutchinson, 1926, nom. cons.
Pharnaceaceae Martinov, 1820.
Glinaceae Link, 1829.
Adenogrammaceae (Fenzl) Nakai, 1942.
Polpodaceae (Fenzl) Nakai, 1942.

78. Caryophyllaceae A. L. de Jussieu, 1789, nom. cons.
Illecebraceae R. Brown, 1810, nom. cons.
Paronychiaceae A. L. de Jussieu, 1815.
Cerastiaceae Vest, 1818.
Dianthaceae Vest, 1818.
Herniariaceae Martinov, 1820.
Ortegaceae Martinov, 1820.
Scleranthaceae Berchtold & J. S. Presl, 1820.
Telephiaceae Martinov, 1820.
Stellariaceae Dumortier, 1822.
Alsinaceae (de Candolle) Bartling, 1825, nom. cons.
Silenaceae (de Candolle) Bartling, 1825.
Corrigiolaceae (Dumortier) Dumortier, 1829.

21. Polygonales Dumortier, 1829.

79. Polygonaceae A. L. de Jussieu, 1789, nom. cons.
Rumicaceae Martinov, 1820.
Eriogonaceae (Dumortier) Meisner, 1841.
Persicariaceae Adanson ex T. Post & Kuntze, 1903.
Calligonaceae Khalkuziev, 1985.

22. Plumbaginales Lindley, 1833.

80. Plumbaginaceae A. L. de Jussieu, 1789, nom. cons.
Staticaceae Cassel, 1817.
Armeriaceae Horaninow, 1834.
Limoniaceae Seringe, 1851, nom. cons. prop.
Aegialitidaceae Linczevski, 1968.

IV. Dilleniidae Takhtajan ex Reveal & Takhtajan, 1993.

23. Dilleniales Hutchinson, 1924.
Paeoniales Heintze, 1927.

81. Dilleniaceae Salisbury, 1807, nom. cons.
Soramiaceae Martinov, 1820.
Hibbertiaceae J. Agardh, 1858.

82. Paeoniaceae F. Rudolphi, 1830, nom. cons.

24. Theales Lindley, 1833.
Elatinales Nakai, 1949.
Medusagynales Brenan, 1952.
Actinidiales Takhtajan ex Reveal, 1992.
Ochnales Hutchinson ex Reveal, 1992.
Paracryphiales Takhtajan ex Reveal, 1992.

83. Ochnaceae de Candolle, 1811, nom. cons.
Sauvagesiaceae (Gingins ex de Candolle) Dumortier, 1829.
Lophiraceae Loudon, 1830.
Gomphiaceae de Candolle ex Schnizlein, 1843–1870.

APPENDIX 15.1 (Continued).

Strasburgeriaceae Engler & Gilg, 1924, nom. cons.
Diegodendraceae Capuron, 1964.
Euthemidaceae Tieghem ex Solereder, 1908.
Luxemburgiaceae Solereder, 1908.

84. Sphaerosepalaceae (Warburg) Tieghem ex Bullock, 1959.
    Rhopalocarpaceae Hemsley ex Takhtajan, 1987.

85. Sarcolaenaceae Caruel, 1881, nom. cons.
    Schizolaenaceae Barnhart, 1895.
    Rhodolaenaceae Bullock, 1958.

86. Dipterocarpaceae Blume, 1825, nom. cons.
    Monotaceae (Gilg) Maury ex Takhtajan, 1987.

87. Caryocaraceae Szyszylowicz, 1893, nom. cons.

88. Theaceae D. Don, 1825, nom. cons.
    Camelliaceae de Candolle, 1816.
    Ternstroemiaceae Mirbel ex de Candolle, 1816.
    Gordoniaceae (de Candolle) Sprengel, 1826.
    Bonnetiaceae (Bartling) L. Beauvisage ex Nakai, 1948.
    Sladeniaceae (Gilg & Werdermann) Airy Shaw, 1965.
    Asteropeiaceae (Szyszylowicz) Takhtajan ex Reveal & Hoogland, 1990.

89. Actinidiaceae Hutchinson, 1926, nom. cons.
    Saurauiaceae J. Agardh, 1858, nom. cons.

90. Scytopetalaceae Engler, 1897, nom. cons.
    Rhaptopetalaceae Solander, 1908.

91. Pentaphylacaceae Engler, 1897, nom. cons.

92. Tetrameristaceae Hutchinson, 1959.

93. Pellicieraceae (Triana & Planchon) L. Beauvisage ex Bullock, 1959.

94. Oncothecaceae Kobuski ex Airy Shaw, 1965.

95. Marcgraviaceae Choisy, 1824, nom. cons.

96. Quiinaceae Choisy ex Engler, 1888, nom. cons.

97. Elatinaceae Dumortier, 1829, nom. cons.
    Cryptaceae Rafinesque, 1820.
    Alsinastraceae Ruprecht ex Bubani, 1901.

98. Paracryphiaceae Airy Shaw, 1965.

99. Medusagynaceae Engler & Gilg, 1924, nom. cons.

100. Clusiaceae Lindley, 1836, nom. cons., nom. alt.: Guttiferae.
    Guttiferae A. L. de Jussieu, 1789, nom. cons., nom. alt.: Clusiaceae.
    Hypericaceae A. L. de Jussieu, 1789, nom. cons.
    Ascyraceae Plenck, 1796.
    Garciniaceae Bartling, 1830.
    Cambogiaceae Horaninow, 1834.
    Calophyllaceae J. Agardh, 1858.

25. Malvales Dumortier, 1829

101. Elaeocarpaceae A. L. de Jussieu ex de Candolle, 1824, nom. cons.
    Aristoteliaceae Dumortier, 1829.

102. Tiliaceae A. L. de Jussieu, 1789, nom. cons.
    Sparmanniaceae J. Agardh, 1858.

103. Sterculiaceae (de Candolle) Bartling, 1830, nom. cons.
    Byttneriaceae R. Brown, 1814, nom. cons.
    Hermanniaceae Berchtold & J. S. Presl, 1820.
    Lasiopetalaceae Reichenbach, 1823.
    Dombeyaceae (de Candolle) Bartling, 1830.
    Triplobaceae Rafinesque, 1838.
    Helicteraceae J. Agardh, 1858.
    Melochiaceae J. Agardh, 1858.
    Theobromataceae J. Agardh, 1858.
    Chiranthodendraceae A. Gray, 1887.
    Cacaoaceae Augier ex T. Post & Kuntze, 1903.

104. Bombacaceae Kunth, 1822, nom. cons.

105. Malvaceae A. L. de Jussieu, 1789, nom. cons.
    Philippodendraceae Endlicher, 1841.
    Hibiscaceae J. Agardh, 1858.
    Plagianthaceae J. Agardh, 1858.

26. Lecythidales Cronquist, 1957.

106. Lecythidaceae Poiteau, 1825, nom. cons.
    Napoleonaeaceae A. Richard, 1827.
    Barringtoniaceae F. Rudolphi, 1830, nom. cons.
    Gustaviaceae Burnett, 1835.
    Asteranthaceae Knuth, 1939, nom. cons.
    Foetidaceae (Nidenzu) Airy Shaw, 1965.

27. Nepenthales Dumortier, 1829.
    Sarraceniales Bromhead, 1838.
    Droserales Grisebach, 1854.

107. Sarraceniaceae Dumortier, 1829, nom. cons.
    Heliamphoraceae Chrtek, Slavíková, & Studicka, 1992.

108. Nepenthaceae Dumortier, 1829, nom. cons.

109. Droseraceae Salisbury, 1808, nom. cons.
    Dionaeaceae Rafinesque, 1837.
    Aldrovandaceae Nakai, 1949.
    Drosophyllaceae Chrtek, Slavíková, & Studicka, 1989.

28. Violales Perleb, 1826.

APPENDIX 15.1 (Continued).

Begoniales Dumortier, 1829.
Cucurbitales Dumortier, 1829.
Cistales Reichenbach, 1828.
Loasales Bessey, 1907.
Tamaricales Hutchinson, 1924.
Ancistrocladales Takhtajan ex Reveal, 1992.
Dioncophyllales Takhtajan ex Reveal, 1992.
Fouquieriales Takhtajan ex Reveal, 1992.

110. Flacourtiaceae Richard ex de Candolle, 1824, nom. cons.
  Prockiaceae Bertuch, 1801.
  Samydaceae Ventenat, 1808, nom. cons.
  Homaliaceae R. Brown, 1818.
  Paropsiaceae Dumortier, 1829.
  Kiggelariaceae Link, 1829.
  Pangiaceae Endlicher, 1841.
  Plagiopteraceae Airy Shaw, 1965.
  Aphloiaceae Takhtajan, 1985.
  Berberidopsidaceae (Veldkamp) Takhtajan, 1985.

111. Peridiscaceae Kuhlmann, 1950, nom. cons.

112. Bixaceae Link, 1831, nom. cons.
  Cochlospermaceae Planchon, 1847, nom. cons.

113. Cistaceae A. L. de Jussieu, 1789, nom. cons.
  Helianthemaceae G. Meyer, 1836.

114. Huaceae A. Chevalier, 1947.

115. Lacistemataceae Martius, 1826, nom. cons.

116. Scyphostegiaceae Hutchinson, 1926, nom. cons.

117. Stachyuraceae J. Agardh, 1858, nom. cons.

118. Violaceae Batsch, 1802, nom. cons.
  Ionidiaceae Mertens & W. Koch, 1823.
  Leoniaceae A. H. L. de Candolle, 1844
  Alsodeiaceae J. Agardh, 1858.

119. Tamaricaceae Link, 1821, nom. cons.
  Reaumuriaceae Ehrenberg ex Lindley, 1830.

120. Frankeniaceae A. de Saint-Hilaire ex Gray, 1821, nom. cons.

121. Dioncophyllaceae (Gilg) Airy Shaw, 1952, nom. cons.

122. Ancistrocladaceae Planchon ex Walpers, 1851, nom. cons.

123. Turneraceae Kunth ex de Candolle, 1828, nom. cons.
  Piriquetaceae Martinov, 1820.

124. Malesherbiaceae D. Don, 1827, nom. cons.

125. Passifloraceae A. L. de Jussieu ex Kunth, 1817, nom. cons.
  Smeathmanniaceae Martius ex Perleb, 1838.
  Modeccaceae J. Agardh, 1858.

126. Achariaceae Harms, 1897, nom. cons.

127. Caricaceae Dumortier, 1829, nom. cons.
  Papayaceae Blume, 1823.

128. Fouquieriaceae de Candolle, 1828, nom. cons.

129. Hoplestigmataceae Gilg, 1924, nom. cons.

130. Cucurbitaceae A. L. de Jussieu, 1789, nom. cons.
  Nhandirobaceae T. G. Lestiboudois, 1826.
  Zanoniaceae Dumortier, 1829.
  Bryoniaceae G. Meyer, 1836.

131. Datiscaceae R. Brown ex Lindley, 1830, nom. cons.
  Tetramelaceae (Warburg) Airy Shaw, 1965.

132. Begoniaceae C. Agardh, 1824, nom. cons.

133. Loasaceae Dumortier, 1822, nom cons.
  Gronoviaceae Endlicher, 1841.
  Cevalliaceae Grisebach, 1854.

29. Salicales Lindley, 1833.

134. Salicaceae Mirbel, 1815, nom. cons.

30. Capparales Hutchinson, 1926.

Brassicales Burnett, 1835.

135. Tovariaceae Pax, 1891, nom. cons.

136. Capparaceae A. L. de Jussieu, 1789, nom. cons.
  Cleomaceae Horaninow, 1834.
  Koeberliniaceae Engler, 1895, nom. cons.
  Pentadiplandraceae Hutchinson & Dalziel, 1928.
  Oxystylidaceae Hutchinson, 1969.

137. Brassicaceae Burnett, 1835, nom. cons., nom. alt.: Cruciferae.
  Cruciferae A. L. de Jussieu, 1789, nom. cons., nom. alt.: Brassicaceae.
  Drabaceae Martinov, 1820.
  Erysimaceae Martinov, 1820.
  Sisymbriaceae Martinov, 1820.
  Thlaspiaceae Martinov, 1820.
  Stanleyaceae Nuttall, 1834.
  Raphanaceae Horaninow, 1847.

138. Moringaceae R. Brown ex Dumortier, 1829, nom. cons.
  Hyperantheraceae Link, 1829.

139. Resedaceae de Candolle ex Gray, 1821, nom. cons.
  Astrocarpaceae A. Kerner, 1891.

31. Batales Engler, 1907.

140. Gyrostemonaceae Endlicher, 1841, nom. cons.

141. Bataceae Martius ex Meisner, 1842, nom. cons.

32. Ericales Dumortier, 1829.

142. Cyrillaceae Endlicher, 1841, nom. cons.

APPENDIX 15.1 (Continued).

143. Clethraceae Klotzsch, 1851, nom. cons.
144. Grubbiaceae Endlicher, 1839, nom. cons.
    Ophiraceae Arnott, 1841.
145. Empetraceae Gray, 1821, nom. cons.
146. Epacridaceae R. Brown, 1810, nom. cons.
    Stypheliaceae Horaninow, 1834.
    Prionotaceae Hutchinson, 1969.
147. Ericaceae A. L. de Jussieu, 1789, nom. cons.
    Rhododendraceae A. L. de Jussieu, 1789.
    Rhodoraceae Ventenat, 1799.
    Azaleaceae Vest, 1818.
    Ledaceae Link, 1821.
    Vacciniaceae de Candolle ex Gray, 1821, nom. cons.
    Andromedaceae (Endlicher) Schnizlein, 1843–1870.
    Menziesiaceae Klotzsch, 1851.
    Arbutaceae J. Agardh, 1858.
    Arctostaphylaceae J. Agardh, 1858.
    Salaxidaceae J. Agardh, 1858.
    Diplarchaceae Klotzsch, 1860.
148. Pyrolaceae Dumortier, 1829, nom. cons.
149. Monotropaceae Nuttall, 1818, nom. cons.
    Hypopitydaceae Link, 1829.

33. Diapensiales Engler & Gilg, 1924.
150. Diapensiaceae (Link) Lindley, 1836, nom. cons.
    Galacaceae D. Don, 1828.

34. Ebenales Engler, 1892.
    Styracales Burnett, 1835.
151. Sapotaceae A. L. de Jussieu, 1789, nom. cons.
    Achradaceae Vest, 1818.
    Bumeliaceae Barnhart, 1895.
    Boerlagellaceae H. J. Lam, 1925.
    Sarcospermataceae H. J. Lam, 1925, nom. cons.
152. Ebenaceae Gürcke, 1891, nom. cons.
    Diospyraceae Vest, 1818.
153. Styracaceae Dumortier, 1829, nom. cons.
    Halesiaceae D. Don, 1828.
154. Lissocarpaceae Gilg, 1924, nom. cons.
155. Symplocaceae Desfontaines, 1820, nom. cons.

35. Primulales Dumortier, 1829.
156. Theophrastaceae Link, 1829, nom. cons.
157. Myrsinaceae R. Brown, 1810, nom. cons.
    Ardisiaceae A. L. de Jussieu, 1810.
    Aegicerataceae Blume, 1833.
    Embeliaceae J. Agardh, 1858.
158. Primulaceae Ventenat, 1799, nom. cons.
    Lysimachiaceae A. L. de Jussieu, 1789.
    Anagallidaceae Batsch ex Borckhausen, 1797.
    Samolaceae Rafinesque, 1820.
    Coridaceae J. Agardh, 1858.

V. Rosidae Takhtajan, 1967.

36. Rosales Perleb, 1826.
    Bruniales Dumortier, 1829.
    Saxifragales Dumortier, 1829.
    Brexiales Lindley, 1833.
    Pittosporales Lindley, 1833.
    Connarales Burnett, 1835.
    Cunoniales Hutchinson, 1924.
    Hydrangeales Nakai, 1943.
    Parnassiales Nakai, 1943.
    Byblidales Nakai ex Reveal, 1992.
    Crossosomatales Takhtajan ex Reveal, 1992.
159. Brunelliaceae Engler, 1897, nom. cons.
160. Connaraceae R. Brown, 1818, nom. cons.
    Cnestidaceae (Rafinesque) Rafinesque, 1830.
161. Eucryphiaceae Endlicher, 1841, nom. cons.
162. Cunoniaceae R. Brown, 1814, nom. cons.
    Baueraceae Lindley, 1830.
    Belangeraceae J. Agardh, 1858.
    Callicomaceae J. Agardh, 1858.
163. Davidsoniaceae Bange, 1952.
164. Dialypetalanthaceae Rizzini & Occhioni, 1948.
165. Pittosporaceae R. Brown, 1814, nom. cons.
166. Byblidaceae (Engler) Domin, 1922, nom. cons.
    Roridulaceae Engler & Gilg, 1924, nom. cons.
167. Hydrangeaceae Dumortier, 1829, nom. cons.
    Hortensiaceae Berchtold & J. S. Presl, 1820.
    Philadelphaceae Martinov, 1820.
    Kirengeshomaceae Nakai, 1943.
168. Columelliaceae D. Don, 1828, nom. cons.
169. Grossulariaceae de Candolle, 1805, nom. cons.
    Ribesiaceae Marquis, 1820.
    Escalloniaceae R. Brown ex Dumortier, 1829, nom. cons.
    Brexiaceae Loudon, 1830.
    Rousseaceae de Candolle, 1839.
    Carpodetaceae Fenzl, 1841.
    Polyosmaceae Blume, 1851.
    Ixerbaceae Grisebach, 1854.
    Iteaceae J. Agardh, 1858, nom. cons.
    Phyllonomaceae Small, 1905.
    Pterostemonaceae Small, 1905, nom. cons.
    Abrophyllaceae Nakai, 1943.
    Montiniaceae Nakai, 1943, nom. cons.
    Tetracarpaeaceae Nakai, 1943.
    Tribelaceae (Engler) Airy Shaw, 1965.

APPENDIX 15.1 (Continued).

Argophyllaceae (Engler) Takhtajan, 1987.
170. Greyiaceae Hutchinson, 1926, nom. cons.
171. Bruniaceae R. Brown ex de Candolle, 1825, nom. cons.
Berzeliaceae Nakai, 1943.
172. Anisophylleaceae Ridley, 1922.
Polygonanthaceae Croizat, 1943.
173. Alseuosmiaceae Airy Shaw, 1965.
174. Crassulaceae de Candolle, 1805, nom. cons.
Sempervivaceae A. L. de Jussieu, 1789.
Sedaceae Roussel, 1806.
Cotyledonaceae Martinov, 1820.
Rhodiolaceae Martinov, 1820.
Tillaeaceae Martinov, 1820.
175. Cephalotaceae Dumortier, 1829, nom. cons.
176. Saxifragaceae A. L. de Jussieu, 1789, nom. cons.
Parnassiaceae Gray, 1821, nom. cons.
Francoaceae A. de Jussieu, 1832, nom. cons.
Pectiantiaceae Rafinesque, 1837.
Penthoraceae Rydberg ex Britton, 1901, nom. cons.
Lepuropetalaceae (Engler) Nakai, 1943.
Eremosynaceae Dandy, 1959.
Vahliaceae Dandy, 1959.
177. Rosaceae A. L. de Jussieu, 1789, nom. cons.
Spiraeaceae Bertuch, 1801.
Poteriaceae Rafinesque, 1815.
Fragariaceae Richard ex Nestler, 1816.
Pyraceae Vest, 1818.
Alchemillaceae Martinov, 1820.
Prunaceae Berchtold & J. S. Presl, 1820.
Sanguisorbaceae Marquis, 1820.
Tormentillaceae Martinov, 1820.
Agrimoniaceae Gray, 1821.
Dryadaceae Gray, 1821.
Ulmariaceae Gray, 1821.
Amygdalaceae (A. L. de Jussieu) D. Don, 1825, nom. cons.
Potentillaceae (A. L. de Jussieu) Perleb, 1826.
Quillajaceae D. Don, 1831.
Mespilaceae Schultz-Schultzenstein, 1832.
Neilliaceae Miquel, 1855.
Cercocarpaceae J. Agardh, 1858.
Coleogynaceae J. Agardh, 1858.
Lindleyaceae J. Agardh, 1858.
Rhodotypaceae J. Agardh, 1858.
Cydoniaceae Schnizlein, 1858.
Sorbaceae Brenner, 1886.
Malaceae Small ex Britton, 1903, nom. cons.
178. Neuradaceae Link, 1829, nom. cons. prop.
Grielaceae Martinov, 1820, nom. rej. prop.
179. Crossosomataceae Engler, 1897, nom. cons.
180. Chrysobalanaceae R. Brown, 1818, nom. cons.
Licaniaceae Martinov, 1820.
Hirtellaceae Horaninow, 1847.
181. Surianaceae Arnott, 1834, nom. cons.
Stylobasiaceae J. Agardh, 1858.
182. Rhabdodendraceae (Huber) Prance, 1968.
37. Fabales Bromhead, 1838.
183. Mimosaceae R. Brown, 1814, nom. cons.
184. Caesalpiniaceae R. Brown, 1814, nom. cons.
Cassiaceae Vest, 1818.
Bauhiniaceae Martinov, 1820.
Swartziaceae (de Candolle) Bartling, 1830.
Ceratoniaceae Link, 1831.
Detariaceae (de Candolle) Hess, 1832.
185. Fabaceae Lindley, 1836, nom. cons., nom. alt.: Leguminosae vel Papilionaceae.
Leguminosae A. L. de Jussieu, 1789, nom. cons., nom. alt.: Fabaceae
Papilionaceae Giseke, 1792, nom. cons., nom. alt.: Fabaceae.
Robiniaceae Vest, 1818.
Aspalathaceae Martinov, 1820.
Astragalaceae Martinov, 1820.
Coronillaceae Martinov, 1820.
Tamarindaceae Berchtold & J. S. Presl, 1820.
Viciaceae Berchtold & J. S. Presl, 1820.
Sophoraceae Sprengel ex J. A. Weinmann, 1824.
Hedysaraceae Oken, 1826.
Lotaceae Oken, 1826.
Phaseolaceae Schnizlein, 1834–1870.
Lathyraceae Burnett, 1835.
Ciceraceae W. Steele, 1847.
38. Proteales Dumortier, 1829.
Elaeagnales Bromhead, 1838.
186. Elaeagnaceae A. L. de Jussieu, 1789, nom. cons.
Hippophaeaceae G. Meyer, 1836.
187. Proteaceae A. L. de Jussieu, 1789, nom. cons.
39. Podostemales Lindley, 1833.
188. Podostemaceae Richard ex C. Agardh, 1822, nom. cons.
Marathraceae Dumortier, 1829.
Philocrenaceae Bongard, 1834.
Tristichaceae J. C. Willis, 1915.
40. Haloragales Bromhead, 1838.
Gunnerales Takhtajan ex Reveal, 1992.
189. Haloragaceae R. Brown, 1814, nom. cons.
Cercodiaceae A. L. de Jussieu, 1817.
Myriophyllaceae Schultz-Schultzenstein, 1832.

APPENDIX 15.1 (Continued).

190. Gunneraceae Meisner, 1842, nom. cons.

41. Myrtales Reichenbach, 1828.
    Thymelaeales Willkomm, 1854.

191. Sonneratiaceae Engler & Gilg, 1924, nom. cons.
    Blattiaceae Niedenzu, 1892.
    Duabangaceae Takhtajan, 1986.

192. Lythraceae Jaume Saint-Hilaire, 1805, nom. cons.
    Salicariaceae A. L. de Jussieu, 1789.
    Ammanniaceae Horaninow, 1834.
    Lagerstroemiaceae J. Agardh, 1858.
    Lawsoniaceae J. Agardh, 1858.

193. Rhynchocalycaceae L. S. Johnson & Briggs, 1985.

194. Alzateaceae S. Graham, 1985.

195. Penaeaceae Sweet ex Guillemin, 1828, nom. cons.

196. Crypteroniaceae A. de Candolle, 1868, nom. cons.

197. Thymelaeaceae A. L. de Jussieu, 1789, nom. cons.
    Daphnaceae Ventenat, 1799.
    Aquilariaceae R. Brown, 1818.
    Phaleriaceae Meisner, 1841.
    Gonystylaceae Gilg, 1897, nom. cons.

198. Trapaceae Dumortier, 1829, nom. cons.

199. Myrtaceae A. L. de Jussieu, 1789, nom. cons.
    Melaleucaceae Vest, 1818.
    Chamelauciaceae de Candolle ex F. Rudolphi, 1830.
    Leptospermaceae F. Rudolphi, 1830.
    Myrrhiniaceae Arnott, 1839.
    Heteropyxidaceae Engler & Gilg, 1920, nom. cons.
    Kaniaceae Nakai, 1943.
    Psiloxylaceae Croizat, 1960.

200. Punicaceae Horaninow, 1834, nom. cons.

201. Onagraceae A. L. de Jussieu, 1789, nom. cons.
    Epilobiaceae Ventenat, 1799.
    Oenotheraceae Robin, 1807.
    Isnardiaceae Martinov, 1820.
    Jussiaeaceae Martinov, 1820.
    Circaeaceae Lindley, 1829.

202. Oliniaceae Harvey & Sonder, 1862, nom. cons., emend prop.

203. Melastomataceae A. L. de Jussieu, 1789, nom. cons.
    Rhexiaceae Dumortier, 1822.
    Memecylaceae de Candolle, 1828.
    Miconiaceae C. Koch, 1857.
    Blakeaceae Reichenbach ex Barnhart, 1895.

204. Combretaceae R. Brown, 1810, nom. cons.
    Terminaliaceae Jaume Saint-Hilaire, 1805.
    Myrobalanaceae Martinov, 1820.
    Bucidaceae Voigt, 1827.
    Sheadendraceae G. Bertoloni, 1850.

42. Rhizophorales Tieghem ex Reveal, 1993.

205. Rhizophoraceae R. Brown, 1814, nom. cons.
    Mangiaceae Rafinesque, 1837.
    Cassipoureaceae J. Agardh, 1858.
    Macarisiaceae J. Agardh, 1858.

43. Cornales Dumortier, 1829.
    Aralidiales Takhtajan ex Reveal, 1992.
    Toricelliales Takhtajan ex Reveal, 1992.

206. Alangiaceae de Candolle, 1828, nom. cons.
    Metteniusaceae Karsten ex Schnizlein, 1843–1870.

207. Cornaceae (Dumortier) Dumortier, 1829, nom. cons.
    Nyssaceae A. L. de Jussieu ex Dumortier, 1829, nom. cons.
    Helwingiaceae Decaisne, 1836.
    Aucubaceae J. Agardh, 1858.
    Mastixiaceae Calestani, 1905.
    Toricelliaceae (Wangerin) Hu, 1934.
    Davidiaceae (Harms) H. L. Li, 1955.
    Melanophyllaceae Takhtajan ex Airy Shaw, 1972.
    Aralidiaceae Philipson & Stone, 1980.
    Curtisiaceae (Harms) Takhtajan, 1987.
    Griseliniaceae (Wangerin) Takhtajan, 1987.

208. Garryaceae Lindley, 1834, nom. cons.

44. Santalales Dumortier, 1829.
    Balanophorales Dumortier, 1829.
    Cynomoriales Burnett, 1835.

209. Medusandraceae Brenan, 1952, nom. cons.

210. Dipentodontaceae Merrill, 1941, nom. cons.

211. Olacaceae Mirbel ex de Candolle, 1824, nom. cons.
    Schoepfiaceae Blume, 1850.
    Aptandraceae Miers, 1853.
    Ximeniaceae Martinet, 1873.
    Tetrastylidiaceae Calestani, 1905.
    Octoknemaceae Tieghem ex Engler, 1909, nom. cons.
    Erythropalaceae Sleumer, 1942, nom. cons.

212. Opiliaceae (Bentham) Valeton, 1886, nom. cons.
    Cansjeraceae J. Agardh, 1858.

213. Santalaceae R. Brown, 1810, nom. cons.
    Thesiaceae Vest, 1818.
    Osyridaceae Martinov, 1820.
    Anthobolaceae Dumortier, 1829.
    Henslowiaceae Lindley, 1835.

APPENDIX 15.1 (Continued).

Canopodaceae C. Presl, 1851.
Exocarpaceae J. Agardh, 1858.

214. Misodendraceae J. Agardh, 1858, nom. cons.

215. Loranthaceae A. L. de Jussieu, 1808, nom. cons.
Dendrophthoaceae Tieghem ex Nakai, 1952.
Elytranthaceae Tieghem ex Nakai, 1952.
Gaiadendraceae Tieghem ex Nakai, 1952.
Nuytsiaceae Tieghem ex Nakai, 1952.
Psittacanthaceae Nakai, 1952.

216. Viscaceae Batsch, 1802.
Phoradendraceae H. Karsten, 1860.
Arceuthobiaceae Tieghem ex Nakai, 1952.
Bifariaceae Nakai, 1952.
Ginalloaceae Tieghem ex Nakai, 1952.
Lepidocerataceae Nakai, 1952.

217. Eremolepidaceae Tieghem ex Nakai, 1952.

218. Balanophoraceae Richard, 1822, nom. cons.
Cynomoriaceae (C. Agardh) Endlicher ex Lindley, 1833, nom. cons.
Lophophytaceae Horaninow, 1847.
Mystropetalaceae J. D. Hooker, 1853.
Sarcophytaceae A. Kerner, 1891.
Scybaliaceae A. Kerner, 1891.
Langsdorffiaceae Tieghem ex Pilger & Krause, 1914.
Dactylanthaceae (Engler) Takhtajan, 1987.
Heloseaceae (Schott & Endlicher) Tieghem ex Reveal & Hoogland, 1990.

45. Rafflesiales W. R. B. Oliver, 1895.

Hydnorales Takhtajan ex Reveal, 1992.

219. Hydnoraceae C. Agardh, 1821, nom. cons.

220. Mitrastemonaceae Makino, 1911, nom. cons.

221. Rafflesiaceae Dumortier, 1829, nom. cons.
Cytinaceae (Brongniart) A. Richard, 1824.
Apodanthaceae (R. Brown) Tieghem ex Takhtajan, 1987.

46. Celastrales Baskerville, 1839.

Geissolomotales Takhtajan ex Reveal, 1992.
Icacinales Tieghem ex Reveal, 1993.

222. Geissolomataceae Endlicher, 1841, nom. cons.

223. Celastraceae R. Brown, 1814, nom. cons.
Euonymaceae A. L. de Jussieu ex Berchtold & J. S. Presl, 1820.
Goupiaceae Miers, 1862.
Chingithamnaceae Handel-Mazzetti, 1932.
Lophopyxidaceae (Engler) H. Pfeiffer, 1951.
Siphonodontaceae (Croizat) Gagnepain & Tardieu-Blot ex Tardieu-Blot, 1951, nom. cons.
Canotiaceae Airy Shaw, 1965.
Pottingeriaceae (Engler) Takhtajan, 1987.

224. Hippocrateaceae A. L. de Jussieu, 1811, nom. cons.
Salaciaceae Rafinesque, 1838.

225. Stackhousiaceae R. Brown, 1814, nom. cons.

226. Salvadoraceae Lindley, 1836, nom. cons.
Azimaceae Wight & Gardner, 1845.

227. Tepuianthaceae Maguire & Steyermark, 1981.

228. Aquifoliaceae Bartling, 1830, nom. cons.
Ilicaceae Berchtold & J. S. Presl, 1820.
Phellinaceae (Loesener) Takhtajan, 1967.
Sphenostemonaceae P. Royen & Airy Shaw, 1972.

229. Icacinaceae (Bentham) Miers, 1851, nom. cons.
Phytocrenaceae Arnott ex R. Brown, 1852.
Pennantiaceae J. Agardh, 1858.

230. Aextoxicaceae Engler & Gilg, 1920, nom. cons.

231. Cardiopteridaceae Blume, 1847, nom. cons.
Peripterygiaceae F. Williams, 1905.

232. Corynocarpaceae Engler, 1897, nom. cons.

233. Dichapetalaceae Baillon, 1886, nom. cons.
Chailletiaceae R. Brown, 1818.

47. Euphorbiales Lindley, 1833.

Buxales Takhtajan ex Reveal, 1992.
Simmondsiales Reveal, 1992.

234. Buxaceae Dumortier, 1822, nom. cons.
Pachysandraceae J. Agardh, 1858.
Stylocerataceae (Pax) Baillon ex Reveal & Hoogland, 1990.

235. Simmondsiaceae (Müller Argoviensis) Tieghem ex Reveal & Hoogland, 1990.

236. Pandaceae Engler & Gilg, 1913, nom. cons.

237. Euphorbiaceae A. L. de Jussieu, 1789, nom. cons.
Tithymalaceae Ventenat, 1799.
Mercurialaceae Martinov, 1820.
Ricinaceae Martinov, 1820.
Stilaginaceae Agardh, 1824.
Antidesmataceae Loudon, 1830.
Scepaceae Lindley, 1836.
Trewiaceae Lindley, 1836.
Tragiaceae Rafinesque, 1838.
Pseudanthaceae Endlicher, 1839.
Putranjivaceae Endlicher, 1841.
Acalyphaceae J. Agardh, 1858.
Bertyaceae J. Agardh, 1858.
Crotonaceae J. Agardh, 1858.
Hippomanaceae J. Agardh, 1858.
Micrantheaceae J. Agardh, 1858.
Phyllanthaceae J. Agardh, 1858.

APPENDIX 15.1 (Continued).

Aporusaceae Lindley ex Miquel, 1859.
Ziziphaceae Adanson ex T. Post & Kuntze, 1903.
Peraceae Klotzsch, 1859.
Picrodendraceae Small ex Britton & Millspaugh, 1920, nom. cons.
Porantheraceae (Pax) Hurusawa, 1954.
Ricinocarpaceae (Müller Argoviensis) Hurusawa, 1954.
Androstachyaceae Airy Shaw, 1965.
Bischofiaceae (Müller Argoviensis) Airy Shaw, 1965.
Hymenocardiaceae Airy Shaw, 1965.
Uapacaceae (Müller Argoviensis) Airy Shaw, 1965.
Paivaeusaceae A. D. J. Meeuse, 1990.

48. Rhamnales Dumortier, 1829.

Vitales Burnett, 1835.

238. Rhamnaceae A. L. de Jussieu, 1789, nom. cons.
Frangulaceae de Candolle, 1805.
Gouaniaceae Rafinesque, 1837.
Phylicaceae J. Agardh, 1858.

239. Leeaceae (de Candolle) Dumortier, 1829, nom. cons.

240. Vitaceae A. L. de Jussieu, 1789, nom. cons.
Ampelopsidaceae Kosteletzky, 1835.
Cissaceae Horaninow, 1847.
Pterisanthaceae J. Agardh, 1858.

49. Linales Baskerville, 1839.

241. Erythroxylaceae Kunth, 1822, nom. cons.
Nectaropetalaceae (H. Winkler) Exell & Mendonça, 1951.

242. Humiriaceae A. de Jussieu, 1829, nom. cons.

243. Ixonanthaceae (Bentham) Exell & Mendonça, 1951, nom. cons.

244. Hugoniaceae Arnott, 1834.
Ctenolophonaceae (H. Winkler) Exell & Mendonça, 1951.

245. Linaceae de Candolle ex Gray, 1821, nom. cons.

50. Polygalales Dumortier, 1829.

246. Malpighiaceae A. L. de Jussieu, 1789, nom. cons.

247. Vochysiaceae A. Saint-Hilaire, 1820, nom. cons.
Euphroniaceae Marcano-Berti, 1989.

248. Trigoniaceae Endlicher, 1841, nom. cons.

249. Tremandraceae R. Brown ex de Candolle, 1824, nom. cons.

250. Polygalaceae R. Brown, 1814, nom. cons.
Moutabeaceae Endlicher, 1841.
Diclidantheraceae J. Agardh, 1858, nom. cons.
Emblingiaceae (Pax) Airy Shaw, 1965.

251. Xanthophyllaceae (Chodat) Gagnepain ex Reveal & Hoogland, 1990.

252. Krameriaceae Dumortier, 1829, nom. cons.

51. Sapindales Dumortier, 1829.

Rutales Perleb, 1826.

253. Staphyleaceae (de Candolle) Lindley, 1829, nom. cons.
Ochranthaceae Lindley ex Endlicher, 1841.
Tapisciaceae (Pax) Takhtajan, 1987.

254. Melianthaceae Link, 1829, nom. cons.

255. Bretschneideraceae Engler & Gilg, 1924, nom. cons.

256. Akaniaceae O. Stapf, 1912, nom. cons.

257. Sapindaceae A. L. de Jussieu, 1789, nom. cons.
Allophylaceae Martinov, 1820.
Ornithropaceae Martinov, 1820.
Dodonaeaceae Link, 1829, nom. cons.
Koelreuteriaceae J. Agardh, 1858.
Ptaeroxylaceae J. Leroy, 1960.

258. Hippocastanaceae de Candolle, 1824, nom. cons.
Aesculaceae Berchtold & J. S. Presl, 1820.
Paviaceae Horaninow, 1834.

259. Aceraceae A. L. de Jussieu, 1789, nom. cons.

260. Burseraceae Kunth, 1824, nom. cons.
Balsameaceae Dumortier, 1829.
Terebinthaceae A. L. Jussieu, 1789

261. Anacardiaceae Lindley, 1830, nom. cons.
Comocladiaceae Martinov, 1820.
Spondiadaceae Martinov, 1820.
Vernicaceae Link, 1829.
Schinaceae Rafinesque, 1837.
Pistaciaceae Martius ex Perleb, 1838.
Sumachiaceae Perleb, 1838.
Podoaceae Baillon ex Franchet, 1889.
Blepharocaryaceae Airy Shaw, 1965.

262. Julianiaceae Hemsley, 1906, nom. cons.

263. Simaroubaceae de Candolle, 1811, nom. cons.
Quassiaceae Bertoloni, 1827.
Soulameaceae Endlicher, 1841.
Simabaceae Horaninow, 1847.
Ailanthaceae (Planchon) J. Agardh, 1858.
Castelaceae J. Agardh, 1858.
Holacanthaceae Jadin ex Engler, 1906.
Irvingiaceae (Engler) Exell & Mendonça, 1951, nom. cons.
Kirkiaceae (Engler) Takhtaian, 1967.

264. Cneoraceae Link, 1829, nom. cons.

265. Meliaceae A. L. de Jussieu, 1789, nom. cons.
Cedrelaceae R. Brown, 1814.

APPENDIX 15.1 (Continued).

Swieteniaceae Berchtold & J. S. Presl, 1820.
Aitoniaceae (Harvey) Harvey ex Reveal & Hoogland, 1992.

266. Rutaceae A. L. de Jussieu, 1789, nom. cons.
Aurantiaceae A. L. de Jussieu, 1789.
Citraceae Roussel, 1806.
Diosmaceae R. Brown, 1814.
Amyridaceae R. Brown, 1818.
Dictamnaceae Vest, 1818.
Jamboliferaceae Martinov, 1820.
Zanthoxylaceae Berchtold & J. S. Presl, 1820.
Fraxinellaceae Nees & Martius, 1823.
Pteleaceae Kunth, 1824.
Boroniaceae J. Agardh, 1858.
Correaceae J. Agardh, 1858.
Diplolaenaceae J. Agardh, 1858.
Pilocarpaceae J. Agardh, 1858.
Spatheliaceae J. Agardh, 1858.
Flindersiaceae (Engler) C. White ex Airy Shaw, 1965.

267. Zygophyllaceae R. Brown, 1814, nom. cons.
Nitrariaceae Berchtold & J. S. Presl, 1820.
Balanitaceae Endlicher, 1841, nom. cons.
Tribulaceae Trautvetter, 1853.
Tetradiclidaceae (Engler) Takhtajan, 1986.
Peganaceae (Engler) Tieghem ex Takhtajan, 1987.

52. Geraniales Dumortier, 1829.

Balsaminales Lindley, 1833.
Limnanthales Nakai, 1930.
Tropaeolales Takhtajan ex Reveal, 1992.

268. Oxalidaceae R. Brown, 1818, nom. cons.
Hypseocharitaceae Weddell, 1861.
Lepidobotryaceae Léonard, 1950, nom. cons.
Averrhoaceae Hutchinson, 1959.
Dirachmaceae Hutchinson, 1959.

269. Geraniaceae A. L. de Jussieu, 1789, nom. cons.
Ledocarpaceae Meyen, 1834.
Vivianiaceae Klotzsch, 1836.
Biebersteiniaceae Endlicher, 1841.
Rhynchothecaceae Endlicher, 1841.
Erodiaceae Horaninow, 1847.

270. Limnanthaceae R. Brown, 1833, nom. cons.

271. Tropaeolaceae A. L. de Jussieu ex de Candolle, 1824, nom. cons.
Cardamindaceae Link, 1831.

272. Balsaminaceae A. Richard, 1822, nom. cons.
Impatientaceae Barnhart, 1895.

53. Apiales Nakai, 1930.

Araliales Burnett, 1835.

273. Araliaceae A. L. de Jussieu, 1789, nom. cons.
Hederaceae Giseke, 1792.
Botryodendraceae J. Agardh, 1858.

274. Apiaceae Lindley, 1836, nom. cons., nom. alt.: Umbelliferae.
Umbelliferae A. L. de Jussieu, 1789, nom. cons., nom. alt.: Apiaceae.
Angelicaceae Martinov, 1820.
Bupleuraceae Martinov, 1820.
Daucaceae Martinov, 1820.
Imperatoriaceae Martinov, 1820.
Pastinacaceae Martinov, 1820.
Coriandraceae Burnett, 1835.
Smyrniaceae Burnett, 1835.
Eryngiaceae Rafinesque, 1838.
Ammiaceae Small, 1903.
Hydrocotylaceae (Drude) Hylander, 1945, nom. cons.
Saniculaceae (Drude) A. & D. Löve, 1974.

VI. Asteridae Takhtajan, 1967.

Lamiidae Takhtajan ex Reveal, 1992.

54. Gentianales Lindley, 1833.

275. Loganiaceae R. Brown ex C. Martius, 1827, nom. cons.
Strychnaceae de Candolle ex Perleb, 1826.
Potaliaceae Martius, 1827.
Spigeliaceae Martius, 1827.
Gardneriaceae Wallich ex Perleb, 1838.
Desfontainiaceae Endlicher, 1841, nom. cons.
Antoniaceae J. Agardh, 1858.

276. Gentianaceae A. L. de Jussieu, 1789, nom. cons.
Coutoubeaceae Martinov, 1820.
Obolariaceae Martinov, 1820.
Chironiaceae Horaninow, 1847.

277. Saccifoliaceae Maguire & Pires, 1978.

278. Apocynaceae A. L. de Jussieu, 1789, nom. cons.
Vincaceae Vest, 1818.
Cerberaceae Martinov, 1820.
Pacouriaceae Martinov, 1820.
Plumeriaceae Horaninow, 1834.
Ophioxylaceae Martius ex Perleb, 1838.
Willughbeiaceae J. Agardh, 1858.
Plocospermataceae Hutchinson, 1973.

279. Asclepiadaceae R. Brown, 1810, nom. cons. emend. prop.
Stapeliaceae Horaninow, 1834.
Cynanchaceae G. Meyer, 1836.
Periplocaceae Schlechter, 1905, nom. cons.

55. Solanales Dumortier, 1829.

Convolvulales Dumortier, 1829.
Polemoniales Bromhead, 1838.

280. Duckeodendraceae Kuhlmann, 1950.

APPENDIX 15.1 (Continued).

281. Nolanaceae Dumortier, 1829, nom. cons.
282. Solanaceae A. L. de Jussieu, 1789, nom. cons.
    Hyoscyamaceae Vest, 1818.
    Atropaceae Martinov, 1820.
    Nicotianaceae Martinov, 1820.
    Daturaceae Rafinesque, 1828.
    Cestraceae Schlechtendal, 1833.
    Lyciaceae Rafinesque, 1840.
    Sclerophylacaceae Miers, 1848.
    Goetzeaceae Miers ex Airy Shaw, 1965.
    Salpiglossidaceae (Bentham) Hutchinson, 1969.
283. Convolvulaceae A. L. de Jussieu, 1789, nom. cons.
    Cressaceae Rafinesque, 1821.
    Dichondraceae Dumortier, 1829, nom. cons.
    Erycibaceae Endlicher, 1840.
    Poranaceae J. Agardh, 1858.
    Humbertiaceae Pichon, 1947, nom. cons.
284. Cuscutaceae (Dumortier) Dumortier, 1829, nom. cons.
285. Retziaceae Bartling, 1830.
286. Menyanthaceae (Dumortier) Dumortier, 1829, nom. cons.
287. Polemoniaceae A. L. de Jussieu, 1789, nom. cons.
    Cobaeaceae D. Don, 1824.
288. Hydrophyllaceae R. Brown, 1817, nom. cons.
    Ellisiaceae Berchtold & J. S. Presl, 1820.
    Hydroleaceae Berchtold & J. S. Presl, 1820.
    Sagoneaceae Martinov, 1820.

56. Lamiales Bromhead, 1838.
    Boraginales Dumortier, 1829.
289. Lennoaceae Solms-Laubach, 1870, nom. cons.
290. Boraginaceae A. L. de Jussieu, 1789, nom. cons.
    Sebestenaceae Ventenat, 1799.
    Buglossaceae Hoffmannsegg & Link, 1810.
    Anchusaceae Vest, 1818.
    Cerinthaceae Martinov, 1820.
    Heliotropiaceae Schrader, 1819, nom. cons.
    Onosmaceae Martinov, 1820.
    Cordiaceae R. Brown ex Dumortier, 1829, nom. cons.
    Ehretiaceae Martius ex Lindley, 1830, nom. cons.
    Echiaceae Rafinesque, 1837.
    Wellstediaceae (Pilger) Novák, 1943.
291. Verbenaceae Jaume Saint-Hilaire, 1805, nom. cons.
    Viticaceae A. L. de Jussieu, 1789.
    Lantanaceae Martinov, 1820.
    Tinaceae Martinov, 1820.
    Stilbaceae Kunth, 1831, nom. cons.
    Aegiphilaceae Rafinesque, 1838.
    Siphonanthaceae Rafinesque, 1838.
    Avicenniaceae Endlicher, 1841, nom. cons.
    Phrymaceae Schauer, 1847, nom. cons.
    Durantaceae J. Agardh, 1858.
    Nyctanthaceae J. Agardh, 1858.
    Petreaceae J. Agardh, 1858.
    Chloanthaceae Hutchinson, 1959.
    Cyclocheilaceae Marais, 1981.
    Nesogenaceae Marais, 1981.
    Symphoremataceae (Meisner) Moldenke ex Reveal & Hoogland, 1991.
292. Lamiaceae Lindley, 1836, nom. cons., nom. alt.: Labiatae.
    Labiatae A. L. de Jussieu, 1789, nom. cons., nom. alt.: Lamiaceae.
    Glechomaceae Martinov, 1820.
    Melittaceae Martinov, 1820.
    Nepetaceae Horaninow, 1834.
    Menthaceae Burnett, 1835.
    Salviaceae Rafinesque, 1837.
    Scutellariaceae Caruel, 1894.
    Tetrachondraceae Wettstein, 1924.
    Salazariaceae F. Barkley, 1975.

57. Callitrichales Dumortier, 1829.
    Hippuridales Burnett, 1835.
    Hydrostachyales Diels ex Reveal, 1992.
293. Hippuridaceae Link, 1821, nom. cons.
294. Callitrichaceae Link, 1821, nom. cons.
295. Hydrostachyaceae Engler, 1898, nom. cons.

58. Plantaginales Lindley, 1833.
296. Plantaginaceae A. L. de Jussieu, 1789, nom. cons.
    Littorellaceae Gray, 1821.
    Psylliaceae Horaninow, 1834.

59. Scrophulariales Lindley, 1833.
    Oleales Lindley, 1833.
297. Buddlejaceae K. Wilhelm, 1910, nom. cons.
298. Oleaceae Hoffmannsegg & Link, 1813–1820, nom. cons.
    Jasminaceae A. L. de Jussieu, 1789.
    Lilacaceae Ventenat, 1799.
    Fraxinaceae Vest, 1818.
    Ligustraceae G. Meyer, 1836.
    Bolivariaceae Grisebach, 1838.
    Forestieraceae Endlicher, 1841.
    Schreberaceae (Wight) Schnizlein, 1843–1870.
    Syringaceae Horaninow, 1847.
299. Scrophulariaceae A. L. de Jussieu, 1789, nom. cons.

APPENDIX 15.1 (Continued).

Pedicularidaceae A. L. de Jussieu, 1789.
Rhinanthaceae Ventenat, 1799.
Antirrhinaceae Persoon, 1807.
Carpanthaceae Rafinesque, 1808.
Veronicaceae Cassel, 1817.
Caprariaceae Martinov, 1820.
Chelonaceae Martinov, 1820.
Digitalidaceae Martinov, 1820.
Euphrasiaceae Martinov, 1820.
Gratiolaceae Martinov, 1820.
Linariaceae Martinov, 1820.
Melampyraceae Richard ex Hooker & Lindley, 1821.
Verbascaceae Rafinesque, 1821.
Hebenstretiaceae Horaninow, 1834.
Aragoaceae D. Don, 1835.
Sibthorpiaceae D. Don, 1835.
Erinaceae (Link) Trinius, 1835.
Halleriaceae (Link) Trinius, 1835.
Scoperiaceae (Link) Trinius, 1835.
Calceolariaceae Rafinesque, 1838.
Oxycladaceae (Miers) Schnizlein, 1843–1870.
Limosellaceae J. Agardh, 1858.
Ellisiophyllaceae Honda, 1930.

300. Globulariaceae de Candolle, 1805, nom. cons.
Selaginaceae Choisy, 1823, nom. cons.

301. Myoporaceae R. Brown, 1810, nom. cons.
Bontiaceae Horaninow, 1834.
Oftiaceae Takhtajan & Reveal, 1993.

302. Orobanchaceae Ventenat, 1799, nom. cons.
Aeginetiaceae Livera, 1927.
Phelypaeaceae Horaninow, 1834.

303. Gesneriaceae Dumortier, 1822, nom. cons.
Belloniaceae Martinov, 1820.
Didymocarpaceae D. Don, 1822.
Cyrtandraceae Jack, 1823.
Besleriaceae Rafinesque, 1838.
Ramondaceae Godron, 1850.

304. Acanthaceae A. L. de Jussieu, 1789, nom. cons.
Justiciaceae Rafinesque, 1838.
Thunbergiaceae Bremekamp, 1954.
Meyeniaceae Sreemadhavan, 1977.
Nelsoniaceae (Nees) Sreemadhavan, 1977.
Thomandersiaceae Sreemadhavan, 1977.

305. Pedaliaceae R. Brown, 1810, nom. cons.
Sesamaceae R. Brown ex Berchtold & J. S. Presl, 1820.
Martyniaceae O. Stapf, 1895, nom. cons.
Trapellaceae Honda & Sakisaka, 1930.

306. Bignoniaceae A. L. de Jussieu, 1789, nom. cons.
Crescentiaceae Dumortier, 1829.
Paulowniaceae Nakai, 1949.

307. Mendonciaceae Bremekamp, 1954.

308. Lentibulariaceae Richard, 1808, nom. cons.
Utriculariaceae Hoffmannsegg & Link, 1813–1820.
Pinguiculaceae Dumortier, 1829.

60. Campanulales H.G.L. Reichenbach, 1828.

Goodeniales Lindley, 1833.
Stylidiales Takhtajan ex Reveal, 1992.

309. Pentaphragmataceae J. Agardh, 1858, nom. cons.

310. Sphenocleaceae Martius ex A.P. de Candolle, 1839, nom. cons.

311. Campanulaceae A. L. de Jussieu, 1789, nom. cons.
Lobeliaceae R. Brown, 1817, nom. cons.
Jasionaceae Dumortier, 1829.
Cyphiaceae A. de Candolle, 1839.
Nemacladaceae Nuttall, 1842.
Cyphocarpaceae Meirs, 1848.
Cyananthaceae J. Agardh, 1858.

312. Stylidiaceae R. Brown, 1810, nom. cons.

313. Donatiaceae Hutchinson, 1959, nom. cons., emend. prop.

314. Brunoniaceae Dumortier, 1829, nom. cons.

315. Goodeniaceae R. Brown, 1810, nom. cons.
Scaevolaceae Lindley, 1830.

61. Rubiales Dumortier, 1829.

316. Rubiaceae A. L. de Jussieu, 1789, nom. cons.
Cinchonaceae Batsch, 1802.
Coffeaceae Batsch, 1802.
Guettardaceae Batsch, 1802.
Aparinaceae Hoffmannsegg & Link, 1813–1820.
Operculariaceae A. L. de Jussieu ex Perleb, 1818.
Catesbaeaceae Martinov, 1820.
Coutareaceae Martinov, 1820.
Hydrophylacaceae Martinov, 1820.
Nonateliaceae Martinov, 1820.
Pagamaeaceae Martinov, 1820.
Randiaceae Martinov, 1820.
Sabiceaceae Martinov, 1820.
Cephalanthaceae Rafinesque, 1820.
Hedyotidaceae Dumortier, 1822.
Gardeniaceae Dumortier, 1829.
Lygodisodeaceae Bartling, 1830.
Psychotriaceae F. Rudolphi, 1830.
Asperulaceae Spenner, 1835.
Galiaceae Lindley, 1836.
Lippayaceae Meisner, 1838.
Houstoniaceae Rafinesque, 1840.
Naucleaceae (de Candolle) Wernham, 1911.
Henriqueziaceae Bremekamp, 1957.

APPENDIX 15.1 (Continued).

317. Theligonaceae Dumortier, 1829, nom. cons.

62. Dipsacales Dumortier, 1829.
    Adoxales Nakai, 1949.
    318. Caprifoliaceae A. L. de Jussieu, 1789, nom. cons.
        Sambucaceae Batsch ex Borckhausen, 1797.
        Loniceraceae Vest, 1818.
        Viburnaceae Rafinesque, 1820.
        Carlemanniaceae Airy Shaw, 1965.
    319. Adoxaceae Trautvetter, 1853, nom. cons.
    320. Valerianaceae Batsch, 1802, nom. cons.
        Triplostegiaceae (Höck) Bobrov ex Airy Shaw, 1965.
    321. Dipsacaceae A. L. de Jussieu, 1789, nom. cons.
        Morinaceae Rafinesque, 1820.
        Scabiosaceae Adanson ex T. Post & Kuntze, 1903.

63. Calycerales Burnett, 1835.
    322. Calyceraceae R. Brown ex Richard, 1820, nom. cons.
        Boopidaceae Cassini, 1816.

64. Asterales Lindley, 1833.
    323. Asteraceae Dumortier, 1822, nom. cons., nom. alt.: Compositae.
        Cichoriaceae A. L. de Jussieu, 1789, nom. cons.
        Cynaraceae A. L. de Jussieu, 1789.
        Compositae Giseke, 1792, nom. cons., nom. alt.: Asteraceae.
        Cnicaceae Vest, 1818.
        Tanacetaceae Vest, 1818.
        Xanthiaceae Vest, 1818.
        Anthemidaceae Martinov, 1820.
        Artemisiaceae Martinov, 1820.
        Athanasiaceae Martinov, 1820.
        Centaureaceae Martinov, 1820.
        Eupatoriaceae Martinov, 1820.
        Lapsanaceae Martinov, 1820.
        Picridaceae Martinov, 1820.
        Santolinaceae Martinov, 1820.
        Serratulaceae Martinov, 1820.
        Carduaceae Dumortier, 1822.
        Echinopaceae Dumortier, 1822.
        Helianthaceae Dumortier, 1822.
        Heleniaceae Rafinesque, 1824.
        Acarnaceae Link, 1829.
        Ambrosiaceae Link, 1829, nom. cons., emend. prop.
        Calendulaceae Link, 1829.
        Coreopsidaceae Link, 1829.
        Helichrysaceae Link, 1829.
        Partheniaceae Link, 1829.
        Perdiciaceae Link, 1829.
        Gnaphaliaceae F. Rudolphi, 1830.
        Senecionaceae Spenner, 1834.
        Mutisiaceae Burnett, 1835.
        Nassauviaceae Burmeister, 1837.
        Vernoniaceae Burmeister, 1837.
        Matricariaceae J. Voigt, 1845.
        Lactucaceae Drude, 1879.
        Arctotidaceae Bessey, 1914.
        Inulaceae Bessey, 1914.

B. Liliopsida Batsch, 1802.

VII. Alismatidae Takhtajan, 1967.
    Triurididae Takhtajan ex Reveal, 1992.

65. Alismatales Dumortier, 1829.
    Butomales Hutchinson, 1934.
    324. Butomaceae Richard, 1815–1816, nom. cons.
    325. Limnocharitaceae Takhtajan ex Cronquist, 1981.
    326. Alismataceae Ventenat, 1799, nom. cons.
        Damasoniaceae Nakai, 1943.

66. Hydrocharitales Dumortier, 1829.
    327. Hydrocharitaceae A. L. de Jussieu, 1789, nom. cons.
        Elodeaceae Dumortier, 1829.
        Stratiotaceae Link, 1829.
        Vallisneriaceae Link, 1829.
        Halophilaceae J. Agardh, 1858.
        Enhalaceae Nakai, 1943.
        Thalassiaceae Nakai, 1943.
        Blyxaceae (Ascherson & Gürke) Nakai, 1949.

67. Najadales Reichenbach, 1828.
    Potamogetonales Dumortier, 1829.
    Aponogetonales Hutchinson, 1934.
    Juncaginales Hutchinson, 1934.
    Cymodoceales Nakai, 1943.
    Posidoniales Nakai, 1943.
    Zosterales Nakai, 1943.
    Scheuchzeriales B. Boivin, 1956.
    328. Aponogetonaceae J. Agardh, 1858, nom. cons.
    329. Scheuchzeriaceae F. Rudolphi, 1830, nom. cons.
    330. Juncaginaceae Richard, 1808, nom. cons.
        Triglochinaceae Chevallier, 1827.
        Lilaeaceae Dumortier, 1829, nom. cons.
        Heterostylaceae Hutchinson, 1934.
        Maundiaceae Nakai, 1943.
    331. Potamogetonaceae Dumortier, 1829, nom. cons.
        Hydrogetonaceae Link, 1829.

APPENDIX 15.1 (Continued).

332. Ruppiaceae Horaninow ex Hutchinson, 1934, nom. cons.
333. Najadaceae A. L. de Jussieu, 1789, nom. cons.
334. Zannichelliaceae Dumortier, 1829, nom. cons.
335. Posidoniaceae Hutchinson, 1934, nom. cons., emend. prop.
336. Cymodoceaceae N. Taylor, 1909, nom. cons.
337. Zosteraceae Dumortier, 1829, nom. cons.

68. Triuridales J. D. Hooker, 1876.
338. Petrosaviaceae Hutchinson, 1934, nom. cons.
339. Triuridaceae Gardner, 1843, nom. cons.
Lacandoniaceae E. Martiñes & C. H. Ramos, 1989.

VIII. Arecidae Takhtajan, 1967.
69. Arecales Bromhead, 1840.
340. Arecaceae Schultz-Schultzenstein, 1832, nom. cons., nom. alt.: Palmae.
Palmae A. L. de Jussieu, 1789, nom. cons., nom. alt.: Arecaceae.
Borassaceae Schultz-Schultzenstein, 1832.
Cocosaceae Schultz-Schultzenstein, 1832.
Coryphaceae Schultz-Schultzenstein, 1832.
Phoenicaceae Schutz-Schultzenstein, 1832.
Sabalaceae Schultz-Schultzenstein, 1832.
Sagaceae Schultz-Schultzenstein, 1832.
Phytelephantaceae Martius ex Perleb, 1838.
Nypaceae Brongniart ex LeMaout & Descaisne, 1868.
Manicariaceae Cook, 1910.
Acristaceae Cook, 1913.
Ceroxylaceae Cook, 1913.
Chamaedoreaceae Cook, 1913.
Geonomataceae Cook, 1913.
Iriarteaceae Cook, 1913.
Lepidocaryaceae Cook, 1913.
Malortieaceae Cook, 1913.
Pseudophoenicaceae Cook, 1913.
Synechanthaceae Cook, 1913.

70. Cyclanthales Nakai, 1930.
341. Cyclanthaceae Poiteau ex A. Richard, 1824, nom. cons., emend. prop.

71. Pandanales Lindley, 1833.
342. Pandanaceae R. Brown, 1810, nom. cons.
Freycinetiaceae Brongniart ex LeMaout & Descaisne, 1868.

72. Arales Dumortier, 1829.
343. Acoraceae Martinov, 1820.
344. Araceae A. L. de Jussieu, 1789, nom. cons.
Pistiaceae Richard ex C. Agardh, 1822.
Callaceae Reichenbach ex Bartling, 1830.
Orontiaceae Bartling, 1830.
Arisaraceae Rafinesque, 1838.
Pothoaceae Rafinesque, 1838.
Lentiscaceae Horaninow, 1843.
Cryptocorynaceae J. Agardh, 1858.
Caladiaceae Salisbury, 1866.
Dracontiaceae Salisbury, 1866.
345. Lemnaceae Gray, 1821, nom. cons.
Wolffiaceae Bubani, 1901-1902.

IX. Commelinidae Takhtajan, 1967.
73. Commelinales Dumortier, 1829.
Xyridales Lindley, 1846.
Mayacales Nakai, 1943.
346. Rapateaceae Dumortier, 1829, nom. cons.
347. Xyridaceae C. Agardh, 1823, nom. cons.
Abolbodaceae Nakai, 1943.
348. Mayacaceae Kunth, 1842, nom. cons.
349. Commelinaceae R. Brown, 1810, nom. cons.
Ephemeraceae Batsch, 1802, nom. rej. prop.
Cartonemataceae Pichon, 1946, nom. cons.

74. Eriocaulales Nakai, 1930.
350. Eriocaulaceae Palisot de Beauvois ex Desvaux, 1828, nom. cons.

75. Restionales Perleb, 1838.
351. Flagellariaceae Dumortier, 1829, nom. cons.
352. Joinvilleaceae Tomlinson & A. C. Smith, 1970.
353. Restionaceae R. Brown, 1810, nom. cons.
Elegiaceae Rafinesque, 1837.
Anarthriaceae D. Cutler & Airy Shaw, 1965.
Ecdeiocoleaceae D. Cutler & Airy Shaw, 1965.
354. Centrolepidaceae Endlicher, 1836, nom. cons.

76. Juncales Dumortier, 1829.
355. Juncaceae A. L. de Jussieu, 1789, nom. cons.
356. Thurniaceae Engler, 1907, nom. cons.

77. Cyperales Burnett, 1835.
Poales Burnett, 1835.
357. Cyperaceae A. L. de Jussieu, 1789, nom. cons.
Scirpaceae Batsch ex Borckhausen, 1797.
Papyraceae Burnett, 1835.
Kobresiaceae Gilly, 1952.
358. Poaceae (R. Brown) Barnhart, 1895, nom. cons., nom. alt.: Gramineae.

APPENDIX 15.1 (Continued).

Gramineae A. L. de Jussieu, 1789, nom. cons., nom. alt.: Poaceae.
Aegilopaceae Martinov, 1820.
Alopecuraceae Martinov, 1820.
Andropogonaceae Martinov, 1820.
Avenaceae Martinov, 1820.
Maydaceae Martinov, 1820.
Melicaceae Martinov, 1820.
Nardaceae Martinov, 1820.
Saccharaceae Martinov, 1820.
Agrostidaceae Burnett, 1835.
Hordaceae Burnett, 1835.
Miliaceae Burnett, 1835.
Oryzaceae Burnett, 1835.
Phalaridaceae Burnett, 1835.
Stipaceae Burnett, 1835.
Spartinaceae Burnett, 1835.
Panicaceae Voigt, 1845.
Arundinaceae Brenner, 1886.
Zeaceae A. Kerner, 1891.
Arundinellaceae (O. Stapf) Herter, 1940.
Chloridaceae (Reichenbach) Herter, 1940.
Eragrostidaceae (O. Stapf) Herter, 1940.
Lepturaceae (Holmberg) Herter, 1940.
Pappophoraceae (Kunth) Herter, 1940.
Pharaceae (O. Stapf) Herter, 1940.
Tristeginaceae (Link) Herter, 1940.
Sporobolaceae Herter, 1941.
Anomochloaceae Nakai, 1943.
Parianaceae Nakai, 1943.
Streptochaetaceae Nakai, 1943.
Festucaceae (Dumortier) Herter, 1952.

78. Hydatellales Cronquist, 1980.

359. Hydatellaceae Hamann, 1976.

79. Typhales Dumortier, 1829.

360. Sparganiaceae F. Rudolphi, 1830, nom. cons.

361. Typhaceae A. L. de Jussieu, 1789, nom. cons.

X. Zingiberidae Cronquist, 1978.

80. Bromeliales Lindley, 1833.

362. Bromeliaceae A. L. de Jussieu, 1789, nom. cons.
Tillandsiaceae Wilbrand, 1834.

81. Zingiberales Grisebach, 1854.
Cannales Dumortier, 1829.
Musales Burnett, 1835.

363. Strelitziaceae (Schumann) Hutchinson, 1934. nom. cons.

364. Heliconiaceae (A. Richard) Nakai, 1941.

365. Musaceae A. L. de Jussieu, 1789, nom. cons.

366. Lowiaceae Ridley, 1924, nom. cons.

367. Zingiberaceae Lindley, 1835, nom. cons.
Amomaceae Jaume Saint-Hilaire, 1805.
Curcumaceae Dumortier, 1829.
Alpiniaceae R. Brown ex F. Rudolphi, 1830.

368. Costaceae (Meisner) Nakai, 1941.

369. Cannaceae A. L. de Jussieu, 1789, nom. cons.

370. Marantaceae Petersen, 1888, nom. cons.

XI. Liliidae Takhtajan, 1967.

82. Liliales Perleb, 1826.
Asteliales Dumortier, 1829.
Iridales Dumortier, 1829.
Taccales Dumortier, 1829.
Smilacales Lindley, 1833.
Amaryllidales Bromhead, 1838.
Asparagales Bromhead, 1838.
Dioscoreales J. D. Hooker, 1876.
Agavales Hutchinson, 1934.
Alstroemeriales Hutchinson, 1934.
Alliales Traub, 1972.
Hanguanales R. Dahlgren ex Reveal, 1992.
Melanthiales R. Dahlgren ex Reveal, 1992.
Velloziales R. Dahlgren ex Reveal, 1992.
Tecophilaeales Traub ex Reveal, 1993.

371. Philydraceae Link, 1821, nom. cons.

372. Pontederiaceae Kunth, 1816, nom. cons.
Heterantheraceae J. Agardh, 1858.

373. Haemodoraceae R. Brown, 1810, nom. cons.
Xiphidiaceae Dumortier, 1829.
Dilatridaceae M. Roemer, 1840.
Lophiolaceae Nakai, 1943.
Conostylidaceae (Pax) Takhtajan, 1987.

374. Cyanastraceae Engler, 1900, nom. cons.

375. Liliaceae A. L. de Jussieu, 1789, nom. cons.
Asparagaceae A. L. de Jussieu, 1789, nom. cons.
Asphodelaceae A. L. de Jussieu, 1789.
Narcissaceae A. L. de Jussieu, 1789.
Hyacinthaceae Batsch ex Borckhausen, 1797.
Leucojaceae Batsch ex Borckhausen, 1797.
Liriaceae Batsch ex Borckhausen, 1797.
Tulipaceae Batsch ex Borckhausen, 1797.
Melanthiaceae Batsch, 1802, nom. cons.
Merenderaceae de Mirbel, 1804.
Amaryllidaceae Jaume Saint-Hilaire, 1805, nom. cons.
Colchicaceae de Candolle, 1805, nom. cons.
Hemerocallidaceae R. Brown, 1810.
Hypoxidaceae R. Brown, 1814, nom. cons.

APPENDIX 15.1 (Continued).

Crinaceae Vest, 1818.
Scillaceae Vest, 1818.
Veratraceae Vest, 1818.
Erythroniaceae Martinov, 1820.
Gilliesiaceae Lindley, 1826.
Paridaceae Dumortier, 1827.
Alstroemeriaceae Dumortier, 1829, nom. cons.
Asteliaceae Dumortier, 1829.
Calochortaceae Dumortier, 1829.
Campynemataceae Dumortier, 1829.
Brunsvigiaceae Horaninow, 1834.
Compsoaceae Horaninow, 1834.
Convallariaceae Horaninow, 1834.
Funkiaceae Horaninow, 1834, nom. rej. prop.
Aphyllanthaceae Burnett, 1835.
Galanthaceae G. Meyer, 1836.
Gethyllidaceae Rafinesque, 1838.
Aspidistraceae Endlicher, 1841.
Eriospermaceae Endlicher, 1841.
Herreriaceae Endlicher, 1841.
Ophiopogonaceae Endlicher, 1841.
Uvulariaceae A. Gray ex Kunth, 1843, nom. cons. prop.
Trilliaceae Lindley, 1846, nom. cons.
Nartheciaceae E. M. Fries ex Bjurzon, 1846.
Pancratiaceae Horaninow, 1847.
Agapanthaceae Voigt, 1850.
Alliaceae J. Agardh, 1858, nom. cons.
Anthericaceae J. Agardh, 1858.
Heloniadaceae J. Agardh, 1858.
Tecophilaeaceae F. Leybold, 1862, nom. cons.
Androsynaceae Salisbury, 1866.
Bulbocodiaceae Salisbury, 1866.
Cepaceae Salisbury, 1866.
Cyanellaceae Salisbury, 1866.
Cyrtanthaceae Salisbury, 1866.
Dianellaceae Salisbury, 1866.
Eucomidaceae Salisbury, 1866.
Fritillariaceae Salisbury, 1866.
Haemanthaceae Salisbury, 1866.
Lachenaliaceae Salisbury, 1866.
Oporanthaceae Salisbury, 1866.
Ornithogalaceae Salisbury, 1866.
Peliosanthaceae Salisbury, 1866.
Polygonataceae Salisbury, 1866.
Strumariaceae Salisbury, 1866.
Themidaceae Salisbury, 1866.
Tulbaghiaceae Salisbury, 1866.
Zephyranthaceae Salisbury, 1866.
Conantheraceae (D. Don) J. D. Hooker, 1873.
Laxmanniaceae Bubani, 1901–1902.
Ruscaceae Sprengel ex Hutchinson, 1934, nom. cons.
Ixiolirionaceae (Pax) Nakai, 1943.
Hesperocallidaceae Traub, 1972.
Milulaceae Traub, 1972.
Blandfordiaceae R. Dahlgren & Clifford, 1985.
Medeolaceae (S. Watson) Takhtajan, 1987.
Hostaceae B. Mathew, 1988, nom. cons. prop.
Lanariaceae H. F. J. Huber ex R. Dahlgren, 1988.

376. Iridaceae A. L. de Jussieu, 1789, nom. cons.
  Crocaceae Vest, 1818.
  Ixiaceae Horaninow, 1834.
  Galaxiaceae Rafinesque, 1836.
  Gladiolaceae Rafinesque, 1838.
  Isophysidaceae (Hutchinson) Barkley, 1948.

377. Velloziaceae Endlicher, 1841, nom. cons.
  Barbaceniaceae Walker-Arnott, 1842.
  Acanthochlamydaceae (S. C. Chen) P. C. Kao, 1989.

378. Aloaceae Batsch, 1802.

379. Agavaceae Endlicher, 1841, nom. cons.
  Phormiaceae J. Agardh, 1858.
  Yuccaceae J. Agardh, 1858.
  Dracaenaceae Salisbury, 1866, nom. cons.
  Nolinaceae Nakai, 1943.
  Sansevieriaceae Nakai, 1936.
  Doryanthaceae R. Dahlgren & Clifford, 1985.

380. Xanthorrhoeaceae Dumortier, 1829, nom. cons.
  Dasypogonaceae Dumortier, 1829.
  Calectasiaceae Endlicher, 1838.
  Kingiaceae Endlicher, 1838.
  Lomandraceae J. P. Lotsy, 1911.

381. Hanguanaceae Airy Shaw, 1965.

382. Taccaceae Dumortier, 1829, nom. cons.

383. Stemonaceae Engler, 1887, nom. cons.
  Roxburghiaceae Wallich, 1832.
  Croomiaceae Nakai, 1937.
  Pentastemonaceae Duyfjes, 1992.

384. Smilacaceae Ventenat, 1799, nom. cons.
  Philesiaceae Dumortier, 1829, nom. cons.
  Lapageriaceae Kunth, 1850.
  Luzuriagaceae J. P. Lotsy, 1911.
  Petermanniaceae Hutchinson, 1934, nom. cons.
  Rhipogonaceae Conran & Clifford, 1985.

385. Dioscoreaceae R. Brown, 1810, nom. cons.

APPENDIX 15.1 (Continued).

Tamaceae Martinov, 1820.
Tamnaceae J. Kickx, 1826.
Stenomeridaceae J. Agardh, 1858, nom. cons.
Trichopodaceae Hutchinson, 1934, nom. cons.

83. Orchidales Dumortier, 1829.

Burmanniales Heintze, 1927.

386. Geosiridaceae Jonker, 1939, nom. cons.

387. Burmanniaceae Blume, 1827, nom. cons.
Tripterellaceae Dumortier, 1829.
Thismiaceae J. Agardh, 1858, nom. cons.

388. Corsiaceae Beccari, 1878, nom. cons.

389. Orchidaceae A. L. de Jussieu, 1789, nom. cons.
Apostasiaceae Lindley, 1833, nom. cons., emend. prop.
Cypripediaceae Lindley, 1833.
Neottiaceae Horaninow, 1834.
Vanillaceae Lindley, 1835.
Limodoraceae Horaninow, 1847.
Neuwiediaceae (Burns-Balogh & Funk) R. Dahlgren ex Reveal & Hoogland, 1991.

The sequence of flowering plants follows A. Cronquist (1981, 1988).

A number of nomenclatural problems regarding family and suprafamilial names remain unsolved. The XVth Botanical Congress (September 1993) may decide to "protect" family names and their bibliographic references (R. D. Hoogland and J. L. Reveal 1993). Should protection be granted, the above citations will be retained; if not, more than 100 will be altered. Ongoing studies no doubt will discover new names and uncover earlier references for unprotected names.

A treatment of all names is under preparation by Hoogland and Reveal.

Names above the rank of family are even more chaotic. Numerous suprafamilial names were recently validated (J. L. Reveal 1992, 1992b, 1993, 1993b, 1993c; Reveal and A. L. Takhtajan 1993). The authorship of Liliopsida is altered here from that given by Cronquist et al. (1966), and while there are numerous earlier class names for dicots (e.g., Rosiopsida Batsch, 1802), a pre-1966 validation of Magnoliopsida has not been found. Essentially no effort has been made to review the pre-1966 literature for subclass names.

The Botanical Congress will also address the question of banning certain publications. The discovery of some 310 generic names validly published by John Hill in the 1750s (Reveal 1991) means that some family names long regarded as illegitimate (e.g., Operculariaceae, Phelypaeaceae, Terebinthaceae) because their generic stem was illegitimate are now available because the Hill names were not superfluous when published. If Hill's works are banned, then several synonyms listed above should be deleted.

APPENDIX 15.2. Concordance of Family Names Accepted by Cronquist, Takhtajan, and Thorne

Name	*Cronquist*	*Takhtajan*	*Thorne*
Abrophyllaceae	Grossulariaceae	Abrophyllaceae	Escalloniaceae
Acanthaceae	Acanthaceae	Acanthaceae	Acanthaceae
Aceraceae	Aceraceae	Aceraceae	Sapindaceae
Achariaceae	Achariaceae	Achariaceae	Achariaceae
Achatocarpaceae	Achatocarpaceae	Achatocarpaceae	Achatocarpaceae
Acoraceae	Acoraceae	Araceae	Acoraceae
Actinidiaceae	Actinidiaceae	Actinidiaceae	Actinidiaceae
Adoxaceae	Adoxaceae	Adoxaceae	Adoxaceae
Aegicerataceae	Myrsinaceae	Aegicerataceae	Myrsinaceae
Aextoxicaceae	Aextoxicaceae	Aextoxicaceae	Aextoxicaceae
Agavaceae	Agavaceae	Agavaceae	Agavaceae
Agdestidaceae	Phytolaccaceae	Agdestidaceae	Agdestidaceae
Aizoaceae	Aizoaceae	Aizoaceae	Aizoaceae
Akaniaceae	Akaniaceae	Akaniaceae	Akaniaceae
Alangiaceae	Alangiaceae	Alangiaceae	Alangiaceae
Alismataceae	Alismataceae	Alismataceae	Alismataceae
Alliaceae	Liliaceae	Alliaceae	Alliaceae
Aloaceae	Aloaceae	Asphodelaceae	Asphodelaceae
Alseuosmiaceae	Alseuosmiaceae	Alseuosmiaceae	Alseuosmiaceae
Alstroemeriaceae	Liliaceae	Alstroemeriaceae	Alstroemeriaceae
Altingiaceae	Hamamelidaceae	Altingiaceae	Hamamelidaceae
Alzateaceae	Alzateaceae	Alzateaceae	Alzateaceae
Amaranthaceae	Amaranthaceae	Amaranthaceae	Amaranthaceae
Amaryllidaceae	Liliaceae	Amaryllidaceae	Amaryllidaceae
Amborellaceae	Amborellaceae	Amborellaceae	Amborellaceae
Anacardiaceae	Anacardiaceae	Anacardiaceae	Anacardiaceae
Anarthriaceae	Restionaceae	Anarthriaceae	Restionaceae
Ancistrocladaceae	Ancistrocladaceae	Ancistrocladaceae	Ancistrocladaceae
Anisophylleaceae	Anisophylleaceae	Anisophylleaceae	Anisophylleaceae
Annonaceae	Annonaceae	Annonaceae	Annonaceae
Antoniaceae	Loganicaceae	Antoniaceae	Loganiaceae
Aphloiaceae	Flacourtiaceae	Aphloiaceae	Flacourtiaceae
Aphyllanthaceae	Liliaceae	Aphyllanthaceae	Aphyllanthaceae
Apiaceae	Apiaceae	Apiaceae	Apiaceae
Apocynaceae	Apocynaceae	Apocynaceae	Apocynaceae
Apodanthaceae	Rafflesiaceae	Apodanthaceae	Rafflesiaceae
Aponogetonaceae	Aponogetonaceae	Aponogetonaceae	Aponogetonaceae
Aptandraceae	Olacaceae	Aptandraceae	Olacaceae
Aquifoliaceae	Aquifoliaceae	Aquifoliaceae	Aquifoliaceae
Araceae	Araceae	Araceae	Araceae
Araliaceae	Araliaceae	Araliaceae	Araliaceae

The name-column on the left cites all of the family names recognized by Cronquist, Takhtajan, and Thorne in their respective systems. The other columns indicate the families under which each author disposed of the names in the first column.

APPENDIX 15.2 (Continued).

Name	*Cronquist*	*Takhtajan*	*Thorne*
Aralidiaceae	Cornaceae	Aralidiaceae	Aralidiaceae
Arecaceae	Arecaceae	Arecaceae	Arecaceae
Argophyllaceae	Grossulariaceae	Argophyllaceae	Escalloniaceae
Aristolochiaceae	Aristolochiaceae	Aristolochiaceae	Aristolochiaceae
Asclepiadaceae	Asclepiadaceae	Asclepiadaceae	Apocynaceae
Asparagaceae	Liliaceae	Asparagaceae	Asparagaceae
Asphodelaceae	Liliaceae	Asphodelaceae	Asphodelaceae
Asteliaceae	Liliaceae	Asteliaceae	Dracaenaceae
Asteraceae	Asteraceae	Asteraceae	Asteraceae
Asteropeiaceae	Theaceae	Asteropeiaceae	Asteropeiaceae
Atherospermataceae	Monimiaceae	Atherospermataceae	Monimiaceae
Aucubaceae	Cornaceae	Aucubaceae	Aucubaceae
Austrobaileyaceae	Austrobaileyaceae	Austrobaileyaceae	Austrobaileyaceae
Avicenniaceae	Verbenaceae	Avicenniaceae	Avicenniaceae
Balanitaceae	Zygophyllaceae	Balanitaceae	Balanitaceae
Balanopaceae	Balanopaceae	Balanopaceae	Balanopaceae
Balanophoraceae	Balanophoraceae	Balanophoraceae	Balanophoraceae
Balsaminaceae	Balsaminaceae	Balsaminaceae	Balsaminaceae
Barbeuiaceae	Phytolaccaceae	Barbeuiaceae	Barbeuiaceae
Barbeyaceae	Barbeyaceae	Barbeyaceae	Barbeyaceae
Barclayaceae	Barclayaceae	Barclayaceae	Nymphaeaceae
Basellaceae	Basellaceae	Basellaceae	Basellaceae
Bataceae	Bataceae	Bataceae	Bataceae
Begoniaceae	Begoniaceae	Begoniaceae	Begoniaceae
Berberidaceae	Berberidaceae	Berberidaceae	Berberidaceae
Berberidopsidaceae	Flacourtiaceae	Berberidopsidaceae	Flacourtiaceae
Betulaceae	Betulaceae	Betulaceae	Betulaceae
Biebersteiniaceae	Geraniaceae	Biebersteiniaceae	Geraniaceae
Bignoniaceae	Bignoniaceae	Bignoniaceae	Bignoniaceae
Bixaceae	Bixaceae	Bixaceae	Bixaceae
Blandfordiaceae	Liliaceae	Blandfordiaceae	Blandfordiaceae
Bombacaceae	Bombacaceae	Bombacaceae	Bombacaceae
Bonnetiaceae	Theaceae	Bonnetiaceae	Bonnetiaceae
Boraginaceae	Boraginaceae	Boraginaceae	Boraginaceae
Brassicaceae	Brassicaceae	Brassicaceae	Brassicaceae
Bretschneideraceae	Bretschneideraceae	Bretschneideraceae	Bretschneideraceae
Brexiaceae	Grossulariaceae	Brexiaceae	Brexiaceae
Bromeliaceae	Bromeliaceae	Bromeliaceae	Bromeliaceae
Brunelliaceae	Brunelliaceae	Brunelliaceae	Cunoniaceae
Bruniaceae	Bruniaceae	Bruniaceae	Bruniaceae
Brunoniaceae	Brunoniaceae	Brunoniaceae	Brunoniaceae

The name-column on the left cites all of the family names recognized by Cronquist, Takhtajan, and Thorne in their respective systems. The other columns indicate the families under which each author disposed of the names in the first column.

APPENDIX 15.2 (Continued).

Name	*Cronquist*	*Takhtajan*	*Thorne*
Buddlejaceae	Buddlejaceae	Buddlejaceae	Buddlejaceae
Burmanniaceae	Burmanniaceae	Burmanniaceae	Burmanniaceae
Burseraceae	Burseraceae	Burseraceae	Burseraceae
Butomaceae	Butomaceae	Butomaceae	Butomaceae
Buxaceae	Buxaceae	Buxaceae	Buxaceae
Byblidaceae	Byblidaceae	Byblidaceae	Byblidaceae
Cabombaceae	Cabombaceae	Cabombaceae	Cabombaceae
Cactaceae	Cactaceae	Cactaceae	Cactaceae
Caesalpiniaceae	Caesalpiniaceae	Fabaceae	Fabaceae
Callitrichaceae	Callitrichaceae	Callitrichaceae	Callitrichaceae
Calochortaceae	Liliaceae	Calochortaceae	Liliaceae
Calycanthaceae	Calycanthaceae	Calycanthaceae	Calycanthaceae
Calyceraceae	Calyceraceae	Calyceraceae	Calyceraceae
Campanulaceae	Campanulaceae	Campanulaceae	Campanulaceae
Campynemataceae	Liliaceae	Melanthiaceae	Campynemataceae
Canellaceae	Canellaceae	Canellaceae	Canellaceae
Cannabaceae	Cannabaceae	Cannabaceae	Cannabaceae
Cannaceae	Cannaceae	Cannaceae	Cannaceae
Capparaceae	Capparaceae	Capparaceae	Capparaceae
Caprifoliaceae	Caprifoliaceae	Caprifoliaceae	Caprifoliaceae
Cardiopteridaceae	Cardiopteridaceae	Cardiopteridaceae	Cardiopteridaceae
Caricaceae	Caricaceae	Caricaceae	Caricaceae
Carlemanniaceae	Caprifoliaceae	Carlemanniaceae	Caprifoliaceae
Carpodetaceae	Grossulariaceae	Carpodetaceae	Carpodetaceae
Caryocaraceae	Caryocaraceae	Caryocaraceae	Caryocaraceae
Caryophyllaceae	Caryophyllaceae	Caryophyllaceae	Caryophyllaceae
Casuarinaceae	Casuarinaceae	Casuarinaceae	Casuarinaceae
Cecropiaceae	Cecropiaceae	Cecropiaceae	Cecropiaceae
Celastraceae	Celastraceae	Celastraceae	Celastraceae
Centrolepidaceae	Centrolepidaceae	Centrolepidaceae	Centrolepidaceae
Cephalotaceae	Cephalotaceae	Cephalotaceae	Cephalotaceae
Ceratophyllaceae	Ceratophyllaceae	Ceratophyllaceae	Ceratophyllaceae
Cercidiphyllaceae	Cercidiphyllaceae	Cercidiphyllaceae	Cercidiphyllaceae
Chenopodiaceae	Chenopodiaceae	Chenopodiaceae	Chenopodiaceae
Chloanthaceae	Verbenaceae	Chloanthaceae	Lamiaceae
Chloranthaceae	Chloranthaceae	Chloranthaceae	Chloranthaceae
Chrysobalanaceae	Chrysobalanaceae	Chrysobalanaceae	Chrysobalanaceae
Circaeasteraceae	Circaeasteraceae	Circaeasteraceae	Circaeasteraceae
Cistaceae	Cistaceae	Cistaceae	Cistaceae
Clethraceae	Clethraceae	Clethraceae	Clethraceae
Clusiaceae	Clusiaceae	Clusiaceae	Clusiaceae
Cneoraceae	Cneoraceae	Cneoraceae	Cneoraceae
Cobaeaceae	Polemoniaceae	Cobaeaceae	Polemoniaceae
Cochlospermaceae	Bixaceae	Cochlospermaceae	Cochlospermaceae

APPENDIX 15.2 (Continued).

Name	*Cronquist*	*Takhtajan*	*Thorne*
Colchicaceae	Liliaceae	Melanthiaceae	Colchicaceae
Columelliaceae	Columelliaceae	Columelliaceae	Columelliaceae
Combretaceae	Combretaceae	Combretaceae	Combretaceae
Commelinaceae	Commelinaceae	Commelinaceae	Commelinaceae
Connaraceae	Connaraceae	Connaraceae	Connaraceae
Conostylidaceae	Haemodoraceae	Conostylidaceae	Haemodoraceae
Convallariaceae	Liliaceae	Convallariaceae	Asparagaceae
Convolvulaceae	Convolvulaceae	Convolvulaceae	Convolvulaceae
Cordiaceae	Boraginaceae	Cordiaceae	Boraginaceae
Coriariaceae	Coriariaceae	Coriariaceae	Coriariaceae
Cornaceae	Cornaceae	Cornaceae	Cornaceae
Corsiaceae	Corsiaceae	Corsiaceae	Corsiaceae
Corylaceae	Betulaceae	Corylaceae	Betulaceae
Corynocarpaceae	Corynocarpaceae	Corynocarpaceae	Corynocarpaceae
Costaceae	Costaceae	Costaceae	Costaceae
Crassulaceae	Crassulaceae	Crassulaceae	Crassulaceae
Crossosomataceae	Crossosomataceae	Crossosomataceae	Crossosomataceae
Crypteroniaceae	Crypteroniaceae	Crypteroniaceae	Crypteroniaceae
Ctenolophonaceae	Hugoniaceae	Ctenolophonaceae	Ctenolophonaceae
Cucurbitaceae	Cucurbitaceae	Cucurbitaceae	Cucurbitaceae
Cunoniaceae	Cunoniaceae	Cunoniaceae	Cunoniaceae
Curtisiaceae	Cornaceae	Curtisiaceae	Curtisiaceae
Cuscutaceae	Cuscutaceae	Cuscutaceae	Convolvulaceae
Cyanastraceae	Cyanastraceae	Cyanastraceae	Cyanastraceae
Cyclanthaceae	Cyclanthaceae	Cyclanthaceae	Cyclanthaceae
Cyclocheilaceae	Verbenaceae	Cyclocheilaceae	Nesogenaceae
Cymodoceaceae	Cymodoceaceae	Cymodoceaceae	Cymodoceaceae
Cynomoriaceae	Balanophoraceae	Cynomoriaceae	Cynomoriaceae
Cyperaceae	Cyperaceae	Cyperaceae	Cyperaceae
Cyphiaceae	Campanulaceae	Cyphiaceae	Campanulaceae
Cyphocarpaceae	Campanulaceae	Cyphocarpaceae	Campanulaceae
Cyrillaceae	Cyrillaceae	Cyrillaceae	Cyrillaceae
Cytinaceae	Rafflesiaceae	Cytinaceae	Rafflesiaceae
Dactylanthaceae	Balanophoraceae	Dactylanthaceae	Balanophoraceae
Daphniphyllaceae	Daphniphyllaceae	Daphniphyllaceae	Daphniphyllaceae
Dasypogonaceae	Xanthorrhoeaceae	Dasypogonaceae	Dasypogonaceae
Datiscaceae	Datiscaceae	Datiscaceae	Datiscaceae
Davidiaceae	Cornaceae	Davidiaceae	Cornaceae
Davidsoniaceae	Davidsoniaceae	Davidsoniaceae	Davidsoniaceae
Degeneriaceae	Degeneriaceae	Degeneriaceae	Degeneriaceae
Desfontainiaceae	Loganiaceae	Desfontainiaceae	Desfontainiaceae

The name-column on the left cites all of the family names recognized by Cronquist, Takhtajan, and Thorne in their respective systems. The other columns indicate the families under which each author disposed of the names in the first column.

APPENDIX 15.2 (Continued).

Name	*Cronquist*	*Takhtajan*	*Thorne*
Dialypetalanthaceae	Dialypetalanthaceae	Dialypetalanthaceae	Dialypetalanthaceae
Diapensiaceae	Diapensiaceae	Diapensiaceae	Diapensiaceae
Dichapetalaceae	Dichapetalaceae	Dichapetalaceae	Dichapetalaceae
Didiereaceae	Didiereaceae	Didiereaceae	Didiereaceae
Didymelaceae	Didymelaceae	Didymelaceae	Didymelaceae
Diegodendraceae	Ochnaceae	Diegodendraceae	Diegodendraceae
Dilleniaceae	Dilleniaceae	Dilleniaceae	Dilleniaceae
Dioncophyllaceae	Dioncophyllaceae	Dioncophyllaceae	Dioncophyllaceae
Dioscoreaceae	Dioscoreaceae	Dioscoreaceae	Dioscoreaceae
Dipentodontaceae	Dipentodontaceae	Dipentodontaceae	Dipentodontaceae
Dipsacaceae	Dipsacaceae	Dipsacaceae	Dipsacaceae
Dipterocarpaceae	Dipterocarpaceae	Dipterocarpaceae	Dipterocarpaceae
Dirachmaceae	Oxalidaceae	Dirachmaceae	Geraniaceae
Donatiaceae	Donatiaceae	Donatiaceae	Stylidiaceae
Doryanthaceae	Agavaceae	Doryanthaceae	Phormiaceae
Dracaenaceae	Agavaceae	Dracaenaceae	Dracaenaceae
Droseraceae	Droseraceae	Droseraceae	Droseraceae
Duabangaceae	Sonneratiaceae	Duabangaceae	Lythraceae
Duckeodendraceae	Duckeodendraceae	Duckeodendraceae	Duckeodendraceae
Ebenaceae	Ebenaceae	Ebenaceae	Ebenaceae
Ecdeiocoleaceae	Restionaceae	Ecdeiocoleaceae	Ecdeiocoleaceae
Ehretiaceae	Boraginaceae	Ehretiaceae	Boraginaceae
Elaeagnaceae	Elaeagnaceae	Elaeagnaceae	Elaeagnaceae
Elaeocarpaceae	Elaeocarpaceae	Elaeocarpaceae	Elaeocarpaceae
Elatinaceae	Elatinaceae	Elatinaceae	Elatinaceae
Emblingiaceae	Polygalaceae	Emblingiaceae	Sapindaceae
Empetraceae	Empetraceae	Empetraceae	Empetraceae
Epacridaceae	Epacridaceae	Epacridaceae	Epacridaceae
Eremolepidaceae	Eremolepidaceae	Eremolepidaceae	Eremolepidaceae
Eremosynaceae	Saxifragaceae	Eremosynaceae	Eremosynaceae
Ericaceae	Ericaceae	Ericaceae	Ericaceae
Eriocaulaceae	Eriocaulaceae	Eriocaulaceae	Eriocaulaceae
Eriospermaceae	Liliaceae	Eriospermaceae	Eriospermaceae
Erythroxylaceae	Erythroxylaceae	Erythroxylaceae	Erythroxylaceae
Escalloniaceae	Grossulariaceae	Escalloniaceae	Escalloniaceae
Eucommiaceae	Eucommiaceae	Eucommiaceae	Eucommiaceae
Eucryphiaceae	Eucryphiaceae	Eucryphiaceae	Eucryphiaceae
Euphorbiaceae	Euphorbiaceae	Euphorbiaceae	Euphorbiaceae
Eupomatiaceae	Eupomatiaceae	Eupomatiaceae	Eupomatiaceae
Eupteleaceae	Eupteleaceae	Eupteleaceae	Eupteleaceae
Fabaceae	Fabaceae	Fabaceae	Fabaceae
Fagaceae	Fagaceae	Fagaceae	Fagaceae
Flacourtiaceae	Flacourtiaceae	Flacourtiaceae	Flacourtiaceae
Flagellariaceae	Flagellariaceae	Flagellariaceae	Flagellariaceae

APPENDIX 15.2 (Continued).

Name	*Cronquist*	*Takhtajan*	*Thorne*
Fouquieriaceae	Fouquieriaceae	Fouquieriaceae	Fouquieriaceae
Francoaceae	Saxifragaceae	Francoaceae	Francoaceae
Frankeniaceae	Frankeniaceae	Frankeniaceae	Frankeniaceae
Fumariaceae	Fumariaceae	Fumariaceae	Papaveraceae
Garryaceae	Garryaceae	Garryaceae	Garryaceae
Geissolomataceae	Geissolomataceae	Geissolomataceae	Geissolomataceae
Gentianaceae	Gentianaceae	Gentianaceae	Gentianaceae
Geosiridaceae	Geosiridaceae	Geosiridaceae	Iridaceae
Geraniaceae	Geraniaceae	Geraniaceae	Geraniaceae
Gesneriaceae	Gesneriaceae	Gesneriaceae	Gesneriaceae
Gisekiaceae	Phytolaccacea	Gisekiaceae	Phytolaccacea
Glaucidiaceae	Ranunculaceae	Glaucidiaceae	Glaucidiaceae
Globulariaceae	Globulariaceae	Globulariaceae	Globulariaceae
Goetzeaceae	Solanaceae	Goetzeaceae	Goetzeaceae
Gomortegaceae	Gomortegaceae	Gomortegaceae	Gomortegaceae
Gonystylaceae	Thymelaeaceae	Gonystylaceae	Gonystylaceae
Goodeniaceae	Goodeniaceae	Goodeniaceae	Goodeniaceae
Goupiaceae	Celastraceae	Goupiaceae	Goupiaceae
Greyiaceae	Greyiaceae	Greyiaceae	Greyiaceae
Griseliniaceae	Cornaceae	Griseliniaceae	Griseliniaceae
Grossulariaceae	Grossulariaceae	Grossulariaceae	Grossulariaceae
Grubbiaceae	Grubbiaceae	Grubbiaceae	Grubbiaceae
Gunneraceae	Gunneraceae	Gunneraceae	Gunneraceae
Gyrocarpaceae	Hernandiaceae	Gyrocarpaceae	Hernandiaceae
Gyrostemonaceae	Gyrostemonaceae	Gyrostemonaceae	Gyrostemonaceae
Haemodoraceae	Haemodoraceae	Haemodoraceae	Haemodoraceae
Halophilaceae	Hydrocharitaceae	Halophilaceae	Hydrocharitaceae
Halophytaceae	Chenopodiaceae	Halophytaceae	Halophytaceae
Haloragaceae	Haloragaceae	Haloragaceae	Haloragaceae
Hamamelidaceae	Hamamelidaceae	Hamamelidaceae	Hamamelidaceae
Hanguanaceae	Hanguanaceae	Hanguanaceae	Hanguanaceae
Hectorellaceae	Portulacaceae	Hectorellaceae	Hectorellaceae
Heliconiaceae	Heliconiaceae	Heliconiaceae	Heliconiaceae
Heloseaceae	Balanophoraceae	Heloseaceae	Balanophoraceae
Helwingiaceae	Cornaceae	Helwingiaceae	Helwingiaceae
Hemerocallidaceae	Liliaceae	Hemerocallidaceae	Hemerocallidaceae
Hernandiaceae	Hernandiaceae	Hernandiaceae	Hernandiaceae
Herreriaceae	Liliaceae	Herreriaceae	Asparagaceae
Hesperocallidaceae	Liliaceae	Hesperocallidaceae	Hyacinthaceae
Heteropyxidaceae	Myrtaceae	Heteropyxidaceae	Myrtaceae
Himantandraceae	Himantandraceae	Himantandraceae	Himantandraceae

The name-column on the left cites all of the family names recognized by Cronquist, Takhtajan, and Thorne in their respective systems. The other columns indicate the families under which each author disposed of the names in the first column.

APPENDIX 15.2 (Continued).

Name	*Cronquist*	*Takhtajan*	*Thorne*
Hippocastanaceae	Hippocastanaceae	Hippocastanaceae	Hippocastanaceae
Hippocrateaceae	Hippocrateaceae	Celastraceae	Celastraceae
Hippuridaceae	Hippuridaceae	Hippuridaceae	Hippuridaceae
Hoplestigmataceae	Hoplestigmataceae	Hoplestigmataceae	Hoplestigmataceae
Hostaceae	Liliaceae	Hostaceae	Hostaceae
Huaceae	Huaceae	Huaceae	Huaceae
Hugoniaceae	Hugoniaceae	Hugoniaceae	Hugoniaceae
Humiriaceae	Humiriaceae	Humiriaceae	Humiriaceae
Hyacinthaceae	Liliaceae	Hyacinthaceae	Hyacinthaceae
Hydatellaceae	Hydatellaceae	Hydatellaceae	Hydatellaceae
Hydnoraceae	Hydnoraceae	Hydnoraceae	Hydnoraceae
Hydrangeaceae	Hydrangeaceae	Hydrangeaceae	Hydrangeaceae
Hydrastidaceae	Ranunculaceae	Hydrastidaceae	Hydrastidaceae
Hydrocharitaceae	Hydrocharitaceae	Hydrocharitaceae	Hydrocharitaceae
Hydrocotylaceae	Apiaceae	Hydrocotylaceae	Hydrocotylaceae
Hydropeltidaceae	Cabombaceae	Hydropeltidaceae	Cabombaceae
Hydrophyllaceae	Hydrophyllaceae	Hydrophyllaceae	Hydrophyllaceae
Hydrostachyaceae	Hydrostachyaceae	Hydrostachyaceae	Hydrostachyaceae
Hypecoaceae	Fumariaceae	Hypecoaceae	Papaveraceae
Hypericaceae	Clusiaceae	Hypericaceae	Clusiaceae
Hypoxidaceae	Liliaceae	Hypoxidaceae	Hypoxidaceae
Hypseocharitaceae	Oxalidaceae	Hypseocharitaceae	Geraniaceae
Icacinaceae	Icacinaceae	Icacinaceae	Icacinaceae
Idiospermaceae	Idiospermaceae	Idiospermaceae	Calycanthaceae
Illiciaceae	Illiciaceae	Illiciaceae	Illiciaceae
Iridaceae	Iridaceae	Iridaceae	Iridaceae
Irvingiaceae	Simaroubaceae	Irvingiaceae	Simaroubaceae
Iteaceae	Grossulariaceae	Iteaceae	Escalloniaceae
Ixerbaceae	Grossulariaceae	Ixerbaceae	Brexiaceae
Ixiolirionaceae	Liliaceae	Ixiolirionaceae	Ixiolirionaceae
Ixonanthaceae	Ixonanthaceae	Ixonanthaceae	Ixonanthaceae
Joinvilleaceae	Joinvilleaceae	Joinvilleaceae	Joinvilleaceae
Juglandaceae	Juglandaceae	Juglandaceae	Juglandaceae
Julianiaceae	Julianiaceae	Anacardiaceae	Anacardiaceae
Juncaceae	Juncaceae	Juncaceae	Juncaceae
Juncaginaceae	Juncaginaceae	Juncaginaceae	Juncaginaceae
Kingdoniaceae	Circaeasteraceae	Kingdoniaceae	Ranunculaceae
Kiggelariaceae	Flacourtiaceae	Kiggelariaceae	Flacourtiaceae
Kirkiaceae	Simaroubaceae	Kirkiaceae	Simaroubaceae
Krameriaceae	Krameriaceae	Krameriaceae	Krameriaceae
Lacistemataceae	Lacistemataceae	Lacistemataceae	Lacistemataceae
Lactoridaceae	Lactoridaceae	Lactoridaceae	Lactoridaceae
Lamiaceae	Lamiaceae	Lamiaceae	Lamiaceae
Lanariaceae	Liliaceae	Tecophilaeaceae	Lanariaceae

APPENDIX 15.2 (Continued).

Name	*Cronquist*	*Takhtajan*	*Thorne*
Langsdorffiaceae	Balanophoraceae	Langsdorffiaceae	Balanophoraceae
Lardizabalaceae	Lardizabalaceae	Lardizabalaceae	Lardizabalaceae
Lauraceae	Lauraceae	Lauraceae	Lauraceae
Lecythidaceae	Lecythidaceae	Lecythidaceae	Lecythidaceae
Ledocarpaceae	Geraniaceae	Ledocarpaceae	Geraniaceae
Leeaceae	Leeaceae	Leeaceae	Vitaceae
Leitneriaceae	Leitneriaceae	Leitneriaceae	Leitneriaceae
Lemnaceae	Lemnaceae	Lemnaceae	Lemnaceae
Lennoaceae	Lennoaceae	Lennoaceae	Lennoaceae
Lentibulariaceae	Lentibulariaceae	Lentibulariaceae	Lentibulariaceae
Lepidobotryaceae	Oxalidaceae	Lepidobotryaceae	Oxalidaceae
Lepuropetalaceae	Saxifragaceae	Lepuropetalaceae	Lepuropetalaceae
Lilaeaceae	Juncaginaceae	Lilaeaceae	Juncaginaceae
Liliaceae	Liliaceae	Liliaceae	Liliaceae
Limnanthaceae	Limnanthaceae	Limnanthaceae	Limnanthaceae
Limnocharitaceae	Limnocharitaceae	Limnocharitaceae	Alismataceae
Linaceae	Linaceae	Linaceae	Linaceae
Lissocarpaceae	Lissocarpaceae	Lissocarpaceae	Lissocarpaceae
Loasaceae	Loasaceae	Loasaceae	Loasaceae
Lobeliaceae	Campanulaceae	Lobeliaceae	Campanulaceae
Loganiaceae	Loganiaceae	Loganiaceae	Loganiaceae
Lophiraceae	Ochnaceae	Lophiraceae	Ochnaceae
Lophophytaceae	Balanophoraceae	Lophophytaceae	Balanophoraceae
Lophopyxidaceae	Celastraceae	Lophopyxidaceae	Lophopyxidaceae
Loranthaceae	Loranthaceae	Loranthaceae	Loranthaceae
Lowiaceae	Lowiaceae	Lowiaceae	Lowiaceae
Luzuriagaceae	Smilacaceae	Luzuriagaceae	Luzuriagaceae
Lythraceae	Lythraceae	Lythraceae	Lythraceae
Magnoliaceae	Magnoliaceae	Magnoliaceae	Magnoliaceae
Malesherbiaceae	Malesherbiaceae	Malesherbiaceae	Malesherbiaceae
Malpighiaceae	Malpighiaceae	Malpighiaceae	Malpighiaceae
Malvaceae	Malvaceae	Malvaceae	Malvaceae
Marantaceae	Marantaceae	Marantaceae	Marantaceae
Marcgraviaceae	Marcgraviaceae	Marcgraviaceae	Marcgraviaceae
Martyniaceae	Pedaliaceae	Martyniaceae	Martyniaceae
Mastixiaceae	Cornaceae	Mastixiaceae	Cornaceae
Maundiaceae	Juncaginaceae	Maundiaceae	Juncaginaceae
Mayacaceae	Mayacaceae	Mayacaceae	Mayacaceae
Medeolaceae	Liliaceae	Medeolaceae	Liliaceae
Medusagynaceae	Medusagynaceae	Medusagynaceae	Medusagynaceae
Medusandraceae	Medusandraceae	Medusandraceae	Medusandraceae

The name-column on the left cites all of the family names recognized by Cronquist, Takhtajan, and Thorne in their respective systems. The other columns indicate the families under which each author disposed of the names in the first column.

APPENDIX 15.2 (Continued).

Name	*Cronquist*	*Takhtajan*	*Thorne*
Melanophyllaceae	Cornaceae	Melanophyllaceae	Montiniaceae
Melanthiaceae	Liliaceae	Melanthiaceae	Melanthiaceae
Melastomataceae	Melastomataceae	Melastomataceae	Melastomataceae
Meliaceae	Meliaceae	Meliaceae	Meliaceae
Melianthaceae	Melianthaceae	Melianthaceae	Melianthaceae
Meliosmaceae	Sabiaceae	Meliosmaceae	Sabiaceae
Memecylaceae	Melastomataceae	Meliosmaceae	Melastomataceae
Mendonciaceae	Mendonciaceae	Mendonciaceae	Acanthaceae
Menispermaceae	Menispermaceae	Menispermaceae	Menispermaceae
Menyanthaceae	Menyanthaceae	Menyanthaceae	Menyanthaceae
Metteniusaceae	Alangiaceae	Metteniusaceae	Metteniusaceae
Mimosaceae	Mimosaceae	Fabaceae	Fabaceae
Misodendraceae	Misodendraceae	Misodendraceae	Misodendraceae
Mitrastemonaceae	Mitrastemonaceae	Mitrastemonaceae	Rafflesiaceae
Molluginaceae	Molluginaceae	Molluginaceae	Molluginaceae
Monimiaceae	Monimiaceae	Monimiaceae	Monimiaceae
Monotaceae	Dipterocarpaceae	Monotaceae	Monotaceae
Monotropaceae	Monotropaceae	Ericaceae	Ericaceae
Montiniaceae	Grossulariaceae	Montiniaceae	Montiniaceae
Moraceae	Moraceae	Moraceae	Moraceae
Morinaceae	Dipsacaceae	Morinaceae	Morinaceae
Moringaceae	Moringaceae	Moringaceae	Moringaceae
Musaceae	Musaceae	Musaceae	Musaceae
Myoporaceae	Myoporaceae	Myoporaceae	Myoporaceae
Myricaceae	Myricaceae	Myricaceae	Myricaceae
Myristicaceae	Myristicaceae	Myristicaceae	Myristicaceae
Myrothamnaceae	Myrothamnaceae	Myrothamnaceae	Myrothamnaceae
Myrsinaceae	Myrsinaceae	Myrsinaceae	Myrsinaceae
Myrtaceae	Myrtaceae	Myrtaceae	Myrtaceae
Mystropetalaceae	Balanophoraceae	Mystropetalaceae	Balanophoraceae
Najadaceae	Najadaceae	Najadaceae	Hydrocharitaceae
Nandinaceae	Berberidaceae	Nandinaceae	Berberidaceae
Nelumbonaceae	Nelumbonaceae	Nelumbonaceae	Nelumbonaceae
Nemacladaceae	Campanulaceae	Nemacladaceae	Campanulaceae
Nepenthaceae	Nepenthaceae	Nepenthaceae	Nepenthaceae
Nesogenaceae	Verbenaceae	Nesogenaceae	Nesogenaceae
Neuradaceae	Neuradaceae	Neuradaceae	Neuradaceae
Nitrariaceae	Zygophyllaceae	Nitrariaceae	Zygophyllaceae
Nolanaceae	Nolanaceae	Nolanaceae	Nolanaceae
Nolinaceae	Agavaceae	Nolinaceae	Dracaenaceae
Nothofagaceae	Nothofagaceae	Nothofagaceae	Nothofagacea
Nyctaginaceae	Nyctaginaceae	Nyctaginaceae	Nyctaginaceae
Nymphaeaceae	Nymphaeaceae	Nymphaeaceae	Nymphaeaceae
Nyssaceae	Cornaceae	Nyssaceae	Cornaceae

APPENDIX 15.2 (Continued).

Name	*Cronquist*	*Takhtajan*	*Thorne*
Ochnaceae	Ochnaceae	Ochnaceae	Ochnaceae
Octoknemaceae	Olacaceae	Octoknemaceae	Olacaceae
Oftiaceae	Myoporaceae	Oftiaceae	Myoporaceae
Olacaceae	Olacaceae	Olacaceae	Olacaceae
Oleaceae	Oleaceae	Oleaceae	Oleaceae
Oliniaceae	Oliniaceae	Oliniaceae	Oliniaceae
Onagraceae	Onagraceae	Onagraceae	Onagraceae
Oncothecaceae	Oncothecaceae	Oncothecaceae	Oncothecaceae
Opiliaceae	Opiliaceae	Opiliaceae	Opiliaceae
Orchidaceae	Orchidaceae	Hypoxidaceae	Orchidaceae
Orobanchaceae	Orobanchaceae	Scrophulariaceae	Scrophulariaceae
Oxalidaceae	Oxalidaceae	Oxalidaceae	Oxalidaceae
Paeoniaceae	Paeoniaceae	Paeoniaceae	Paeoniaceae
Pandaceae	Pandaceae	Pandaceae	Euphorbiaceae
Pandanaceae	Pandanaceae	Pandanaceae	Pandanaceae
Papaveraceae	Papaveraceae	Papaveraceae	Papaveraceae
Paracryphiaceae	Paracryphiaceae	Paracryphiaceae	Paracryphiaceae
Parnassiaceae	Saxifragaceae	Parnassiaceae	Parnassiaceae
Passifloraceae	Passifloraceae	Passifloraceae	Passifloraceae
Pedaliaceae	Pedaliaceae	Pedaliaceae	Pedaliaceae
Peganaceae	Zygophyllaceae	Peganaceae	Zygophyllaceae
Pellicieraceae	Pellicieraceae	Pellicieraceae	Pellicieraceae
Penaeaceae	Penaeaceae	Penaeaceae	Penaeaceae
Pentaphragmataceae	Pentaphragmataceae	Pentaphragmataceae	Pentaphragmataceae
Pentaphylacaceae	Pentaphylacaceae	Pentaphylacaceae	Pentaphylacaceae
Penthoraceae	Saxifragaceae	Penthoraceae	Penthoraceae
Peperomiaceae	Piperaceae	Peperomiaceae	Piperaceae
Peridiscaceae	Peridiscaceae	Peridiscaceae	Peridiscaceae
Petermanniaceae	Smilacaceae	Petermanniaceae	Petermanniaceae
Petiveriaceae	Phytolaccaceae	Petiveriaceae	Petiveriaceae
Petrosaviaceae	Petrosaviaceae	Melanthiaceae	Melanthiaceae
Phellinaceae	Aquifoliaceae	Phellinaceae	Phellinaceae
Philesiaceae	Smilacaceae	Philesiaceae	Philesiaceae
Philydraceae	Philydraceae	Philydraceae	Philydraceae
Phormiaceae	Agavaceae	Phormiaceae	Phormiaceae
Phrymaceae	Verbenaceae	Phrymaceae	Phrymaceae
Phyllonomaceae	Grossulariaceae	Phyllonomaceae	Escalloniaceae
Physenaceae	Physenaceae	Physenaceae	Physenaceae
Phytolaccaceae	Phytolaccaceae	Phytolaccaceae	Phytolaccaceae
Piperaceae	Piperaceae	Piperaceae	Piperaceae
Pittosporaceae	Pittosporaceae	Pittosporaceae	Pittosporaceae

The name-column on the left cites all of the family names recognized by Cronquist, Takhtajan, and Thorne in their respective systems. The other columns indicate the families under which each author disposed of the names in the first column.

APPENDIX 15.2 (Continued).

Name	*Cronquist*	*Takhtajan*	*Thorne*
Plagiopteraceae	Flacourtiaceae	Plagiopteraceae	Plagiopteraceae
Plantaginaceae	Plantaginaceae	Plantaginaceae	Plantaginaceae
Platanaceae	Platanaceae	Platanaceae	Platanaceae
Plocospermataceae	Apocynaceae	Plocospermataceae	Loganiaceae
Plumbaginaceae	Plumbaginaceae	Plumbaginaceae	Plumbaginaceae
Poaceae	Poaceae	Poaceae	Poaceae
Podoaceae	Anacardiaceae	Podoaceae	Anacardiaceae
Podophyllaceae	Berberidaceae	Podophyllaceae	Berberidaceae
Podostemaceae	Podostemaceae	Podostemaceae	Podostemaceae
Polemoniaceae	Polemoniaceae	Polemoniaceae	Polemoniaceae
Polygalaceae	Polygalaceae	Polygalaceae	Polygalaceae
Polygonaceae	Polygonaceae	Polygonaceae	Polygonaceae
Polyosmaceae	Grossulariaceae	Polyosmaceae	Escalloniaceae
Pontederiaceae	Pontederiaceae	Pontederiaceae	Pontederiaceae
Portulacaceae	Portulacaceae	Portulacaceae	Portulacaceae
Posidoniaceae	Posidoniaceae	Posidoniaceae	Posidoniaceae
Potamogetonaceae	Potamogetonaceae	Potamogetonaceae	Potamogetonaceae
Pottingeriaceae	Celastraceae	Pottingeriaceae	Celastraceae
Primulaceae	Primulaceae	Primulaceae	Primulaceae
Proteaceae	Proteaceae	Proteaceae	Proteaceae
Psiloxylaceae	Myrtaceae	Psiloxylaceae	Myrtaceae
Ptaeroxylaceae	Sapindaceae	Ptaeroxylaceae	Ptaeroxylaceae
Pteridophyllaceae	Fumariaceae	Pteridophyllaceae	Papaveraceae
Pterostemonaceae	Grossulariaceae	Pterostemonaceae	Escalloniaceae
Punicaceae	Punicaceae	Punicaceae	Lythraceae
Pyrolaceae	Pyrolaceae	Ericaceae	Ericaceae
Quiinaceae	Quiinaceae	Quiinaceae	Quiinaceae
Rafflesiaceae	Rafflesiaceae	Rafflesiaceae	Rafflesiaceae
Ranunculaceae	Ranunculaceae	Ranunculaceae	Ranunculaceae
Rapateaceae	Rapateaceae	Rapateaceae	Rapateaceae
Reaumuriaceae	Tamaricaceae	Reaumuriaceae	Tamaricaceae
Resedaceae	Resedaceae	Resedaceae	Resedaceae
Restionaceae	Restionaceae	Restionaceae	Restionaceae
Retziaceae	Retziaceae	Retziaceae	Stilbaceae
Rhabdodendraceae	Rhabdodendraceae	Rhabdodendraceae	Rhabdodendraceae
Rhamnaceae	Rhamnaceae	Rhamnaceae	Rhamnaceae
Rhipogonaceae	Smilacaceae	Rhipogonaceae	Rhipogonaceae
Rhizophoraceae	Rhizophoraceae	Rhizophoraceae	Rhizophoraceae
Rhodoleiaceae	Hamamelidaceae	Rhodoleiaceae	Hamamelidaceae
Rhoipteleaceae	Rhoipteleaceae	Rhoipteleaceae	Rhoipteleaceae
Rhynchocalycaceae	Rhynchocalycaceae	Rhynchocalycaceae	Rhynchocalyaceae
Rhynchothecaceae	Geraniaceae	Rhynchothecaceae	Geraniaceae
Roridulaceae	Byblidaceae	Roridulaceae	Roridulaceae
Rosaceae	Rosaceae	Rosaceae	Rosaceae

APPENDIX 15.2 (Continued).

Name	*Cronquist*	*Takhtajan*	*Thorne*
Rousseaceae	Grossulariaceae	Rousseaceae	Brexiaceae
Rubiaceae	Rubiaceae	Rubiaceae	Rubiaceae
Ruppiaceae	Ruppiaceae	Ruppiaceae	Potamogetonaceae
Ruscaceae	Liliaceae	Ruscaceae	Asparagaceae
Rutaceae	Rutaceae	Rutaceae	Rutaceae
Sabiaceae	Sabiaceae	Sabiaceae	Sabiaceae
Saccifoliaceae	Saccifoliaceae	Saccifoliaceae	Saccifoliaceae
Salicaceae	Salicaceae	Salicaceae	Salicaceae
Salvadoraceae	Salvadoraceae	Salvadoraceae	Salvadoraceae
Sambucaceae	Caprifoliaceae	Sambucaceae	Adoxaceae
Santalaceae	Santalaceae	Santalaceae	Santalaceae
Sapindaceae	Sapindaceae	Sapindaceae	Sapindaceae
Sapotaceae	Sapotaceae	Sapotaceae	Sapotaceae
Sarcolaenaceae	Sarcolaenaceae	Sarcolaenaceae	Sarcolaenaceae
Sarcophytaceae	Balanophoraceae	Sarcophytaceae	Balanophoraceae
Sargentodoxaceae	Sargentodoxaceae	Sargentodoxaceae	Sargentodoxaceae
Sarraceniaceae	Sarraceniaceae	Sarraceniaceae	Sarraceniaceae
Saururaceae	Saururaceae	Saururaceae	Saururaceae
Sauvagesiaceae	Ochnaceae	Sauvagesiaceae	Ochnaceae
Saxifragaceae	Saxifragaceae	Saxifragaceae	Saxifragaceae
Scheuchzeriaceae	Scheuchzeriaceae	Scheuchzeriaceae	Scheuchzeriaceae
Schisandraceae	Schisandraceae	Schisandraceae	Schisandraceae
Sclerophylacaceae	Solanaceae	Sclerophylacaceae	Solanaceae
Scrophulariaceae	Scrophulariaceae	Scrophulariaceae	Scrophulariaceae
Scybaliaceae	Balanophoraceae	Scybaliaceae	Balanophoraceae
Scyphostegiaceae	Scyphostegiaceae	Scyphostegiaceae	Scyphostegiaceae
Scytopetalaceae	Scytopetalaceae	Scytopetalaceae	Scytopetalaceae
Selaginaceae	Globulariaceae	Selaginaceae	Scrophulariaceae
Sesuviaceae	Aizoaceae	Sesuviaceae	Aizoaceae
Simaroubaceae	Simaroubaceae	Simaroubaceae	Simaroubaceae
Simmondsiaceae	Simmondsiaceae	Simmondsiaceae	Simmondsiaceae
Siparunaceae	Monimiaceae	Siparunaceae	Monimiaceae
Sladeniaceae	Theaceae	Sladeniaceae	Theaceae
Smilacaceae	Smilacaceae	Smilacaceae	Smilacaceae
Solanaceae	Solanaceae	Solanaceae	Solanaceae
Sonneratiaceae	Sonneratiaceae	Sonneratiaceae	Lythraceae
Sparganiaceae	Sparganiaceae	Sparganiaceae	Typhaceae
Sphaerosepalaceae	Sphaerosepalaceae	Sphaerosepalaceae	Diegodendraceae
Sphenocleaceae	Sphenocleaceae	Sphenocleaceae	Sphenocleaceae
Sphenostemonaceae	Aquifoliaceae	Sphenostemonaceae	Sphenostemonaceae
Spigeliaceae	Loganiaceae	Spigeliaceae	Loganiaceae

The name-column on the left cites all of the family names recognized by Cronquist, Takhtajan, and Thorne in their respective systems. The other columns indicate the families under which each author disposed of the names in the first column.

APPENDIX 15.2 (Continued).

Name	*Cronquist*	*Takhtajan*	*Thorne*
Stachyuraceae	Stachyuraceae	Stachyuraceae	Stachyuraceae
Stackhousiaceae	Stackhousiaceae	Stackhousiaceae	Stackhousiaceae
Staphyleaceae	Staphyleaceae	Staphyleaceae	Staphyleaceae
Stegnospermataceae	Phytolaccaceae	Stegnospermataceae	Stegnospermataceae
Stemonaceae	Stemonaceae	Stemonaceae	Stemonaceae
Stenomeridaceae	Dioscoreaceae	Stenomeridaceae	Dioscoreaceae
Sterculiaceae	Sterculiaceae	Sterculiaceae	Sterculiaceae
Stilbaceae	Verbenaceae	Stilbaceae	Stilbaceae
Strasburgeriaceae	Ochnaceae	Strasburgeriaceae	Strasburgeriaceae
Strelitziaceae	Strelitziaceae	Strelitziaceae	Strelitziaceae
Stylidiaceae	Stylidiaceae	Stylidiaceae	Stylidiaceae
Stylobasiaceae	Surianaceae	Stylobasiaceae	Sapindaceae
Styloceratаceae	Buxaceae	Stylocerataceae	Buxaceae
Styracaceae	Styracaceae	Styracaceae	Styracaceae
Surianaceae	Surianaceae	Surianaceae	Surianaceae
Symphoremataceae	Verbenaceae	Symphoremataceae	Symphoremataceae
Symplocaceae	Symplocaceae	Symplocaceae	Symplocaceae
Taccaceae	Taccaceae	Taccaceae	Taccaceae
Tamaricaceae	Tamaricaceae	Tamaricaceae	Tamaricaceae
Tapisciaceae	Staphyleaceae	Tapisciaceae	Staphyleaceae
Tecophilaeaceae	Liliaceae	Tecophilaeaceae	Tecophilaeaceae
Tepuianthaceae	Tepuianthaceae	Tepuianthaceae	Tepuianthaceae
Tetracarpaeaceae	Grossulariaceae	Tetracarpaeaceae	Tetracarpaeaceae
Tetracentraceae	Tetracentraceae	Tetracentraceae	Trochodendraceae
Tetrachondraceae	Lamiaceae	Tetrachondraceae	Tetrachondraceae
Tetradiclidaceae	Zygophyllaceae	Tetradiclidaceae	Zygophyllaceae
Tetragoniaceae	Aizoaceae	Tetragoniaceae	Aizoaceae
Tetrameristaceae	Tetrameristaceae	Tetrameristaceae	Tetrameristaceae
Thalassiaceae	Hydrocharitaceae	Thalassiaceae	Hydrocharitaceae
Theaceae	Theaceae	Theaceae	Theaceae
Theligonaceae	Theligonaceae	Theligonaceae	Rubiaceae
Theophrastaceae	Theophrastaceae	Theophrastaceae	Theophrastaceae
Thurniaceae	Thurniaceae	Thurniaceae	Thurniaceae
Thymelaeaceae	Thymelaeaceae	Thymelaeaceae	Thymelaeaceae
Ticodendraceae	Ticodendraceae	Ticodendraceae	Ticodendraceae
Tiliaceae	Tiliaceae	Tiliaceae	Tiliaceae
Toricelliaceae	Cornaceae	Toricelliaceae	Toricelliaceae
Tovariaceae	Tovariaceae	Tovariaceae	Capparaceae
Trapaceae	Trapaceae	Trapaceae	Trapaceae
Trapellaceae	Pedaliaceae	Trapellaceae	Pedaliaceae
Tremandraceae	Tremandraceae	Tremandraceae	Tremandraceae
Tribelaceae	Grossulariaceae	Tribelaceae	Escalloniaceae
Trichopodaceae	Dioscoreaceae	Trichopodaceae	Trichopodaceae
Trigoniaceae	Trigoniaceae	Trigoniaceae	Trigoniaceae

APPENDIX 15.2 (Continued).

Name	*Cronquist*	*Takhtajan*	*Thorne*
Trilliaceae	Liliaceae	Trilliaceae	Trilliaceae
Trimeniaceae	Trimeniaceae	Trimeniaceae	Trimeniaceae
Triplostegiaceae	Valerianaceae	Triplostegiaceae	Triplostegiaceae
Triuridaceae	Triuridaceae	Triuridaceae	Triuridaceae
Trochodendraceae	Trochodendraceae	Trochodendraceae	Trochodendraceae
Tropaeolaceae	Tropaeolaceae	Tropaeolaceae	Tropaeolaceae
Turneraceae	Turneraceae	Turneraceae	Turneraceae
Typhaceae	Typhaceae	Typhaceae	Typhaceae
Ulmaceae	Ulmaceae	Ulmaceae	Ulmaceae
Urticaceae	Urticaceae	Urticaceae	Urticaceae
Vahliaceae	Saxifragaceae	Vahliaceae	Vahliaceae
Valerianaceae	Valerianaceae	Valerianaceae	Valerianaceae
Velloziaceae	Velloziaceae	Velloziaceae	Velloziaceae
Verbenaceae	Verbenaceae	Verbenaceae	Verbenaceae
Viburnaceae	Caprifoliaceae	Viburnaceae	Adoxaceae
Violaceae	Violaceae	Violaceae	Violaceae
Viscaceae	Viscaceae	Viscaceae	Viscaceae
Vitaceae	Vitaceae	Vitaceae	Vitaceae
Viticaceae	Verbenaceae	Viticaceae	Lamiaceae
Vivianiaceae	Geraniaceae	Vivianiaceae	Geraniaceae
Vochysiaceae	Vochysiaceae	Vochysiaceae	Vochysiaceae
Winteraceae	Winteraceae	Winteraceae	Winteraceae
Xanthophyllaceae	Xanthophyllaceae	Polygalaceae	Polygalaceae
Xanthorrhoeaceae	Xanthorrhoeaceae	Xanthorrhoeaceae	Xanthorrhoeaceae
Xyridaceae	Xyridaceae	Xyridaceae	Xyridaceae
Zannichelliaceae	Zannichelliaceae	Zannichelliaceae	Zannichelliaceae
Zingiberaceae	Zingiberaceae	Zingiberaceae	Zingiberaceae
Zosteraceae	Zosteraceae	Zosteraceae	Zosteraceae
Zygophyllaceae	Zygophyllaceae	Zygophyllaceae	Zygophyllaceae

The name-column on the left cites all of the family names recognized by Cronquist, Takhtajan, and Thorne in their respective systems. The other columns indicate the families under which each author disposed of the names in the first column.

Literature Cited

Robert W. Kiger, Editor

This is a consolidated list of all works cited in volume 1. In the entries for articles, serial titles are cited by the abbreviated forms recommended in G. D. R. Bridson and E. R. Smith (1991). Cross references to the corresponding full serial titles are interpolated in the list alphabetically by abbreviated form. Two or more works published in the same year by the same author or group of coauthors will be distinguished uniquely and consistently throughout all volumes of *Flora of North America* by lowercase letters (b, c, d, . . .) suffixed to the date for the second and subsequent works in the set. The suffixes are assigned in order of editorial encounter, and do not reflect actual chronological sequence of publication. The first work by any particular author or group from any given year carries the implicit date suffix "a"; thus, the sequence of explicit suffixes begins with "b". In two cases, this list includes citations with dates suffixed "b" that are not preceded by citations of "a" works (i.e., ones with no date suffix) for the same year. This does not reflect omissions here but rather that there are corresponding "a" works cited (and encountered first from) elsewhere in the *Flora* that are not pertinent here.

A. A. P. G. Mem. = A A P G Memoir. [American Association of Petroleum Geologists.]

Abh. K. K. Zool.-Bot. Ges. Wien = Abhandlungen der Kaiserlich-königlichen zoologisch-botanischen Gesellschaft in Wien.

Abh. Naturwiss. Naturwiss. Verein Hamburg = Abhandlungen aus dem Gebiete der Naturwissenschaften herausgegeben von dem Naturwissenschaftlichen Verein in Hamburg.

Abrams, L. and R. S. Ferris. 1923–1960. Illustrated Flora of the Pacific States: Washington, Oregon, and California. 4 vols. Stanford.

Achuff, P. 1989. Old-growth forests of the Canadian Rocky Mountain national parks. Nat. Areas J. 9: 12–26.

Acta Ecol. = Acta Ecologica.

Acta Horti Berg. = Acta Horti Bergiani.

Acta Phytogeogr. Suec. = Acta Phytogeographica Suecica.

Advancem. Sci. = Advancement of Science; Report of the British Association for the Advancement of Science.

Advances Bot. Res. = Advances in Botanical Research.

Ager, T. A. 1983. Holocene vegetational history of Alaska. In: H. E. Wright Jr., ed. 1983. Late-Quaternary Environments of the United States. Vol. 2. The Holocene. Minneapolis. Pp. 128–141.

Akin, W. E. 1991. Global Patterns: Climate, Vegetation, and Soils. Norman.

Albee, B. J., L. M. Shultz, and S. Goodrich. 1988. Atlas of the Vascular Plants of Utah. Salt Lake City.

Allard, R. W., M. A. Saghai-Maroof, Zhang Q., and R. A. Jorgenson. 1990. Genetic and molecular organization of ribosomal DNA (rDNA) variants in wild and cultivated barley. Genetics 126: 743–751.

Allen, G. M., P. F. J. Eagles, and S. D. Price, eds. 1990. Conserving Carolinian Canada. Waterloo.

Alston, R. E. and B. L. Turner. 1963. Biochemical Systematics. Englewood Cliffs.

Alston, R. E. and B. L. Turner. 1963b. Natural hybridization among four species of *Baptisia* Leguminosae. Amer. J. Bot. 50: 159–173.

Amer. Antiquity = American Antiquity.

Amer. Biol. Teacher = American Biology Teacher.
Amer. Fern J. = American Fern Journal; a Quarterly Devoted to Ferns.
Amer. J. Bot. = American Journal of Botany.
Amer. J. Sci. Arts = American Journal of Science, and Arts.
Amer. Midl. Naturalist = American Midland Naturalist; Devoted to Natural History, Primarily That of the Prairie States.
Amer. Naturalist = American Naturalist...
Amer. Sci. = American Scientist... [Subtitle varies.]
Anderson, C. M. and M. Treshow. 1980. A review of environmental and genetic factors that affect height in *Spartina alterniflora* Loisel (salt marsh cordgrass). Estuaries 3: 168–176.
Anderson, E. 1936. The species problem in *Iris*. Ann. Missouri Bot. Gard. 23: 457–509.
Anderson, R. C. and L. E. Brown. 1986. Stability and instability in plant communities following fire. Amer. J. Bot. 78: 364–368.
Andrews, J. T., ed. 1985. Quaternary Environments: Eastern Canadian Arctic, Baffin Bay, and Western Greenland. Boston.
Andrews, J. T. 1987. The late Wisconsin glaciation and deglaciation of the Laurentide Ice Sheet. In: W. F. Ruddiman and H. E. Wright Jr., eds. 1987. North America and Adjacent Oceans during the Last Deglaciation. Boulder. Pp. 13–37.
Andrews, J. T. and R. G. Barry, eds. 1976. Abstracts of the Fourth Biennial Meeting of the American Quaternary Association. Tempe.
Ann. Assoc. Amer. Geogr. = Annals of the Association of American Geographers.
Ann. Bot. (London) = Annals of Botany. (London.)
Ann. Missouri Bot. Gard. = Annals of the Missouri Botanical Garden.
Annual Rev. Ecol. Syst. = Annual Review of Ecology and Systematics.
Anonymous. 1967. Sasakubo Yasuo Kyoju Kanreki Kinen Ronbunshu. Jubilee Publication in the Commemoration of Professor Yasuo Sasa, Dr. Sc., Sixtieth Birthday. Sapporo.
Arber, A. 1925. Monocotyledons. A Morphological Study. Cambridge.
Archer, S., C. Scifres, C. R. Bassham, and R. Maggio. 1988. Autogenic succession in a subtropical savanna: Conversion of grassland to thorn woodland. Ecol. Monogr. 58: 111–127.
Arctic Alpine Res. = Arctic and Alpine Research.
Argus, G. W. 1977. Threatened and endangered species problems in North America. Canada. In: G. T. Prance and T. S. Elias, eds. 1977. Extinction Is Forever. Bronx. Pp. 17–29.
Argus, G. W. and K. M. Pryer. 1990. Rare Vascular Plants in Canada—Our Natural Heritage. Ottawa.
Arnason, T., R. J. Hebda, and T. Johns. 1981. Use of plants for food and medicine by native peoples of eastern Canada. Canad. J. Bot. 59: 2189–2325.
Arno, S. F. and J. R. Habeck. 1972. Ecology of alpine larch (*Larix lyallii* Parl.) in the Pacific Northwest. Ecol. Monogr. 42: 417–450.
Asch, D. L. and N. E. Asch. 1977. Chenopod as cultigen: A re-evaluation of some prehistoric collections from eastern North America. Midcontinental J. Archaeol. 2: 3–45.
Avdulov, N. P. 1931. Karyosystematische Untersuchung der Familie Gramineen. Trudy Prikl. Bot., prilož. 44.
Axelrod, D. I. 1950. Evolution of desert vegetation. Publ. Carnegie Inst. Wash. 590: 215–306.
Axelrod, D. I. 1958. Evolution of the Madro-Tertiary Geoflora. Bot. Rev. (Lancaster) 24: 433–509.
Axelrod, D. I. 1966. The Eocene Copper Basin flora of northeastern Nevada. Univ. Calif. Publ. Geol. Sci. 59: 1–124.
Axelrod, D. I. 1975. Evolution of the Madrean-Tethyan sclerophyll vegetation. Ann. Missouri Bot. Gard. 62: 280–334.
Axelrod, D. I. 1976. History of the coniferous forests, California and Nevada. Univ. Calif. Publ. Bot. 70: 1–62.
Axelrod, D. I. 1979. Age and origin of Sonoran Desert vegetation. Occas. Pap. Calif. Acad. Sci. 132: 1–74.
Axelrod, D. I. 1985. Rise of the grassland biome, central North America. Bot. Rev. (Lancaster) 51: 163–201.
Axelrod, D. I. 1986. Analysis of some palaeogeographic and palaeoecologic problems of palaeobotany. Palaeobotanist 35: 115–129.
Axelrod, D. I. 1986b. Cenozoic history of some western American pines. Ann. Missouri Bot. Gard. 73: 565–641.
Axelrod, D. I. 1986c. The Sierra redwood (*Sequoiadendron*) forest: End of a dynasty. Geophytology 16: 25–36.
Axelrod, D. I. 1987. The late Oligocene Creede flora, Colorado. Univ. Calif. Publ. Geol. Sci. 130: 1–235.
Axelrod, D. I. 1988. An interpretation of high montane conifers in western Tertiary floras. Paleobiology 14: 301–306.
Axelrod, D. I. 1989. Age and origin of chaparral. In: S. C. Keeley, ed. 1989. The California Chaparral: Paradigms Reexamined. Los Angeles. Pp. 7–19.
Axelrod, D. I. 1990. Age and origin of subalpine forest zone. Paleobiology 16: 360–369.
Axelrod, D. I., M. T. Kalin Arroyo, and P. H. Raven. 1991. Historical development of temperate vegetation in the Americas. Revista Chilena Hist. Nat. 64: 413–446.
Axelrod, D. I. and P. H. Raven. 1985. Origins of the Cordilleran flora. J. Biogeogr. 12: 21–47.
Ayensu, E. S. 1981. Assessment of threatened plant species in the United States. In: H. Synge, ed. 1981. The Biological Aspects of Rare Plant Conservation. Chichester. Pp. 19–58.
Ayensu, E. S. and R. A. DeFilipps. 1978. Endangered and Threatened Plants of the United States. Washington.

Babcock, E. B. 1947. The Genus *Crepis*. 2 vols. Berkeley.
Babcock, E. B. and G. L. Stebbins. 1938. The American species of *Crepis*. Their interrelationships and distribution

as affected by polyploidy and apomixis. Publ. Carnegie Inst. Wash. 504.

Bailey, D. K. 1970. Phytogeography and taxonomy of *Pinus* subsection *Balfourianae*. Ann. Missouri Bot. Gard. 57: 210–249.

Bailey, D. K. 1987. A study of *Pinus* subsection *Cembroides* I: The single-needle pinyons of the Californias and the Great Basin. Notes Roy. Bot. Gard. Edinburgh 44: 275–310.

Bailey, D. K. and F. G. Hawksworth. 1979. Pinyons of the Chihuahuan Desert region. Phytologia 44: 129–133.

Bailey, L. H. 1943. Species of *Cucurbita*. Gentes Herb. 6: 267–322, fig. 143.

Bailey, L. H., E. Z. Bailey, and Bailey Hortorium Staff. 1976. Hortus Third. A Concise Dictionary of Plants Cultivated in the United States and Canada. New York.

Bakeless, J. 1961. The Eyes of Discovery. New York.

Baker, H. G. 1948. Dimorphism and monomorphism in the Plumbaginaceae I. A survey of the family. Ann. Bot. (London), n. s. 12: 207–209.

Baker, H. G. 1953. Dimorphism and monomorphism in the Plumbaginaceae III. Correlation of distribution patterns with dimorphism and monomorphism in *Limonium*. Ann. Bot. (London), n. s. 17: 615–627.

Baker, H. G. 1955. Self-compatibility and establishment after long-distance dispersal. Evolution 9: 347–348.

Baker, H. G. 1965. Characteristics and modes of origin of weeds. In: H. G. Baker and G. L. Stebbins, eds. 1965. The Genetics of Colonizing Species. New York. Pp. 147–172.

Baker, H. G. 1974. The evolution of weeds. Annual Rev. Ecol. Syst. 5: 1–24.

Baker, H. G. and G. L. Stebbins, eds. 1965. The Genetics of Colonizing Species. New York.

Baker, R. G. 1983. Holocene vegetational history of the western United States. In: H. E. Wright Jr., ed. 1983. Late-Quaternary Environments of the United States. Vol. 2. The Holocene. Minneapolis. Pp. 109–127.

Baker, R. G., R. S. Rhodes II, D. P. Schwert, A. C. Ashworth, T. J. Frest, G. R. Hallberg, and J. A. Janssens. 1986. A full-glacial biota from southeastern Iowa, USA. J. Quatern. Sci. 1: 91–107.

Baker, R. G., J. van Nest, and G. Woodworth. 1989. Dissimilarity coefficients for fossil pollen spectra from Iowa and western Illinois during the last 30,000 years. Palynology 13: 63–77.

Ball, P. W. 1968. *Vicia*. In: T. G. Tutin et al., eds. 1964–1980. Flora Europaea. 5 vols. Cambridge. Vol. 2, pp. 129–136.

Bally, A. W. and R. A. Palmer, eds. 1989. The Geology of North America—An Overview. Boulder. [Geology of North America. Vol. A.]

Bally, A. W, C. R. Scotese, and M. I. Ross. 1989. North America; plate tectonic setting and tectonic elements. In: A. W. Bally and R. A. Palmer, eds. 1989. The Geology of North America—An Overview. Boulder. Pp. 1–15.

Barabé, D., ed. 1985. Colleque du centennaire du Frère Marie-Victorin. Bull. Soc. Animat. Jard. Inst. Bot., Montreal 9(3): 1–94.

Barbour, M. G. 1969. Patterns of genetic similarity between *Larrea divaricata* of North and South America. Amer. Midl. Naturalist 81: 54–67.

Barbour, M. G. 1988. Californian upland forests and woodlands. In: M. G. Barbour and W. D. Billings, eds. 1988. North American Terrestrial Vegetation. New York. Pp. 131–164.

Barbour, M. G. 1992. Life at the leading edge: The beach plant syndrome. In: U. Seeliger, ed. 1992. Coastal Plant Communities of Latin America. San Diego. Pp. 291–307.

Barbour, M. G., N. H. Berg, G. F. Kittel, and M. E. Kunz. 1990. Snowpack and the distribution of a major vegetation ecotone in the Sierra Nevada of California. J. Biogeogr. 18: 141–149.

Barbour, M. G. and W. D. Billings, eds. 1988. North American Terrestrial Vegetation. New York.

Barbour, M. G., J. H. Burk, and W. D. Pitts. 1987. Terrestrial Plant Ecology, ed. 2. Palo Alto.

Barbour, M. G., T. M. De Jong, and A. F. Johnson. 1975. Additions and corrections to a review of North American Pacific coast beach vegetation. Madroño 23: 130–134.

Barbour, M. G., T. M. De Jong, and B. M. Pavlik. 1985. Marine beach and dune plant communities. In: B. F. Chabot and H. A. Mooney, eds. 1985. Physiological Ecology of North American Plant Communities. New York. Pp. 296–322.

Barbour, M. G. and A. F. Johnson. 1988. Beach and dune. In: M. G. Barbour and J. Major, eds. 1988. Terrestrial Vegetation of California, ed. 2. Sacramento. Pp. 223–261.

Barbour, M. G. and J. Major, eds. 1988. Terrestrial Vegetation of California, ed. 2. Sacramento.

Barbour, M. G. and R. A. Minnich. 1990. The myth of chaparral conversion. Israel J. Bot. 39: 453–463.

Barbour, M. G., M. Rejmanek, A. F. Johnson, and B. M. Pavlik. 1987. Beach vegetation and plant distribution patterns along the northern Gulf of Mexico. Phytocoenologia 15: 201–233.

Barbour, M. G. and R. H. Robichaux. 1976. Beach phytomass along the California coast. Bull. Torrey Bot. Club 103: 16–20.

Barbour, M. G. and R. A. Woodward. 1985. The Shasta red fir forest of California. Canad. J. Forest Res. 15: 570–576.

Barkley, T. M. 1983. Field Guide to the Common Weeds of Kansas. Lawrence, Kans.

Barkworth, M. E. and D. R. Dewey. 1985. Genomically based genera in the perennial Triticeae of North America: Identification and membership. Amer. J. Bot. 72: 767–776.

Barnosky, C. W. 1987. Response of vegetation to climatic changes of different duration in the late Neogene. Trends Ecol. Evol. 2: 247–250.

Barnosky, C. W., P. M. Anderson, and P. J. Bartlein. 1987.

The northwestern U.S. during deglaciation; vegetational history and paleoclimatic implications. In: W. F. Ruddiman and H. E. Wright Jr., eds. 1987. North America and Adjacent Oceans during the Last Deglaciation. Boulder. Pp. 289–321.

Barry, J. M. 1980. Natural Vegetation of South Carolina. Columbia.

Barry, W. J. 1972. The Central Valley Prairie. Vol. 1. California Prairie Ecosystem. Sacramento.

Bartel, J. A. 1987. The federal listing of rare and endangered plants: What is involved and what does it mean? In: T. S. Elias, ed. 1987. Conservation and Management of Rare and Endangered Plants. Sacramento. Pp. 15–22.

Bartlein, P. J. 1988. Late-Tertiary and Quaternary palaeoenvironments. In: B. Huntley and T. Webb III, eds. 1988. Vegetation History. Dordrecht. Pp. 113–152.

Batten, D. J. 1984. Palynology, climate and the development of late Cretaceous floral provinces in the Northern Hemisphere; a review. In: P. Brenchley, ed. 1984. Fossils and Climate. New York. Pp. 127–164.

Baur, E. 1932. Artumgrenzung und Artbildung in der Gattung *Antirrhinum* Sektion *Antirrhinastrum*. Z. Indukt. Abstammungs-Vererbungsl. 63: 256–302.

Baye, P. R. 1990. Comparative Growth Responses and Population Ecology of European and American Beach Grasses (*Ammophila* spp.) in Relation to Sand Accretion and Salinity. Ph.D. dissertation. University of Western Ontario.

Bayer, R. J. 1985. Investigations into the evolutionary history of the polyploid complexes in *Antennaria* (Asteraceae Inuleae) I. The *A. neodioica* complex. Pl. Syst. Evol. 150: 143–163.

Bayer, R. J. 1985b. Investigations into the evolutionary history of the polyploid complexes in *Antennaria* (Asteraceae Inuleae) II. The *A. parlinii* complex. Rhodora 87: 321–329.

Bazzaz, F. A. 1986. Life history of colonizing plants: Some demographic, genetic, and physiologic features. In: H. A. Mooney and J. A. Drake, eds. 1986. Ecology of Biological Invasions of North America and Hawaii. New York. Pp. 96–110.

Bean, R. C., C. D. Richards, and F. Hyland. 1966. Revised checklist of the vascular plants of Maine. Bull. Josselyn Bot. Soc. Maine 8.

Beatley, J. C. 1975. Climates and vegetation patterns across the Mojave/Great Basin desert transition of southern Nevada. Amer. Midl. Naturalist 93: 53–70.

Becker, H. F. 1961. Oligocene plants from the upper Ruby River Basin, southwestern Montana. Mem. Geol. Soc. Amer. 82.

Behnke, H.-D. 1976. Ultrastructure of sieve-element plastids in Caryophyllales (Centrospermae), evidence for delimitation and classification of the order. Pl. Syst. Evol. 126: 31–54.

Beiswenger, J. M. 1991. Late Quaternary vegetational history of Graves Lake, Idaho. Ecol. Monogr. 61(2): 165–182.

Bell, W. A. 1949. Uppermost Cretaceous and Paleocene floras of western Canada. Bull. Geol. Surv. Canada 13.

Benson, L. and R. S. Thompson. 1987. The physical record of lakes in the Great Basin. In: W. F. Ruddiman and H. E. Wright Jr., eds. 1987. North America and Adjacent Oceans during the Last Deglaciation. Boulder. Pp. 241–260.

Benson, L. D. 1943. Revisions of status of southwestern trees and shrubs. Amer. J. Bot. 30: 230–240.

Benson, L. D. 1962. Plant Taxonomy, Methods and Principles. New York.

Bentham, G. and J. D. Hooker. 1862–1883. Genera Plantarum ad Exemplaria Imprimis in Herbariis Kewensibus Servata Definita. 3 vols. London.

Berg, C. C. 1977. Urticales, their differentiation and systematic position. Pl. Syst. Evol., Suppl. 1: 349–374.

Berger, A. L. 1978. Long-term variations of caloric insolation resulting from the earth's orbital elements. Quatern. Res. 9: 139–167.

Berggren, W. A. and C. D. Hollister. 1974. Paleogeography, paleobiogeography and the history of circulation in the Atlantic Ocean. In: W. W. Hay, ed. 1974. Studies in Paleooceanography. Tulsa. Pp. 126–186.

Berkeley, E. and D. S. Berkeley. 1963. John Clayton, Pioneer of American Botany. Chapel Hill.

Berkeley, E. and D. S. Berkeley. 1969. Dr. Alexander Garden of Charles Town. Chapel Hill.

Berkeley, E. and D. S. Berkeley. 1982. The Life and Travels of John Bartram, from Lake Ontario to the River St. John. Tallahassee.

Bernabo, J. C. and T. Webb III. 1977. Changing patterns in the Holocene pollen record of northeastern North America: A mapped summary. Quatern. Res. 84: 64–96.

Bertness, M. D. and A. M. Ellison. 1987. Determinants of pattern in a New England salt marsh plant community. Ecol. Monogr. 57: 129–147.

Bessey, C. E. 1897. Phylogeny and taxonomy of the angiosperms. Bot. Gaz. 34: 145–178.

Bessey, C. E. 1915. The phylogenetic taxonomy of flowering plants. Ann. Missouri Bot. Gard. 2: 109–164.

Betancourt, J. L., T. R. Van Devender, and P. S. Martin, eds. 1990. Packrat Middens: The Last 40,000 Years of Biotic Change. Tucson.

Bharadwaj, D. C. et al., eds. [1978–1981.] Proceedings. Fourth International Palynological Conference [Lucknow, India, 1976–1977]. 3 vols. Lucknow.

Bierhorst, D. W. 1971. Morphology of Vascular Plants. New York.

Billings, W. D. 1949. The shadscale vegetation zone of Nevada and eastern California in relation to climate and soils. Amer. Midl. Naturalist 42: 87–109.

Billings, W. D. 1951. Vegetational zonation in the Great Basin of western North America. In: International Union of Biological Sciences. 1951. Les Bases Ecologiques de la Regeneration de la Vegetation des Zones Arides. On the Ecological Foundations of the Regeneration of Vegetation in Arid Zones. [Symposium.] Stockholm, Juillet 1950. Paris. Pp. 101–122.

Billings, W. D. 1974. Adaptations and origins of alpine plants. Arctic Alpine Res. 6: 129–142.

Billings, W. D. 1988. Alpine vegetation. In: M. G. Barbour and W. D. Billings, eds. 1988. North American Terrestrial Vegetation. New York. Pp. 401–420.

Billington, R. A. 1982. Westward Expansion: A History of the American Frontier, ed. 2. New York and London.

Biótica = Biótica; Publicación del Instituto Nacional de Investigaciones sobre Recursos Bióticos.

Bird, J. B. 1980. The Natural Landscapes of Canada, ed. 2. Toronto.

Birkeland, P. W. 1984. Soils and Geomorphology. New York.

Birkeland, P. W., R. M. Burke, and J. C. Yount. 1976. Preliminary comments on late Cenozoic glaciations in the Sierra Nevada. In: W. C. Mahaney, ed. 1976. Quaternary Stratigraphy of North America. Stroudsburg, Pa. Pp. 283–295.

Black, R. A. and L. C. Bliss. 1978. Recovery sequence of *Picea mariana/Vaccinium uliginosum* forests after burning near Inuvik, Northwest Territories, Canada. Canad. J. Bot. 56: 2020–2030.

Black, R. A. and L. C. Bliss. 1980. Reproductive ecology of *Picea mariana* (Mill.) B.S.P. at treeline near Inuvik, Northwest Territories, Canada. Ecol. Monogr. 50: 331–354.

Blackmore, S. and I. K. Ferguson, eds. 1986. Pollen and Spores: Form and Function. London.

Bliss, L. C. 1963. Alpine plant communities of the Presidential Range, New Hampshire. Ecology 44: 678–697.

Bliss, L. C. 1981. North American and Scandanavian tundras and polar deserts. In: L. C. Bliss et al., eds. 1981. Tundra Ecosystems: A Comparative Analysis. New York. Pp. 8–24.

Bliss, L. C. 1988. Arctic tundra and polar desert biome. In: M. G. Barbour and W. D. Billings, eds. 1988. North American Terrestrial Vegetation. New York. Pp. 1–32.

Bliss, L. C., O. W. Heal, and J. J. Moore, eds. 1981. Tundra Ecosystems: A Comparative Analysis. New York.

Bliss, L. C. and J. Svoboda. 1984. Plant communities and plant production in the western Queen Elizabeth Islands. Holarc. Ecol. 7: 324–344.

Blunt, W. and W. T. Stearn. 1971. The Compleat Naturalist: A Life of Linnaeus. London.

Böcher, T. W., K. Holmen, and K. Jakobsen. 1968. The Flora of Greenland, ed. 2. Copenhagen.

Boellstorff, J. 1978. North American Pleistocene stages reconsidered in light of probable Pliocene-Pleistocene continental glaciation. Science 202: 305–307.

Boivin, B. 1967. Enumeration des plantes du Canada. VII—Resume statistique et regions adjacentes. Naturaliste Canad. 94: 625–655.

Boivin, B. 1968. Enumeration des plantes du Canada. Provancheria 6.

Booth, D. B. 1987. Timing and processes of deglaciation along the southern margin of the Cordilleran ice sheet. In: W. F. Ruddiman and H. E. Wright Jr., eds. 1987. North America and Adjacent Oceans during the Last Deglaciation. Boulder. Pp. 71–90.

Borchert, M. T., F. W. Davis, J. Michaelsen, and L. D. Oyler. 1989. Interactions of factors affecting seedling recruitment of blue oak *(Quercus douglasii)* in California. Ecology 70: 389–404.

Bot. Gaz. = Botanical Gazette; Paper of Botanical Notes.

Bot. Helv. = Botanica Helvetica.

Bot. J. Linn. Soc. = Botanical Journal of the Linnean Society.

Bot. Jahrb. Syst. = Botanische Jahrbücher für Systematik, Pflanzengeschichte und Pflanzengeographie.

Bot. Not. = Botaniska Notiser.

Bot. Rev. (Lancaster) = Botanical Review, Interpreting Botanical Progress.

Bovis, M. J. 1987. The interior mountains and plateaus. In: W. L. Graf, ed. 1987. Geomorphic Systems of North America. Boulder. Pp. 469–515.

Bowen, D. Q. 1985. Quaternary Geology, a Stratigraphic Framework for Multidisciplinary Work. Oxford.

Bower, F. O. 1894. A theory of the strobilus in archegoniate plants. Ann. Bot. (London) 8: 343–365.

Bowers, J. E. 1984. Plant geography of southwestern sand dunes. Desert Pl. 6: 21–42.

Box, E. O. 1981. Macroclimate and Plant Forms: An Introduction to Predictive Modeling in Phytogeography. The Hague, Boston, and Hingham, Mass. [Tasks for Vegetation Science. Vol. 1.]

Boyd, R. S. and M. G. Barbour. 1986. Relative salt tolerance of *Cakile edentula* from lacustrine and marine beaches. Amer. J. Bot. 73: 236–241.

Brakenridge, G. R. 1978. Evidence for a cold, dry full-glacial climate in the American Southwest. Quatern. Res. 9: 22–40.

Bramwell, D., O. Hamann, V. H. Heywood, and H. Synge, eds. 1987. Botanic Gardens and World Conservation Strategy. London.

Braun, E. L. 1950. Deciduous Forests of Eastern North America. Philadelphia.

Breckton, G. J. and M. G. Barbour. 1974. Review of North American Pacific coast beach vegetation. Madroño 22: 333–360.

Brenchley, P., ed. 1984. Fossils and Climate. New York.

Bretting, P. K. and G. P. Nabhan. 1986. Ethnobotany of devil's claw (*Proboscidea parviflora* ssp. *parviflora*) in the greater Southwest. J. Calif. & Great Basin Anthropol. 8: 226–237.

Bridson, G. D. R. and E. R. Smith. 1991. B-P-H/S. Botanico-Periodicum-Huntianum/Supplementum. Pittsburgh.

British Columbia Ministry of Forests. 1988. Biogeoclimatic Zones of British Columbia. Victoria.

Britton, N. L. 1901. Manual of the Flora of the Northern States and Canada. New York.

Britton, N. L., L. M. Underwood, W. A. Murrill, J. H. Barnhart, and H. W. Rickett, eds. 1905–1972. North American Flora.... 42 vols. New York. [Vols. 1–34, 1905–1957; Ser. 2, Vols. 1–8, 1954–1972.]

Brittonia = Brittonia; a Journal of Systematic Botany....

Broecker, W. S. and G. H. Denton. 1989. The role of ocean-atmosphere reorganizations in glacial cycles. Geochim. Cosmochim. Acta 53: 2465–2501.

Broecker, W. S. and J. van Donk. 1970. Insolation changes, ice volumes, and O_{18} record in deep sea cores. Rev. Geophys. Space Phys. 8: 169–198.

Brooks, R. E. 1983. *Trifolium stoloniferum,* running buffalo clover: Description, distribution and current status. Rhodora 85: 343–354.

Brophy, W. B. and D. R. Parnell. 1974. Hybridization between *Quercus agrifolia* and *Quercus wislizenii* (Fagaceae). Madroño 22: 290–302.

Brown, D. E. 1982. Great Basin conifer woodland. Desert Pl. 4: 52–57.

Brown, D. E. 1982b. Great Basin montane scrubland. Desert Pl. 4: 83–84.

Brown, D. E., ed. 1982c. Biotic communities of the American Southwest—United States and Mexico. Desert Pl. 4: 1–341.

Brown, D. M. 1941. Vegetation of Roan Mountain: A phytosociological and successional study. Ecol. Monogr. 11: 61–97.

Brown, S. 1981. A comparison of the structure, primary productivity, and transpiration of cypress ecosystems in Florida. Ecol. Monogr. 51: 403–427.

Brown, W. H. 1938. The bearing of nectaries on the phylogeny of flowering plants. Proc. Amer. Philos. Soc. 79: 549–595.

Bryant, V. M. Jr. and R. G. Holloway, eds. 1985. Pollen Records of Late-Quaternary North American Sediments. Dallas.

Bryson, R. A. 1966. Air masses, streamlines, and the boreal forest. Geogr. Bull. 8: 228–269.

Bryson, R. A., D. A. Baerreis, and W. M. Wendland. 1970. The character of late-glacial and post-glacial climatic changes. In: W. Dort Jr. and J. K. Jones Jr., eds. 1970. Pleistocene and Recent Environments of the Central Great Plains. Lawrence, Kans. Pp. 53–74.

Bryson, R. A. and F. K. Hare, eds. 1974. Climates of North America. Amsterdam and New York. [World Survey of Climatology. Vol. 11.]

Bryson, R. A. and F. K. Hare. 1974b. The climates of North America. In: R. A. Bryson and F. K. Hare, eds. 1974. Climates of North America. Amsterdam and New York. Pp. 1–47.

Bryson, R. A. and W. M. Wendland. 1967. Tentative climatic patterns for some late glacial and post-glacial episodes in central North America. In: W. J. Mayer-Oakes, ed. 1967. Life, Land, Water. Winnipeg. Pp. 271–298.

Buchholz, J. T. 1939. The generic segregation of the sequoias. Amer. J. Bot. 26: 535–538.

Buffington, L. C. and C. H. Herbel. 1965. Vegetational changes of a semi-desert grassland range from 1858 to 1963. Ecol. Monogr. 35: 139–164.

Bull. Canad. Petrol. Geol. = Bulletin of Canadian Petroleum Geology.

Bull. Entomol. Soc. Amer. = Bulletin of the Entomological Society of America.

Bull. Geol. Soc. Amer. = Bulletin of the Geological Society of America.

Bull. Geol. Surv. Canada = Bulletin of the Geological Survey of Canada.

Bull. Josselyn Bot. Soc. Maine = Bulletin of the Josselyn Botanical Society of Maine.

Bull. Kansas Geol. Surv. = Bulletin of the Kansas Geological Survey.

Bull. Mus. Natl. Hist. Nat., B, Adansonia = Bulletin du Muséum National d'Histoire Naturelle. Section B, Adansonia: Botanique Phytochimie.

Bull. Natl. Mus. Canada = Bulletin of the National Museum of Canada.

Bull. New York State Mus. Sci. Serv. = Bulletin of the New York State Museum and Science Service.

Bull. S. Calif. Acad. Sci. = Bulletin of the Southern California Academy of Sciences.

Bull. Soc. Animat. Jard. Inst. Bot., Montreal = Bulletin de la Société d'Animation du Jardin et de l'Institut Botaniques.

Bull. Soc. Bot. France = Bulletin de la Société Botanique de France.

Bull. Soc. Hist. Nat. Toulouse = Bulletin de la Société d'Histoire Naturelle de Toulouse.

Bull. Torrey Bot. Club = Bulletin of the Torrey Botanical Club.

Bull. U.S. Geol. Surv. = Bulletin of the United States Geological Survey.

Buol, S. W., F. D. Hole, and R. J. McCracken. 1980. Soil Genesis and Classification, ed. 2. Ames.

Burk, J. H. 1988. Sonoran Desert. In: M. G. Barbour and J. Major, eds. 1988. Terrestrial Vegetation of California, ed. 2. Sacramento. Pp. 869–889.

Burns, R. M., ed. 1983. Silvicultural Systems for the Major Forest Types of the United States. Washington. [Agric. Handb. 445.]

Butler Univ. Bot. Stud. = Butler University Botanical Studies.

C. R. C. Crit. Rev. Pl. Sci. = C R C Critical Reviews in Plant Sciences.

Cain, S. A. 1930. An ecological study of the heath balds of the Great Smoky Mountains. Butler Univ. Bot. Stud. 1: 177–208.

Cain, S. A. 1944. Foundations of Plant Geography. New York.

Calder, J. A. and R. L. Taylor. 1968. Flora of the Queen Charlotte Islands. 2 vols. Ottawa.

Caldwell, M. M. 1985. Cold desert. In: B. F. Chabot and H. A. Mooney, eds. 1985. Physiological Ecology of North American Plant Communities. New York. Pp. 198–212.

Callicot, J. B. 1986. On the intrinsic value of nonhuman species. In: B. G. Norton, ed. 1986. The Preservation of Species. The Value of Biological Diversity. Princeton. Pp. 138–172.

Camp, W. H. and C. L. Gilly. 1943. The structure and origin of species. Brittonia 4: 323–385.

Campbell, J. J. N., M. Evans, M. E. Medley, and N. L. Taylor. 1988. Buffalo clovers in Kentucky (*Trifolium stoloniferum* and *T. reflexum*): Historical records, presettlement environment, rediscovery, endangered status, cultivation and chromosome number. Rhodora 90: 399–418.

Canad. Field-Naturalist = Canadian Field-Naturalist.

Canad. J. Anthropol. = Canadian Journal of Anthropology.

Canad. J. Bot. = Canadian Journal of Botany.

Canad. J. Earth Sci. = Canadian Journal of Earth Sciences.

Canad. J. Forest Res. = Canadian Journal of Forest Research.

Canada Soil Survey Committee. 1978. The Canadian System of Soil Classification. Ottawa. [Canada Dept. Agric. Publ. 1946.]

Candolle, A. P. de. 1813. Théorie Élementaire de la Botanique. Paris.

Candolle, A. P. de. 1827. Organographie Végétal. 2 vols. Paris.

Carlson, A. W. 1986. Ginseng: America's botanical drug connection to the Orient. Econ. Bot. 40: 233–249.

Carpenter, D. E., M. G. Barbour, and C. J. Bahre. 1986. Old field succession in Mojave Desert scrub. Madroño 33: 111–122.

Carpenter, J. R. 1940. The grassland biome. Ecol. Monogr. 10: 617–684.

Carter, H. B. 1988. Sir Joseph Banks, 1743–1820. London.

Castanea = Castanea; Journal of the Southern Appalachian Botanical Club.

Cevallos-Ferriz, S. R. S. and R. A. Stockey. 1989. Permineralized fruits and seeds from the Princeton chert (middle Eocene) of British Columbia: Nymphaeaceae. Bot. Gaz. 150: 207–217.

Cevallos-Ferriz, S. R. S. and R. A. Stockey. 1990. Permineralized fruits and seeds from the Princeton chert (middle Eocene) of British Columbia: Vitaceae. Canad. J. Bot. 68: 288–295.

Cevallos-Ferriz, S. R. S. and R. A. Stockey. 1990b. Vegetative remains of the Magnoliaceae from the Princeton chert (middle Eocene) of British Columbia. Canad. J. Bot. 68: 1327–1339.

Chabot, B. F. and H. A. Mooney, eds. 1985. Physiological Ecology of North American Plant Communities. New York.

Chabreck, R. H. 1972. Vegetation, water, and soil characteristics of the Louisiana coastal region. Louisiana Agric. Exp. Sta. Bull. 664.

Chaney, R. W. 1959. Miocene floras of the Columbia Plateau. Part I, composition and interpretation. Publ. Carnegie Inst. Wash. 617: 1–134.

Chaney, R. W. 1967. Miocene forests of the Pacific Basin; their ancestors and their descendents. In: Anonymous. 1967. Sasakubo Yasuo Kyoju Kanreki Kinen Ronbunshu. Jubilee Publication in the Commemoration of Professor Yasuo Sasa, Dr. Sc., Sixtieth Birthday. Sapporo. Pp. 209–239.

Chaney, R. W. and D. I. Axelrod. 1959. Miocene floras of the Columbia Plateau. Part II, systematic considerations. Publ. Carnegie Inst. Wash. 617: 135–237.

Chaney, R. W. and M. K. Elias. 1936. Late Tertiary floras from the High Plains. Publ. Carnegie Inst. Wash. 476(1): 1–72.

Chapin, F. S. III. 1980. The mineral nutrition of wild plants. Annual Rev. Ecol. Syst. 11: 233–260.

Chapin, F. S. III and G. R. Shaver. 1985. Arctic. In: B. F. Chabot and H. A. Mooney, eds. 1985. Physiological Ecology of North American Plant Communities. New York. Pp. 16–40.

Chapman, V. J. 1976. Coastal Vegetation, ed. 2. New York.

Chapman, V. J., ed. 1977. Wet Coastal Ecosystems. Amsterdam.

Chappell, J. 1978. Theories of upper Quaternary ice ages. In: A. B. Pittock et al., eds. 1978. Climatic Change and Variability, a Southern Perspective. Cambridge. Pp. 211–225.

Cheadle, V. I. 1953. Independent origin of vessels in the monocotyledons and dicotyledons. Phytomorphology 3: 23–44.

Christensen, N. L. 1988. Vegetation of the southeastern coastal plain. In: M. G. Barbour and W. D. Billings, eds. 1988. North American Terrestrial Vegetation. New York. Pp. 317–363.

Christensen, N. L. 1989. Landscape history and ecological change. J. Forest Hist. 33: 116–124.

Christensen, N. L., R. B. Burchell, A. Liggett, and E. L. Simms. 1981. The structure and development of pocosin vegetation. In: C. J. Richardson, ed. 1981. Pocosin Wetlands. Stroudsburg, Pa. Pp. 43–61.

Christie, R. L. and G. E. Rouse. 1976. Eocene beds at Lake Hazen, northern Ellesmere Island. Rep. Activities Geol. Surv. Canada 76-1C: 153–156.

Clague, J. J. 1989. Quaternary geology of the Canadian Cordillera. In: R. J. Fulton, ed. 1989. Quaternary Geology of Canada and Greenland. Ottawa. Pp. 16–95.

Clark, J. R. and J. Benforado, eds. 1981. Wetlands of Bottomland Hardwood Forests. Amsterdam.

Clark, J. S. 1986. Dynamism in the barrier-beach vegetation of Great South Beach, New York. Ecol. Monogr. 56: 97–126.

Clausen, J. 1926. Genetical and cytological investigations on *Viola tricolor* and *V. arvensis* Murr. Hereditas (Lund) 8: 1–156.

Clausen, J. 1951. Stages in the Evolution of Plant Species. Ithaca, N.Y.

Clausen, J., D. D. Keck, and W. M. Hiesey. 1940. Experimental studies on the nature of species. I. Effect of varied environments on western American plants. Publ. Carnegie Inst. Wash. 520.

Clausen, R. E. 1932. Interspecific hybridization in *Nicotiana* XIII. Further data as to the origin and constitution of *Nicotiana tabacum*. Svensk Bot. Tidskr. 26: 123–136.

Clausen, R. E. and T. H. Goodspeed. 1925. Interspecific hybridization in *Nicotiana* II. A tetraploid *glutinosa-tabacum* hybrid, a verification of Winge's hypothesis. Genetics 10: 219–284.

CLIMAP Project. 1981. Seasonal Reconstructions of the Earth's Surface at the Last Glacial Maximum. Boulder. [Geol. Soc. Amer. Map Chart Ser. MC-36.]

Climat. Change = Climatic Change.

Cody, M. L. and H. A. Mooney. 1978. Convergence versus nonconvergence in Mediterranean-climate ecosystems. Annual Rev. Ecol. Syst. 9: 265–321.

Cody, W. J. and D. M. Britton. 1989. Ferns and Fern Allies of Canada. Ottawa.

Coffin, B. and L. Pfanmuller, eds. 1988. Minnesota's Endangered Flora and Fauna. Minneapolis.

COHMAP Members. 1988. Climatic changes of the last 18,000 years: Observations and model simulations. Science 241: 1043–1052.

Cole, F. W. 1980. Introduction to Meteorology, ed. 3. New York.

Cole, K. 1985. Past rates of change, species richness, and a model of vegetational inertia in the Grand Canyon, Arizona. Amer. Naturalist 125: 289–303.

Collinson, M. E. 1990. Plant evolution and ecology during the early Cainozoic diversification. Advances Bot. Res. 17: 1–98.

Conservation Biol. = Conservation Biology; Journal of the Society for Conservation Biology.

Contr. Gray Herb. = Contributions from the Gray Herbarium of Harvard University.

Contr. Paleontol. Carnegie Inst. Wash. = Contributions to Paleontology (from the Carnegie Institution of Washington). [Forms part of: Publ. Carnegie Inst. Washington.]

Contr. Ser. Amer. Assoc. Stratigr. Palynologists = Contributions Series, American Association of Stratigraphic Palynologists.

Contr. U.S. Natl. Herb. = Contributions from the United States National Herbarium.

Cook, F. R. and D. Muir. 1984. The Committee on the Status of Endangered Wildlife in Canada (COSEWIC): History and progress. Canad. Field-Naturalist 98: 63–70.

Cooper, W. S. 1922. The broad-sclerophyll vegetation of California. Publ. Carnegie Inst. Wash. 319.

Cooperrider, T. S. 1957. Introgressive hybridization between *Quercus marilandica* and *Quercus velutina* in Iowa. Amer. J. Bot. 44: 804–810.

Core, E. L. 1966. The Natural Vegetation of West Virginia. Parsons, W.Va.

Corns, I. G. W. 1974. Arctic plant communities east of the Mackenzie Delta. Canad. J. Bot. 52: 1730–1745.

Corps of Engineers [U.S. Army]. 1973. National Shoreline Study. 5 vols. Washington.

Correll, D. S. and M. C. Johnston. 1970. Manual of the Vascular Plants of Texas. Renner, Tex.

Coulter, J. M. and A. Nelson. 1909. New Manual of Botany of the Central Rocky Mountains (Vascular Plants). New York.

Countryman, W. D. 1977. Threatened and endangered species problems in North America. The northeastern United States. In: G. T. Prance and T. S. Elias, eds. 1977. Extinction Is Forever. Bronx. Pp. 30–35.

Coupland, R. T., ed. 1979. Grassland Ecosystems of the World: Analysis of Grasslands and Their Uses. Cambridge and New York.

Cowan, C. W. 1985. Understanding the evolution of plant husbandry in eastern North America: Lessons from botany, ethnography and archaeology. In: R. I. Ford, ed. 1985. Prehistoric Food Production in North America. Ann Arbor. Pp. 205–243.

Cowdrey, A. E. 1983. This Land This South: An Environmental History. Lexington, Ky.

Crabbe, J. A., A. C. Jermy, and J. T. Mickel. 1975. A new generic sequence for the pteridophyte herbarium. Fern Gaz. 11: 141–162.

Craighead, F. C. Sr. 1971. The Trees of South Florida. Vol. 1. The Natural Environments and Their Succession. Coral Gables.

Crandall, D. L. 1958. Ground vegetation patterns of the spruce-fir area of the Great Smoky Mountains National Park. Ecol. Monogr. 28: 337–360.

Crane, P. R. and S. Blackmore, eds. 1989. Evolution, Systematics, and Fossil History of the Hamamelidae. Vol. 1. Introduction and "Lower" Hamamelidae. Oxford. [Syst. Assoc. Special Vol. 40A.]

Crane, P. R., S. R. Manchester, and D. L. Dilcher, 1990. A preliminary survey of fossil leaves and well-preserved reproductive structures from the Sentinel Butte Formation (Paleocene) near Almont, North Dakota. Fieldiana, Geol. 20: 1–63.

Crane, P. R. and R. A. Stockey. 1987. *Betula* leaves and reproductive structures from the middle Eocene of British Columbia, Canada. Canad. J. Bot. 65: 2490–2500.

Crawford, D. J. 1990. Plant Molecular Systematics: Macromolecular Approaches. New York.

Creighton, H. B. and B. McClintock. 1931. A correlation of cytological and genetical crossingover in *Zea mays.* Proc. Natl. Acad. Sci. U.S.A. 17: 492–497.

Crepet, W. L., C. P. Daghlian, and M. Zavada. 1980. Investigations of angiosperms from the Eocene of North America: A new juglandaceous catkin. Rev. Palaeobot. Palynol. 30: 361–370.

Crepet, W. L. and G. D. Feldman. 1991. The earliest remains of grasses in the fossil record. Amer. J. Bot. 78: 1010–1014.

Critchfield, W. B. 1984. Impact of the Pleistocene on the genetic structure of North American conifers. In: R. M. Lanner, ed. 1984. Proceedings of the Eighth North American Forest Biology Workshop. Logan. Pp. 70–118.

Cronin, T. M., B. J. Szabo, T. A. Ager, J. E. Hazel, and J. P. Owens. 1981. Quaternary climates and sea levels of the U.S. Atlantic coastal plain. Science 211: 233–240.

Cronon, W. 1983. Changes in the Land: Indians, Colonists and the Ecology of New England. New York.

Cronquist, A. 1968. The Evolution and Classification of Flowering Plants. Boston.

Cronquist, A. 1977. The Compositae revisited. Brittonia 29: 137–153.

Cronquist, A. 1977b. On the taxonomic significance of secondary metabolites in angiosperms. Pl. Syst. Evol., Suppl. 1: 179–189.

Cronquist, A. 1978. Once again, what is a species? In: L. V. Knutson, ed. 1978. Biosystematics in Agriculture. Montclair, N.J. Pp. 3–20.

Cronquist, A. 1981. An Integrated System of Classification of Flowering Plants. New York.
Cronquist, A. 1982. Floristic provinces of North America. Brittonia 34: 143–145.
Cronquist, A. 1987. A botanical critique of cladism. Bot. Rev. (Lancaster) 53: 1–52.
Cronquist, A. 1988. The Evolution and Classification of Flowering Plants, ed. 2. Bronx.
Cronquist, A., A. H. Holmgren, N. H. Holmgren, J. L. Reveal, P. K. Holmgren, and R. C. Barneby. 1972+. Intermountain Flora. Vascular Plants of the Intermountain West, U.S.A. 4+ vols. New York and London. [Vol. 1, 1972; Vol. 3, Part B, 1989; Vol. 4, 1984; Vol. 6, 1977.]
Cronquist, A., A. L. Takhtajan, and W. Zimmermann. 1966. On the higher taxa of Embryobionta. Taxon 15: 129–134.
Crop Sci. (Madison) = Crop Science. (Madison, Wis.)
Crosby, J. 1949. Selection of an unfavorable gene complex. Evolution 3: 212–230.
Cross, A. T. and R. E. Taggart. 1982. Causes of short-term sequential changes in fossil plant assemblages: Some considerations based on a Miocene flora of the northwest United States. Ann. Missouri Bot. Gard. 69: 679–734.
Crosswhite, F. S. and C. D. Crosswhite. 1984. A classification of life forms of the Sonoran Desert, with emphasis on the seed plants and their survival strategies. Desert Pl. 5: 131–161.
Crotz, K. 1989. Ewaniana: The Writings of Joe and Nesta Ewan; with a Preface by Ian MacPhail; and Introduction by Emanuel D. Rudolph. Chillicothe, Ill.
Crowell, J. C. and W. Berger, eds. 1982. Pre-Pleistocene Climates. Washington.
Cwynar, L. C. and G. M MacDonald. 1987. Geographical variation of lodgepole pine in relation to population history. Amer. Naturalist 129: 463–469.
Cwynar, L. C. and J. C. Ritchie. 1980. Arctic steppe-tundra: A Yukon perspective. Science 208: 1375–1377.

Dahlgren, G. 1989. An updated angiosperm classification. Bot. J. Linn. Soc. 100: 197–203.
Dahlgren, R. M. T. 1983. General aspects of angiosperm evolution and macrosystematics. Nordic J. Bot. 3: 119–149.
Dahlgren, R. M. T., H. T. Clifford, and P. F. Yeo. 1985. The Families of the Monocotyledons. Structure, Evolution, and Taxonomy. Berlin etc.
Daubenmire, R. 1943. Vegetation zonation in the Rocky Mountains. Bot. Rev. (Lancaster) 9: 325–393.
Daubenmire, R. 1978. Plant Geography, with Special Reference to North America. New York.
Davidse, G., D. E. Boufford, S. A. Spongberg, W. Hamilton, M. C. McKenna, Hsü J., Hou H. Y., M. B. Davis, D. H. S. Chang, Cheng Z., Wu Z., A. R. Kruckeberg, Ying T. S., E. L. Little Jr., R. T. Allen, D. I. Axelrod, C. R. Parks, N. G. Miller, J. F. Wendel, K. M. McDougal, J. B. Phipps, Hong D. Y., Chen S. C., M. Kato, K. Iwatsuki, P. S. White, He S. A., and F. S. Santamour Jr. 1983. Biogeographical relationships between temperate eastern Asia and temperate eastern North America: The Twenty-Ninth Annual Systematics Symposium. Ann. Missouri Bot. Gard. 70(3–4): 421–749.
Davis, E. B. 1987. Guide to Information Sources in the Botanical Sciences. Littleton, Colo.
Davis, M. B. 1976. Pleistocene biogeography of temperate deciduous forests. Geosci. & Man 13: 13–26.
Davis, M. B. 1981. Quaternary history and the stability of forest communities. In: D. C. West et al., eds. 1981. Forest Succession: Concepts and Application. New York. Pp. 132–153.
Davis, M. B. 1983. Holocene vegetational history of the eastern United States. In: H. E. Wright Jr., ed. 1983. Late-Quaternary Environments of the United States. Vol. 2. The Holocene. Minneapolis. Pp. 166–181.
Davis, M. B. 1986. Climatic instability, time lags, and community disequilibrium. In: J. Diamond and T. J. Case, eds. 1986. Community Ecology. New York. Pp. 269–284.
Davis, M. B. 1987. Invasion of forest communities during the Holocene: Beech and hemlock in the Great Lakes region. In: A. J. Gran et al., eds. 1987. Colonization, Succession, and Stability. Oxford. Pp. 373–393.
Davis, M. B. and G. L. Jacobson Jr. 1985. Late glacial and early Holocene landscapes in northern New England and adjacent areas of Canada. Quatern. Res. 23: 341–368.
Davis, S. D., S. J. M. Droop, P. Gregerson, L. Henson, C. J. Leon, J. L. Villa-Lobos, H. Synge, and J. Zantovska. 1986. Plants in Danger. What Do We Know? Gland.
De Beer, G. 1954. *Archaeopteryx* and evolution. Advancem. Sci. 11: 160–170.
Deam, C. C. 1940. Flora of Indiana. Indianapolis.
DeAngelis, D. L., R. H. Gardner, and H. H. Shugart. 1981. Productivity of forest ecosystems studied during IBP: The woodlands data set. In: D. E. Reichle, ed. 1981. Dynamic Properties of Forest Ecosystems. Cambridge. Pp. 567–673.
Decker, D. S. 1988. Origin(s), evolution and systematics of *Cucurbita pepo* (Cucurbitaceae). Econ. Bot. 42: 3–15.
Delcourt, H. R. 1987. The impact of prehistoric agriculture and land occupation on natural vegetation. Trends Ecol. Evol. 2: 39–44.
Delcourt, H. R. and P. A. Delcourt. 1977. The Tunica Hills, Louisiana-Mississippi: Late glacial locality for spruce and deciduous forest species. Quatern. Res. 7: 218–237.
Delcourt, H. R. and P. A. Delcourt. 1983. Late-Quaternary vegetational dynamics and community stability reconsidered. Quatern. Res. 19: 265–271.
Delcourt, H. R. and P. A. Delcourt. 1984. Late-Quaternary paleoclimates and biotic responses across eastern North America and the northwestern Atlantic Ocean. Palaeogeogr. Palaeoclimatol. Palaeoecol. 48: 263–284.
Delcourt, H. R. and P. A. Delcourt. 1991. Quaternary Ecology, a Paleoecological Perspective. London.

Delcourt, P. A. and H. R. Delcourt. 1980. Pollen preservation and Quaternary environmental history in the southeastern United States. Palynology 4: 215–231.

Delcourt, P. A. and H. R. Delcourt. 1987. Long-term Forest Dynamics of the Temperate Zone. New York. [Ecological Studies. Vol. 63.]

Delcourt, P. A. and H. R. Delcourt. 1987b. Late-Quaternary dynamics of temperate forests: Applications of paleoecology to issues of global environmental change. Quatern. Sci. Rev. 6: 129–146.

Delcourt, P. A., H. R. Delcourt, R. C. Brister, and L. E. Lackey. 1980. Quaternary vegetation history of the Mississippi Embayment. Quatern. Res. 13: 111–132.

Delcourt, P. A., H. R. Delcourt, P. A. Cridlebaugh, and J. Chapman. 1986. Holocene ethnobotanical and paleoecological record of human impact on vegetation in the Little Tennessee River Valley, Tennessee, U.S.A. Quatern. Res. 25: 330–349.

Denton, G. H. and T. J. Hughes. 1983. Milankovitch theory of ice ages: Hypothesis of ice-sheet linkage between regional insolation and global climate. Quatern. Res. 20: 125–144.

Desert Pl. = Desert Plants.

di Castri, F. and H. A. Mooney, eds. 1973. Mediterranean Type Ecosystems: Origin and Structure. Berlin and New York. [Ecological Studies. Vol. 7.]

Diamond, D. D. and F. E. Smiens. 1985. Composition, classification, and species response patterns of remnant tallgrass prairie in Texas. Amer. Midl. Naturalist 113: 294–309.

Diamond, J. and T. J. Case, eds. 1986. Community Ecology. New York.

Dilcher, D. L. 1973. A palaeoclimatic interpretation of the Eocene floras of southeastern North America. In: A. Graham, ed. 1973. Vegetation and Vegetational History of Northern Latin America. Amsterdam and New York. Pp. 39–59.

Dilcher, D. L. and T. N. Taylor, eds. 1980. Biostratigraphy of Fossil Plants. Stroudsburg, Pa.

Dilcher, D. L., M. S. Zavada, A. Cronquist, J. H. Jones, E. J. Romero, R. B. Kaul, P. K. Endress, A. L. Bogle, P. R. Crane, R. A. Stockey, B. H. Tiffney, D. E. Giannasi, and D. Macklin. 1986. Phylogeny of the Hamamelidae. [Symposium.] Ann. Missouri Bot. Gard. 73(2): 225–441.

Dolan, R., P. J. Godfrey, and W. E. Odum. 1973. Man's impact on the barrier islands of North Carolina. Amer. Sci. 61: 152–162.

Dolan, R., B. Hayden, and H. Lins. 1980. Barrier islands. Amer. Sci. 68: 16–25.

Dort, W. Jr. and J. K. Jones Jr., eds. 1970. Pleistocene and Recent Environments of the Central Great Plains. Lawrence, Kans.

Doyle, J. and M. O'Leary. 1935. Pollination in *Tsuga, Cedrus, Pseudotsuga, Larix.* Sci. Proc. Roy. Dublin Soc. 21: 191–204.

Driver, H. E. 1961. Indians of North America. Chicago.

Drury, W. H. Jr. 1956. Bog flats and physiographic processes in the upper Kuskokwin River region, Alaska. Contr. Gray Herb. 178: 1–30.

Dupree, A. H. 1959. Asa Gray, 1810–1888. Cambridge, Mass.

Duvigneaud, P., ed. 1971. Productivity of Forest Ecosystems. Productivite des Ecosystemes Forestiers. Proceedings of the Brussels Symposium Organized by Unesco and the International Biological Programme (27–31 October 1969). Paris.

Dyke, A. S. and V. K. Prest. 1987. Paleogeography of northern North America, 18,000–5,000 years ago. Geological Survey of Canada, Map 1703A, scale 1:12,500,000. With: R. J. Fulton, ed. 1989. Quaternary Geology of Canada and Greenland. Ottawa. Annex.

Eagleman, J. R. 1985. Meteorology. The Atmosphere in Action, ed. 2. Belmont, Calif.

Eames, A. J. 1961. Morphology of Angiosperms. New York.

Earth Planet. Sci. Lett. = Earth and Planetary Science Letters.

Eckenwalder, J. E. 1976. Re-evaluation of Cupressaceae and Taxodiaceae: A proposed merger. Madroño 23: 237–256.

Eckenwalder, J. E. 1980b. Taxonomy of the West Indian cycads. J. Arnold Arbor. 61: 701–722.

Ecol. Monogr. = Ecological Monographs.

Ecology = Ecology, a Quarterly Journal Devoted to All Phases of Ecological Biology.

Econ. Bot. = Economic Botany; Devoted to Applied Botany and Plant Utilization.

Ehleringer, J. R. 1985. Annuals and perennials of warm deserts. In: B. F. Chabot and H. A. Mooney, eds. 1985. Physiological Ecology of North American Plant Communities. New York. Pp. 162–180.

Ehrenfeld, J. G. 1990. Dynamics and process of barrier island vegetation. Rev. Aquatic Sci. 2: 437–480.

Ehrlich, P. R. 1980. The strategy of conservation, 1980–2000. In: M. E. Soule and B. A. Wilcox, eds. 1980. Conservation Biology. An Evolutionary-Ecological Perspective. Sunderland, Mass. Pp. 329–344.

Ehrlich, P. R. 1986. Extinction: What is happening now and what needs to be done. In: D. K. Elliott, ed. 1986. Dynamics of Extinction. New York. Pp. 157–164.

Ehrlich, P. R. and P. H. Raven. 1964. Butterflies and plants—A study of coevolution. Evolution 18: 586–608.

Elias, M. K. 1942. Tertiary prairie grasses and other herbs from the High Plains. Special Pap. Geol. Soc. Amer. 41.

Elias, T. S. 1977. Threatened and endangered species problems in North America. An overview. In: G. T. Prance and T. S. Elias, eds. 1977. Extinction Is Forever. Bronx. Pp. 13–16.

Elias, T. S. 1987. Conservation and Management of Rare and Endangered Plants. Sacramento.

Ellefson, C. L., T. L. Stephens, and D. Welsh. 1992. Xeriscape™ Gardening: Water Conservation for the American Landscape. New York.

Elliott, D. K., ed. 1986. Dynamics of Extinction. New York.

Elliott, D. L. and S. K. Short. 1979. The current regenerative capacity of the northern Canadian trees, Keewatin, N.W.T., Canada: Some preliminary observations. Arctic Alpine Res. 11: 243–251.

Elliott-Fisk, D. L. 1983. The stability of the northern Canadian tree limit. Ann. Assoc. Amer. Geogr. 73: 560–576.

Elliott-Fisk, D. L. 1988. The Boreal Forest. In: M. G. Barbour and W. D. Billings, eds. 1988. North American Terrestrial Vegetation. New York. Pp. 33–62.

Elston, R., ed. 1976. Holocene Environmental Change in the Great Basin. Reno. [Nevada Archeol. Surv., Res. Pap. 6.]

Eluterius, L. N. 1980. Tidal Marsh Plants of Mississippi and Adjacent States. Ocean Springs, Miss. [Gulf Coast Res. Lab., Publ. MASGP-77-039.]

Endang. Spec. Update = Endangered Species Update.

Engler, H. G. A., ed. 1900–1953. Das Pflanzenreich.... 107 vols. Berlin. [Sequence of volume (Heft) numbers (order of publication) is independent of the sequence of series and family (Roman and Arabic) numbers (taxonomic order).]

Engler, H. G. A., H. Harms, J. Mattfeld, H. Melchior, and E. Werdermann, eds. 1924+. Die natürlichen Pflanzenfamilien...., ed. 2. 26+ vols. Leipzig and Berlin.

Engler, H. G. A. and K. Prantl, eds. 1887–1915. Die natürlichen Pflanzenfamilien.... 254 fasc. Leipzig. [In this work's complex and inconsistently applied numbering scheme, the sequence of fascicle (Lieferung) numbers (order of publication) is independent of the sequence of division (Teil) and subdivision (Abteilung) numbers (taxonomic order).]

Environm. Conservation = Environmental Conservation.

Erdkunde = Erdkunde. Archiv für wissenschaftliche Geographie.

Erwin, D. M. and R. A. Stockey. 1990. Sapindaceous flowers from the middle Eocene (Allenby Fm.) of British Columbia. Canad. J. Bot. 68: 2025–2034.

Erwin, T. L. 1983. Tropical forest canopies, the last biotic frontier. Bull. Entomol. Soc. Amer. 29: 14–19.

Erwin, T. L. 1988. The tropical forest canopy: The heart of biotic diversity. In: E. O. Wilson, ed. 1988. Biodiversity. Washington. Pp. 123–129.

Escher, A. and W. S. Watt, eds. 1976. Geology of Greenland. Copenhagen.

Escher, A. and W. S. Watt. 1976b. Summary of the geology of Greenland. In: A. Escher and W. S. Watt, eds. 1976. Geology of Greenland. Copenhagen. Pp. 13–16.

Espejel, I. 1986. Studies on Coastal and Sand Dune Vegetation of the Yucatan Peninsula. Ph.D. dissertation. Uppsala University.

Estuaries = Estuaries; Journal of Research on Any Aspect of Natural Science Applied to Estuaries.

Evenari, M., I. Noy-Meir, and D. W. Goodall, eds. 1985. Hot Deserts and Arid Shrublands. Vol. 1. Amsterdam.

Evolution = Evolution, International Journal of Organic Evolution.

Ewan, J. 1955. San Francisco as a mecca for nineteenth century naturalists. In: E. L. Kessel, ed. 1955. A Century of Progress in the Natural Sciences, 1853–1953. San Francisco. Pp. 1–64.

Ewan, J., ed. 1969. A Short History of Botany in the United States. New York and London.

Ewan, J. and N. Ewan. 1970. John Banister and His Natural History of Virginia 1678–1692. Urbana.

Ewan, J. and N. Ewan. 1981. Biographical Dictionary of Rocky Mountain Naturalists: A Guide to the Writings and Collections of Botanists, Zoologists, Geologists, Artists, and Photographers, 1682–1932. Utrecht etc. [Regnum Veg. 107.]

Eyre, F. H., ed. 1980. Forest Cover Types of the United States and Canada. Washington.

Eyre, S. R. 1971. Vegetation and Soils. Chicago.

Farabee, M. J. 1990. Triprojectate fossil pollen genera. Rev. Palaeobot. Palynol. 65: 341–347.

Farabee, M. J. and J. J. Skvarla. 1988. Examination of a pollen tetrad *Integricorpus reticulatus* (Mtchedlishvili) Standley from the Maastrichtian of North Dakota, U.S.A. Palynology 12: 43–48.

Farrar, D. R. 1967. Gametophytes of four tropical fern genera reproducing independently of their sporophytes in the southern Appalachians. Science 155: 1266–1267.

Farrar, D. R. 1985. Independent fern gametophytes in the wild. Proc. Roy. Soc. Edinburgh, B 86: 361–369.

Favre, D. S. 1989. International Trade in Endangered Species. A Guide to CITES. Dordrecht.

Fay, J. F. 1981. The endangered species program and plant reserves in the United States. In: H. Synge, ed. 1981. The Biological Aspects of Rare Plant Conservation. Chichester. Pp. 447–452.

Fed. Reg. = Federal Register. [U.S. Government.]

Fenneman, N. M. 1931. Physiography of Western United States. New York and London.

Fenneman, N. M. 1938. Physiography of Eastern United States. New York and London.

Ferlatte, W. J. 1974. A Flora of the Trinity Alps of Northern California. Berkeley.

Fern Gaz. = Fern Gazette; Journal of the British Pteridological Society.

Fernald, M. L. 1950. Gray's Manual of Botany, ed. 8. New York.

Fernald, M. L., A. C. Kinsey, and R. C. Rollins. 1958. Edible Wild Plants of Eastern North America. New York.

Fieldiana, Geol. = Fieldiana: Geology.

Finkelstein, M. 1990. National Parks System Plan. Ottawa.

Fish and Wildlife Service [U.S.D.I.]. 1954. Gulf of Mexico: Its Origin, Waters, and Marine Life. Washington. [U.S.D.I. Fish Wildlife Serv., Fish. Bull. 55.]

Fish and Wildlife Service [U.S.D.I.]. 1990. Endangered and threatened wildlife and plants: Review of plant taxa for listing as endangered or threatened species; notice of review. Fed. Reg. 55(35): 6184–6229. [50 CFR 17.]

Fish and Wildlife Service [U.S.D.I.]. 1992. Endangered and Threatened Wildlife and Plants: 50 CFR 17.11 and 17.12. Washington.

Fish Wildlife Res. = Fish and Wildlife Research.

Flint, R. F. 1971. Glacial and Quaternary Geology. New York.
Florin, R. 1951. Evolution in cordaites and conifers. Acta Horti Berg. 15: 285–388.
Florin, R. 1956. Nomenclatural notes on genera of living gymnosperms. Taxon 5: 188–192.
Flous, F. 1937. Revision du genre *Pseudotsuga*. Bull. Soc. Hist. Nat. Toulouse 71: 33–164.
Ford, R. I., ed. 1985. Prehistoric Food Production in North America. Ann Arbor.
Ford, R. I. 1985b. Patterns of prehistoric food production in North America. In: R. I. Ford, ed. 1985. Prehistoric Food Production in North America. Ann Arbor. Pp. 341–364.
Ford, R. I., ed. 1986. An Ethnobiology Source Book: The Use of Plants and Animals by American Indians. New York.
Forest Sci. = Forest Science.
Forman, R. T. T. and R. E. Boerner. 1981. Fire frequency and the pine barrens of New Jersey. Bull. Torrey Bot. Club 108: 34–50.
Fowells, H. A. 1965. Sylvics of Forest Trees of the United States. Washington. [Agric. Handb. 271.]
Fox, S. 1981. John Muir and His Legacy: The American Conservation Movement. Boston.
Frankis, M. P. 1988. Generic interrelationships in Pinaceae. Notes Roy. Bot. Gard. Edinburgh 45: 527–548.
Franklin, J. F. and C. T. Dyrness. 1973. Natural Vegetation of Oregon and Washington. Portland. [U.S.D.A. Forest Serv., Gen. Techn. Rep. PNW-8.]
Frankton, C. and G. A. Mulligan. 1970. Weeds of Canada. [Ottawa.]
Frederiksen, N. O. 1980. Mid-Tertiary climate of southeastern United States: The sporomorph evidence. J. Paleontol. 54: 728–739.
Frederiksen, N. O. 1984. Stratigraphic, paleoclimatic, and paleobiogeographic significance of Tertiary sporomorphs from Massachusetts. Profess. Pap. U.S. Geol. Surv. 1308: 1–25.
Frederiksen, N. O. 1987. Tectonic and paleogeographic setting of a new latest Cretaceous floristic province in North America. Palaios 2: 533–542.
Frederiksen, N. O. 1988. Sporomorph biostratigraphy, floral changes, and paleoclimatology, Eocene and earliest Oligocene of the eastern Gulf coast. Profess. Pap. U.S. Geol. Surv. 1448: 1–68.
Fredskild, B. 1973. Studies in the vegetational history of Greenland, palaeobotanical investigations of some Holocene lake and bog deposits. Meddel. Grønland 198: 1–245.
Fredskild, B. 1985. Holocene pollen records from West Greenland. In: J. T. Andrews, ed. 1985. Quaternary Environments: Eastern Canadian Arctic, Baffin Bay, and Western Greenland. Boston. Pp. 643–681.
Frick, G. F. and R. P. Stearns. 1961. Mark Catesby: The Colonial Audubon. Urbana.
Friis, E. M., W. G. Chaloner, and P. R. Crane, eds. 1987. The Origins of Angiosperms and Their Biological Consequences. Cambridge and New York.
Friis, E. M., W. G. Chaloner, and P. R. Crane. 1987b. Introduction to angiosperms. In: E. M. Friis et al., eds. 1987. The Origins of Angiosperms and Their Biological Consequences. Cambridge and New York. Pp. 1–15.
Fulton, R. J., ed. 1984. Quaternary Stratigraphy of Canada: A Canadian Contribution to IGCP Project 24. Ottawa. [Geol. Surv. Canada, Pap. 84-10.]
Fulton, R. J., ed. 1989. Quaternary Geology of Canada and Greenland. Ottawa. [Geology of Canada. No. 1. Geology of North America. Vol. K-1.]
Funder, S., coord. 1989. Quaternary geology of Canada and Greenland. In: R. J. Fulton, ed. 1989. Quaternary Geology of Canada and Greenland. Ottawa. Pp. 741–792.

Ganders, F. R. 1989. Adaptive radiation in Hawaiian *Bidens*. In: L. V. Giddings et al., eds. 1989. Genetics and the Founder Principle. New York. Pp. 99–112.
Gard. Bull. Straits Settlem. = Gardens' Bulletin. Straits Settlements.
Gastony, G. J. and L. D. Gottlieb. 1985. Genetic variation in the homosporous fern *Pellaea andromedifolia*. Amer. J. Bot. 72: 257–267.
Gates, D. H., L. A. Stoddart, and C. W. Cook. 1956. Soil as a factor influencing plant distribution on salt deserts of Utah. Ecol. Monogr. 26: 155–175.
Gates, F. C. 1940. Annotated List of the Plants of Kansas: Ferns and Flowering Plants, with Maps Showing Distribution of Species. [Topeka.]
Genetics = Genetics; a Periodical Record of Investigations Bearing on Heredity and Variation.
Gentes Herb. = Gentes Herbarum; Occasional Papers on the Kinds of Plants.
Geochim. Cosmochim. Acta = Geochimica et Cosmochimica Acta.
Geogr. Bull. = Geographical Bulletin.
Geophytology = Geophytology; an International Journal of Palaeobotany, Palynology and Allied Sciences.
Geosci. & Man = Geoscience and Man.
Gersmehl, P. J. 1971. Factors involved in the persistence of southern Appalachian treeless balds: An experimental study. Proc. Assoc. Amer. Geogr. 3: 56–61.
Giddings, L. V., K. Y. Kaneshiro, and W. W. Anderson, eds. 1989. Genetics and the Founder Principle. New York.
Gleason, H. A. and A. Cronquist. 1963. Manual of Vascular Plants of Northeastern United States and Adjacent Canada. Princeton.
Gleason, H. A. and A. Cronquist. 1964. The Natural Geography of Plants. New York.
Glooschenko, W. A. 1980. Coastal salt marshes in Canada. In: C. D. A. Rubec and F. C. Follet, eds. 1980. Workshop on Canadian Wetlands. Toronto. Pp. 39–47.
Glooschenko, W. A. 1980b. Coastal ecosystems of the James/Hudson Bay area of Ontario, Canada. Zhurn. Geomorph. 34: 214–224.
Glooschenko, W. A., I. P. Martini, and K. Clarke-Whistler. 1988. Salt marshes of Canada. In: National Wetlands Working Group [Canada]. 1988. Wetlands of Canada. Montreal. Pp. 347–377.

Godfrey, P. J. and M. M. Godfrey. 1976. Barrier Island Ecology of Cape Lookout National Seashore and Vicinity, North Carolina. Washington. [U.S. Natl. Park Serv., Sci. Monogr. 9.]

Goetzmann, W. H. 1966. Exploration and Empire: The Explorer and the Scientist in the Winning of the American West. New York.

Goldberg, A. 1986. Classification, evolution, and phylogeny of the families of dicotyledons. Smithsonian Contr. Bot. 58.

Goldberg, A. 1989. Classification, evolution, and phylogeny of the families of monocotyledons. Smithsonian Contr. Bot. 71.

Goldberg, D. E. 1982. The distribution of evergreen and deciduous trees relative to soil type: An example from the Sierra Madre, Mexico, and a general model. Ecology 63: 942–951.

Good, R. 1974. The Geography of Flowering Plants, ed. 4. London.

Goodall, D. W., R. A. Perry, and K. M. W. Howes, eds. 1979. Arid Land Ecosystems: Structure, Functioning, and Management. Vol. 1. Cambridge and New York.

Goodspeed, T. H. 1934. *Nicotiana* phylesis in the light of chromosome number, morphology and behavior. Univ. Calif. Publ. Bot. 17: 369–398.

Goodspeed, T. H. and R. E. Clausen. 1928. Interspecific hybridization in *Nicotiana* VIII. The *sylvestris-tomentosa-tabacum* hybrid triangle and its bearing on the origin of *tabacum*. Univ. Calif. Publ. Bot. 11: 243–254.

Goss, R. W. 1960. Mycorrhizae of ponderosa pine in Nebraska grassland soil. Nebraska Agric. Exp. Sta. Res. Bull. 192.

Gottlieb, L. D. 1973. Enzyme differentiation and phylogeny in *Clarkia franciscana, C. rubicunda* and *C. amoena*. Evolution 27: 205–214.

Gottlieb, L. D. 1974. Genetic confirmation of the origin of *Clarkia lingulata*. Evolution 28: 244–250.

Gottlieb, L. D. 1974b. Gene duplication and fixed heterozygosity for alcohol dehydrogenase in the diploid plant *Clarkia franciscana*. Proc. Natl. Acad. Sci. U.S.A. 71: 1816–1818.

Gottlieb, L. D. 1986. Genetic differentiation, speciation in *Clarkia* (Onagraceae). In: K. Iwatsuki et al., eds. 1986. Modern Aspects of Species. Tokyo. Pp. 145–160.

Gottlieb, L. D. and S. W. Edwards. 1992. An electrophoretic test of the genetic independence of a newly discovered population of *Clarkia franciscana*. Madroño 39: 1–7.

Graf, W. L., ed. 1987. Geomorphic Systems of North America. Boulder. [Geol. Soc. Amer., Centen. Special Vol. 2.]

Graham, A. 1965. The Sucker Creek and Trout Creek Miocene Floras of Southeastern Oregon. Kent, Ohio. [Kent State Univ., Res. Ser. 9.]

Graham, A. 1966. Plantae Rariores Camschatcenses: A translation of the dissertation of Jonas P. Helenius, 1750. Brittonia 18: 131–139.

Graham, A., ed. 1972. Floristics and Paleofloristics of Asia and Eastern North America. Amsterdam and New York.

Graham, A. 1972b. Outline of the origin and historical recognition of floristic affinities between Asia and eastern North America. In: A. Graham, ed. 1972. Floristics and Paleofloristics of Asia and Eastern North America. Amsterdam and New York. Pp. 1–18.

Graham, A., ed. 1973. Vegetation and Vegetational History of Northern Latin America. Amsterdam and New York.

Graham, A. 1973b. History of the arborescent temperate element in the northern Latin American biota. In: A. Graham, ed. 1973. Vegetation and Vegetational History of Northern Latin America. Amsterdam and New York. Pp. 301–314.

Gran, A. J., M. J. Crawley, and P. J. Edwards, eds. 1987. Colonization, Succession, and Stability. Oxford.

Grana = Grana; an International Journal of Palynology Including World Pollen and Spore Flora.

Grant, V. 1971. Plant Speciation. New York.

Grant, V. 1981. Plant Speciation, ed. 2. New York.

Grant, V. and K. A. Grant. 1965. Flower Pollination in the Phlox Family. New York.

Graustein, J. E. 1967. Thomas Nuttall, Naturalist; Explorations in America, 1808–1841. Cambridge, Mass.

Gray, A. 1840. Dr. Seybold, Flora Japonica. [Review.] Amer. J. Sci. Arts 39: 175–176.

Gray, A. 1846. Analogy between the flora of Japan and that of the United States. Amer. J. Sci. Arts, ser. 2, 2: 135–136.

Gray, J. T. and W. H. Schlesinger. 1981. Biomass, production, and litterfall in the coastal sage scrub of southern California. Amer. J. Bot. 68: 24–33.

Grayum, M. H. 1987. A summary of the evidence and arguments supporting the removal of *Acorus* from Araceae. Taxon 36: 723–729.

Great Basin Naturalist Mem. = Great Basin Naturalist Memoirs.

Great Plains Flora Association. 1977. Atlas of the Flora of the Great Plains. Ames.

Great Plains Flora Association. 1986. Flora of the Great Plains. Lawrence, Kans.

Gregg, W. P. Jr., S. L. Krugman, and J. D. Wood Jr., eds. 1989. Proceedings of the Symposium on Biosphere Reserves, Fourth World Wilderness Congress, September 14–17, 1987, YMCA of the Rockies, Estes Park, Colorado, USA. Atlanta.

Gregory, W. C. 1941. Phylogenetic and cytological studies in the Ranunculaceae. Trans. Amer. Philos. Soc., n. s. 31: 443–521.

Greller, A. M. 1980. Correlation of some climate statistics with distribution of broadleaved forest zones in Florida, U.S.A. Bull. Torrey Bot. Club 107: 189–219.

Greller, A. M. 1988. Deciduous forest. In: M. G. Barbour and W. D. Billings, eds. 1988. North American Terrestrial Vegetation. New York. Pp. 287–316.

Greuter, W., H. M. Burdet, W. G. Chaloner, V. Demoulin, R. Grolle, D. L. Hawksworth, D. H. Nicolson, P. C. Silva, F. A. Stafleu, E. G. Voss, and J. McNeill, eds. 1988. International Code of Botanical Nomenclature:

Adopted by the Fourteenth International Botanical Congress, Berlin, July–August 1987. Königstein. [Regnum Veg. 118.]

Gribbin, J. 1991. Climate now. (Inside science. No. 44. Pp. 1–4.) New Sci. 129(16 Mar.): centerfold.

Grichuk, V. P. 1984. Late Pleistocene vegetation history. In: A. A. Velichko et al., eds. 1984. Late Quaternary Environments of the Soviet Union. Minneapolis. Pp. 155–178.

Griffin, J. R. 1988. Oak woodland. In: M. G. Barbour and J. Major, eds. 1988. Terrestrial Vegetation of California, ed. 2. Sacramento. Pp. 383–415.

Grimm, E. C. 1983. Chronology and dynamics of vegetation change in the prairie-woodland region of southern Minnesota, USA. New Phytol. 93: 311–350.

Groves, R. H. and F. di Castri, eds. 1991. Biogeography of Mediterranean Invasions. Cambridge and New York.

Guetter, P. J. and J. E. Kutzbach. 1990. A modified Köppen classification applied to model simulations of glacial and interglacial climates. Climat. Change 16: 193–215.

Guthrie, R. D. 1984. Mosaics, allelochemics and nutrients, and ecological theory of late Pleistocene megafaunal extinctions. In: P. S. Martin and R. G. Klein, eds. 1984. Quaternary Extinctions. Tucson. Pp. 259–298.

Habeck, J. R. 1988. Present-day vegetation in the northern Rocky Mountains. Ann. Missouri Bot. Gard. 74: 804–840.

Hafen, L. R., W. E. Hollon, and C. C. Rister. 1970. Western America: The Exploration, Settlement, and Development of the Region beyond the Mississippi, ed. 3. Englewood Cliffs.

Haines, B. L. and E. L. Dunn. 1985. Coastal marshes. In: B. F. Chabot and H. A. Mooney, eds. 1985. Physiological Ecology of North American Plant Communities. New York. Pp. 323–347.

Haines, R. W. and K. A. Lye. 1975. Seedlings of Nymphaeaceae. Bot. J. Linn. Soc. 70: 255–265.

Halliday, G. 1989. The vegetation and flora of Greenland. Pl. Today 1989(Nov.–Dec.): 197–202.

Hallier, H. 1901. Über die Verwandtschaftsverhältnisse der Tubifloren und Ebenalen, den polyphyletischen Ursprung der Sympetalen und Apetalen und die Anordnung der Angiospermen Überhaupt. Vorstudien zum Entwurf eines Stammbaums der Blütenpflanzen. Abh. Naturwiss. Naturwiss. Verein Hamburg 16: 1–112.

Hamilton, T. D., K. M. Reed, and R. M. Thorson, eds. 1986. Glaciation in Alaska, the Geologic Record. Anchorage.

Hamilton, T. D., K. M. Reed, and R. M. Thorson. 1986b. Glaciation in Alaska—Introduction and overview. In: T. D. Hamilton et al., eds. 1986. Glaciation in Alaska, the Geologic Record. Anchorage. Pp. 1–8.

Hamilton, T. D. and R. M. Thorson. 1983. The Cordilleran Ice Sheet in Alaska. In: S. C. Porter, ed. 1983. Late-Quaternary Environments of the United States. Vol. 1. The Late Pleistocene. Minneapolis. Pp. 38–70.

Hanes, T. L. 1988. California chaparral. In: M. G. Barbour and J. Major, eds. 1988. Terrestrial Vegetation of California, ed. 2. Sacramento. Pp. 417–470.

Hansen, H. L., V. Kurmis, and D. D. Ness. 1974. The ecology of upland forest communities and implications for management of Itasca State Park, Minnesota. Minnesota Agric. Exp. Sta. Techn. Bull. 298.

Hara, H. 1972. Patterns of differentiation in flowering plants. In: A. Graham, ed. 1972. Floristics and Paleofloristics of Asia and Eastern North America. Amsterdam and New York. Pp. 55–60.

Harborne, J. B. and B. L. Turner. 1984. Plant Chemosystematics. New York.

Hardin, J. W. 1977. Threatened and endangered species problems in North America. The southeastern United States. In: G. T. Prance and T. S. Elias, eds. 1977. Extinction Is Forever. Bronx. Pp. 36–40.

Harlan, J. R. 1975. Crops and Man. Madison.

Harper, R. M. 1911. The relation of climax vegetation to islands and peninsulas. Bull. Torrey Bot. Club 38: 515–525.

Harrington, H. D. 1954. Manual of the Plants of Colorado. Denver.

Harris, D. and G. Hillman, eds. 1989. Foraging and Farming. London.

Harshberger, J. W. 1899. The Botanists of Philadelphia and Their Work. Philadelphia.

Harshberger, J. W. 1911. Phytogeographic Survey of North America. Leipzig and New York.

Hart, J. A. 1987. A cladistic analysis of conifers: Preliminary results. J. Arnold Arbor. 68: 269–307.

Hart, J. A. and R. A. Price. 1990. The genera of Cupressaceae (including Taxodiaceae) in the southeastern United States. J. Arnold Arbor. 71: 275–322.

Haufler, C. H. and D. E. Soltis. 1986. Genetic evidence suggests that homosporous ferns with high chromosome numbers are diploid. Proc. Natl. Acad. Sci. U.S.A. 83: 4389–4393.

Hawkes, J. G. 1983. The Diversity of Crop Plants. Cambridge, Mass.

Hay, W. W., ed. 1974. Studies in Paleooceanography. Tulsa. [Soc. Econ. Paleontol. Mineral. Special Publ. 20.]

Hays, J. D., J. Imbrie, and N. J. Shackelton. 1976. Variations in the Earth's orbit: Pacemaker of the Ice Ages. Science 194: 1121–1132.

Heady, H. F. 1988. Valley grassland. In: M. G. Barbour and J. Major, eds. 1988. Terrestrial Vegetation of California, ed. 2. Sacramento. Pp. 491–514.

Heady, H. F., T. C. Foin, M. J. Hektner, D. W. Taylor, M. G. Barbour, and W. J. Barry. 1988. Coastal prairie and northern coastal scrub. In: M. G. Barbour and J. Major, eds. 1988. Terrestrial Vegetation of California, ed. 2. Sacramento. Pp. 733–760.

Heath, R. C. 1989. Hydrogeologic map of North America showing the major rock units that underlie the surficial layer. In: A. W. Bally and R. A. Palmer, eds. 1989. The Geology of North America—An Overview. Boulder. Plate 11.

Heath, R. C. 1989b. Surficial deposits of North America. (Hydrogeologic map of North America showing the major units that comprise the surficial layer.) In: A. W. Bally and R. A. Palmer, eds. 1989. The Geology of North America—An Overview. Boulder. Plate 12.

Hebda, R. J. and R. W. Mathewes. 1984. Holocene history of cedar and native Indian cultures of the North American Pacific coast. Science 225: 711–713.

Hedrick, U. P. 1950. A History of Horticulture in America to 1860. New York.

Heiser, C. B. 1969. Nightshades, the Paradoxical Plants. San Francisco.

Heiser, C. B. 1989. Domestication of the Cucurbitaceae: *Cucurbita* and *Lagenaria*. In: D. Harris and G. Hillman, eds. 1989. Foraging and Farming. London. Pp. 471–480.

Heiser, C. B. 1990. New perspectives on the origin and evolution of New World domesticated plants: Summary. Econ. Bot. 44(3, suppl.): 111–116.

Helenurm, K. and F. R. Ganders. 1985. Adaptive radiation and genetic differentiation in Hawaiian *Bidens*. Evolution 39: 763–765.

Hellmers, H. 1966. Growth response of redwood seedlings to thermoperiodism. Forest Sci. 12: 276–283.

Hengeveld, R. 1990. Dynamic Biogeography. Cambridge.

Henrey, B. 1975. British Botanical and Horticultural Literature Before 1800; comprising a History and Bibliography of Botanical and Horticultural Books Printed in England, Scotland, and Ireland from the Earliest Times until 1800. 3 vols. London and New York.

Henry, R. D. and A. R. Scott. 1980. Some aspects of the alien component of the spontaneous Illinois vascular flora. Trans. Illinois State Acad. Sci. 73(4).

Heredity = Heredity; an International Journal of Genetics.

Herendeen, P. S. and D. L. Dilcher. 1990. Reproductive and vegetative evidence for the occurrence of *Crudia* (Leguminosae, Caesalpinioideae) in the Eocene of southeastern North America. Bot. Gaz. 151: 402–413.

Herendeen, P. S., D. H. Les, and D. L. Dilcher. 1990. Fossil *Ceratophyllum* (Ceratophyllaceae) from the Tertiary of North America. Amer. J. Bot. 77: 7–16.

Heusser, C. J. 1983. Vegetational history of the northwestern United States including Alaska. In: S. C. Porter, ed. 1983. Late-Quaternary Environments of the United States. Vol. 1. The Late Pleistocene. Minneapolis. Pp. 239–258.

Heusser, C. J. 1983b. Holocene vegetation history of the Prince William Sound region, south-central Alaska. Quatern. Res. 19: 337–355.

Heusser, L. E. and J. E. King. 1988. North America, with special emphasis on the development of the Pacific coastal forest and prairie/forest boundary prior to the last glacial maximum. In: B. Huntley and T. Webb III, eds. 1988. Vegetation History. Dordrecht. Pp. 193–236.

Heywood, V. H., ed. 1978. Flowering Plants of the World. Oxford.

Heywood, V. H. 1989. The Botanic Gardens Conservation Strategy. Gland.

Heywood, V. H., J. B. Harborne, and B. L. Turner, eds. 1977. The Biology and Chemistry of the Compositae. London.

Hickey, L. J. 1977. Stratigraphy and paleobotany of the Golden Valley Formation (early Tertiary) of western North Dakota. Mem. Geol. Soc. Amer. 150.

Hicks, D. J. and B. F. Chabot. 1985. Deciduous forest. In: B. F. Chabot and H. A. Mooney, eds. 1985. Physiological Ecology of North American Plant Communities. New York. Pp. 257–277.

Hinkley, D. 1990. Thinking of yew. Washington Park Arbor. Bull. 53(4): 2–3.

Hitchcock, C. L. and A. Cronquist. 1973. Flora of the Pacific Northwest: An Illustrated Manual. Seattle.

Hitchcock, C. L., A. Cronquist, M. Ownbey, and J. W. Thompson. 1955–1969. Vascular Plants of the Pacific Northwest. 5 vols. Seattle.

Hoey, M. T. and C. R. Parks. 1991. Isozyme divergence between eastern Asian, North American, and Turkish species of *Liquidambar* (Hamamelidaceae). Amer. J. Bot. 78: 938–947.

Holarc. Ecol. = Holarctic Ecology.

Holland, R. F. 1976. The vegetation of vernal pools: A survey. In: S. Jain, ed. 1976. Vernal Pools: Their Ecology and Conservation. Davis. Pp. 11–15.

Holmgren, N. H. 1972. Plant geography of the Intermountain Region. In: A. Cronquist et al. 1972+. Intermountain Flora. Vascular Plants of the Intermountain West, U.S.A. 4+ vols. New York and London. Vol. 1, pp. 77–161.

Holmgren, P. K., N. H. Holmgren, and L. C. Barnett, eds. 1990. Index Herbariorum. Part I: The Herbaria of the World, ed. 8. Bronx. [Regnum Veg. 120.]

Holt, P. C., ed. 1971. The Distributional History of the Biota of the Southern Appalachians. Part 2. Flora. Blacksburg, Va. [Virginia Polytechnic Inst. and State Univ., Res. Div. Monogr. 2.]

Holzner, W. and M. Numata. 1982. Biology and Ecology of Weeds. The Hague.

Hoogland, R. D. and J. L. Reveal. 1993. Vascular plant family names in current use. In: W. Greuter, ed. 1993. NCU-1. Family Names in Current Use for Vascular Plants, Bryophytes, and Fungi. Königstein. Pp. 15–60.

Hoose, P. and S. Crispin. 1990. The status of natural heritage data centres in Canada. In: G. M. Allen et al., eds. 1990. Conserving Carolinian Canada. Waterloo. Pp. 327–331.

Hopkins, D. M., ed. 1967. The Bering Land Bridge. Stanford.

Hopkins, D. M. 1982. Aspects of the paleogeography of Beringia during the late Pleistocene. In: D. M. Hopkins et al., eds. 1982. Paleoecology of Beringia. New York. Pp. 3–28.

Hopkins, D. M., J. V. Matthews Jr., C. E. Schweger, and S. B. Young, eds. 1982. Paleoecology of Beringia. New York.

Hopkins, D. M., J. V. Matthews Jr., J. A. Wolfe, and M. L. Silberman. 1971. A Pliocene flora and insect fauna from the Bering Strait region. Palaeogeogr. Palaeoclimatol. Palaeoecol. 9: 211–231.

Hsu, S. Y. 1980. The *Metasequoia* flora and its phytogeographic significance. J. Arnold Arbor. 61: 41–94.

Hudson, C. 1976. The Southeastern Indians. Knoxville.

Huffman, R. T. and S. W. Forsythe. 1981. Bottomland hardwood forest communities and their relation to anaerobic soil conditions. In: J. R. Clark and J. Benforado, eds. 1981. Wetlands of Bottomland Hardwood Forests. Amsterdam. Pp. 187–196.

Hughes, T. 1987. Ice dynamics and deglaciation models when ice sheets collapsed. In: W. F. Ruddiman and H. E. Wright Jr., eds. 1987. North America and Adjacent Oceans during the Last Deglaciation. Boulder. Pp. 183–220.

Hultén, E. 1940. History of botanical exploration in Alaska and Yukon territories from the time of their discovery to 1940. Bot. Not. 1940: 289–346.

Hultén, E. 1963. Phytogeographical connections of the North Atlantic. In: A. Löve and D. Löve, eds. 1963. North Atlantic Biota and Their History. Oxford. Pp. 45–72.

Humphrey, H. B. 1961. Makers of North American Botany. New York.

Humphrey, R. R. 1974. The Boojum and Its Home. Tucson.

Humphrey, R. R. and L. A. Mehrhoff. 1958. Vegetation changes on a southern Arizona grassland range. Ecology 39: 720–726.

Hunt, C. B. 1974. Natural Regions of the United States and Canada. San Francisco.

Huntley, B. and T. Webb III. 1988. Vegetation History. Dordrecht.

Hutchinson, J. 1926–1934. The Families of Flowering Plants. 2 vols. London.

Hutchinson, J. 1959. The Families of Flowering Plants, ed. 2. 2 vols. Oxford.

Hutchinson, J. 1973. The Families of Flowering Plants Arranged According to a New System Based on Their Probable Phylogeny, ed. 3. Oxford.

Hyde, P. and F. Leydet. 1969. The Last Redwoods. San Francisco.

I.U.C.N. 1985. 1985 United Nations List of National Parks and Protected Areas. Gland. [International Union for Conservation of Nature and Natural Resources, Conservation Monitoring Centre and Commission on National Parks and Protected Areas.]

Illinois Agricultural Experiment Station. 1981. Weeds of the North Central States, rev. ed. Urbana. [Illinois Agric. Exp. Sta. Bull. 772.]

Imbrie, J. and K. P. Imbrie. 1979. Ice Ages, Solving the Mystery. Hillside, N.J.

International Union of Biological Sciences. 1951. Les Bases Ecologiques de la Regeneration de la Vegetation des Zones Arides. On the Ecological Foundations of the Regeneration of Vegetation in Arid Zones. [Symposium.] Stockholm, Juillet 1950. Paris.

Israel J. Bot. = Israel Journal of Botany.

Ives, J. D. and R. G. Barry, eds. 1974. Arctic and Alpine Environments. London.

Iwatsuki, K., P. H. Raven, and W. J. Bock, eds. 1986. Modern Aspects of Species. Tokyo.

J. Air Pollut. Control Assoc. = Journal of the Air Pollution Control Association.

J. Appl. Meteorol. = Journal of Applied Meteorology.

J. Arid Environm. = Journal of Arid Environments.

J. Arizona Acad. Sci. = Journal of the Arizona Academy of Science.

J. Arnold Arbor. = Journal of the Arnold Arboretum.

J. Atmosph. Sci. = Journal of the Atmospheric Sciences.

J. Biogeogr. = Journal of Biogeography.

J. Calif. & Great Basin Anthropol. = Journal of California and Great Basin Anthropology.

J. Ecol. = Journal of Ecology.

J. Fac. Sci. Hokkaido Univ., Ser. 4, Geol. Mineral. = Journal of the Faculty of Science, Hokkaido University. Series 4, Geology and Mineralogy.

J. Forest Hist. = Journal of Forest History.

J. Hattori Bot. Lab. = Journal of the Hattori Botanical Laboratory. [Hattori Shokubutsu Kenkyusho Hokoku.]

J. Paleontol. = Journal of Paleontology.

J. Quatern. Sci. = Journal of Quaternary Science.

J. Range Managem. = Journal of Range Management.

Jacobs, B. F., P. L. Fall, and O. K. Davis, eds. 1985. Late Quaternary vegetation and climates of the American Southwest. Contr. Ser. Amer. Assoc. Stratigr. Palynologists 16: 1–185.

Jacobson, G. L. Jr. 1979. The palaeoecology of white pine (*Pinus strobus*) in Minnesota. J. Ecol. 67: 697–726.

Jacobson, G. L. Jr., T. Webb III, and E. C. Grimm. 1987. Patterns and rates of vegetation change during deglaciation of eastern North America. In: W. F. Ruddiman and H. E. Wright Jr., eds. 1987. North America and Adjacent Oceans during the Last Deglaciation. Boulder. Pp. 277–288.

Jain, S., ed. 1976. Vernal Pools: Their Ecology and Conservation. Davis.

Jain, S. and P. Moyle, eds. 1984. Vernal Pools and Intermittent Streams: A Symposium, May 9 and 10, 1981. Davis.

Jansen, R. K. and J. D. Palmer. 1988. Phylogenetic implications of chloroplast DNA restriction site variation in the Mutisieae (Asteraceae). Amer. J. Bot. 75: 753–766.

Jap. J. Ecol. = Japanese Journal of Ecology. [Nippon Seitaigakkai-shi.]

Jarzen, D. M. 1977. *Aquilapollenites* and some Santalalean genera. Grana 16: 29–39.

Jenkins, R. E. 1975. Endangered plant species: A soluble ecological problem. Nat. Conservancy News (Arlington) 25(4): 20–21.

Jensen, R. J. and W. H. Eshbaugh. 1976. Numerical taxonomic studies of hybridization in *Quercus* I and II. Syst. Bot. 1: 1–19.

Jepson, W. L. [1923–1925.] A Manual of the Flowering Plants of California.... Berkeley.

Johnson, A. F. 1977. A survey of the strand and dune vegetation along the Pacific and southern Gulf coasts of Baja California, Mexico. J. Biogeogr. 7: 83–99.

Johnson, A. F. 1982. Dune vegetation along the eastern shore of the Gulf of California. J. Biogeogr. 9: 317–330.

Johnson, A. F. 1985. Ecologia de *Abronia maritima*, especie

pionera de las dunas del oeste de Mexico. Biótica 10: 19–34.
Johnson, A. F. 1985b. A Guide to the Plant Communities of the Napeague Dunes, Long Island, New York. Mattituck, N.Y.
Johnson, A. F. and M. G. Barbour. 1990. Maritime forests and dunes. In: R. L. Myers and J. J. Ewel, eds. 1990. Ecosystems of Florida. Gainesville. Pp. 429–480.
Jones, G. N. 1936. A Botanical Survey of the Olympic Peninsula. Seattle.
Josselyn, M. 1983. The Ecology of San Francisco Bay Tidal Marshes: A Community Profile. Washington.
Judd, F. W., R. I. Lonard, and S. L. Sides. 1977. The vegetation of South Padre Island, Texas in relation to topography. SouthW. Naturalist 22: 31–48.

Kahler, A. and R. W. Allard. 1970. Genetics of isozyme variants in barley. I. Esterases. Crop Sci. (Madison) 10: 444–448.
Kapp, R. O. 1977. Late Pleistocene and postglacial plant communities of the Great Lakes region. In: R. C. Romans, ed. 1977. Geobotany. New York. Pp. 1–27.
Karrow, P. F. and P. E. Calkin, eds. 1985. Quaternary Evolution of the Great Lakes. [St. John's and Toronto.]
Karrow, P. F. and S. Occhietti. 1989. Quaternary geology of the St. Lawrence Lowlands. In: R. J. Fulton, ed. 1989. Quaternary Geology of Canada and Greenland. Ottawa. Pp. 320–389.
Kartesz, J. T. 1988. A Flora of Nevada. Ph.D. thesis. 4 vols. University of Nevada.
Kato, M. and K. Iwatsuki. 1983. Phytogeographic relationships of pteridophytes between temperate North America and Japan. Ann. Missouri Bot. Gard. 70: 724–733.
Kay, M., F. B. King, and C. K. Robinson. 1980. Cucurbits from Phillips Spring: New evidence and interpretations. Amer. Antiquity 45: 806–822.
Keegan, W. F., ed. 1987. Emergent Horticultural Economics of the Eastern Woodlands. Carbondale. [Center Archaeol. Invest., Occas. Pap. 7.]
Keeley, J. E. and S. C. Keeley. 1988. Chaparral. In: M. G. Barbour and W. D. Billings, eds. 1988. North American Terrestrial Vegetation. New York. Pp. 165–207.
Keeley, S. C., ed. 1989. The California Chaparral: Paradigms Reexamined. Los Angeles.
Keever, C. 1953. Present composition of some stands of former oak-chestnut forest in the southern Blue Ridge Mountains. Ecology 34: 44–54.
Kemp, P. R. 1983. Phenological patterns of Chihuahuan Desert plants in relation to the timing of water availability. J. Ecol. 71: 427–436.
Kershaw, K. A. 1976. The vegetational zonation of the East Pen Island salt marshes, Hudson Bay. Canad. J. Bot. 54: 5–13.
Kershaw, K. A. 1977. Physiological-environmental interactions in lichens. New Phytol. 79: 377–421.
Kessel, E. L., ed. 1955. A Century of Progress in the Natural Sciences, 1853–1953. San Francisco. [Reprint 1974, New York.]
Kihara, H. and T. Ono. 1926. Chromosomenzahlen und systematische Gruppierung der *Rumex*-Arten. Z. Zellf. Mikroskop. Anat. 4: 475–481.
Kilgore, B. M. 1973. The ecological role of fire in Sierran conifer forests. Quatern. Res. 3: 496–513.
King, L. J. 1966. Weeds of the World: Biology and Control. New York.
King, R. M. and H. Robinson. 1970. Studies in the Eupatorieae (Compositae) XIX. New combinations in *Ageratina*. Phytologia 19: 208–229.
King, R. M. and H. Robinson. 1987. The genera of the Eupatorieae (Asteraceae). Monogr. Syst. Bot. Missouri Bot. Gard. 22.
Klekowski, E. J. Jr. 1969. Reproductive biology of the Pteridophyta. II. Theoretical considerations. Bot. J. Linn. Soc. 62: 347–359.
Klopatek, J. M., R. J. Olson, C. J. Emerson, and J. L. Jones. 1979. Land-use conflicts with natural vegetation in the United States. Environm. Conservation 6: 191–199.
Knutson, L. V., ed. 1978. Biosystematics in Agriculture. Montclair, N.J.
Koch, B. E. 1963. Fossil plants from the lower Paleocene of the Agatdalen (Angmartussut) area, central Nugssuaq Peninsula, northwest Greenland. Meddel. Grønland 172.
Koopowitz, H. and H. Kaye. 1990. Plant Extinction: A Global Crisis, ed. 2. London.
Koptur, S., A. R. Smith, and I. Baker. 1982. Nectaries in some neotropical species of *Polypodium* (Polypodiaceae): Preliminary observations and analysis. Biotropica 14: 108–113.
Korstian, C. F. 1937. Perpetuation of spruce on cut-over and burned lands in the higher southern Appalachian Mountains. Ecol. Monogr. 7: 125–167.
Krajina, V. J., ed. 1965. Ecology of Western North America. Vol. 1. Vancouver.
Kramer, K. U. and P. S. Green, eds. 1990. Pteridophytes and gymnosperms. In: K. Kubitzki et al., eds. 1990+. The Families and Genera of Vascular Plants. 1+ vol. Berlin etc. Vol. 1.
Kubitzki, K., K. U. Kramer, and P. S. Green, eds. 1990+. The Families and Genera of Vascular Plants. 1+ vol. Berlin etc.
Kuc, M. 1974. Fossil flora of the Beaufort Formation, Meighen Island, Northwest Territories. Rep. Activities Geol. Surv. Canada 74-1: 193–195.
Küchler, A. W. 1964. Potential Natural Vegetation of the Conterminous United States. New York. [Amer. Geogr. Soc., Special Publ. 36.]
Küchler, A. W. 1972. The oscillations of the mixed prairie in Kansas. Erdkunde 26: 120–129.
Kutzbach, J. E. 1987. Model simulations of the climatic patterns during the deglaciation of North America. In: W. F. Ruddiman and H. E. Wright Jr., eds. 1987. North America and Adjacent Oceans during the Last Deglaciation. Boulder. Pp. 425–446.
Kutzbach, J. E. and P. J. Guetter. 1986. The influence of changing orbital parameters and surface boundary conditions on climate simulations for the past 18,000 years. J. Atmosph. Sci. 43: 1726–1759.

Kutzbach, J. E. and H. E. Wright Jr. 1985. Simulation of the climate of 18,000 years BP: Results for the North American/North Atlantic/European sector and comparison with the geologic record of North America. Quatern. Sci. Rev. 4: 147–187.

Laacke, R. J. and J. F. Fiske. 1983. Sierra Nevada mixed conifers. In: R. M. Burns, ed. 1983. Silvicultural Systems for the Major Forest Types of the United States. Washington. Pp. 44–47.

Lamb, H. F. and M. E. Edwards. 1988. The Arctic. In: B. Huntley and T. Webb III, eds. 1988. Vegetation History. Dordrecht. Pp. 519–555.

Lamoureux, G. and M. M. Grandtner. 1977. Contributions a l'etude ecologique des dunes mobiles. I. Les elements phytosociologiques. Canad. J. Bot. 55: 158–171.

Lamoureux, G. and M. M. Grandtner. 1978. Contributions a l'etude ecologique des dunes mobiles. II. Les elements edaphiques. Canad. J. Bot. 56: 818–832.

Lanner, R. M., ed. 1984. Proceedings of the Eighth North American Forest Biology Workshop. Logan.

Larsen, J. A. 1980. The Boreal Ecosystem. New York.

Larsen, J. A. 1982. Ecology of the Northern Lowland Bogs and Conifer Forests. New York.

Lassoie, J. P., T. M. Hinckley, and C. C. Grier. 1985. Coniferous forests of the Pacific Northwest. In: B. F. Chabot and H. A. Mooney, eds. 1985. Physiological Ecology of North American Plant Communities. New York. Pp. 127–161.

Lawrence, G. H. M. 1951. Taxonomy of Vascular Plants. New York.

Lean, G., D. Hinrichson, and A. Markham. 1990. Atlas of the Environment. London.

Leatherman, S. P. 1982. Barrier Island Handbook, ed. 2. College Park, Md.

Lellinger, D. B. 1985. A Field Manual of the Ferns & Fern-allies of the United States & Canada. Washington.

Lenz, L. W. 1986. Marcus E. Jones: Western Geologist, Mining Engineer & Botanist. Claremont.

Leopold, A. S. 1950. Vegetation zones of Mexico. Ecology 31: 507–518.

Leopold, E. B. and M. F. Denton. 1987. Comparative age of grassland and steppe east and west of the northern Rocky Mountains. Ann. Missouri Bot. Gard. 74: 841–867.

Leopold, E. B. and H. D. MacGinitie. 1972. Development and affinities of Tertiary floras in the Rocky Mountains. In: A. Graham, ed. 1972. Floristics and Paleofloristics of Asia and Eastern North America. Amsterdam and New York. Pp. 147–200.

Leroy, J.-F. 1957. Les Botanistes Français en Amérique du Nord avant 1850. Paris.

Les, D. H. and R. L. Stuckey. 1985. The introduction and spread of *Veronica beccabunga* (Scrophulariaceae) in eastern North America. Rhodora 87: 503–515.

Levin, D. 1979. The nature of plant species. Science 204: 381–384.

Levitzky, G. A. 1931. The karyotype in systematics. Trudy Prikl. Bot. 27: 220–240.

Levy, G. F. 1990. Vegetation dynamics of the Virginia barrier islands. Virginia J. Sci. 41: 300–306.

Lewis, H. 1962. Catastrophic selection as a factor in speciation. Evolution 16: 257–271.

Lewis, H. and M. E. Lewis. 1955. The genus *Clarkia*. Univ. Calif. Publ. Bot. 20: 241–392.

Lewis, H. and P. H. Raven. 1958. *Clarkia franciscana,* a new species from central California. Brittonia 10: 7–13.

Lewis, H. and M. H. Roberts. 1956. The origin of *Clarkia lingulata*. Evolution 10: 126–138.

Lewis, W. H., ed. 1980. Polyploidy: Biological Relevance. New York.

Li, H. L. 1952. Floristic relationships between eastern Asia and eastern North America. Trans. Amer. Philos. Soc., n. s. 41: 371–409.

Li, H. L. 1953. A reclassification of *Libocedrus* and Cupressaceae. J. Arnold Arbor. 34: 17–36.

Li, H. L. 1972. Eastern Asia–eastern North America species-pairs in wide-ranging genera. In: A. Graham, ed. 1972. Floristics and Paleofloristics of Asia and Eastern North America. Amsterdam and New York. Pp. 65–78.

Little, E. L. Jr. 1971b. Endemic, disjunct and northern trees in the Sourthern Appalachians. In: P. C. Holt, ed. 1971. The Distributional History of the Biota of the Southern Appalachians. Part 2. Flora. Blacksburg, Va. Pp. 249–290.

Little, E. L. Jr. 1979. Checklist of United States Trees (Native and Naturalized). Washington. [Agric. Handb. 541.]

Long, R. W. and O. Lakela. 1971. A Flora of Tropical Florida: A Manual of the Seed Plants and Ferns of Southern Peninsular Florida. Coral Gables.

Looman, J. 1983. Distribution of plant species and vegetation in relation to climate. Vegetatio 54: 17–25.

Lorenzi, H. J. and L. S. Jeffery. 1987. Weeds of the United States and Their Control. New York.

Louisiana Agric. Exp. Sta. Bull. = Louisiana Agricultural Experiment Station Bulletin.

Löve, A. and D. Löve, eds. 1963. North Atlantic Biota and Their History. Oxford.

Löve, D. 1970. Subarctic and subalpine: Where and what? Arctic Alpine Res. 2: 63–73.

Löve, D. and A. Löve. 1974. Origin and evaluation of arctic and alpine floras. In: J. D. Ives and R. G. Barry, eds. 1974. Arctic and Alpine Environments. London. Pp. 571–603.

Love, R. M., ed. 1975. The California Annual Grassland Ecosystem: A Symposium Sponsored by the California Chapter, American Society of Agronomy, Anaheim, California, January 30, 1975. Davis. [Inst. Ecol. Publ. 7.]

Lovejoy, T. E. 1979. The epoch of biotic impoverishment. Great Basin Naturalist Mem. 3: 5–10.

Lovejoy, T. E. 1980. Foreword. In: M. E. Soule and B. A. Wilcox, eds. 1980. Conservation Biology. An Evolutionary-Ecological Perspective. Sunderland, Mass. Pp. ix–x.

Lovejoy, T. E. 1986. Species leave the ark one by one. In: B. G. Norton, ed. 1986. The Preservation of Species.

The Value of Biological Diversity. Princeton. Pp. 13–27.

Lowe, D. W., J. R. Matthews, and C. J. Moseley, eds. 1990. The Official World Wildlife Fund Guide to Endangered Species of North America. Vol. 1. Washington.

Lucas, G. Ll. and H. Synge. 1978. The IUCN Plant Red Data Book. Morges.

Lugardon, B. 1971. Contribution à la Connaissance de la Morphogenése et de la Structure des Parois Sporales Chez les Filicinées Isosporées. Thesis. Université Paul Sabatier, Toulouse.

Lugardon, B. 1978. Comparison between pollen and pteridophyte spore walls. In: D. C. Bharadwaj et al., eds. [1978–1981.] Proceedings. Fourth International Palynological Conference [Lucknow, India, 1976–1977]. 3 vols. Lucknow. Vol. 1, pp. 199–206.

Lumaret, R. 1988. Cytology, genetics and evolution in the genus *Dactylis*. C. R. C. Crit. Rev. Pl. Sci. 7: 55–91.

Lutkens, F. K. and E. J. Tarbuck. 1989. The Atmosphere. An Introduction to Meteorology, ed. 4. Englewood Cliffs.

Mabberley, D. J. 1985. Jupiter Botanicus: Robert Brown of the British Museum. Braunschweig.

Mabberley, D. J. 1987. The Plant-book. A Portable Dictionary of the Higher Plants.... Cambridge etc.

Mabry, T. J., J. H. Hunziker, and D. R. DiFeo, eds. 1977. Creosote Bush: Biology and Chemistry of *Larrea* in New World Deserts. Stroudsburg, Pa.

MacDonald, G. M. and L. C. Cwynar. 1985. A fossil pollen based reconstruction of the late Quaternary history of lodgepole pine (*Pinus contorta* ssp. *latifolia*) in the western interior of Canada. Canad. J. Forest Res. 15: 1039–1044.

Macdonald, K. B. and M. G. Barbour. 1974. Beach and salt marsh vegetation of the North American Pacific coast. In: R. J. Reimold and W. H. Queen, eds. 1974. Ecology of Halophytes. New York. Pp. 175–233.

MacGinitie, H. D. 1953. Fossil plants of the Florissant beds, Colorado. Contr. Paleontol. Carnegie Inst. Wash. 599.

MacGinitie, H. D. 1962. The Kilgore flora, a late Miocene flora from northern Nebraska. Univ. Calif. Publ. Geol. Sci. 35: 67–158.

MacGinitie, H. D. 1969. The Eocene Green River flora of northwestern Colorado and northeastern Utah. Univ. Calif. Publ. Geol. Sci. 83: 1–202.

MacGinitie, H. D. 1974. An early middle Eocene flora from the Yellowstone-Absaroka volcanic province, northwestern Wind River Basin, Wyoming. Univ. Calif. Publ. Geol. Sci. 108: 1–103.

Mack, R. N., N. W. Rutter, and S. Valastro. 1983. Holocene vegetational history of the Kootenai River Valley, Montana. Quatern. Res. 20: 177–193.

Mack, R. N. and J. N. Thompson. 1982. Evolution in steppe with few large hooved animals. Amer. Naturalist 119: 757–773.

MacMahon, J. A. 1979. North America's deserts: Their floral and faunal components. In: D. W. Goodall et al., eds. 1979. Arid Land Ecosystems: Structure, Functioning, and Management. Vol. 1. Cambridge and New York. Pp. 21–82.

MacMahon, J. A. 1988. Warm deserts. In: M. G. Barbour and W. D. Billings, eds. 1988. North American Terrestrial Vegetation. New York. Pp. 231–264.

MacMahon, J. A. and F. H. Wagner. 1985. The Mojave, Sonoran, and Chihuahuan deserts of North America. In: M. Evenari et al., eds. 1985. Hot Deserts and Arid Shrublands. Vol. 1. Amsterdam. Pp. 105–202.

Macnair, M. R. and P. Christie. 1983. Reproductive isolation as a pleiotropic effect of copper tolerance in *Mimulus guttatus*. Heredity 50: 295–302.

Macoun, J. 1922. Autobiography of John Macoun, M.A., Canadian Explorer and Naturalist.... With Introduction by Ernest Thompson Seton. [Ottawa.]

Madole, R. F., W. C. Bradley, D. S. Loewenherz, D. F. Ritter, N. W. Rutter, and C. E. Thorn. 1987. Rocky Mountains. In: W. L. Graf, ed. 1987. Geomorphic Systems of North America. Boulder. Pp. 211–257.

Madroño = Madroño; Journal of the California Botanical Society [from vol. 3: a West American Journal of Botany].

Mahall, B. E. and R. B. Park. 1976. The ecotone between *Spartina foliosa* Trin. and *Salicornia virginica* L. in salt marshes of northern San Francisco Bay. J. Ecol. 64: 421–433, 793–809, 811–819.

Mahaney, W. C., ed. 1976. Quaternary Stratigraphy of North America. Stroudsburg, Pa.

Malloch, D. and B. Malloch. 1981. The mycorrhizal status of boreal plants: Species from northern Ontario. Canad. J. Bot. 59: 2167.

Manchester, S. R. 1987. The fossil history of the Juglandaceae. Monogr. Syst. Bot. Missouri Bot. Gard. 21: 1–137.

Manchester, S. R. and P. R. Crane. 1983. Attached leaves, inflorescences, and fruits of *Fagopsis*, an extinct genus of fagaceous affinity from the Oligocene Florissant flora of Colorado, U.S.A. Amer. J. Bot. 70: 1147–1164.

Manton, I. 1950. Problems of Cytology and Evolution in the Pteridophyta. London.

Manum, S. 1962. Studies in the Tertiary flora of Spitsbergen. Norsk Polarinst. Skr. 125: 7–124.

Marie-Victorin, Frère. 1935. Flore Laurentienne. Montreal.

Mark, A. F. 1958. The ecology of southern Appalachian grass balds. Ecol. Monogr. 28: 293–336.

Mark, A. F. 1959. The flora of the grass balds and fields of the southern Appalachian Mountains. Castanea 24: 1–21.

Marshall, N. T. 1993. The Gardener's Guide to Plant Conservation. Washington.

Martin, P. G. and J. M. Dowd. 1986. Phylogenetic studies using protein sequences within the order Myrtales. Ann. Missouri Bot. Gard. 73: 442–448.

Martin, P. S. and B. E. Harrell. 1957. The Pleistocene history of temperate biotas in Mexico and eastern United States. Ecology 38: 468–480.

Martin, P. S. and R. G. Klein, eds. 1984. Quaternary Extinctions. Tucson.

Martin, W. E. 1959. The vegetation of Island Beach State Park. Ecol. Monogr. 21: 1–46.
Mather, A. S. 1990. Global Forest Resources. London.
Mathewes, R. W. 1989. The Queen Charlotte Islands refugium: A paleoecological perspective. In: R. J. Fulton, ed. 1989. Quaternary Geology of Canada and Greenland. Ottawa. Pp. 486–491.
Matthews, J. V. Jr. 1974. Quaternary environments at Cape Deceit (Seward Peninsula, Alaska): Evolution of a tundra ecosystem. Bull. Geol. Soc. Amer. 85: 1353–1384.
Matthews, J. V. Jr. 1976. Arctic steppe—An extinct biome. In: J. T. Andrews and R. G. Barry, eds. 1976. Abstracts of the Fourth Biennial Meeting of the American Quaternary Association. Tempe. Pp. 73–79.
Matthews, J. V. Jr. 1982. East Beringia during late Wisconsin time: A review of the biotic evidence. In: D. M. Hopkins et al., eds. 1982. Paleoecology of Beringia. New York. Pp. 127–150.
Matthews, J. V. Jr., T. W. Anderson, M. Boyko-Diakonow, R. W. Mathewes, J. H. McAndrews, R. J. Mott, P. J. H. Richard, J. C. Ritchie, and C. E. Schweger. 1989. Quaternary environments in Canada as documented by paleobotanical case histories. In: R. J. Fulton, ed. 1989. Quaternary Geology of Canada and Greenland. Ottawa. Pp. 481–539.
Maurer, A. L. 1985. Laws for the Protection of Threatened and Endangered Species in Canada. Burnaby.
Mayer-Oakes, W. J., ed. 1967. Life, Land, Water. Winnipeg. [Proceedings of the 1966 Conference on Environmental Studies of the Glacial Lake Agassiz Region.]
McAndrews, J. H. 1988. Human disturbance of North American forests and grasslands: The fossil pollen record. In: B. Huntley and T. Webb III, eds. 1988. Vegetation History. Dordrecht. Pp. 673–697.
McCaffrey, C. A. and R. D. Dueser. 1990. Plant associations of the Virginia barrier islands. Virginia J. Sci. 41: 282–299.
McCartan, L., B. H. Tiffney, J. A. Wolfe, T. A. Ager, S. L. Wing, L. A. Sirkin, L. W. Ward, and J. Brooks. 1990. Late Tertiary floral assemblage from upland gravel deposits of the southern Maryland coastal plain. Geology 18: 311–314.
McClintock, B. 1932. A correlation of ring-shaped chromosomes with variegation in *Zea mays*. Proc. Natl. Acad. Sci. U.S.A. 18: 677–681.
McCormick, J. F. and R. B. Platt. 1980. Recovery of an Appalachian forest following the chestnut blight, or Catherine Keever—You were right! Amer. Midl. Naturalist 104: 264–273.
McKelvey, S. D. 1955. Botanical Exploration of the Trans-Mississippi West, 1790–1850. Jamaica Plain.
McKenna, M. C. 1983. Holarctic landmass rearrangement, cosmic events, and Cenozoic terrestrial organisms. Ann. Missouri Bot. Gard. 70: 459–489.
McMahan, L. R. 1991. Propagation and reintroduction of imperiled plants, and the role of botanical gardens and arboreta. Endang. Spec. Update 8: 4–7.
McMillan, C. 1959. The role of ecotypic variation in the distribution of the central grassland of North America. Ecol. Monogr. 29: 285–308.
McMinn, H. 1935. An Illustrated Manual of Pacific Coast Trees. Berkeley and Los Angeles.
McMinn, H. 1944. The importance of field hybrids in determining the species in the genus *Ceanothus*. Proc. Calif. Acad. Sci. 25: 323–356.
McMinn, H. 1951. Studies in the genus *Diplacus*. Madroño 11: 33–128.
McMinn, H. 1951b. An Illustrated Manual of California Shrubs. Berkeley and Los Angeles.
Means, J. E., ed. 1982. Forest Succession and Stand Development Research in the Northwest. Corvallis.
Meddel. Grønland = Meddelelser om Grønland, af Kommissionen for Ledelsen af de Geologiske og Geografiske Undersølgeser i Grønland.
Mehringer, P. J. Jr. 1985. Late-Quaternary pollen records from the interior Pacific Northwest and northern Great Basin of the United States. In: V. M. Bryant Jr. and R. G. Holloway, eds. 1985. Pollen Records of Late-Quaternary North American Sediments. Dallas. Pp. 167–189.
Melchior, H., ed. 1964. A. Engler's Syllabus der Pflanzenfamilien..., ed. 12. 2 vols. Berlin.
Mem. Geol. Soc. Amer. = Memoirs of the Geological Society of America.
Mem. New York Bot. Gard. = Memoirs of the New York Botanical Garden.
Merriam, C. H. 1898. Life-zones and Crop-zones of the United States. Washington. [U.S.D.A. Div. Biol. Surv., Bull. 10.]
Mickel, J. T. 1979. How to Know the Ferns and Fern Allies. Dubuque.
Midcontinental J. Archaeol. = Midcontinental Journal of Archaeology.
Milankovič, M. 1941. Canon of Insolation and the Ice-Age Problem (Kanon der Erdbestrahlung und seine Anwendung auf das Eiszeitenproblem). Belgrade. [Roy. Serb. Acad., Special Publ. 132.]
Miller, C. N. Jr., J. A. Wolfe, D. L. Dilcher, D. R. Crabtree, S. L. Wing, J. R. Habeck, E. B. Leopold, and M. F. Denton. 1987. Contributions to a symposium on the evolution of the modern flora of the northern Rocky Mountains. Ann. Missouri Bot. Gard. 74(4): 681–867.
Miller, J. T. 1985. Living in the Environment, ed. 4. Belmont, Calif.
Millspaugh, C. F. 1892. Medicinal Plants. 2 vols. Philadelphia.
Minnesota Agric. Exp. Sta. Techn. Bull. = Minnesota Agricultural Experiment Station. Technical Bulletin.
Minnich, R. A. 1987. The distribution of forest trees in northern Baja California, Mexico. Madroño 34: 98–127.
Minore, D. 1979. Comparative Autecological Characteristics of North-western Tree Species—A Literature Review. Portland. [U.S.D.A. Forest Serv., Gen. Techn. Rep. PNW-87.]
Miranda, F. and A. J. Sharp. 1950. Characteristics of the veg-

etation in certain temperate regions of eastern Mexico. Ecology 31: 313–333.

Mishler, B. D. 1985. The morphological, developmental and phylogenetic basis of species concepts in bryophytes. Bryologist 88: 207–214.

Mishler, B. D. and M. J. Donoghue. 1982. Species concepts: A case for pluralism. Syst. Zool. 31: 491–503.

Mitchell, J. E., N. E. West, and R. W. Miller. 1966. Soil physical properties in relation to plant community patterns in the shadscale zone of north-western Utah. Ecology 47: 627–630.

Mitchell, R. S. 1986. A checklist of New York State plants. Bull. New York State Mus. Sci. Serv. 458.

Moerman, D. E. 1982. Medicinal Plants of America. 2 vols. Ann Arbor.

Monogr. Syst. Bot. Missouri Bot. Gard. = Monographs in Systematic Botany from the Missouri Botanical Garden.

Mooney, H. A. 1988. Southern coastal scrub. In: M. G. Barbour and J. Major, eds. 1988. Terrestrial Vegetation of California, ed. 2. Sacramento. Pp. 471–490.

Mooney, H. A. and J. A. Drake, eds. 1986. Ecology of Biological Invasions of North America and Hawaii. New York.

Mooney, H. A. and P. C. Miller. 1985. Chaparral. In: B. F. Chabot and H. A. Mooney, eds. 1985. Physiological Ecology of North American Plant Communities. New York. Pp. 213–231.

Mooney, H. A. and D. J. Parsons. 1973. Structure and function of the California chaparral. In: F. di Castri and H. A. Mooney, eds. 1973. Mediterranean Type Ecosystems: Origin and Structure. Berlin and New York. Pp. 83–112.

Monk, C. D. 1968. Successional and environmental relationships of the forest vegetation of north central Florida. Amer. Midl. Naturalist 74: 127–140.

Moran, J. M., M. D. Morgan, and J. H. Wiersma. 1986. Introduction to Environmental Science, ed. 2. New York.

Moreno-Casasola, P. 1988. Patterns of plant species distribution on coastal dunes along the Gulf of Mexico. J. Biogeogr. 15: 787–806.

Morin, N. R., R. D. Whetstone, D. Wilken, and K. L. Tomlinson, eds. 1989. Floristics for the 21st century. Proceedings of the workshop sponsored by the American Society of Plant Taxonomists, 4–7 May, 1988, Alexandria, Virginia. Monogr. Syst. Bot. Missouri Bot. Gard. 28.

Morison, R. 1680–1699. Plantarum Historiae Universalis Oxoniensis.... 2 vols. Oxford. [Vols. 2 and 3; vol. 1 never published.]

Morrison, R. G. and G. A. Yarranton. 1973. Diversity, richness, and evenness during a primary sand dune succession at Grand Bend, Ontario. Canad. J. Bot. 51: 2401–2411.

Morrison, R. G. and G. A. Yarranton. 1974. Vegetational heterogeneity during a primary sand dune succession. Canad. J. Bot. 52: 397–410.

Morse, L. E. 1981. The Nature Conservancy and rare plant conservation in the United States. In: H. Synge, ed. 1981. The Biological Aspects of Rare Plant Conservation. Chichester. Pp. 453–457.

Morse, L. E. and M. S. Henifin, eds. 1981. Rare Plant Conservation: Geographical Data Organization. Bronx.

Morzenti, V. M. 1962. A first report of pseudomeiotic sporogenesis, a new type of spore reproduction by which sterile ferns produce gametophytes. Amer. Fern J. 52: 69–78.

Moss, E. H. and J. G. Packer. 1983. Flora of Alberta, ed. 2, Toronto.

Muenscher, W. C. 1941. The Flora of Whatcom County, State of Washington. Ithaca, N.Y.

Muhlenbach, V. 1979. Contributions to the synanthropic (adventive) flora of the railroads in St. Louis, Missouri, U.S.A. Ann. Missouri Bot. Gard. 66: 1–108.

Muir, P. S. and J. E. Lotan. 1985. Disturbance history and serotiny of *Pinus contorta* in western Montana. Ecology 66: 1658–1668.

Müller, M. J. 1982. Selected Climatic Data for a Global Set of Standard Stations for Vegetation Science. The Hague, Boston, and Hingham, Mass. [Tasks for Vegetation Science. Vol. 5.]

Mulroy, T. W. and P. W. Rundel. 1977. Annual plants: Adaptations to desert environments. BioScience 27: 109–114.

Murray, G. E. 1961. Geology of the Atlantic and Gulf Coastal Province of North America. New York.

Myers, N. 1980. Conservation of Moist Tropical Forests. Washington.

Myers, N. 1987. The extinction spasm impending: Synergisms at work. Conservation Biol. 1: 14–21.

Myers, R. L. 1985. Fire and the dynamic relationship between Florida sandhill and sand pine scrub vegetation. Bull. Torrey Bot. Club 112: 241–252.

Myers, R. L. and J. J. Ewel, eds. 1990. Ecosystems of Florida. Gainesville.

Nabhan, G. P. 1985. Native crop diversity in Aridoamerica: Conservation of regional gene pools. Econ. Bot. 39: 387–399.

Nat. Areas J. = Natural Areas Journal; Quarterly Publication of the Natural Areas Association.

Nat. Conservancy News (Arlington) = Nature Conservancy News.

National Research Council [U.S.A.]. 1989. Biologic Markers of Air-pollution Stress and Damage in Forests. Committee on Biologic Markers of Air-pollution Damage in Trees, Board on Environmental Studies and Toxicology, Commission on Life Sciences.... Washington.

National Wetlands Working Group [Canada]. 1988. Wetlands of Canada. Montreal. [Ecol. Land Classific. Ser. 24.]

National Wildflower Research Center. 1992. Wildflower Handbook, ed. 2. Stillwater, Minn.

Naturaliste Canad. = Naturaliste Canadien. Bulletin de Recherches, Observations et Découvertes se Rapportant à l'Histoire Naturelle du Canada.

Nature = Nature; a Weekly Illustrated Journal of Science.
Nebraska Agric. Exp. Sta. Res. Bull. = Nebraska Agricultural Experiment Station Research Bulletin.
Neuenschwander, L., T. H. Thorsted Jr., and R. J. Vogl. 1979. The salt marsh and transitional vegetation of Bahia de San Quintin. Bull. S. Calif. Acad. Sci. 78: 163–182.
New Phytol. = New Phytologist; a British Botanical Journal.
New Sci. = New Scientist.
Niering, W. A., R. H. Whittaker, and C. H. Lowe. 1963. The saguaro: A population in relation to environment. Science 142: 15–23.
Nixon, K. C. and K. P. Steele. 1981. A new species of *Quercus* (Fagaceae) from southern California. Madroño 28: 210–219.
Nobel, P. S. 1985. Desert succulents. In: B. F. Chabot and H. A. Mooney, eds. 1985. Physiological Ecology of North American Plant Communities. New York. Pp. 181–197.
Nobs, M. A. 1963. Experimental studies on species relationships in *Ceanothus*. Publ. Carnegie Inst. Wash. 623.
Nordic J. Bot. = Nordic Journal of Botany.
Norris, G. 1982. Spore-pollen evidence for early Oligocene high-latitude cool climatic episode in northern Canada. Nature 297: 387–389.
Norsk Polarinst. Skr. = Norsk Polarinstitutt Skrifter.
Norton, B. G., ed. 1986. The Preservation of Species. The Value of Biological Diversity. Princeton.
Norton, B. G. 1986b. Epilogue. In: B. G. Norton, ed. 1986. The Preservation of Species. The Value of Biological Diversity. Princeton. Pp. 268–283.
Norton, B. G. 1987. Why Preserve Natural Variety? Princeton.
Notes Roy. Bot. Gard. Edinburgh = Notes from the Royal Botanic Garden, Edinburgh.

Occas. Pap. Calif. Acad. Sci. = Occasional Papers of the California Academy of Sciences.
Oechel, W. C. and W. T. Lawrence. 1985. Taiga. In: B. F. Chabot and H. A. Mooney, eds. 1985. Physiological Ecology of North American Plant Communities. New York. Pp. 66–94.
Oettinger, F. W. 1975. The Vascular Plants of the High Lake Basins in the Vicinity of English Peak, Siskiyou County, California. M.A. thesis. Claremont Graduate School.
Oldfield, M. L. 1984. The Value of Conserving Genetic Resources. Washington.
Olson, J. S. 1958. Rates of succession and soil changes on southern Lake Michigan dunes. Bot. Gaz. 119: 125–170.
Olson, J. S. 1971. Primary productivity: Temperate forests, especially American deciduous types. In: P. Duvigneaud, ed. 1971. Productivity of Forest Ecosystems. Productivite des Ecosystemes Forestiers. Proceedings of the Brussels Symposium Organized by Unesco and the International Biological Programme (27–31 October 1969). Paris. Pp. 235–258.
Olwell, P., M. Singleton, and M. O'Neal. 1992. 1992 Plant Conservation Directory. St. Louis.
Oosting, H. J. 1942. An ecological analysis of the plant communities of piedmont North Carolina. Amer. Midl. Naturalist 28: 1–126.
Oosting, H. J. and W. D. Billings. 1951. A comparison of virgin spruce-fir forest in the northern and southern Appalachian system. Ecology 32: 84–103.
Otte, D. K. S. and J. A. Endler, eds. 1989. Speciation and Its Consequences. Sunderland, Mass.
Ownbey, G. B. and T. Morley. 1991. Vascular Plants of Minnesota: A Checklist and Atlas. Minneapolis.

Packer, J. G. 1969. Polyploidy in the Canadian Arctic Archipelago. Arctic Alpine Res. 1: 15–28.
Page, C. N. 1988. New and maintained genera in the conifer families Podocarpaceae and Pinaceae. Notes Roy. Bot. Gard. Edinburgh 45: 377–395.
Pagney, P. 1973. La Climatologie. Paris.
Palaeogeogr. Palaeoclimatol. Palaeoecol. = Palaeogeography, Palaeoclimatology, Palaeoecology.
Paris, C. A., F. S. Wagner, and W. H. Wagner Jr. 1989. Cryptic species, species delimitation, and taxonomic practice in the homosporous ferns. Amer. Fern J. 70: 46–54.
Parrish, J. T. 1987. Global palaeogeography and palaeoclimate of the late Cretaceous and early Tertiary. In: E. M. Friis et al., eds. 1987. The Origins of Angiosperms and Their Biological Consequences. Cambridge and New York. Pp. 51–74.
Pase, C. P. and D. E. Brown. 1982. Rocky Mountain (Petran) and Madrean montane conifer forest. Desert Pl. 4: 43–48.
Pase, C. P. and D. E. Brown. 1982b. Rocky Mountain (Petran) subalpine conifer forest. Desert Pl. 4: 37–39.
Passini, M.-F., J. Delgado, and M. Salazar. 1989. L'ecosystéme forestier de Basse-Californie: Composition floristique, variables écologiques principales, dynamique. Acta Ecol. 10: 275–293.
Payette, S. and R. Gagnon. 1985. Late Holocene deforestation and tree regeneration in the forest-tundra of Quebec. Nature 313: 570–572.
Payette, S., C. Morneau, L. Sirois, and M. Desponts. 1989. Recent fire history of the northern Quebec biomes. Ecology 70: 656–673.
Peck, M. E. 1961. A Manual of the Higher Plants of Oregon. Corvallis.
Peet, R. K. 1988. Forests of the Rocky Mountains. In: M. G. Barbour and W. D. Billings, eds. 1988. North American Terrestrial Vegetation. New York. Pp. 63–101.
Perlin, J. 1989. A Forest Journey: The Role of Wood in the Development of Civilization. New York.
Peterson, E. B. 1969. Radiosonde data for characterization of a mountain environment in British Columbia. Ecology 50: 200–205.
Péwé, T. L. 1983. The periglacial environment in North America during Wisconsin time. In: S. C. Porter, ed. 1983. Late-Quaternary Environments of the United States. Vol. 1. The Late Pleistocene. Minneapolis. Pp. 157–189.
Péwé, T. L. 1983b. Alpine permafrost in the contiguous U.S., a review. Arctic Alpine Res. 15(2): 145–146.

Phytocoenologia = Phytocoenologia; Journal of the International Society for Plant Geography and Ecology.

Phytologia = Phytologia; Designed to Expedite Botanical Publication.

Phytomorphology = Phytomorphology; an International Journal of Plant Morphology.

Pichi-Sermolli, R. E. G. 1977. Tentamen pteridophytorum in taxonomicum ordinem redigendi. Webbia 31: 313–512.

Piel, K. M. 1971. Palynology of Oligocene sediments from central British Columbia. Canad. J. Bot. 49: 1885–1920.

Pijl, L. van der and C. Dodson. 1966. Orchid Flowers: Their Pollination and Evolution. Coral Gables.

Pilger, R. K. F. 1903. Taxaceae. In: H. G. A. Engler, ed. 1900–1953. Das Pflanzenreich.... 107 vols. Berlin. Vol. 18[IV,5], pp. 1–124.

Pilger, R. K. F. 1926. Coniferae. In: H. G. A. Engler et al., eds. 1924+. Die natürlichen Pflanzenfamilien...., ed. 2. 26+ vols. Leipzig and Berlin. Vol. 13, pp. 121–407.

Piper, C. V. 1906. Flora of the State of Washington. Contr. U.S. Natl. Herb. 11: 1–637.

Piper, C. V. and R. K. Beattie. 1901. The Flora of the Palouse Region. Pullman.

Pirkle, E. C. and W. H. Yoho. 1982. Natural Landscapes of the United States, ed. 3. Dubuque.

Pisias, N. G. and T. C. Moore Jr. 1981. The evolution of Pleistocene climate: A time series approach. Earth Planet. Sci. Lett. 52: 450–458.

Pittock, A. B., L. A. Frakes, D. Jenssen, J. A. Peterson, and J. W. Zillman, eds. 1978. Climatic Change and Variability, a Southern Perspective. Cambridge.

Pl. Syst. Evol. = Plant Systematics and Evolution.

Pl. Syst. Evol., Suppl. = Plant Systematics and Evolution. Supplementum.

Pl. Today = Plants Today.

Plukenet, L. 1691–1705. Phytographia sive Illustriorum & Miniis Cognitarum Icones Tabulis Aeneis Summâ Diligentiâ Elaboratae.... 7 parts. London. [Pars Prior, Pars Altera, 1691; Pars Tertia, 1692; [Pars Quarta], 1694; Almagestum, 1696; Almagesti.... Mantissa, 1700; Almatheum, 1705.]

Polunin, N. 1940. Botany of the Canadian eastern Arctic. Part 1. Pteridophyta and Spermatophyta. Bull. Natl. Mus. Canada 92: 1–408.

Polunin, N. 1959. Circumpolar Arctic Flora. Oxford.

Pomerol, C. and I. Premoli-Silva, eds. 1986. Terminal Eocene Events. Amsterdam.

Porsild, A. E. and W. J. Cody. 1980. Vascular Plants of Continental Northwest Territories, Canada. Ottawa.

Porter, S. C., ed. 1983. Late-Quaternary Environments of the United States. Vol. 1. The Late Pleistocene. Minneapolis.

Porter, S. C., K. L. Pierce, and T. D. Hamilton. 1983. Late Wisconsin mountain glaciation in the western United States. In: S. C. Porter, ed. 1983. Late-Quaternary Environments of the United States. Vol. 1. The Late Pleistocene. Minneapolis. Pp. 71–111.

Prance, G. T. 1977. Introduction. In: G. T. Prance and T. S. Elias, eds. 1977. Extinction Is Forever. Bronx. Pp. 3–4.

Prance, G. T. and T. S. Elias, eds. 1977. Extinction Is Forever. Bronx.

Prescott-Allen, C. and R. Prescott-Allen. 1986. The First Resource: Wild Species in the North American Economy. New Haven.

Price, R. A. 1989. The genera of Pinaceae in the southeastern United States. J. Arnold Arbor. 70: 247–305.

Price, R. A. 1990. The genera of Taxaceae in the southeastern United States. J. Arnold Arbor. 71: 69–91.

Price, R. A. and J. M. Lowenstein. 1989. An immunological comparison of Sciadopityaceae, Taxodiaceae, and Cupressaceae. Syst. Bot. 14: 141–149.

Price, R. A., J. Olsen-Stojkovich, and J. M. Lowenstein. 1987. Relationships among genera of Pinaceae: An immunological comparison. Syst. Bot. 12: 91–97.

Proc. Amer. Philos. Soc. = Proceedings of the American Philosophical Society.

Proc. Assoc. Amer. Geogr. = Proceedings, Association of American Geographers.

Proc. Autumn School Bot., Mahabaleshwar = Proceedings of the Autumn School in Botany, Mahabaleshwar.

Proc. Calif. Acad. Sci. = Proceedings of the California Academy of Science.

Proc. Natl. Acad. Sci. U.S.A. = Proceedings of the National Academy of Sciences of the United States of America.

Proc. Roy. Soc. Edinburgh, B = Proceedings of the Royal Society of Edinburgh. Series B, Biology [later: Biological Sciences].

Profess. Pap. U.S. Geol. Surv. = Professional Papers. United States Geological Survey.

Provancheria = Provancheria; Mémoires de l'Herbier Louis-Marie, Faculté d'Agriculture de l'Université Laval.

Publ. Carnegie Inst. Wash. = Publications of the Carnegie Institution of Washington.

Pursh, F. 1814. Flora Americae Septentrionalis; or, a Systematic Arrangement and Description of the Plants of North America. 2 vols. London.

Pyne, S. J. 1982. Fire in America: A Cultural History of Wildland and Rural Fire. Princeton.

Quart. Rev. Biol. = Quarterly Review of Biology.

Quarterman, E. and C. Keever. 1962. Southern mixed hardwood forest: Climax in the southeastern coastal plain, U.S.A. Ecol. Monogr. 32: 167–185.

Quatern. Res. = Quaternary Research; Interdisciplinary Journal.

Quatern. Sci. Rev. = Quaternary Science Reviews; International Review and Research Journal.

Quinn, J. A. and R. T. Ward. 1969. Ecological differentiation in sand dropseed *(Sporobolus cryptandrus)*. Ecol. Monogr. 39: 61–78.

Radford, A. E., J. W. Hardin, J. R. Massey, E. L. Core, and L. S. Radford, eds. 1980+. Vascular Flora of the Southeastern United States. 2+ vols. Chapel Hill.

Radosevich, S. R. and J. S. Holt. 1984. Weed Ecology: Implications for Vegetation Management. New York.

Ramanujam, C. G. K. and W. N. Stewart. 1969. Fossil woods of Taxodiaceae from the Edmonton Formation (upper Cretaceous) of Alberta. Canad. J. Bot. 47: 115–124.

Rao, V. S. 1970. Floral anatomy in some monocotyledonous taxa. Proc. Autumn School Bot., Mahabaleshwar 1966: 295–302.

Raven, P. H. 1986. Modern aspects of the biological species in plants. In: K. Iwatsuki et al., eds. 1986. Modern Aspects of Species. Tokyo. Pp. 11–29.

Raven, P. H. 1987. The scope of the plant conservation problem world-wide. In: D. Bramwell et al., eds. 1987. Botanic Gardens and World Conservation Strategy. London. Pp. 19–29.

Raven, P. H. and D. I. Axelrod. 1974. Angiosperm phylogeny and past continental movements. Ann. Missouri Bot. Gard. 61: 539–673.

Raven, P. H. and D. I. Axelrod. 1978. Origin and relationships of the California flora. Univ. Calif. Publ. Bot. 72: 1–134.

Ray, J. 1686–1704. Historia Plantarum.... 3 vols. London. [Vols. 1 and 2 paged consecutively.]

Reeder, J. R. 1957. The embryo in grass systematics. Amer. J. Bot. 44: 756–769.

Reichle, D. E., ed. 1981. Dynamic Properties of Forest Ecosystems. Cambridge.

Reimold, R. J. 1977. Mangals and salt marshes of eastern United States. In: V. J. Chapman, ed. 1977. Wet Coastal Ecosystems. Amsterdam. Pp. 157–166.

Reimold, R. J. and W. H. Queen, eds. 1974. Ecology of Halophytes. New York.

Rejmanek, M., C. D. Thomsen, and I. D. Peters. 1991. Invasive vascular plants of California. In: R. H. Groves and F. DiCastri, eds. 1991. Biogeography of Mediterranean Invasions. Cambridge. Pp. 81–101.

Rep. Activities Geol. Surv. Canada = Report of Activities, Geological Survey of Canada.

Rep. (Annual) New Jersey State Mus. = Report (Annual), New Jersey State Museum.

Rep. Invest. Bur. Mines, Washington, DC = Report of Investigations, Bureau of Mines.

Rev. Aquatic Sci. = Reviews in Aquatic Science.

Rev. Geophys. Space Phys. = Reviews of Geophysics and Space Physics.

Rev. Palaeobot. Palynol. = Review of Palaeobotany and Palynology; an International Journal.

Reveal, J. L. 1981. The concept of rarity and population threats in plant communities. In: L. E. Morse and M. S. Henifin, eds. 1981. Rare Plant Conservation: Geographical Data Organization. Bronx. Pp. 41–47.

Reveal, J. L. 1991. Botanical Explorations in the American West—1889–1989: An essay on the last century of a floristic frontier. Ann. Missouri Bot. Gard. 78: 65–80.

Reveal, J. L. 1991b. Two previously unnoticed sources of generic names published by John Hill in 1753 and 1754–1755. Bull. Mus. Natl. Hist. Nat., B, Adansonia 13: 197–239.

Reveal, J. L. 1992. Gentle Conquest: The Botanical Discovery of North America with Illustrations from the Library of Congress. Washington.

Reveal, J. L. 1992b. Validation of subclass and superordinal names in Magnoliophyta Novon 2: 235–237.

Reveal, J. L. 1992c. Validation of ordinal names of extant vascular plants. Novon 2: 238–240.

Reveal, J. L. 1993. A splitter's guide to the higher taxa of the flowering plants (Magnoliophyta) generally arranged to follow the sequence proposed by Thorne (1992) with certain modificaitons. Phytologia 74: 203–263.

Reveal, J. L. 1993b. New ordinal names for extant vascular plants. Phytologia 74: 173–177.

Reveal, J. L. 1993c. New subclass and superordinal names for extant vascular plants. Phytologia 74: 178–179.

Revista Chilena Hist. Nat. = Revista Chilena de Historia Natural.

Rhodora = Rhodora; Journal of the New England Botanical Club.

Richard, P. J. H. 1977. Histoire Post-Wisconsinienne de la Vegetation du Quebec Meridional par l'Analyse Pollinique. 2 vols. Quebec.

Richardson, C. J., ed. 1981. Pocosin Wetlands. Stroudsburg, Pa.

Rickard, W. H. and J. C. Beatley. 1965. Canopy-coverate of the desert shrub vegetation mosaic of the Nevada Test Site. Ecology 46: 524–529.

Riley, R. and V. Chapman. 1958. Genetic control of the cytologically diploid behaviour of hexaploid wheat. Nature 182: 713–715.

Risser, P. G. 1985. Grasslands. In: B. F. Chabot and H. A. Mooney, eds. 1985. Physiological Ecology of North American Plant Communities. New York. Pp. 232–256.

Risser, P. G., E. C. Birney, H. D. Blockner, S. W. May, W. J. Parton, and J. A. Weins. 1981. The True Prairie Ecosystem. Stroudsburg, Pa.

Ritchie, J. C. 1962. A Geobotanical Survey of Northern Manitoba. Montreal. [Arctic Inst. N. Amer., Techn. Pap. 9.]

Ritchie, J. C. 1977. The modern and late Quaternary vegetation of the Campbell-Dolomite uplands near Inuvik, N.W.T., Canada. Ecol. Monogr. 47: 401–423.

Ritchie, J. C. 1980. Towards a late-Quaternary palaeoecology of the ice-free corridor. Canad. J. Anthropol. 1: 15–28.

Ritchie, J. C. 1984. Past and Present Vegetation of the Far Northwest of Canada. Toronto and Buffalo.

Ritchie, J. C. 1987. Postglacial Vegetation of Canada. Cambridge.

Ritchie, J. C. and L. C. Cwynar. 1982. The late Quaternary vegetation of the north Yukon. In: D. M. Hopkins et al., eds. 1982. Paleoecology of Beringia. New York. Pp. 113–126.

Ritchie, J. C., L. C. Cwynar, and R. W. Spear. 1983. Evidence from northwest Canada for an early Holocene Milankovitch thermal maximum. Nature 305: 126–128.

Ritchie, J. C. and G. M. MacDonald. 1986. The patterns of post-glacial spread of white spruce. J. Biogeogr. 13: 527–540.

Ritchie, J. C. and G. A. Yarranton. 1978. Patterns of change in the late-Quaternary vegetation of the western interior of Canada. Canad. J. Bot. 56: 2177–2183.

Roberts, H. A., ed. 1982. Weed Control Handbook: Principles, ed. 7. Oxford.

Roberts, L. 1988. Extinction imminent for native plants. Science 242: 1508.

Roberts, M. R. and N. L. Christensen. 1988. Vegetation variation among mesic successional forest stands in northern lower Michigan. Canad. J. Bot. 66: 1080–1090.

Robichaud, B. and M. F. Buell. 1973. Vegetation of New Jersey. New Brunswick.

Robinson, B. L. and M. L. Fernald. 1908. Gray's New Manual of Botany: A Handbook of the Flowering Plants and Ferns of the Central and Northeastern United States and Adjacent Canada, ed. 7. New York, Cincinnati, and Chicago.

Rodgers, A. D. III. 1942. John Torrey: A Story of North American Botany. Princeton and London.

Rogers, R. S. 1980. Hemlock stands from Wisconsin to Nova Scotia: Transitions in understory composition. Ecology 61: 178–193.

Rogers, R. S. 1981. Mature mesophytic hardwood forest: Community transitions, by layer, from east-central Minnesota to southeastern Michigan. Ecology 62: 1634–1647.

Rohr, B. R., J. I. Lawyer, L. E. Morse, and S. G. Shetler. 1977. The Flora North America Reports—A bibliography and index. Brittonia 29: 419–432.

Romans, R. C., ed. 1977. Geobotany. New York.

Roose, M. L. and L. D. Gottlieb. 1976. Genetic and biochemical consequences of polyploidy in *Tragopogon*. Evolution 30: 818–830.

Rouleau, E. 1981. Guide to the Generic Names Appearing in the Index Kewensis and Its Fifteen Supplements. Lac de Brome.

Rouse, G. E. and S. K. Srivastava. 1972. Palynological zonation of Cretaceous and early Tertiary rocks of the Bonnet Plume Formation, northeastern Yukon, Canada. Canad. J. Earth Sci. 9: 1163–1179.

Rowe, J. S. 1977. Forest Regions of Canada. Ottawa. [Canad. Forest. Serv. Publ. 1300.]

Royall, P. D., P. A. Delcourt, and H. R. Delcourt. 1991. Late Quaternary paleoecology and paleoenvironments of the central Mississippi Alluvial Valley. Bull. Geol. Soc. Amer. 103: 157–170.

Rubec, C. D. A. and F. C. Follet, eds. 1980. Workshop on Canadian Wetlands. Toronto. [Land Directorate, Ser. 12.]

Ruddiman, W. F. and J. E. Kutzbach. 1991. Plateau uplift and climatic change. Sci. Amer. 264(3): 66–75.

Ruddiman, W. F. and A. McIntyre. 1981. The mode and mechanism of the last deglaciation: Oceanic evidence. Quatern. Res. 16: 125–134.

Ruddiman, W. F. and H. E. Wright Jr., eds. 1987. North America and Adjacent Oceans during the Last Deglaciation. Boulder. [Geology of North America. Vol. K-3.]

Rundel, P. W. 1971. Community structure and stability in the giant sequoia groves of the Sierra Nevada, California. Amer. Midl. Naturalist 85: 478–492.

Rundel, P. W. 1972. Habitat restriction in giant sequoia: The environmental control of grove boundaries. Amer. Midl. Naturalist 87: 81–99.

Rundel, P. W., D. J. Parsons, and D. T. Gordon. 1988. Montane and subalpine vegetation of the Sierra Nevada and Cascade ranges. In: M. G. Barbour and J. Major, eds. 1988. Terrestrial Vegetation of California, ed. 2. Sacramento. Pp. 559–599.

Russell, E. W. B. 1983. Indian-set fires in the forests of the northeastern United States. Ecology 64: 78–88.

Russell, E. W. B. 1987. Pre-blight distribution of *Castanea dentata* (Marsh.) Borkh. Bull. Torrey Bot. Club 114: 183–190.

Rutter, N. W. 1984. Pleistocene history of the western Canadian ice-free corridor. In: R. J. Fulton, ed. 1984. Quaternary Stratigraphy of Canada: A Canadian Contribution to IGCP Project 24. Ottawa. Pp. 49–56.

Rydberg, P. A. 1917. Flora of the Rocky Mountains and Adjacent Plains. New York.

Rzedowski, J. 1978. Vegetación de México. México, D.F.

Sauer, J. D. 1967. Geographic Reconnaissance of Seashore Vegetation along the Mexican Gulf Coast. Baton Rouge. [Coastal Stud. Ser. 21.]

Saure, C. O. 1958. Grassland, climax, fire and man. J. Range Managem. 3: 16–22.

Savage, H. Jr. and E. J. Savage. 1986. André and François André Michaux. Charlottesville.

Savile, D. B. O. 1979. Fungi as aids in higher plant classification. Bot. Rev. (Lancaster) 45: 377–503.

Sawyer, J. O. and D. A. Thornburgh. 1988. Montane and subalpine vegetation of the Klamath Mountains. In: M. G. Barbour and J. Major, eds. 1988. Terrestrial Vegetation of California, ed. 2. Sacramento. Pp. 699–732.

Sawyer, J. O., D. A. Thornburgh, and J. R. Griffin. 1988. Mixed evergreen forest. In: M. G. Barbour and J. Major, eds. 1988. Terrestrial Vegetation of California, ed. 2. Sacramento. Pp. 359–415.

Schultes, R. E. and A. Hofmann. 1980. The Botany and Chemistry of Hallucinogens, ed. 2. Springfield, Ill.

Schweger, C. E. 1982. Late Pleistocene vegetation of eastern Beringia: Pollen analysis of dated alluvium. In: D. M. Hopkins et al., eds. 1982. Paleoecology of Beringia. New York. Pp. 95–112.

Schweger, C. E. 1989. Paleoecology of the western Canadian ice-free corridor. In: R. J. Fulton, ed. 1989. Quaternary Geology of Canada and Greenland. Ottawa. Pp. 491–498.

Sci. Amer. = Scientific American.

Sci. Proc. Roy. Dublin Soc. = Scientific Proceedings of the Royal Dublin Society.

Science = Science; an Illustrated Journal [later: a Weekly Journal Devoted to the Advancement of Science]. [American Association for the Advancement of Science.]

Scoggan, H. J. 1978–1979. The Flora of Canada. 4 parts. Ottawa. [Natl. Mus. Nat. Sci. Publ. Bot. 7.]

Seeliger, U., ed. 1992. Coastal Plant Communities of Latin America. San Diego.

Seliskar, D. M. and J. L. Gallagher. 1983. The Ecology of Tidal Marshes of the Pacific Northwest Coast: A Community Profile. Washington.

Seneca, E. D. 1972. Germination and seedling response of Atlantic and Gulf coasts populations of *Uniola paniculata*. Amer. J. Bot. 59: 290–296.

Seymour, F. C. 1969. The Flora of New England: A Manual for the Identification of All Vascular Plants... Growing without Cultivation.... Rutland, Vt.

Sharitz, R. R. and J. W. Gibbons. 1982. The Ecology of Southeastern Shrub Bogs (Pocosins) and Carolina Bays: A Community Profile. Washington.

Sharp, A. J. 1972. The possible significance of some exotic distributions of plants occurring in Japan and/or North America. In: A. Graham, ed. 1972. Floristics and Paleofloristics of Asia and Eastern North America. Amsterdam and New York. Pp. 61–64.

Sharp, A. J. 1972b. Phytogeographical correlations between the bryophytes of eastern Asia and North America. J. Hattori Bot. Lab. 35: 263–268.

Shaver, G. R. and W. D. Billings. 1977. Effects of day length and temperature on root elongation in tundra graminoids. Oecologia 28: 57–65.

Shing, K. H. 1988. Preliminary study on phytogeographic comparison of pteridophytes between China and North America In: Shing K. H. and K. U. Kramer, eds. 1988. Proceedings of the International Symposium on Systematic Pteridology. Beijing. Pp. 203–213.

Shing, K. H. and K. U. Kramer. 1988. Proceedings of the International Symposium on Systematic Pteridology. Beijing.

Shmida, A. 1985. Biogeography of the desert flora. In: M. Evenari et al., eds. 1985. Hot Deserts and Arid Shrublands. Vol. 1. Amsterdam. Pp. 23–77.

Short, S. K., W. N. Mode, and P. T. Davis. 1985. The Holocene record from Baffin Island: Modern and fossil pollen studies. In: J. T. Andrews, ed. 1985. Quaternary Environments: Eastern Canadian Arctic, Baffin Bay, and Western Greenland. Boston. Pp. 608–642.

Shreve, F. 1942. The desert vegetation of North America. Bot. Rev. (Lancaster) 8: 195–246.

Shreve, F. 1951. Vegetation of the Sonoran Desert. Publ. Carnegie Inst. Wash. 591: 1–192.

Shreve, F. and I. L. Wiggins. 1964. Vegetation and Flora of the Sonoran Desert. 2 vols. Stanford.

Silander, J. A. 1979. Microevolution and clone structure in *Spartina patens*. Science 203: 658–660.

Silander, J. A. and J. Antonovics. 1979. The genetic basis of the ecological amplitude of *Spartina patens*. 1. Morphometric and physiological traits. Evolution 33: 1114–1127.

Simberloff, D. 1986. Are we on the verge of a mass extinction in tropical rain forests? In: D. K. Elliott, ed. 1986. Dynamics of Extinction. New York. Pp. 165–180.

Sims, P. L. 1988. Grasslands. In: M. G. Barbour and W. D. Billings, eds. 1988. North American Terrestrial Vegetation. New York. Pp. 265–286.

Sims, P. L., J. S. Singh, and W. K. Lauenroth. 1978. The structure and function of ten western North American grasslands. I. Abiotic and vegetational characteristics. J. Ecol. 66: 251–285.

Singh, J. S., W. K. Lauenroth, R. K. Heitschmidt, and J. L. Dodd. 1983. Structural and functional attributes of the vegetation of northern mixed prairie of North America. Bot. Rev. (Lancaster) 49: 117–149.

Slosson, M. 1902. The origin of *Asplenium ebenoides*. Bull. Torrey Bot. Club 29: 487–495.

Small, J. K. 1933. Manual of the Southeastern Flora, Being Descriptions of the Seed Plants Growing Naturally in Florida, Alabama, Mississippi, Eastern Louisiana, Tennessee, North Carolina, South Carolina and Georgia. New York.

Smith, A. G., A. M. Hurley, and J. C. Briden. 1981. Phanerozoic Paleocontinental World Maps. Cambridge and New York.

Smith, A. R. 1972. Comparison of fern and flowering plant distributions with some evolutionary interpretations for ferns. Biotropica 4: 4–9.

Smith, B. D. 1985. *Chenopodium berlandieri* ssp. *jonesianum*: Evidence for a Hopewellian domesticate from Asch Cave, Ohio. SouthE. Archaeol. 4: 107–133.

Smith, B. D. 1987. The independent domestication of indigenous seed-bearing plants in eastern North America. In: W. F. Keegan, ed. 1987. Emergent Horticultural Economics of the Eastern Woodlands. Carbondale. Pp. 3–47.

Smith, G. I. and F. A. Street-Perrott. 1983. Pluvial lakes of the western United States. In: S. C. Porter, ed. 1983. Late-Quaternary Environments of the United States. Vol. 1. The Late Pleistocene. Minneapolis. Pp. 190–212.

Smith, H. H. and K. Daly. 1959. Discrete populations derived by interspecific hybridization and selection in *Nicotiana*. Evolution 13: 476–487.

Smith, W. K. 1985. Western montane forests. In: B. F. Chabot and H. A. Mooney, eds. 1985. Physiological Ecology of North American Plant Communities. New York. Pp. 95–126.

Smith, W. K. 1988. Vascular plants. In: B. Coffin and L. Pfanmuller, eds. 1988. Minnesota's Endangered Flora and Fauna. Minneapolis. Pp. 33–217.

Smithsonian Contr. Bot. = Smithsonian Contributions to Botany.

Smithsonian Institution. 1975. Report on Endangered and Threatened Species of the United States. Washington. [House Document 94-51, serial no. 94-A.]

Soil Conservation Service [U.S.D.A.]. 1982. National List of Scientific Plant Names. 2 vols. [Washington.]

Soil Survey Staff [U.S.D.A.]. 1975. Soil Taxonomy: A Basic System of Soil Classification for Making and Interpreting Soil Surveys. Washington. [Agric. Handb. 436.]

Soltis, D. E. and P. S. Soltis. 1989. Genetic consequences of autopolyploidy in *Tolmiea* (Saxifragaceae). Evolution 43: 585–594.

Soule, M. E. and B. A. Wilcox, eds. 1980. Conservation Biology. An Evolutionary-Ecological Perspective. Sunderland, Mass.

SouthE. Archaeol. = Southeastern Archaeology.

SouthW. Naturalist = Southwestern Naturalist.

Spaulding, W. G., E. B. Leopold, and T. R. Van Devender. 1983. Late Wisconsin paleoecology of the American Southwest. In: S. C. Porter, ed. 1983. Late-Quaternary Environments of the United States. Vol. 1. The Late Pleistocene. Minneapolis. Pp. 259–293.

Specht, R. L., ed. 1979. Heathlands and Related Shrublands: Descriptive Studies. Amsterdam.

Special Pap. Geol. Soc. Amer. = Special Papers of the Geological Society of America.

Spellenberg, R. 1975. Autogamy and hybridization as evolutionary mechanisms in *Panicum* subgenus *Dichanthelium* (Gramineae). Brittonia 27: 87–95.

Spicer, R. A., J. A. Wolfe, and D. J. Nichols. 1987. Alaskan Cretaceous-Tertiary floras and Arctic origins. Paleobiology 13: 73–83.

Sporne, K. R. 1949. A new approach to the problem of the primitive flower. New Phytol. 85: 419–449.

Sprugel, D. G. 1976. Dynamic structure of wave-generated *Abies balsamea* forests in the northeastern United States. J. Ecol. 64: 899–911.

Stafleu, F. A. 1971. Linnaeus and the Linnaeans: The Spreading of Their Ideas in Systematic Botany, 1735–1789. Utrecht.

Stafleu, F. A. and R. S. Cowan. 1976–1988. Taxonomic Literature: A Selective Guide to Botanical Publications and Collections with Dates, Commentaries and Types, ed. 2. 7 vols. Utrecht, Antwerp, The Hague, and Boston.

Stafleu, F. A. and E. A. Mennega. 1992. Taxonomic Literature: A Selective Guide to Botanical Publications and Collections with Dates, Commentaries and Types. Supplement 1. Königstein.

Stalter, R. 1974. Vegetation in coastal dunes of South Carolina. Castanea 39: 95–103.

Stanley, S. M. 1979. Macroevolution, Pattern and Process. San Francisco.

Staplin, F. L., ed. 1976. Tertiary biostratigraphy, MacKenzie Delta region, Canada. Bull. Canad. Petrol. Geol. 24: 117–136.

Stearns, R. P. 1970. Science in the British Colonies of America. Urbana.

Stebbins, G. L. 1950. Variation and Evolution in Plants. New York.

Stebbins, G. L. 1971. Chromosomal Evolution in Higher Plants. London.

Stebbins, G. L. 1974. Flowering Plants: Evolution Above the Species Level. Cambridge, Mass.

Stebbins, G. L. 1980. Polyploidy in plants: Unsolved problems and prospects. In: W. H. Lewis, ed. 1980. Polyploidy: Biological Relevance. New York. Pp. 495–528.

Stebbins, G. L. 1981. Chromosomes and evolution in the genus *Bromus* (Gramineae). Bot. Jahrb. Syst. 102: 359–379.

Stebbins, G. L. 1983. Mosaic evolution: An integrating principle for the modern synthesis. Experientia (Basel) 39: 823–834.

Stebbins, G. L. 1983b. Mosaic evolution: Mosaic selection and angiosperm phylogeny. Bot. J. Linn. Soc. 88: 149–164.

Stebbins, G. L. 1984. Polyploidy and the distribution of the arcticalpine flora: New evidence and a new approach. Bot. Helv. 94: 1–13.

Stebbins, G. L. 1985. Polyploid hybridization and the invasion of new habitats. Ann. Missouri Bot. Gard. 12: 824–832.

Stebbins, G. L. 1986. The origin and success of polyploids in the circumpolar flora: A new analysis. Trans. & Proc. Bot. Soc. Edinburgh, 150th anniv. suppl.: 17–31.

Stebbins, G. L. and J. Dawe. 1987. Polyploidy and distribution in the European flora: A reappraisal. Bot. Jahrb. Syst. 108: 343–354.

Stebbins, G. L. and J. Major. 1965. Endemism and speciation in the California flora. Ecol. Monogr. 35: 1–35.

Stebbins, R. C. 1974. Off-road vehicles and the fragile desert. Amer. Biol. Teacher 36: 203–208, 294–304.

Steila, D. 1976. The Geography of Soils: Formation, Distribution, and Management. Englewood Cliffs.

Steila, D. and T. E. Pond. 1989. The Geography of Soils: Formation, Distribution, and Management, ed. 2. Totowa, N.J.

Stephenson, N. L. 1990. Climatic control of vegetation distribution: The role of the water balance. Amer. Naturalist 135: 649–670.

Stephenson, S. L. 1986. Changes in a former chestnut-dominated forest after a half century of succession. Amer. Midl. Naturalist 116: 173–179.

Stern, W. L., J. A. Wolfe, M. F. Moseley Jr., D. E. Stone, J. A. Mears, and R. F. Thorne. 1973. What happened to the Amentiferae? [Symposium.] Brittonia 25(4): 315–405.

Stevens, O. A. 1950. Handbook of North Dakota Plants. Fargo.

Steyermark, J. A. 1963. Flora of Missouri. Ames.

Stockey, R. A. 1984. Middle Eocene *Pinus* remains from British Columbia. Bot. Gaz. 145: 262–274.

Stockey, R. A. 1987. A permineralized flower from the middle Eocene of British Columbia. Amer. J. Bot. 74: 1878–1887.

Stone, E. C. and R. B. Vasey. 1968. Preservation of coast redwood on alluvial flats. Science 159: 157–161.

Stone, W. 1912. The plants of southern New Jersey with especial reference to the flora of the Pine Barrens. Rep. (Annual) New Jersey State Mus. 1910: 21–828.

Stout, J. P. 1984. The Ecology of Irregularly Flooded Salt Marshes of the Northeastern Gulf of Mexico: A Community Profile. Washington.

Strausbaugh, P. D. and E. L. Core. 1978. Flora of West Virginia, ed. 2. Grantsville, W.Va.

Sundquist, E. T. and W. S. Broecker, eds. 1985. The Carbon Cycle and Atmospheric CO_2: Natural Variations Archean to Present. Washington. [Amer. Geophys. Union Monogr. 32.]

Svensk Bot. Tidskr. = Svensk Botanisk Tidskrift Utgifven af Svenska Botaniska Föreningen.

Sweitzer, E. M. 1971. The comparative anatomy of Ulmaceae. J. Arnold Arbor. 52: 523–585.

Synge, H., ed. 1981. The Biological Aspects of Rare Plant Conservation. Chichester.

Syst. Bot. = Systematic Botany; Quarterly Journal of the American Society of Plant Taxonomists.

Syst. Zool. = Systematic Zoology.

Sytsma, K. and L. D. Gottlieb. 1986. Chloroplast DNA, evolution and phylogenetic relationships in *Clarkia* Sect. *Peripetasma* (Onagraceae). Evolution 40: 1248–1261.

Syvertson, J. P., G. L. Nickell, R. W. Spellenberg, and G. L. Cunningham. 1976. Carbon reduction pathways and standing crop in three Chihuahuan Desert plant communities. SouthW. Naturalist 21: 311–320.

Szarek, S. R. 1979. Primary production in four North American deserts: Indices of efficiency. J. Arid Environm. 2: 187–255.

Taggart, R. E. and A. T. Cross. 1980. Vegetation change in the Miocene Sucker Creek flora of Oregon and Idaho: A case study in paleosuccession. In: D. L. Dilcher and T. N. Taylor, eds. 1980. Biostratigraphy of Fossil Plants. Stroudsburg, Pa. Pp. 185–210.

Tai, W. and R. K. Vickery Jr. 1970. Cytogenetic relationships of key diploid members of the *Mimulus glabratus* complex (Scrophulariaceae). Evolution 24: 670–679.

Takhtajan, A. L. 1959. Die Evolution der Angiospermen. Jena.

Takhtajan, A. L. 1986. Floristic Regions of the World. Berkeley and Los Angeles.

Takhtajan, A. L. 1987. Systema Magnoliophytorum. Leningrad.

Tatnall, R. R. 1946. Flora of Delaware and the Eastern Shore: An Annotated List of the Ferns and Flowering Plants of the Peninsula of Delaware, Maryland and Virginia. [Wilmington.]

Taxon = Taxon; Journal of the International Association for Plant Taxonomy.

Taylor, D. W. 1977. Floristic relationships along the Cascade-Sierran axis. Amer. Midl. Naturalist 97: 333–349.

Taylor, D. W. 1988. Paleobiogeographic relationships of the Paleogene flora from the southeastern U.S.A.: Implications for west Gondwanaland affinities. Palaeogeogr. Palaeoclimatol. Palaeoecol. 66: 265–275.

Taylor, D. W. 1990. Paleobiogeographic relationships of angiosperms from the Cretaceous and early Tertiary of the North American area. Bot. Rev. (Lancaster) 56: 279–417.

Taylor, D. W. and W. L. Crepet. 1987. Fossil flora evidence of Malpighiaceae and an early plant-pollinator relationship. Amer. J. Bot. 74: 274–286.

Taylor, R. L. and B. MacBryde. 1977. Vascular Plants of British Columbia: A Descriptive Resource Inventory. Vancouver.

Teeri, J. A. and L. G. Stowe. 1976. Climatic pattern and the distribution of C_4 grasses in North America. Oecologia 23: 1–12.

Templeton, A. R. 1989. The meaning of species and speciation: A general perspective. In: D. K. S. Otte and J. A. Endler, eds. 1989. Speciation and Its Consequences. Sunderland, Mass. Pp. 3–27.

Terasmae, J. and T. W. Anderson. 1970. Hypsithermal range extension of white pine (*Pinus strobus* L.) in Quebec, Canada. Canad. J. Earth Sci. 7: 406–413.

Thannheiser, D. 1984. The Coastal Vegetation of Eastern Canada.... Edited and Arranged by Gordon F. Bennett. St. John's. [Mem. Univ. Newfoundland, Occas. Pap. Biol. 8. Translation of: Die Kunstenvegetation Ostkanadas, Paderborn, 1981.]

Thomasson, J. R. 1979. Late Cenozoic grasses and other angiosperms from Kansas, Nebraska, and Colorado: Biostratigraphy and relationships to living taxa. Bull. Kansas Geol. Surv. 218.

Thomasson, J. R. 1987. Late Miocene plants from northeastern Nebraska. J. Paleontol. 61: 1065–1079.

Thompson, D. Q., R. L. Stuckey, and E. B. Thompson. 1987. Spread, impact, and control of purple loosestrife (*Lythrum salicaria*) in North American wetlands. Fish Wildlife Res. 2: i–v, 1–55.

Thompson, R. S. 1988. Western North America: Vegetation dynamics in the western United States: Modes of response to climatic fluctuations. In: B. Huntley and T. Webb III, eds. 1988. Vegetation History. Dordrecht. Pp. 415–458.

Thorne, R. F. 1949. Inland plants on the Gulf coastal plain of Georgia. Castanea 14: 88–97.

Thorne, R. F. 1954. Flowering plants of the waters and shores of the Gulf of Mexico. In: Fish and Wildlife Service [U.S.D.I.]. 1954. Gulf of Mexico: Its Origin, Waters, and Marine Life. Washington. Pp. 193–202.

Thorne, R. F. 1954b. The vascular plants of southwestern Georgia. Amer. Midl. Naturalist 52: 257–327.

Thorne, R. F. 1963. Some problems and guiding principles of angiosperm phylogeny. Amer. Naturalist 97: 287–305.

Thorne, R. F. 1972. Major disjunctions in the geographic ranges of seed plants. Quart. Rev. Biol. 47: 365–411.

Thorne, R. F. 1973. The "Amentiferae" or Hamamelidae as an artificial group: A summary statement. Brittonia 25: 395–405.

Thorne, R. F. 1982. The desert and other transmontane plant communities of southern California. Aliso 10: 219–257.

Thorne, R. F. 1983. Proposed new realignments in the angiosperms. Nordic J. Bot. 3: 85–117.

Thorne, R. F. 1984. Are California's vernal pools unique? In: S. Jain and P. Moyle, eds. 1984. Vernal Pools and Intermittent Streams: A Symposium, May 9 and 10, 1981. Davis. Pp. 1–8.

Thorne, R. F. 1985. Diversity in the Florida flora. [Abstract.] Amer. J. Bot. 72: 938.

Thorne, R. F. 1986. A historical sketch of the vegetation of the Mojave and Colorado deserts of the American Southwest. Ann. Missouri Bot. Gard. 73: 642–651.

Thorne, R. F. 1988. Montane and subalpine forests of the Transverse and Peninsular ranges. In: M. G. Barbour and J. Major, eds. 1988. Terrestrial Vegetation of California, ed. 2. Sacramento. Pp. 537–557.

Thorne, R. F. 1992. An updated classification of the flowering plants. Aliso 13: 365–389.

Thorne, R. F., B. A. Prigge, and J. Henrickson. 1981. A flora of the higher ranges and the Kelso Dunes of the eastern Mojave Desert in California. Aliso 10: 71–186.

Tidwell, W. D. 1972. Physiography of the Intermountain Region. In: A. Cronquist et al. 1972+. Intermountain Flora. Vascular Plants of the Intermountain West, U.S.A. 4+ vols. New York and London. Vol. 1, pp. 10–18.

Tiehm, A. and F. A. Stafleu. 1990. Per Axel Rydberg: A biography, bibliography and list of his taxa. Mem. New York Bot. Gard. 58: 1–75.

Tiffney, B. H. 1985. Perspectives on the origin of the floristic similarity between eastern Asia and eastern North America. J. Arnold Arbor. 66: 73–94.

Tiffney, B. H. 1985b. The Eocene North Atlantic land bridge: Its importance in Tertiary and modern phytogeography of the Northern Hemisphere. J. Arnold Arbor. 66: 243–273.

Tiffney, B. H. and E. S. Barghoorn. 1976. Fruits and seeds of the Brandon lignite. I. Vitaceae. Rev. Palaeobot. Palynol. 22: 169–191.

Tiffney, B. H., J. A. Wolfe, T. A. Ager, and S. L. Wing. 1989. Late Miocene (?) flora of the Brandywine Clay, Maryland. Amer. J. Bot. 76(6, suppl.): 176.

Tippo, O. 1938. Comparative anatomy of the Moraceae and their presumed allies. Bot. Gaz. 100: 1–99.

Tisdale, E. W. 1986. Canyon Grasslands and Associated Shrublands of West-central Idaho and Adjacent Areas. Moscow, Idaho.

Tomlinson, P. B., T. Takaso, and J. A. Rattenbury. 1989. Cone and ovule ontogeny in *Phyllocladus* (Podocarpaceae). Bot. J. Linn. Soc. 99: 209–221.

Trans. Amer. Philos. Soc. = Transactions of the American Philosophical Society Held at Philadelphia for Promoting Useful Knowledge.

Trans. & Proc. Bot. Soc. Edinburgh = Transactions and Proceedings of the Botanical Society Edinburgh.

Trans. Illinois State Acad. Sci. = Transactions of the Illinois State Academy of Science.

Transeau, E. N. 1935. The prairie peninsula. Ecology 16: 423–437.

Traverse, A. 1955. Pollen analysis of the Brandon lignite of Vermont. Rep. Invest. Bur. Mines, Washington, DC 5151.

Traverse, A. 1988. Paleopalynology. London.

Trends Ecol. Evol. = Trends in Ecology and Evolution.

Trettin, H. P. 1989. The Arctic islands. In: A. W. Bally and R. A. Palmer, eds. 1989. The Geology of North America—An Overview. Boulder. Pp. 349–370.

Trewartha, G. T. and L. H. Horn. 1980. An Introduction to Climate, ed. 5. New York.

Trudy Prikl. Bot. = Trudy po Prikladnoi Botanike, Genetike i Selektsii.

Tryon, A. F. 1986. Stasis, diversity and function in spores based on an electron microscope survey of the Pteridophyta. In: S. Blackmore and I. K. Ferguson, eds. 1986. Pollen and Spores: Form and Function. London. Pp. 233–249.

Tryon, A. F. and B. Lugardon. 1991. Spores of the Pteridophyta. New York.

Tryon, R. M. 1969. Taxonomic problems in the geography of North American ferns. BioScience 19: 790–795.

Tryon, R. M. and A. F. Tryon. 1982. Ferns and Allied Plants, with Special Reference to Tropical America. New York, Heidelberg, and Berlin.

Tsukada, M. 1982. *Pseudotsuga menziesii* (Mirb.) Franco: Its pollen dispersal and late Quaternary history in the Pacific Northwest. Jap. J. Ecol. 32: 159–181.

Tucker, J. M. 1952. Taxonomic interrelationships in the *Quercus dumosa* complex. Madroño 11: 234–250.

Tucker, J. M. 1952b. Evolution of the Californian oak *Quercus alvordiana*. Evolution 6: 162–180.

Tuhkanen, S. 1980. Climatic parameters and indices in plant geography. Acta Phytogeogr. Suec. 67: 1–105.

Turesson, G. 1922. The genotypical response of the plant species to the habitat. Hereditas (Lund) 3: 211–350.

Turner, B. L. 1977. Fossil history and geography. In: V. H. Heywood et al., eds. 1977. The Biology and Chemistry of the Compositae. London. Pp. 21–40.

Tutin, T. G., V. H. Heywood, N. A. Burges, D. H. Valentine, S. M. Walters, D. A. Webb, et al., eds. 1964–1980. Flora Europaea. 5 vols. Cambridge.

Univ. Calif. Publ. Bot. = University of California Publications in Botany.

Univ. Calif. Publ. Geol. Sci. = University of California Publications in Geological Sciences.

Upchurch, G. R. and J. A. Wolfe. 1987. Mid-Cretaceous to early Tertiary vegetation and climate: Evidence from fossil leaves and woods. In: E. M. Friis et al., eds. 1987. The Origins of Angiosperms and Their Biological Consequences. Cambridge and New York. Pp. 75–105.

Valiela, I., J. M. Teal, and W. G. Deuser. 1978. The nature of growth forms in the salt marsh grass *Spartina alterniflora*. Amer. Naturalist 112: 461–470.

Van Bruggen, T. 1976. The Vascular Plants of South Dakota. Ames.

Van Cleve, K., C. T. Dyrness, L. A. Viereck, J. Fox, F. S. Chapin III, and W. C. Oechel. 1983. Taiga ecosystems in interior Alaska. BioScience 33: 39–44.

Van Devender, T. R., R. S. Thompson, and J. L. Betancourt. 1987. Vegetation history of the deserts of southwestern North America; the nature and timing of the late Wisconsin-Holocene transition. In: W. F. Ruddiman and H. E. Wright Jr., eds. 1987. North America and Adjacent Oceans during the Last Deglaciation. Boulder. Pp. 323–352.

van Tieghem, P. 1891. Structure et affinités des *Abies* et des genres les plus voisins. Bull. Soc. Bot. France 38: 406–415.

Vandermeer, J. 1980. Saguaros and nurse trees: A new hypothesis to account for population fluctuations. SouthW. Naturalist 25: 357–360.

Vankat, J. L. 1979. The Natural Vegetation of North America. New York.
Vankat, J. L. 1989. Water stress in chaparral shrubs in summer-rain versus summer drought climates: Whither the mediterranean-type climate paradigm? In: S. C. Keeley, ed. 1989. The California Chaparral: Paradigms Reexamined. Los Angeles. Pp. 117–124.
Vankat, J. L. and J. Major. 1978. Vegetation changes in Sequoia National Park, California. J. Biogeogr. 5: 377–402.
Vasek, F. C. 1980. Creosote bush: Long-lived clones in the Mojave Desert. Amer. J. Bot. 67: 246–255.
Vasek, F. C. and M. G. Barbour. 1988. Mojave Desert scrub vegetation. In: M. G. Barbour and J. Major, eds. 1988. Terrestrial Vegetation of California, ed. 2. Sacramento. Pp. 835–867.
Vasek, F. C. and R. F. Thorne. 1988. Transmontane coniferous vegetation. In: M. G. Barbour and J. Major, eds. 1988. Terrestrial Vegetation of California, ed. 2. Sacramento. Pp. 797–832.
Vegetatio = Vegetatio; Acta Geobotanica.
Velichko, A. A., H. E. Wright Jr., and C. W. Barnosky, eds. 1984. Late Quaternary Environments of the Soviet Union. Minneapolis.
Vermeij, G. J. 1986. The biology of human caused extinction. In: B. G. Norton, ed. 1986. The Preservation of Species. The Value of Biological Diversity. Princeton. Pp. 28–49.
Vernhes, J. R. 1989. Biosphere reserves: The beginnings, the present, and future challenges. In: W. P. Gregg Jr. et al., eds. 1989. Proceedings of the Symposium on Biosphere Reserves, Fourth World Wilderness Congress, September 14–17, 1987, YMCA of the Rockies, Estes Park, Colorado, USA. Atlanta. Pp. 7–20.
Vierhapper, F. 1910. Entwurf eines neuen Systemes der Coniferen. Abh. K. K. Zool.-Bot. Ges. Wien 5: 1–56.
Viers, S. D. Jr. 1982. Coast redwood forest: Stand dynamics, successional status, and the role of fire. In: J. E. Means, ed. 1982. Forest Succession and Stand Development Research in the Northwest. Corvallis. Pp. 119–141.
Virginia J. Sci. = Virginia Journal of Science.
Vitousek, P. M., P. R. Ehrlich, A. H. Ehrlich, and P. A. Matson. 1986. Human appropriation of the products of photosynthesis. BioScience 36: 368–373.
Vogel, V. J. 1970. American Indian Medicine. Norman.
Vogl, R. J. and L. T. McHargue. 1966. Vegetation of California fan palm oases on the San Andreas Fault. Ecology 47: 532–540.
Vowinckel, T., W. C. Oechel, and W. G. Boll. 1975. The effect of climate on the photosynthesis of *Picea mariana* at the sub-arctic tree line. I. Field measurements. Canad. J. Bot. 53: 604–620.

Wade, D., J. J. Ewel, and R. Hofstetter. 1980. Fire in South Florida Ecosystems. Washington. [U.S.D.A. Forest Serv., Gen. Techn. Rep. SE-17.]
Wagner, W. H. Jr. 1952. Types of foliar dichotomy in living ferns. Amer. J. Bot. 39: 578–592.
Wagner, W. H. Jr. 1960. Evergreen grapeferns and the meanings of infraspecific categories as used in North American pteridophytes. Amer. Fern J. 50: 32–45.
Wagner, W. H. Jr. 1972. Disjunctions in homosporous vascular plants. Ann. Missouri Bot. Gard. 59: 203–217.
Wagner, W. H. Jr. 1979. Reticulate veins in the systematics of modern ferns. Taxon 28: 87–95.
Wagner, W. H. Jr. and F. S. Wagner. 1977. Fertile-sterile leaf dimorphy in ferns. Gard. Bull. Straits Settlem. 30: 251–267.
Wagner, W. H. Jr. and F. S. Wagner. 1983. Genus communities as a systematic tool in the study of New World *Botrychium* (Ophioglossaceae). Taxon 32: 51–63.
Wagner, W. H. Jr., F. S. Wagner, and J. M. Beitel. 1985. Evidence for interspecific hybridization in pteridophytes with subterranean mycoparasitic gametophytes. Proc. Roy. Soc. Edinburgh, B 86: 273–281.
Wagner, W. H. Jr., F. S. Wagner, and W. C. Taylor. 1986. Detecting abortive spores in herbarium specimens of sterile hybrids. Amer. Fern J. 76: 129–140.
Walter, H. 1979. Vegetation of the Earth, ed. 2. New York.
Walter, H., E. Harnickell, and D. Mueller-Dombois. 1975. Climate Diagram Maps of the Individual Continents and Ecological Regions of the Earth. Berlin.
Walter, H. and H. Lieth. 1967. Klimadiagramm-Weltatlas. Jena.
Walter, K. S., W. H. Wagner Jr., and F. S. Wagner. 1982. Ecological, biosystematic, and nomenclatural notes on Scott's spleenwort, ×*Asplenosorus ebenoides*. Amer. Fern J. 72: 65–75.
Waring, R. H. and J. F. Franklin. 1979. Evergreen coniferous forests of the Pacific Northwest. Science 204: 1380–1386.
Warner, B. G., R. W. Mathewes, and J. J. Clague. 1982. Ice-free conditions on the Queen Charlotte Islands, British Columbia, at the height of late Wisconsin glaciation. Science 218: 675–677.
Warwick, S. I. and L. D. Gottlieb. 1985. Genetic divergence and geographic speciation in *Layia* (Compositae). Evolution 39: 1236–1241.
Washington Park Arbor. Bull. = Washington Park Arboretum Bulletin.
Watts, W. A. 1980. The late Quaternary vegetation history of the southeastern United States. Annual Rev. Ecol. Syst. 11: 387–409.
Watts, W. A. 1980b. Late-Quaternary vegetation history at White Pond on the inner coastal plain of South Carolina. Quatern. Res. 13: 187–199.
Watts, W. A. 1983. Vegetational history of the eastern United States 25,000 to 10,000 years ago. In: S. C. Porter, ed. 1983. Late-Quaternary Environments of the United States. Vol. 1. The Late Pleistocene. Minneapolis. Pp. 294–310.
Watts, W. A. 1988. Europe. In: B. Huntley and T. Webb III, eds. 1988. Vegetation History. Dordrecht. Pp. 155–192.
Weaver, J. E. and T. J. Fitzpatrick. 1934. The prairie. Ecol. Monogr. 4: 109–295.
Webb, T. III. 1988. Eastern North America. In: B. Huntley

and T. Webb III, eds. 1988. Vegetation History. Dordrecht. Pp. 385–414.
Webb, T. III, P. J. Bartlein, and J. E. Kutzbach. 1987. Climatic change in eastern North America during the past 18,000 years: Comparisons of pollen data with model results. In: W. F. Ruddiman and H. E. Wright Jr., eds. 1987. North America and Adjacent Oceans during the Last Deglaciation. Boulder. Pp. 447–462.
Webbia = Webbia; Raccolta di Scritti Botanici.
Weber, W. A. 1953. Handbook of Plants of the Colorado Front Range: Keys for the Identification of the Ferns, Conifers, and Flowering Plants of the Central Rocky Mountains from Pikes Peak to Rocky Mountain National Park, and from the Plains to the Continental Divide. Boulder.
Weber, W. A. 1965. Plant geography in the southern Rocky Mountains. In: H. E. Wright Jr. and D. G. Frey, eds. 1965. The Quaternary of the United States. Princeton. Pp. 433–468.
Weber, W. A. 1987. Colorado Flora: Western Slope. Boulder.
Weber, W. A. and R. C. Wittmann. 1992. Catalog of the Colorado Flora: A Biodiversity Baseline. Niwot, Colo.
Weed Sci. = Weed Science.
Weed Science Society of America. 1984. Composite list of weeds. Weed Sci. 32(suppl. 2): 1–137.
Weed Science Society of America. 1988. Supplement to WSSA composite list of weeds. Weed Sci. 36: 850–851.
Weidick, A. 1976. Glaciation and the Quaternary of Greenland. In: A. Escher and W. S. Watt, eds. 1976. Geology of Greenland. Copenhagen. Pp. 431–458.
Weishaupt, C. G. 1971. Vascular Plants of Ohio: A Manual for Use in Field and Laboratory, ed. 3. Dubuque.
Wells, P. V. 1983. Paleobiogeography of montane islands in the Great Basin since the last glaciopluvial. Ecol. Monogr. 53: 341–382.
Wendland, W. M. 1977. Tropical storm frequencies related to sea surface temperatures. J. Appl. Meteorol. 16: 477–481.
Wernham, H. F. 1912. Floral evolution: With particular reference to the sympetalous dicotyledons. New Phytol. 11: 373–397.
Werth, C. R. and W. H. Wagner Jr. 1990. Proposal to designate reproductively competent species of hybrid origin by an × placed in brackets (reword Article H.3, Note 1). Taxon 39: 699–702.
West, D. C., H. H. Shugart, and D. B. Botkin, eds. 1981. Forest Succession: Concepts and Application. New York.
West, N. E. 1988. Intermountain deserts, shrub steppes, and woodlands. In: M. G. Barbour and W. D. Billings, eds. 1988. North American Terrestrial Vegetation. New York. Pp. 209–230.
Westgate, J. W. and C. T. Gee. 1990. Paleoecology of a middle Eocene mangrove biota (vertebrates, plants, and invertebrates) from southwest Texas. Palaeogeogr. Palaeoclimatol. Palaeoecol. 78: 163–177.
Westman, W. E. 1981. Diversity relations and succession in California coastal sage scrub. Ecology 62: 170–184.
Westman, W. E. 1983. Xeric mediterranean-type shrubland association of Alta and Baja California and the community/continuum debate. Vegetatio 52: 3–19.
White, D. 1941. Prairie soil as a medium for tree growth. Ecology 22: 399–407.
White, D. J. 1987. Ecological Study and Status Report on American Ginseng, *Panax quinquefolium* L., a Threatened Species in Canada. Ottawa. [Unpublished report, submitted to the Subcommittee on Plants, Committee on the Status of Endangered Wildlife in Canada.]
White, P. S., ed. 1984. The Southern Appalachian Spruce-Fir Ecosystem: Its Biology and Threats. Atlanta. [Natl. Park Serv., SE Region, Res./Resources Managem. Rep. SER-71.]
White, P. S. 1984b. The southern Appalachian spruce-fir ecosystem: An introduction. In: P. S. White, ed. 1984. The Southern Appalachian Spruce-Fir Ecosystem: Its Biology and Threats. Atlanta. Pp. 1–21.
Whittaker, R. H. 1960. Vegetation of the Siskiyou Mountains, Oregon and California. Ecol. Monogr. 30: 279–338.
Whittaker, R. H. 1979. Appalachian balds and other North American heathlands. In: R. L. Specht, ed. 1979. Heathlands and Related Shrublands: Descriptive Studies. Amsterdam. Pp. 427–440.
Whittier, D. 1972. Gametophytes of *Botrychium dissectum* as grown in sterile culture. Bot. Gaz. 133: 336–339.
Williams, M. 1989. Americans and Their Forests: A Historical Geography. Cambridge and New York.
Williams, R. L. 1984. Aven Nelson of Wyoming. Boulder.
Williams, W. T., M. Brady, and S. C. Willison. 1977. Air pollution damage to the forests of the Sierra Nevada Mountains of California. J. Air Pollut. Control Assoc. 27: 230–234.
Willis, J. C. 1973. A Dictionary of the Flowering Plants and Ferns, ed. 8, revised by H. K. Airy Shaw. Cambridge.
Wilson, E. O., ed. 1988. Biodiversity. Washington.
Wilson, E. O. 1988b. The current state of biological diversity. In: E. O. Wilson, ed. 1988. Biodiversity. Washington. Pp. 3–18.
Wilson, K. A. 1959. Sporangia of the fern genera allied with *Polypodium* and *Vittaria*. Contr. Gray Herb. 185: 97–127.
Wing, S. L. 1987. Eocene and Oligocene floras and vegetation of the Rocky Mountains. Ann. Missouri Bot. Gard. 74: 748–784.
Wolfe, J. A. 1972. An interpretation of Alaskan Tertiary floras. In: A. Graham, ed. 1972. Floristics and Paleofloristics of Asia and Eastern North America. Amsterdam and New York. Pp. 201–233.
Wolfe, J. A. 1973. Fossil forms of Amentiferae. Brittonia 25: 334–355.
Wolfe, J. A. 1975. Some aspects of the plant geography of the Northern Hemisphere during the late Cretaceous and Tertiary. Ann. Missouri Bot. Gard. 62: 264–279.
Wolfe, J. A. 1977. Paleogene floras from the Gulf of Alaska region. Profess. Pap. U.S. Geol. Surv. 997.
Wolfe, J. A. 1978. A paleobotanical interpretation of Tertiary climates in the Northern Hemisphere. Amer. Sci. 66: 694–704.

Wolfe, J. A. 1985. Distribution of major vegetation types during the Tertiary. In: E. T. Sundquist and W. S. Broecker, eds. 1985. The Carbon Cycle and Atmospheric CO_2: Natural Variations Archean to Present. Washington. Pp. 357–375.

Wolfe, J. A. 1987. Late Cretaceous-Cenozoic history of deciduousness and the terminal Cretaceous event. Paleobiology 13: 215–226.

Wolfe, J. A. 1987b. An overview of the origins of the modern vegetation and flora of the northern Rocky Mountains. Ann. Missouri Bot. Gard. 74: 785–803.

Wolfe, J. A. and R. Z. Poore. 1982. Tertiary marine and nonmarine climatic trends. In: J. C. Crowell and W. Berger, eds. 1982. Pre-Pleistocene Climates. Washington. Pp. 154–158.

Wolfe, J. A. and H. E. Schorn. 1989. Paleoecologic, paleoclimatic, and evolutionary significance of the Oligocene Creede flora, Colorado. Paleobiology 14: 180–198.

Wolfe, J. A. and T. Tanai. 1987. Systematics, phylogeny, and distribution of *Acer* (maples) in the Cenozoic of western North America. J. Fac. Sci. Hokkaido Univ., Ser. 4, Geol. Mineral. 22: 1–246.

Wolfe, J. A. and G. R. Upchurch. 1987. North American nonmarine climates and vegetation during the late Cretaceous. Palaeogeogr. Palaeoclimatol. Palaeoecol. 61: 33–77.

Wolfe, J. A. and G. R. Upchurch. 1987b. Leaf assemblages across the Cretaceous-Tertiary boundary in the Raton Basin, New Mexico and Colorado. Proc. Natl. Acad. Sci. U.S.A. 84: 5096–5100.

Wolfe, J. A. and W. Wehr. 1987. Middle Eocene dicotyledonous plants from Republic, northeastern Washington. Bull. U.S. Geol. Surv. 1597.

Wood, C. E. Jr., ed. 1958+. Generic flora of the southeastern United States. J. Arnold Arbor. 39+. [A long-running series of taxonomic treatments by various authors, each article titled "The genera of...of the southeastern United States."]

Wood, C. E. Jr. 1971. Some floristic relationships between the southern Appalachians and western North America. In: P. C. Holt, ed. 1971. The Distributional History of the Biota of the Southern Appalachians. Part 2. Flora. Blacksburg, Va. Pp. 331–404.

Woodward, F. I. 1987. Climate and Plant Distribution. Cambridge and New York.

Woodwell, G. M. 1977. The challenge of endangered species. In: G. T. Prance and T. S. Elias, eds. 1977. Extinction Is Forever. Bronx. Pp. 5–10.

Wright, H. E. Jr. 1976. The dynamic nature of Holocene vegetation, a problem in paleoclimatology, biogeography, and stratigraphic nomenclature. Quatern. Res. 6: 581–596.

Wright, H. E. Jr. 1981. Vegetation east of the Rocky Mountains 18,000 years ago. Quatern. Res. 15: 113–125.

Wright, H. E. Jr., ed. 1983. Late-Quaternary Environments of the United States. Vol. 2. The Holocene. Minneapolis.

Wright, H. E. Jr. 1989. The Quaternary. In: A. W. Bally and R. A. Palmer, eds. 1989. The Geology of North America—An Overview. Boulder. Pp. 513–536.

Wright, H. E. Jr. and D. G. Frey, eds. 1965. The Quaternary of the United States. Princeton.

Wright, S. 1931. Evolution in Mendelian populations. Genetics 16: 97–159.

Wunderlin, R. P. 1982. Guide to the Vascular Plants of Central Florida. Tampa.

Yang, T. W. 1970. Major chromosome races of *Larrea divaricata* in North America. J. Arizona Acad. Sci. 6: 41–45.

Yanovsky, E. 1936. Food Plants of the North American Indians. Washington. [U.S.D.A., Misc. Publ. 237.]

Young, J. A., R. A. Evans, and J. Major. 1988. Sagebrush steppe. In: M. G. Barbour and J. Major, eds. 1988. Terrestrial Vegetation of California, ed. 2. Sacramento. Pp. 763–796.

Young, J. A., R. A. Evans, and P. T. Tueller. 1976. Great Basin plant communities—Pristine and grazed. In: R. Elston, ed. 1976. Holocene Environmental Change in the Great Basin. Reno. Pp. 186–215.

Z. Indukt. Abstammungs- Vererbungsl. = Zeitschrift für induktive Abstammungs- und Vererbungslehre.

Z. Zellf. Mikroskop. Anat. = Zeitschrift für Zellforschung und mikroskopische Anatomie.

Zedler, J. B. 1982. The Ecology of Southern California Coastal Salt Marshes: A Community Profile. Washington.

Zhurn. Geomorph. = Zhurnal Geomorphologii.

Ziegler, P. A. 1988. Evolution of the Arctic–North Atlantic and the western Tethys. A. A. P. G. Mem. 43.

Zinke, P. J. 1988. The redwood forest and associated north coast forests. In: M. G. Barbour and J. Major, eds. 1988. Terrestrial Vegetation of California, ed. 2. Sacramento. Pp. 678–698.

Zohary, D. and U. Plitman. 1979. Chromosome polymorphism, hybridization and colonization in the *Vicia sativa* group (Fabaceae). Pl. Syst. Evol. 131: 143–156.

Index

Page numbers in *italics* designate the location of an illustration or photograph.